Problem Solving

in

Analytical Chemistry

A Practical Handbook Containing
Over 1000 Worked Examples, Problems and Answers

Themistocles P. Hadjiioannou
University of Athens

Gary D. Christian
University of Washington

Constantinos E. Efstathiou
University of Athens

Demetrios P. Nikolelis
University of Athens

PERGAMON PRESS

OXFORD · NEW YORK · BEIJING · FRANKFURT
SÃO PAULO · SYDNEY · TOKYO · TORONTO

U.K. Pergamon Press plc, Headington Hill Hall, Oxford OX3 0BW, England

U.S.A. Pergamon Press, Inc., Maxwell House, Fairview Park, Elmsford, New York 10523, U.S.A.

PEOPLE'S REPUBLIC OF CHINA Pergamon Press, Room 4037, Qianmen Hotel, Beijing, People's Republic of China

FEDERAL REPUBLIC OF GERMANY Pergamon Press GmbH, Hammerweg 6, D-6242 Kronberg, Federal Republic of Germany

BRAZIL Pergamon Editora Ltda, Rua Eça de Queiros, 346, CEP 04011, Paraiso, São Paulo, Brazil

AUSTRALIA Pergamon Press Australia Pty Ltd., P.O. Box 544, Potts Point, N.S.W. 2011, Australia

JAPAN Pergamon Press, 5th Floor, Matsuoka Central Building, 1-7-1 Nishishinjuku, Shinjuku-ku, Tokyo 160, Japan

CANADA Pergamon Press Canada Ltd., Suite No. 271, 253 College Street, Toronto, Ontario, Canada M5T 1R5

First edition 1988

Library of Congress Cataloging in Publication Data

Problem solving in analytical chemistry.
Includes index.
1. Chemistry, Analytic—Problems, exercises, etc.
I. Chatzēiōannou, Themistoklēs P.
QD75.9.P76 1988 543'.0076 88–25479

British Library Cataloguing in Publication Data

Problem solving in analytical chemistry: a practical handbook containing over 1000 worked examples, problems and answers.
1. Chemical analysis — Questions & answers
I. Hadjiioannou, Themistocles P.
543'.007

ISBN 0–08–036968–5 Hardcover
ISBN 0–08–036967–7 Flexicover

Printed in Great Britain by A. Wheaton & Co. Ltd., Exeter

Contents

Preface

The solution of problems in analytical chemistry is usually a prerequisite for the calculation of the composition of various samples. The present book, "Problem Solving in Analytical Chemistry," aims at familiarizing the student with the calculations performed in analytical chemistry, and in chemistry in general, and at consolidating theoretical knowledge by applying it to the solution of concrete or real problems. This problem solving book may be used as a companion to a more conventional textbook in quantitative analysis courses to provide numerous supplemental problems and expose the student to various approaches for problem solving. It is sufficiently complete that it may serve as the sole text in a problem solving course. It is also useful in basic chemistry, chemical equilibrium or instrumental analysis courses. Topics in each chapter are arranged so that assignments may be made that are relevant to the particular course level or material. Thus, for example, in Chapter 5, dealing with acid-base equilibria, separate topics include simple monoprotic acids, mass balance and electroneutrality, the diverse salt effect (thermodynamic dissociation constant), indicator theory, polyprotic acids (α-values), logarithmic acid-base concentration diagrams, buffers, and hydrolysis. Often two or three approaches are presented for solving a particular problem. Instrumentation subject matter has been restricted to potentiometric methods, spectrophotometry, and chromatographic methods, in order to provide a book of reasonable length and cost. This book can also be used by anyone who deals with chemical calculations and especially with chemical analysis, such as chemical engineers, clinical chemists, pharmacists, biologists, geologists, microbiologists, and so forth.

The book contains 18 chapters, which deal with the most common analytical methods. In each chapter there is a short introduction of the relevant theory, and equations are given to facilitate the comprehension of the theoretical principle and the solution of the relevant problems. About 340 solved examples are given throughout the book. In addition, about 640 unsolved problems are included, some of which deal with new material so as to cover the field of Analytical Chemistry as much as possible. Answers to odd-numbered problems are given in the Appendix. In addition, a Solutions Manual is available with detailed solutions of each problem (see below). Some of the examples constitute part of the theory of the respective chapters, such as the calculation of titration curves or the multiple equilibria in systems involving acids-bases or complexes. The data given in several problems are actual data obtained either by students during their training in the laboratory or by researchers in the Laboratory of Analytical Chemistry during their research. Many of the problems have been given in various exams for pretesting.

In solving a problem, the student should understand the principles of the analytical methods involved and should be able to formulate the proper equation, comprehending the meaning of any term in the equation. Therefore, before solving a problem, the student should familiarize herself or himself with the relevant material of analytical chemistry, either from the material in this book or from a companion text. At the end of the book, tables containing constants needed for the solution of problems are given.

This text represents translation from the Greek version by Hadjiioannou, Efstathiou, and Nikolelis (1984). It has been used by students in Greece in manuscript or textbook form since 1982, and several modifications and improvements to problems have been made in the interim.

We would be grateful to our readers for bringing to our attention any oversight or error in the text.

We wish to express our gratitude to all who assisted in this edition of the book, starting with our students, who, with their interest and zeal in the courses of Analytical Chemistry, were the driving force of the whole endeavor. We thank our colleagues for valuable criticism and advice, in particular, M. A. Koupparis, E. P. Diamandis, A. C. Calokerinos, M. Timotheou-Potamia, and A. Makri, University of Athens. We are also grateful to the students T. Tsekouras, N. Psaroudakis, and F. Papadimitracopoulos for checking the Solutions Manual.

Finally we wish to express our deep appreciation to our wives for their forbearance and assistance during the writing of the book. Sue Christian's efficient and timely typing of the English draft of the manuscript is especially appreciated. John Shipman undertook the difficult task of providing the final version in computer format that adds greatly to its readability. The support of the Center for Process Analytical Chemistry (CPAC) at the University of Washington in final manuscript production is gratefully acknowledged.

T. P. Hadjiioannou

G. D. Christian

C. E. Efstathiou

D. P. Nikolelis

Laboratory of Analytical Chemistry
University of Athens

and

Department of Chemistry
University of Washington

Solutions Manual. A manual with detailed solutions to each problem is available for use by students.

Chapter 1

Mathematical Calculations and Statistical Treatment of Analytical Data

1.1 The Branches of Analytical Chemistry

Analytical chemistry is concerned with the chemical characterization of matter. The principal objective is the determination of the qualitative and quantitative composition of a chemical system. *Qualitative analysis* is concerned with the identification of the elements, ions or compounds present in an unknown sample, while *quantitative analysis* is concerned with the determination of the quantity of one or more components of the sample.

This book will deal with the computations required in performing quantitative analysis, or needed for a quantitative understanding of the measurement technique used (e.g., construction of titration curves).

The mathematical (quantitative) calculations of analytical chemistry are based on experimental data obtained with accurate measurements of various physical parameters, such as mass, volume, electrode potential, light absorption, and so forth. The treatment of data is mainly done on the basis of simple stoichiometric relations and chemical equilibrium constants. In instrumental methods of analysis the quantity of a component is calculated from measurement of a physical property which is related to the mass or the concentration of the component. The statistical treatment of data characterizes their reliability.

1.2 Errors in Quantitative Analysis

Every physical measurement is subject to errors, and therefore there is an uncertainty in the results that can be minimized but never eliminated completely. The error is represented by the equation

$$E = x_i - \mu \tag{1.1}$$

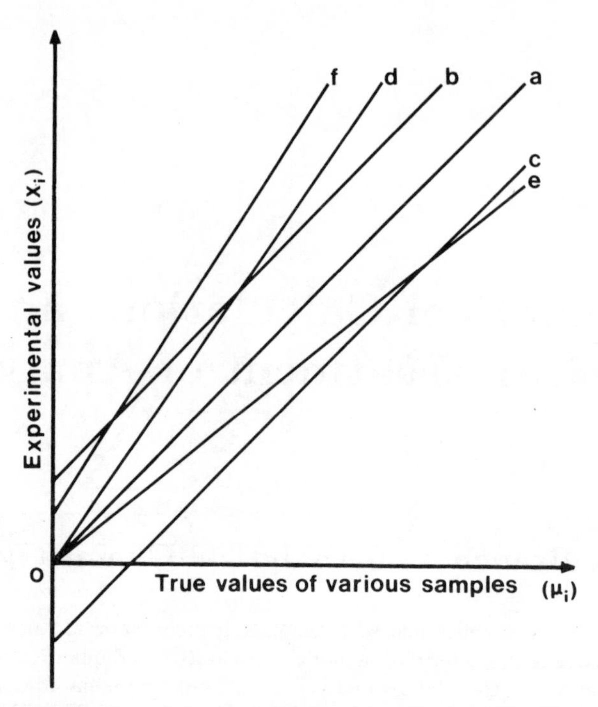

Figure 1.1. *Correlation between experimental and true values in various types of determinate errors: a) Ideal correlation, absence of determinate error, (b) constant positive determinate error, (c) constant negative determinate error, (d) proportional positive error, (e) proportional negative error, and f) composite (constant proportional) error.*

where x_i is the experimental value and μ is the true value.

Errors are classified into two categories, depending on their characteristics. *Determinate (systematic) errors* can be attributed to concrete causes (instrumental defects, reagent impurities, personal errors, method errors, etc.). They are unidirectional (Example 1.1 below) and are fixed for a series of measurements under the same experimental conditions. These errors usually can be corrected in various ways, e.g., by theoretical calculation of corrections, calibration of an instrument, analysis of standards, running of a blank, and so forth. Determinate errors may be *constant*, when the *absolute* value of E is the same for all examples. For example, when a fixed volume of reagent added to all samples contains a fixed amount of the analyte as a contaminant, the result of the analysis will always be in error by that fixed amount. Or the error may be *proportional* when the *relative* error (E/μ) is constant (i.e., the absolute error E is proportional to the quantity μ of the determined substance). For example, if the concentration of a standard base solution used for titration of acids is in error by +0.1%, then the apparent amount of acid reacted with it will always be in error by −0.1% since the volume of base required for reaction will be 0.1% low. In addition, there are *composite* systematic errors, which are a combination of constant and proportional errors. Representative examples of systematic errors are given in Figure 1.1.

The *standard deviation* for a *small* number of measurements, s, is defined as

$$s = \sqrt{\frac{\sum_{i=1}^{N}(x_i - \bar{x})^2}{N - 1}} \tag{1.7}$$

The quantity N−1 in Equation (1.7) is called the *degrees of freedom* of the system.

The quantity s^2 is called the *variance*. For faster calculation of the variance (and s) with calculators, the following relations are used:

$$s^2 = \frac{\sum x_i^2 - N\bar{x}^2}{N - 1} = \frac{\sum x_1^2 - \frac{(\sum x_i)^2}{N}}{N - 1} \qquad (1.8a)(1.8b)$$

For a *large* number of measurements, the standard deviation, σ, is defined as

$$\sigma = \sqrt{\frac{\sum_{i=1}^{N}(x_i - \mu)^2}{N}} \tag{1.9}$$

Strictly speaking, only the standard deviation, s, is determined experimentally, whereas the population standard deviation, σ, cannot be determined with a finite number of measurements. The value s is useful as an estimate of σ, and will more nearly approach σ as the number of measurements increases.

The mean of a series of N measurements taken from an infinite population will show a smaller scatter from the true value, μ, than will a single measurement. The scatter of the mean from μ will decrease as N is increased; as N gets very large the sample mean will approach μ. The mean derived from N measurements can be shown to be $\sqrt{N}$ times more reliable than a single measurement. Hence, the standard deviation of the mean, often referred to as the *standard error*, is

$$s_{\bar{x}} = \frac{s}{\sqrt{N}} = \sqrt{\frac{\sum_{i=1}^{N}(x_i - \bar{x})^2}{N(N - 1)}} \tag{1.10}$$

Instead of using the *absolute* values of the error, range and deviations, which are expressed in the same units as the measurements themselves, it is often more useful to express these terms as quantities *relative* to the magnitude of the quantity being measured. Thus

$$\text{Relative error} = E_r = E/\mu \tag{1.11}$$

$$\text{Relative range} = R_r = R/\bar{x} \tag{1.12}$$

$$\text{Relative average deviation} = \bar{d}_r = \bar{d}/\bar{x} \tag{1.13}$$

$$\text{Relative standard deviation} = s_r = s/\bar{x} \tag{1.14}$$

It can be concluded from the definitions that these relative quantities have no units. When these quantities are multiplied by 10^2, 10^3, 10^6 or 10^9, they become expressed in terms of percentage (%), parts per thousand (‰ or ppt), parts per million (ppm), or parts per billion (ppb), respectively. The relative standard deviation (standard deviation divided by the mean), when expressed as a percent, is also known as the *coefficient of variation.* This gives an indication of how good or poor the precision is relative to the magnitude of the measured value.

Example 1.2. Four measurements of the weight of an object whose correct weight is 0.1026 g are 0.1021, 0.1025, 0.1019, 0.1023 g. Calculate the mean, the median, the range, the relative range (%), the average deviation, the relative average deviation (%), the standard deviation, the relative standard deviation (%), the error of the mean, and the relative error of the mean (%).

Solution. We have

Mean : $\bar{x} = (0.1021 + 0.1025 + 0.1019 + 0.1023)/4 = \mathbf{0.1022}$ g

Median : $M = (0.1021 + 0.1023)/2 = \mathbf{0.1022}$ g

Range : $R = 0.1025 - 0.1019 = \mathbf{0.0006}$ g

Relative range (%) : $R_r(\%) = (0.0006/0.1022) \times 100 = \mathbf{0.59\%}$

Average deviation :
$\bar{d} = (0.0001 + 0.0003 + 0.0003 + 0.0001)/4 = \mathbf{0.0002}$ g

Relative average deviation (%) :
$\bar{d}_r(\%) = (0.0002/0.1022) \times 100 = \mathbf{0.2\%}$

Standard deviation :
$s = \sqrt{[(0.0001)^2 + (0.0003)^2 + (0.0003)^2 + (0.0001)^2]/(4-1)} = \mathbf{0.0003}$ g

Relative standard deviation (%) :
$s_r(\%) = (0.0003/0.1022) \times 100 = \mathbf{0.3\%}$

Error of the mean :
$E = 0.1022 - 0.1026 = \mathbf{-0.0004}$ g

Relative error of the mean (%):
$E_r(\%) = (-0.0004/0.1026) \times 100 = \mathbf{-0.4\%}$.

As the number of measurements increases, the experimental standard deviation, s, becomes a better estimate of the true standard deviation, σ ($s \to \sigma$, when $N \to \infty$).

Whenever there are different *sets* of precision data collected on separate samples of differing composition, it is preferable to calculate the *pooled standard deviation* of the method, s_p, from the pooled data. For this, one uses the equation

$$s_p = \sqrt{\frac{\sum_{i=1}^{N_1}(x_i - \bar{x}_1)^2 + \sum_{i=1}^{N_2}(x_i - \bar{x}_2)^2 + \ldots + \sum_{i=1}^{N_k}(x_i - \bar{x}_k)^2}{N - k}} \quad (1.15)$$

where $\bar{x}_1$, $\bar{x}_2 \ldots \bar{x}_k$ are the means of N_1, $N_2 \ldots N_k$ analyses for each of k different samples and N is the total number of measurements.

Example 1.3. The manganese in samples of six drinking waters was determined by a catalytic method. The results are given below. Calculate the standard deviation of the method, based on the pooled data.

Sample number	Number of replications	Mn, ppb	Mean, ppb Mn
1	4	1.92, 1.96, 2.01, 2.20	2.03
2	5	4.70, 4.88, 5.40, 4.96, 4.88	4.96
3	3	3.60, 3.16, 3.28	3.35
4	6	4.12, 3.86, 4.24, 4.32, 3.78, 3.90	4.04
5	4	1.14, 1.16, 1.28, 0.98	1.14
6	4	2.22, 2.30, 2.44, 2.08	2.26

Solution. The sum of the squares of the deviations from the mean, that is, the term $\sum(x_i - \bar{x}_j)^2 (j = 1, 2, \ldots 6)$ for sample 1 is calculated as follows:

x_i	$\mid x_i - \bar{x}_1 \mid$	$(x_i - \bar{x}_1)^2$
1.92	0.11	0.0121
1.96	0.07	0.0049
2.04	0.01	0.0001
2.20	0.17	0.0289
$\bar{x}_1 = 2.03$		$\sum(x_i - \bar{x}_1)^2 = 0.0460$

In the same way, the term $(x_i - \bar{x}_j)^2$ is calculated for the other samples. Therefore, we have

$$s_p = \sqrt{\frac{0.0460 + 0.2740 + 0.1035 + 0.2443 + 0.0456 + 0.0680}{26 - 6}} = \mathbf{0.20} \text{ ppb Mn}$$

Note. If the number of degrees of freedom (N−1) is at least 20, the calculated value of s can be considered to be a good approximation of σ.

1.4 Application of Statistics to Small Sets of Data

Statistics are used for the quantitative study of random errors. In quantitative analysis, we usually deal with only a few measurements per sample or set of samples. Statistics can nevertheless be appropriately applied. They allow questions like the following, arising from a set of measurements, to be answered.

1. On the basis of the mean, $\overline{x}$, and the standard deviation, s, what is the probability that the true value, μ, will fall within a certain interval about $\overline{x}$?

2. If the standard deviation of a single measurement is known, how many times should a measurement be replicated so that μ will fall, with a large probability (high confidence level), within a certain interval about $\overline{x}$ (Example 1.5)?

3. If two sets of N_1 and N_2 measurements give the means $\overline{x}_1$ and $\overline{x}_2$, respectively, how large should the difference $\overline{x}_1 - \overline{x}_2$ be in order to be considered as *significant* (i.e., not caused by random errors)? Such an answer is necessary in any comparison of results that were obtained with two different methods (Section 1.5, Equation (1.19), and Example 1.8 below) or by two analysts (Example 1.9).

4. If in a series of measurements of the same reliability, one result or more is questionable because of its large difference from the rest of the results, how large should this difference be so that the questionable result can be rejected in calculating a mean for the set?

5. If the standard deviations s_1 and s_2 of two sets of measurements obtained using two separate methods are different, how large should their difference be, so as to conclude with a high confidence level, that there is a significant difference in the precision of the two methods (the F-test, Equation (1.22))?

Typical examples of the applications of statistics to the solutions of these questions are given in the following sections.

Confidence limits and confidence interval. The *confidence limits* are the extreme values about $\overline{x}$, that define the *confidence interval*, that is, the interval within which the true value, μ, can be expected to fall with a certain probability or degree of confidence. This probability is usually expressed as a percentage and is defined as the *confidence level.* The higher the confidence level (i.e., the closer to 100% confidence), the greater necessarily will be the confidence interval to encompass μ.

A confidence level of 99% or 95% is usually considered as satisfactory, and in some cases, even 90% is used. The difference (100 − confidence level) is defined as the *error probability* and gives the probability that μ falls outside the confidence interval.

For a known σ and N measurements, the confidence limits are computed by the expression

$$\mu = \overline{x} \pm z\sigma/\sqrt{N} \tag{1.16}$$

Table 1.1. *Value of t for Various Degrees of Freedom and Confidence Levels.*

Confidence level, % $\nu = N - 1$	50	90	95	99	99.9
1	1.000	6.314	12.706	63.657	636.619
2	0.816	2.920	4.303	9.925	31.598
3	0.765	2.353	3.182	5.841	12.941
4	0.741	2.132	2.776	4.604	8.610
5	0.727	2.015	2.571	4.032	6.859
6	0.718	1.943	2.447	3.707	5.959
7	0.711	1.895	2.365	3.500	5.405
8	0.706	1.860	2.306	3.355	5.041
9	0.703	1.833	2.262	3.250	4.781
10	0.700	1.812	2.228	3.169	4.587
11	0.697	1.796	2.201	3.106	4.437
12	0.695	1.782	2.179	3.055	4.318
13	0.694	1.771	2.160	3.012	4.221
14	0.692	1.761	2.145	2.977	4.140
15	0.691	1.753	2.131	2.947	4.073
20	0.687	1.725	2.086	2.845	3.850
25	0.684	1.708	2.060	2.787	3.725
30	0.683	1.697	2.042	2.750	3.646
∞[a]	**0.674**	**1.645**	**1.960**	**2.576**	**3.291**

[a] For infinite measurements, t = z.

where z is a variable which gives the deviation of $\bar{x}$ from μ in units of standard deviation σ, that is, $z = (\bar{x} - \mu)/\sigma$. If σ is not known, the confidence limits are computed by the expression

$$\mu = \bar{x} \pm (ts)/\sqrt{N} \qquad (1.17)$$

where t is a variable which increases with increasing confidence level and decreasing degrees of freedom, ν ($\nu = N - 1$) (Table 1.1).

Example 1.4. The data obtained for the volumetric determination of Fe in an ore correspond to a normal distribution curve. The mean of a large number of determinations is 18.12%, and the method has a standard deviation $s \to \sigma = 0.04\%$. Calculate the 95% confidence interval for the result, if it was based upon a single analysis.

Solution. It is deduced from Figure 1.3 (or more accurately from statistical tables) that 95% of the population falls in the interval $\pm 1.960\sigma$ from the true value. Therefore, 95% of the experimental data will fall in the range

$18.12 \pm (1.960)(0.04)/\sqrt{1}$, that is, between **18.04%** and **18.20%**.

Example 1.5. The standard deviation σ of a single determination of copper is 0.10 μg/mL. a) What is the probability that a mean of 4 determinations will fall within $\pm 0.15 \mu$g/mL of the true value? b) From the result of (a), what is the probability that a mean of four determinations will fall more than 0.15 μg/mL above the true value? c) Within what range of the true value will 95% of the means of four determinations fall? d) What is the minimum number of determinations that should be made in order that their mean would fall within $\pm 0.07 \mu$g/mL of the true value 95% of the time?

Solution. a) the standard deviation of the mean, σ_x, for four determinations is $\sigma/\sqrt{N} = 0.10/\sqrt{4} = 0.05 \mu$g/mL. Hence, 0.15 μg/mL is $(0.15/0.05)\sigma_{\bar{x}} = 3.0\sigma_{\bar{x}}$. From Figure 1.3 (or more accurately from statistical tables), the probability of a mean falling within the range $\pm 3.0\sigma_{\bar{x}}$, that is, from -0.15 to $+0.15 \mu$g/mL about the true value, is **99.74%**.

b) From part (a), the probability of the mean of four determinations falling within the range 0 to +0.15 μg/mL of the true value is 99.75%/2 = 49.87%. Hence, the probability that the mean of four determinations is at least 0.15 μg/mL larger than the true value is 50.00 − 49.87 = **0.13%**.

c) From Figure 1.3, 95% of the means of four determinations fall within the range $\pm 1.960\sigma_{\bar{x}}$, that is, within $\pm(1.960)(0.05) = \pm$**0.10** μg/mL of the true value.

d) For a 95% probability that the mean of N determinations will fall within the range $\pm 0.07 \mu$g/mL about the true value, $|\ \bar{x} - \mu\ | = 1.960\sigma_{\bar{x}} = 0.07$, or $\sigma_{\bar{x}} = 0.07/1.960 = 0.0357$. Now $\sigma_{\bar{x}} = \sigma/\sqrt{N}$. Hence, $N = \sigma^2/\sigma_{\bar{x}}^2 = (0.10)^2/(0.0357)^2 = 7.8$, that is, at least **8** determinations should be made.

Example 1.6. To test the performance of a spectrophotometer, 60 replicate readings of the percent transmittance of a solution were taken. The standard deviation of these data was ±0.12% T. How many replicate readings should be taken for each subsequent measurement if the error of the mean due to the instrument is to be kept below a) ±0.15% T with 99% certainty? b) ±0.15% T with 95% certainty? c) ±0.10% T with 99% certainty? d) ±0.10% T with 95% certainty? e) What are the 95% and 99% confidence limits for a single measurement?

Solution. (a) Since N = 60, $s \simeq \sigma = 0.12\%$ T. From Table 1.1, for 99% confidence level:

$$(z\sigma)/\sqrt{N} = (2.576 \times 0.12)/\sqrt{N} = 0.15, \text{ or } N = 4.2 \rightarrow \mathbf{5} \text{ measurements.}$$

(a) For 95% confidence level:

$$(1.960 \times 0.12)/\sqrt{N} = 0.15, \text{ or } N = 2.5 \rightarrow \mathbf{3} \text{ measurements.}$$

(b) For 99% confidence level:

$$(2.576 \times 0.12)/\sqrt{N} = 0.10, \text{ or } N = 9.6 \rightarrow \mathbf{10} \text{ measurements.}$$

(c) For 95% confidence level:

$$(1.960 \times 0.12)/\sqrt{N} = 0.10, \text{ or } N = 5.5 \rightarrow \mathbf{6} \text{ measurements.}$$

(d) 95% confidence limits:

$$(\pm z\sigma)/\sqrt{N} = \pm(1.960 \times 0.12)/\sqrt{1} = \pm\mathbf{0.24}\%\ T.$$

99% confidence limits:

$$(2.576 \times 0.12)/\sqrt{1} = \pm\mathbf{0.31}\%\ T.$$

1.5 Statistical Tests of Significance

Experimental data rarely agree completely with those expected on the basis of a theoretical model. In addition, during the comparison of two sets of experimental data there is, as a rule, disagreement of values. In such cases, the arithmetic difference between two values can be attributed either to determinate errors, in which case it is significant, or to indeterminate errors. The checking of the significance of the observed differences is done with special statistical tests.

In all statistical tests, certain assumptions and hypotheses are made. For example, an indispensable assumption in each statistical test is that all samples used for making a judgment are random. A basis hypothesis is the *null hypothesis*, in which we presume that any differences in the numerical quantities being compared, e.g., $\bar{x}$ and μ, or $\bar{x}_1$ and $\bar{x}_2$, or s_1 and s_2, are due only to random errors. To check the correctness of the null hypothesis, we compute the probability that the observed differences are the result of random errors. Usually, if the observed difference is equal to or less than the difference expected in 95% of the cases (the 95% probability level), the null hypothesis is considered valid, and the difference is judged to be statistically insignificant. Some of the more common tests are described below.

Students' test, or t test. Comparison of the experimental mean $\bar{x}$ with the true value μ. The testing of the reliability of a method of analysis is usually done by employing that method in the replicate analysis of a standard sample, and comparing the value of $\bar{x}$ with μ. If $|\bar{x}-\mu| > ts/\sqrt{N}$ (Equation 1.17) or $|\bar{x}-\mu| > z\sigma/\sqrt{N}$ (Equation 1.16), $\bar{x}$ is judged significantly different from μ, at a predetermined confidence level. On the other hand, if $|\bar{x}-\mu| < ts/\sqrt{N}$ or $|\bar{x}-\mu| < z\sigma/\sqrt{N}$, it is judged that there is no significant difference between $\bar{x}$ and μ, although they may not be equal. The mean $\bar{x}$ is "equal" to μ in the broad limits of Table 1.1, that is, $\bar{x}$ can be appreciably different from μ and still be judged not significantly different from μ. A significant difference at the confidence level chosen is assumed to be due to determinate errors and not random errors.

The comparison of $\bar{x}$ with μ can also be done as follows. The value of t is computed by the expression

$$t = (|\mu - \bar{x}|\sqrt{N})/s \tag{1.17a}$$

and then is compared with the theoretical value of t, which is computed statistically (Table 1.1) for a predetermined confidence level. If $t_{exp} > t_{theor}$, it is concluded that there is significant difference (determinate error) between $\bar{x}$ and μ. On the other hand, if $t_{exp} < t_{theor}$, the difference is considered insignificant and is attributed to random errors. (Note that the absolute value of t is used without regard to the sign.)

Example 1.7. An emission spectroscopic method for the determination of Mn, Ni, and Cr in steel was tested by analyzing several samples of a National Bureau of Standards (NBS) standard reference material (SRM). From the data given below, decide if a determinate error in any of the analyses is indicated at the 95% confidence level. Assume that the NBS values are correct.

Element	Number of Analyses, A	Mean, %	Relative Standard Deviation, %	NBS Value, %
a) Mn	6	0.52	4.1	0.59
b) Ni	7	0.38	3.2	0.37
c) Cr	8	0.084	5.2	0.078

Solution. a) $\bar{x} - \mu = 0.52 - 0.59 = -0.07\%$ and $s = 0.52 \times 0.041 = 0.0213$. From Table 1.1, for $\nu = N - 1 = 6 - 1 = 5$, $t = 2.571$. Hence,

$$\frac{ts}{\sqrt{N}} = \frac{2.571 \times 0.0213}{\sqrt{6}} = 0.0224.$$

In 95% of cases, the mean $\bar{x}$ will differ from the true value μ by 0.0224% or less. Hence, at the 95% confidence level, the difference −0.07% is significant, and the existence of a negative determinate error is very probable.

We reach the same conclusion as follows:

$$\pm t = (\mu - \bar{x})\sqrt{N}/s = (0.59 - 0.52)\sqrt{6}/0.0213 = 8.05.$$

From Table 1.1, we find that at the 95% confidence level t has a value of 2.571 for five degrees of freedom. Since $8.05 > 2.571$, it is concluded that the existence of a (negative) determinate error is very probable. Note that by either method, the experimental value is 3.13 times that of the theoretical value.

b) $\bar{x} - \mu = 0.38 - 0.37 = 0.01\%$ and $s = 0.38 \times 0.032 = 0.0122$. From Table 1.1, for $\nu = 6$, we have $t = 2.447$. Hence,

$$ts/\sqrt{N} = (2.447 \times 0.0122)/\sqrt{7} = 0.0113 > 0.01 = \bar{x} - \mu.$$

Therefore, we conclude at the 95% confidence level that there is no determinate error.

c) $\bar{x} - \mu = 0.084 - 0.078 = 0.006$ and $s = 0.084 \times 0.052 = 0.0044$. From Table 1.1, for $\nu = 7$, we have $t = 2.365$. Hence,

$$ts/\sqrt{N} = (2.365 \times 0.0044)/\sqrt{8} = 0.0037 < 0.006 = \bar{x} - \mu.$$

Hence, at the 95% confidence level the difference 0.006% is significant and the presence of a (positive) determinate error is very probable.

Comparison of experimental means $\bar{x}_1$ and $\bar{x}_2$ (t test). The comparison of data obtained with two methods or by two persons or by the same person but under different experimental conditions is often indispensable in analytical chemistry for testing the validity of an analysis. For this purpose, the two experimental means $\bar{x}_1$ and $\bar{x}_2$ are compared using the t test. A similar comparison is done to establish whether two materials are identical or different. t is computed by the expression

$$t = \frac{\bar{x}_1 - \bar{x}_2}{s_{1-2}} \sqrt{\frac{N_1 N_2}{N_1 + N_2}}, \tag{1.18}$$

where s_{1-2} is the standard deviation of any individual value in each of the two sets of data, which is computed by the expression

$$s_{1-2} = \sqrt{\frac{\sum(x_{i_1} - \bar{x}_1)^2 + \sum(x_{i_2} - \bar{x}_2)^2}{N_1 + N_2 - 2}} \tag{1.19}$$

(Example 1.8). Equation (1.19) is a partial case of the general equation (1.15). In the special case where $N_1 = N_2 = N$, Equation (1.18) is simplified to

$$t = \frac{\bar{x}_1 - \bar{x}_2}{s_{1-2}} \sqrt{\frac{N}{2}} \tag{1.20}$$

Also, in the special case where $N_1 = N$ and $N_2 = \infty$, Equation (1.18) is simplified to Equation (1.17*a*). Then the computed value t is compared with the theoretical value t (Table 1.1), for $N_1 + N_2 - 2$ degrees of freedom and a predetermined confidence level, exactly as in the comparison of the mean $\bar{x}$ with the true value μ (Example 1.7).

Note. The employment of the t test for the comparison of the means of two sets of data (distributions) presupposes that the standard deviations of these two distributions do not differ significantly. To compare the two standard deviations, the F test is employed (Equation (1.22) below).

Paired t test. Two methods are often compared by analyzing several different samples. In this case, the comparison refers to paired data, and t is computed by the expression

$$t = |\bar{x}_A - \bar{x}_B| \sqrt{\frac{N(N-1)}{\sum(d_i - \bar{d})^2}} = \frac{\bar{d}\sqrt{N}}{s_d} \tag{1.21a)(1.21b}$$

where N is the number of paired samples, which are analyzed by methods A and B, d_i the positive or negative value of the differences $x_{i_A} - x_{i_B}$, $\bar{d}$ the mean of all individual differences, d_i, and s_d is the standard deviation of an individual difference.

Example 1.8. In the determination of calcium by two methods, standard method A and tested potentiometric method B, the following two sets of results were obtained with 6 individual samples. Determine whether the two methods differ significantly from each other at the 95% confidence level. Assume that the two methods have similar precisions.

Sample Number	Ca, mg/100 mL Standard Method A	Tested Method B	d_i	$d_i - \bar{d}$	$(d_i - \bar{d})^2$
1	21.0	20.4	−0.6	−1.2	1.44
2	23.8	25.4	1.6	1.0	1.00
3	17.4	17.2	−0.2	−0.8	0.64
4	33.8	35.0	1.2	0.6	0.36
5	21.8	22.4	0.6	0.0	0.00
6	22.2	23.0	0.8	0.2	0.04
	$\bar{x}_B = 23.3$	$\bar{x}_A = 23.9$	$\sum = 3.4$ $\bar{d} = 0.57$		$\sum = 3.48$

Solution. Substituting the computed values of $\bar{x}_A$, $\bar{x}_B$, $d_i - \bar{d}$, and $(d_i - \bar{d})^2$ in Equation (1.21*a*), we have

$$t_{exp} = | 23.90 - 23.33 | \sqrt{(6 \times 5)/3.48} = 1.67.$$

The same value of t_{exp} is also found using Equation (1.21*b*). From Table 1.1, t_{theor} = 2.571 at the 95% confidence level. Since $t_{exp} = 1.67 < t_{theor} = 2.571$, it is concluded that the two methods do not differ significantly from each other.

Comparison of variances, s^2 (F test). The F test is used to compare whether there is a significant difference in the precisions of two series of measurements. If the precisions are significantly different, then we cannot make comparisons of the means of the two measurements, for example, by the t-test. F is defined by

$$F = s_1{}^2/s_2{}^2 \tag{1.22}$$

where s_1 and s_2 are the standard deviations of the two sets of data ($s_1 > s_2$). Values of F are available in statistical tables for ν_1 and ν_2 degrees of freedom (i.e., for various numbers of measurements by each method) and for various confidence levels. Part of such a table is given in Table 1.2. When $F_{exp} > F_{theor}$, it is concluded that s_1 is statistically different from s_2, whereas when $F_{exp} \leq F_{theor}$, there is no significant difference between s_1 and s_2.

Example 1.9. The results obtained by two analysts for the carbon content of an organic compound are given below.

Analyst 1: 49.32, 49.41, 49.66, 49.45

Analyst 2: 49.09, 49.08, 49.25, 49.13, 49.10, 49.19

Compare a) the standard deviations and b) the mean values of the two sets of data.

Table 1.2. *Values of F for Various Degrees of Freedom and Confidence Levels.*

ν_2 \ ν_1	2	3	4	5	6	10	∞	Confidence levels, %[a]
2	9.00	9.16	9.24	9.29	9.33	9.39	9.49	90
	19.00	19.16	19.25	19.30	19.33	19.37	19.50	95
	99.00	99.17	99.25	99.30	99.33	99.36	99.50	99
3	5.46	5.39	5.34	5.31	5.28	5.23	5.13	90
	9.55	9.28	9.12	9.01	8.98	8.78	8.53	95
	30.82	29.46	28.71	28.24	27.91	27.23	26.12	99
4	4.32	4.19	4.11	4.06	4.01	3.92	3.76	90
	6.94	6.59	6.39	6.26	6.16	5.96	5.63	95
	18.00	16.69	15.98	15.52	15.21	14.54	13.46	99
5	3.78	3.62	3.52	3.45	3.40	3.30	3.10	90
	5.79	5.41	5.19	5.05	4.95	4.74	4.36	95
	13.27	12.06	11.39	10.97	10.67	10.05	9.02	99
6	3.46	3.29	3.18	3.11	3.05	2.94	2.72	90
	5.14	4.76	4.53	4.39	4.28	4.06	3.67	95
	10.92	9.78	9.15	8.75	8.47	7.87	6.88	99
10	2.92	2.73	2.61	2.52	2.46	2.32	2.06	90
	4.10	3.71	3.48	3.33	3.22	2.97	2.54	95
	7.56	6.55	5.99	5.64	5.39	4.85	3.91	99
∞	2.30	2.08	1.94	1.85	1.77	1.60	1.00	90
	3.00	2.60	2.37	2.21	2.10	1.83	1.00	95
	4.61	3.78	3.32	3.02	2.80	2.32	1.00	99

[a] Confidence level is the probability that the function $F = s_1^2/s_2^2$ will be smaller than the tabular value. The term (100 - confidence level) is the *error probability*, that is, the probability that the function $F = s_1^2/s_2^2$ will be larger than the tabular value.

Solution.

a) The standard deviations of the two sets of data are computed as follows:

x_{i_1}	$\mid x_{i_1} - \overline{x}_1 \mid$	$(x_{i_1} - \overline{x}_1)^2$	x_{i_2}	$\mid x_{i_2} - \overline{x}_2 \mid$	$(x_{i_2} - \overline{x}_2)^2$
49.32	0.14	0.0196	49.09	0.05	0.0025
49.41	0.05	0.0025	49.08	0.06	0.0036
49.66	0.20	0.0400	49.25	0.11	0.0121
49.45	0.01	0.0001	49.13	0.01	0.0001
			49.10	0.04	0.0016
			49.19	0.05	0.0025
$\overline{x}_1 = 49.46$		$\sum(x_{i_1} - \overline{x}_1)^2 = 0.0622$			
			$\overline{x}_2 = 49.14$		$\sum(x_{i_2} - \overline{x}_2)^2 = 0.0224$

$s_1^2 = 0.0622/3 = 0.0207$ and $s_1 = 0.14$, $s_2^2 = 0.0224/5 = 0.0045$ and $s_2 = 0.07$. From Table 1.2, $F = 5.41$ for $\nu_1 = 3$ and $\nu_2 = 5$ at the 95% confidence level. We have

$$F_{exp} = s_1^2/s_2^2 = 0.0207/0.0045 = 4.60$$

The value of F will exceed 5.41 (Table 1.2) only 5% of the time. Hence, it is concluded that the value of F is not statistically significant and, therefore, the t test can be applied (Equation (1.18)).

b) We have

$$s_{1-2} = \sqrt{(0.0622 + 0.0224)/(4 + 6 - 2)} = 0.103$$

and

$$t = [(49.46 - 49.14)/(0.103)]\sqrt{(4 \times 6)/(4 + 6)} = 4.8$$

The values of t for $\nu = 4 + 6 - 2 = 8$ are 2.306 and 3.355 at the 95% and 99% confidence levels, respectively. Since $t_{exp} > t_{theor}$ at the 99% confidence level, the difference $\overline{x}_1 - \overline{x}_2 = 49.46 - 49.14 = 0.32$ is statistically significant.

The Chi-square (χ^2) test. The χ^2 test is used to determine whether the frequency of occurrence of an event is statistically different from the expected frequency. The quantity χ^2 is defined by

$$\chi^2 = \sum \frac{(f_i - F_i)^2}{F_i} \tag{1.23}$$

where f_i is the observed frequency, F_i is the expected frequency, and the summation refers to all classes of observations. Each column heading in Table 1.3 gives the probability, P, that the tabular value for ν degrees of freedom will be exceeded.

Example 1.10. The following numbers denote the frequency of occurrence of each of the digits in the last significant figure of a buret reading, with an estimate of 0.01 mL. Determine whether these data show a bias during the reading of the buret.

Table 1.3. *Values of χ^2 for Various Degrees of Freedom.*

ν \ P	0.995	0.99	0.95	0.90	0.50	0.10	0.05	0.01	0.005
1	0.00004	0.00016	0.0039	0.0158	0.45	2.71	3.84	6.63	7.88
2	0.0100	0.0201	0.1026	0.211	1.39	4.61	5.99	9.21	10.60
3	0.1717	0.115	0.352	0.584	2.37	6.25	7.81	11.34	12.84
4	0.207	0.297	0.711	1.064	3.36	7.78	9.49	13.28	14.86
5	0.412	0.554	1.15	1.61	4.35	9.24	11.07	15.09	16.75
6	0.676	0.872	1.64	2.20	5.35	10.64	12.59	16.81	18.55
7	0.99	1.239	2.17	2.83	6.35	12.02	14.07	18.48	20.28
8	1.34	1.65	2.73	3.49	7.34	13.36	15.51	20.09	21.96
9	1.73	2.09	3.33	4.17	8.34	14.68	16.92	21.67	23.59
10	2.16	2.56	3.94	4.87	9.34	15.98	18.31	23.21	25.19
11	2.60	3.05	4.57	5.58	10.34	17.28	19.68	24.72	26.76
12	3.07	3.57	5.23	6.30	11.34	18.55	21.03	26.22	28.30
13	3.57	4.11	5.89	7.04	12.34	19.81	22.36	27.69	29.82
14	4.07	4.66	6.57	7.79	13.34	21.06	23.68	29.14	31.32
15	4.60	5.23	7.26	8.55	14.24	22.31	25.00	30.58	32.80
20	7.43	8.26	10.85	12.44	19.34	28.41	31.41	37.57	40.00
25	10.52	11.52	14.61	16.47	24.34	34.38	37.65	44.31	46.93
30	13.79	14.95	18.49	20.60	29.34	40.26	43.77	50.89	53.67

Digit	Frequency	Digit	Frequency
0	208	5	156
1	73	6	81
2	76	7	83
3	87	8	97
4	82	9	57

Solution. The sum of all readings is 1000. Hence, the expected frequency of occurrence of each of the 10 digits, F_i, is $1000/10 = 100$. Therefore, we have the following table:

Last Digit	f_i	$\lvert f_i - F_i \rvert$	$(f_i - F_i)^2$
0	208	108	11664
1	73	27	729
2	76	24	576
3	87	13	169
4	82	18	324
5	156	56	3136
6	81	19	361
7	83	17	289
8	97	3	9
9	57	43	1849
	$\sum f_i = 1000$		$\sum (f_i - F_i)^2 = 19106$

Hence, $\chi^2 = 19106/100 = 191.1$. From Table 1.3, $\chi^2 = 23.59$ for 9 degrees of freedom and $P = 0.005$. Since $191.1 >> 23.59$, it is concluded that during the reading of the buret there is a bias in favor of digits 0 and 5. It should be noted that this bias decreases the accuracy of the reading by an individual. Such a bias is different for each person.

1.6 Tests for Rejection of Data

When a set of data contains a value that differs considerably from the other values, the rejection or retention of such a value can be made on the basis of the **Q-test**. (It should be noted that a result is rejected *a priori* whenever it is known that a determinate error was made in the respective measurement.)

The Q-test is the most often used test for small numbers of measurements. Q is defined as the ratio given by the difference of the suspect value from its nearest neighbor divided by the range of the results. Theoretical values of Q are available in statistical tables (Table 1.4). If $Q_{exp} > Q_{theor}$, the suspect result can be rejected. If there is more than one suspect value, they are arranged in increasing order of numbers, $x_1, x_2 \ldots x_{N-1}, x_N$, the smallest value x_1 is checked, then the largest, X_N, and so on. The value of Q_{exp} is always less than unity. The Q test can not be applied in the case of three replicates, when two coincide, because in such a case $Q_{exp} = 1$, and the suspect value would always be rejected, regardless of the magnitude of its deviation from the two identical results.

Example 1.11. Apply the Q test in order to determine whether any of the following data should be rejected.

a) 70.10, 69.62, 69.70

Table 1.4. *Values of Q for Various Numbers of Replicates and Confidence Levels.*

Suspect Result	Formula for the test	N	$Q_{0.90}^{a}$	$Q_{0.95}$	$Q_{0.99}$
Smallest value, x_1	$Q = \frac{x_2 - x_1}{x_N - x_1}$	3	0.94	0.98	0.99
		4	0.76	0.85	0.93
		5	0.64	0.73	0.82
		6	0.56	0.64	0.74
Largest value, x_N	$Q = \frac{x_N - x_{N-1}}{x_N - x_1}$	7	0.51	0.59	0.68
		8	0.47	0.54	0.63
		9	0.44	0.51	0.60
		10	0.41	0.48	0.57

[a] If $Q_{exp} > Q_{theor}$, the suspect result is rejected.

b) 70.10, 69.62, 69.70, 69.65

Solution. a) At first glance, the small difference between the values 69.62 and 69.70 justifies the rejection of the suspect value 70.10. However, with the Q test the suspect result is retained because

$$Q_{exp} = \frac{70.10 - 69.70}{70.10 - 69.62} = 0.83 < 0.94 = Q_{0.90} \text{ (for N = 3, Table 1.4).}$$

b) $Q_{exp} = \frac{70.10 - 69.70}{70.10 - 69.62} = 0.83 > 0.76 = Q_{0.90}$ (for N = 4, Table 1.4). Hence, the value 70.10 should be rejected.

Notes. 1) The series of data in b contains the same data as in a, and in addition the value 69.65. Because this value falls near (and between) the other two valid numbers, more credence is added to those values, and the suspect value becomes relatively less valid, as indicated by the Q test.

2) If the suspect value is retained with the Q test, it is often better to report the median, M, of the set instead of the mean, because M is insensitive to the divergent value and all data are taken into account for its computation. On the other hand, $\bar{x}$ is very sensitive to a divergent value, especially when N is very small.

1.7 Propagation of Errors in Computations

Table 1.5. *Error of a Computed Result.*

Computation	Determinate Error	Random Error
Addition or subtraction, $y = a + b - c$	$E_y = E_a + E_b - E_c$	$s_y^2 = s_a^2 + s_b^2 + s_c^2$
Multiplication or division, $y = ab/c$	$\frac{E_y}{y} = \frac{E_a}{a} + \frac{E_b}{b} - \frac{E_c}{c}$	$\left(\frac{s_y}{y}\right)^2 = \left(\frac{s_a}{a}\right)^2 + \left(\frac{s_b}{b}\right)^2 + \left(\frac{s_c}{c}\right)^2$
General, $y = f(a, b, c \ldots)$	$E_y = \frac{\partial y}{\partial a}E_a + \frac{\partial y}{\partial b}E_b + \ldots$	$s_y^2 = \left(\frac{\partial y}{\partial a}\right)^2 s_a^2 + \left(\frac{\partial y}{\partial b}\right)^2 s_b^2 + \ldots$

It is often necessary to compute the error in the result of a composite measurement, that is, in a result which was computed from more than one datum, a, b, c..., each of which has an error associated with it. In such cases, the effect of the individual errors associated with a, b, c... on the computed quantity y depends 1) on how y is related to a, b, c... and 2) on the types of errors (determinate or indeterminate). Table 1.5 summarizes how the error of a computed result is related to the errors of the individual data. In the case of determinate errors, for addition or subtraction, the *absolute* error of the final result, E_y, is determined by the individual *absolute* determinate errors. For multiplication or division, the *relative* error of the final result, E_y/y, is determined by the *relative* determinate errors, associated with a, b, c... In the case of random errors, for addition or subtraction, the variance of the final result, s_y^2, is determined by the variances of the terms a, b, c... In multiplication or division, the square of the relative standard deviation of the final result, $(s_y/y)^2$, is determined by the squares of the individual relative deviations.

Example 1.12. Calculate a) the error in the result and b) the standard deviation of the result of the following calculation, where the numbers in parentheses are absolute determinate errors (case a) or absolute standard deviations (case b, neglect sign):

$$\begin{array}{r} +1.00(+0.03) \\ +5.40(-0.04) \\ \underline{-3.24}(-0.06) \\ 3.16 \end{array}$$

Solution. a) From Table 1.5, the error in the result is E_y (or Δ_y) $= 0.03 + (-0.04) - (0.06) = \mathbf{+0.05}$.

b) Since the numbers in parentheses refer to absolute standard deviations, from Table 1.5,

$$s_y = \sqrt{(0.03)^2 + (0.04)^2 + (0.06)^2} = \mathbf{0.08}.$$

Hence, $y = 3.16(\pm 0.08)$.

Example 1.13. Calculate the standard deviation of the result of the following computation:

$$y = \frac{(8.20 \pm 0.04)(0.0100 \pm 0.0002)}{(3.94 \pm 0.08)} = 0.0208.$$

Solution. Since the computation involves multiplication and division, from Table 1.5,

$$\frac{s_y}{y} = \sqrt{\left(\frac{0.04}{8.20}\right)^2 + \left(\frac{0.0002}{0.0100}\right)^2 + \left(\frac{0.08}{3.94}\right)^2} = 0.029.$$

Hence, $s_y = 0.029 \times 0.0208 = \mathbf{0.0006}$ and $y = 0.0208(\pm 0.0006)$.

1.8 Significant Figures

When a computation is based on experimental data, the error or the uncertainty of the final result can be computed as previously described. A rough estimate of the uncertainty is obtained from the number of significant figures. *Significant figures* are considered the digits of a number known with certainty plus the first uncertain one. In general, the last digit of a number is considered to have an uncertainty of ± 1, except if information is given for the existence of greater uncertainty. For example, a weight of 1.1925 g means not exactly 1.1925 g, but a value in the range 1.1925 ± 0.0001 g, or $1.1924 - 1.1926$ g. On the other hand, when it is stated that a rock is 5.44 ± 0.05 million years old, there is an uncertainty of 0.05 million years, that is, fifty thousand years.

As a rule, only significant figures should be included in the value given for a physical quantity, neither more nor fewer than the order of magnitude of the probable error of this value. For example, if the accuracy of a method for the determination of copper is 99.9% (relative error 0.1%), and the results of an analysis are given as 9.574%, 9.6%, and 9.57%, only the value 9.57% is correct. The value 9.574% denotes a higher accuracy (relative error of only about 0.01%) than the real one, whereas the value 9.6% denotes a lower accuracy (relative error about 1%).

Special attention is required in deciding whether the zeroes bounded by digits on both sides are significant figures. Zeroes at the beginning of a number are never significant figures, because they simply determine the position of the decimal point. For example, the weight of a precipitate can be expressed as 0.0713 g or 71.3 mg, without any change in the uncertainty (± 0.0001 g or ± 0.1 mg) or the number of significant figures (three). Zeroes at the end of a number are considered significant figures. If the last zero of a number is not a significant figure, it should be eliminated and the number expressed using exponential notation. For example, the age of the earth is written as 4.6 billion years or 4.6×10^9 years, because it is known with a certainty of 0.1 billion years and not 1 year.

In carrying out numerical computations, the accuracy of the final result should be of the same order of magnitude as the least precise number. Specifically, computations are carried out as follows:

1. Non-significant digits are dropped. In rounding numbers, if the residue is larger than 5, the last retained digit is increased by one, whereas if the residue is smaller than 5, the last digit is retained unchanged. If the residue is 5, the last digit is retained unchanged if even, and increased by one if odd. For better consistency, one or two more digits than necessary are generally retained in rounding intermediate factors during a multistep calculation, and all uncertain digits are dropped in the final result.

2. In addition or subtraction, the result cannot be more precise than the value with the least number of digits beyond the decimal place, and consequently the largest *absolute* uncertainty. Therefore, as many digits beyond the decimal place are retained in the result as there are in the number with the least number of digits beyond the decimal. In addition and subtraction, the number of significant figures or digits beyond the decimal in the answer may be drastically reduced from those of some of the individual numbers.

For example,

$$\begin{array}{rr} & 22.1 \\ + & 2.21 \\ & 0.221 \\ \hline & 24.5 \end{array} \qquad \begin{array}{r} 23.1572 \\ -23.1539 \\ \hline 0.0033 \end{array}$$

In the addition, each number added has 3 significant figures, but the number 22.1 has only 1 decimal place, and therefore, the sum 24.5 also has only 1 decimal place. In the subtraction, the difference has only 2 significant figures, although the minuend and the subtrahend have 6 significant figures, because these two numbers differ very little in magnitude. Therefore, the accuracy of the difference is very small (1/33 or 3/100).

If an addition or subtraction includes numbers with positive or negative exponents, the numbers should be written as powers with the same exponent. For example, suppose that the following concentrations of hydrogen ions have to be added: 3.00×10^{-2} $\underline{M}$, 2.28×10^{-3} $\underline{M}$, and 1×10^{-6} $\underline{M}$. The numbers are written and added, as follows:

$$\begin{array}{c} 3.00 \times 10^{-2}\underline{M} \\ 0.228 \times 10^{-2}\underline{M} \\ 0.0001 \times 10^{-2}\underline{M} \end{array}$$

Sum = 3.2281×10^{-2} $\underline{M}$ = 3.23×10^{-2} $\underline{M}$. The accuracy of the sum cannot be better than $\pm 0.01 \times 10^{-2}$ $\underline{M}$; therefore, the sum is written as 3.23×10^{-2} $\underline{M}$ and not as 3.2281 $\times 10^{-2}$ $\underline{M}$.

3. In multiplication or division, the number of significant figures in the product or the quotient depends on the *relative* uncertainty of the multiplied or divided numbers,

and it should be of the same order as that of the least precise component. That is, it should lie in the range 0.2x–2x, where x is the largest relative uncertainty. For example, suppose the product of $0.0145 \times 5.25 \times 2.0265$ is requested. The relative uncertainties of the terms of the product are $(1/145) \times 100 = 0.69\%$, $(1/525) \times 100 = 0.19\%$, and $(1/20265) \times 100 = 0.0049\%$. Since the first term of the product has the largest relative uncertainty (key number), this component determines the relative uncertainty of the product, and therefore the number of significant figures. Hence, $0.0145 \times 5.25 \times 2.065 = 0.154$. The percent uncertainty of the product is $(1/154) \times 100 = 0.65\%$, that is, of the same order as the 0.69% uncertainty of the least precise term.

Whenever the answer in a multiplication or division lies close to 1 times a power of 10, care should be taken in determining the number of significant figures. For example, if the normalities of two solutions are 0.0999 and 0.1001, they practically have the same uncertainty, although they have 3 and 4 significant figures, respectively. In a multiplication or division, the number with the least number of significant figures is called the *key number* because it is known the least precisely. If the answer is of a smaller magnitude than the key number, ***without regard to the decimal point***, then one more digit may be carried. This extra digit usually is carried as a subscript to indicate it is no more precise than the last digit in the larger (key) number. **For example, in the multiplication $0.782 \times 365.3 = 285._7$, the key number 0.782 is known to $(1/782) \times 1000 = 1.3‰$, and the answer is known no more precisely than this, and the subscript indicates this. The subscript (extra digit) is carried because the answer is known better than $(1/286) \times 1000 = 3.5‰$.**

4. Logarithmic terms, e.g., pH or pM, as a rule, should have as many digits to the right of the decimal as there are significant figures in the original number. Similarly, the antilog should have as many significant figures as there are digits to the right of the decimal. For example, if $[H^+] = 3.4 \times 10^{-4}$ $\underline{M}$, pH = 3.47. The number 3 (the characteristic of the logarithm) simply denotes the order of magnitude and only 0.47 is a measure of the accuracy of $[H^+]$ (the relative accuracy of the decimal part of the logarithm, i.e., $(1/47) \times 100 = 2.1\%$, is a good measure of the relative accuracy of $[H^+]$ (2.9%)). Respectively, if $pK_a = 4.757$, $K_a = 1.75 \times 10^{-5}$.

In solving problems in this text, it is assumed that the data agree with the aforementioned rules regarding significant figures just discussed. In problems where such expressions are required as "1 g sample," "0.1 $\underline{M}$ solution," "25 mL pipet," and so forth, it can be assumed that the weight of the sample, the molarity of the solution, and the volume of the pipet are known with a precision at least equal to the precision of the other terms in the problem.

The estimate of an analytical error. The accuracy of a multistep analysis can not be better than the accuracy of the least accurate step. The propagation of errors should be taken into account in estimating the accuracy of an analysis.

Example 1.14. In the standardization of NaOH, 38.76 mL of NaOH was required for 0.8169 g of potassium acid phthalate. Estimate the accuracy we should expect in the molarity of the NaOH solution.

Solution. Let the weighing error be ±0.0001 g and the buret reading error be ±0.02 mL, and assume an additional uncertainty of ±0.03 mL, associated with the determination of the indicator color at the end point. Since two weighings and two buret readings are involved, the uncertainty (standard deviation) in the weight of potassium acid phthalate is the square root of the sum of the variances (s^2) of the two readings, i.e.,

$$s_1 = \sqrt{(0.0001)^2 + (0.0001)^2} = \pm 0.00014 \text{ g},$$

whereas the uncertainty in the volume of NaOH is

$$s_2 = \sqrt{(0.02)^2 + (0.02)^2 + (0.03)^2} = \pm 0.041 \text{ mL}.$$

The corresponding relative standard deviations are $(0.00014/0.8169) \times 1000 = 0.17‰$ and $(0.041/38.76) \times 1000 = 1.1‰$. Since the formula weight of potassium acid phthalate is known very accurately, the uncertainty in its value can be ignored. Hence, the standard deviation, and consequently the expected accuracy in determining the molarity of the NaOH solution is

$$s_3 = \sqrt{(0.17)^2 + (1.1)^2} = \mathbf{1.1‰}.$$

Note. The value of s_3 increases as the weight of potassium acid phthalate decreases. For example, if the above weight of potassium acid phthalate is reduced to one tenth, the volume of NaOH will also be reduced to one tenth, and (using percentages)

$$s_3 = \sqrt{(0.17)^2 + \left(\frac{0.041 \times 100}{3.88}\right)^2} = 1.1\%.$$

1.9 Statistics of Linear Relationships

Linear least squares. In analytical chemistry, a calibration curve is often prepared by plotting a parameter y as a function of variable x, for example, the absorbance of a certain wavelength of light by a solution as a function of concentration. This relation is found with a series of measurements. Then we have the relation

$$y = f(x) \tag{1.24}$$

In practice, the relation is often linear. Then Equation (1.24) becomes

$$y = ax + b \tag{1.25}$$

where a is the slope of the linear curve (the change of y for a change in x equal to unity) and b is the intercept on the y axis (the value of y for x = 0). For the objective of plotting of the best straight line, usually the *method of least squares* is used. The method is based on the principle that the best line is that for which the sum of the squares of the deviations is minimum. While the method is applicable to non-linear relations as well, we will consider the linear case only. For the case of Equation (1.25), this sum, S, is given by

$$S = \sum_{i=1}^{N} (ax_i + b - y_i)^2 \tag{1.26}$$

From Equation (1.26), by differential calculus we find

$$a = \frac{N\sum x_i y_i - \sum x_i \sum y_i}{N\sum x_i^2 - (\sum x_i)^2} \tag{1.27}$$

$$b = \frac{\sum x_i^2 \sum y_i - \sum x_i \sum x_i y_i}{N\sum x_i^2 - (\sum x_i)^2} \tag{1.28}$$

On the basis of the calculated values of the coefficients a and b, a *regression line* is plotted which gives the best linear fitting of the data.

Example 1.15. From the following data determine the linear least squares line. Assume a linear relation between y and x.

x	1	2	3	4	5	6	7	8
y	3.1	6.0	8.7	12.9	15.3	17.9	22.0	23.7

Solution. We compute the values of x_i^2 and $x_i y_i$ and write the results in the following table:

N	x_i	y_i	x_i^2	$x_i y_i$
1	1	3.1	1	3.1
2	2	6.0	4	12.0
3	3	8.7	9	26.1
4	4	12.9	16	51.6
5	5	15.3	25	76.5
6	6	17.9	36	107.4
7	7	22.0	49	154.4
8	8	23.7	64	189.6
	$\sum x_i = 36$	$\sum y_i = 109.6$	$\sum x_i^2 = 204$	$\sum x_i y_i = 620.3$

Substituting the values of the above table in Equations (1.27) and (1.28), we have

$$a = \frac{(8 \times 620.3) - (36 \times 109.6)}{(8 \times 204) - (36)^2} = 3.03$$

$$b = \frac{(204 \times 109.6) - (36 \times 620.3)}{(8 \times 204) - (36)^2} = 0.082$$

Therefore, the equation of the regression line is

$$y = 3.03x + 0.082$$

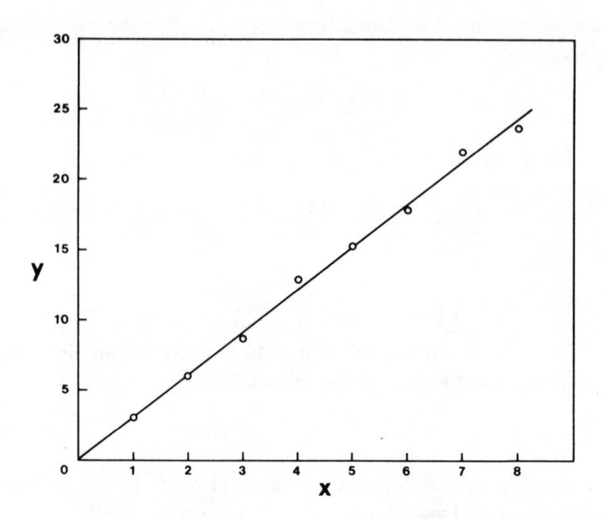

Figure 1.5. *Regression line for Example 1.15.*

The regression line and the experimental points are given in Figure 1.5.

Note. It should be noticed that the method of least squares is applied under the following presuppositions: 1) The measured values of y correspond to determined values of x (free of error). 2) The error in the values of y (expressed as the variance of y) is independent of the magnitude of x. 3) The straight line should be a satisfactory approximation of the way variable y depends on variable x.

The correlation coefficient. The *Pearson correlation coefficient*, r, is used as a measure of the linear correlation between two variables, x and y, and is given by

$$r = \frac{\sum x_i y_i - N\bar{x}\bar{y}}{\sqrt{(\sum x_i^2 - N\bar{x}^2)(\sum y_i^2 - N\bar{y}^2)}} = \frac{N\sum x_i y_i - \sum x_i \sum y_i}{\sqrt{[N\sum x_i^2 - (\sum x_i)^2][N\sum y_i^2 - (\sum y_i)^2]}} \quad (1.29a), (1.29b)$$

The correlation coefficient can vary from -1 to $+1$. A positive value of r indicates a positive correlation between the two variables (high values of x correspond to high values of y), whereas a negative value indicates a negative correlation. When $r = 0$, there is no correlation between x and y.

The correlation coefficient can be calculated for a calibration curve to ascertain the degree of correlation between the measured value (e.g., the absorbance of a solution) and the concentration of the solution, that is, to determine the linearity of the calibration curve.

As a rule, $|r| > 0.99$ indicates very good linearity. An $r > 0.999$ can sometimes be obtained with special care (usually, in the value of r we retain at least one digit beyond the last 9). the correlation coefficient is useful for the comparison of two methods (Example 1.8).

Example 1.16. Calculate the correlation coefficient, r, for the data in Example 1.8.

Solution. Using the data of Example 1.8, we construct the following table.

x_i	y_i	$x_i y_i$	x_i^2	y_i^2
21.0	20.4	428.40	441.00	416.16
23.8	25.4	604.52	566.44	645.16
17.4	17.2	299.28	302.76	295.84
33.8	35.0	1183.00	1142.44	1225.00
21.8	22.4	488.32	475.24	501.76
22.2	23.0	510.60	492.84	529.00
$\sum x_i = 140.0$	$\sum y_i = 143.4$	$\sum x_i y_i = 3514.12$	$\sum x_i^2 = 3420.72$	$\sum y_i^2 = 3612.92$

Substituting the above values in Equation (1.29*b*), we have

$$r = \frac{(6)(3514.12) - (140.0)(143.4)}{\sqrt{[(6)(3420.72) - (140.0)^2][(6)(3612.92) - (143.4)^2]}} = \mathbf{0.994}$$

Note. The high value of r indicates that the tested potentiometric method is practically as good as the standard method, i.e., there is a high degree of correlation between the results obtained by either method.

1.10 Problems

Errors in quantitative analysis

1.1. The following results were obtained for a sample that contains 25.00% KCl. Does the method contain a determinate error?

Sample Taken, g	KCl Found, g
0.4000	0.0978
0.3600	0.0878
0.3800	0.0927
0.4400	0.1079

Accuracy and precision

1.2. A 2.6450 g sample of iron ore that contains 53.51% Fe is dissolved and diluted to 250 mL. A spectrophotometric method gave the following results for the solution: $x_1 = 5.84$ mg Fe/mL, $x_2 = 5.77$ mg Fe/mL, $x_3 = 5.65$ mg Fe/mL, $x_4 = 5.66$ mg Fe/mL. a) Estimate whether the accuracy of the method is satisfactory, if the acceptable error is smaller than 1%, and b) compute the standard deviation.

1.3. A sample contains about 25% of the ion A^{2-}. A^{2-} is determined by precipitating the compound M_2A. M has an atomic weight about 3.5 times that of A. If the uncertainty in determining the weight of the precipitate is not to exceed 1% on a balance accurate to 0.1 mg, what is the minimum size of the sample that should be taken for analysis?

1.4. The following absorbance readings were obtained in the spectrophotometric determination of glucose in successive serum samples of a patient. Compute the pooled standard deviation.

Day 1 (Sample A)	Day 2 (Sample B)	Day 3 (Sample C)
0.886	0.777	0.704
0.890	0.722	0.675
0.900	0.709	0.687
0.885	0.744	0.697

1.5. To check the accuracy of a new method, a standard copper alloy containing 23.24% Cu was analyzed and calculated to contain 23.17% Cu. Calculate the relative error of the method in ppt.

Application of statistics to small sets of data

1.6. The following results were obtained for the weight of an object: 55.10, 55.30, 55.50 mg. The true weight is 54.90 mg. a) Calculate the 95% confidence limits for the mean, assuming 1) no additional knowledge about the precision and 2) that σ for a single measurement is 0.24 mg. b) Is there a significant difference between the mean $\bar{x}$ and the true value μ at the 95% confidence level? c) What is the probability of the above conclusion being incorrect?

1.7. The following results were obtained during the gravimetric and volumetric determination of iron in sample A. Gravimetric: 26.53, 26.45, 26.44, 26.52, 26.40; Volumetric: 26.46, 26.48, 26.56, 26.41, 26.59. Compute the 90, 95, and 99% confidence intervals for the means of the gravimetric and volumetric methods.

1.8. If the true percentage of Fe in a sample is 21.32 and the standard deviation of the method is 0.15%, what is the probability that the result for a single measurement will fall in the range a) 21.17–21.47%, b) 21.00–21.50%?

1.9. The standard deviation, σ, of a single determination of glucose in serum by an automated method is 3 mg/dL (dL = 0.1 L). a) What is the probability that a mean of 2, 4, or 9 determinations will fall within more than ± 2 mg/dL of the true value? b) What is the probability that a mean of 4 determinations will fall more than 2 mg/dL above the true value? c) Within what range of the true value will 95% of the means of 2, 4, or 9 determinations fall? d) What is the minimum number of determinations that should be made in order that their mean will fall within ± 6, ± 3, or ± 1 mg/dL of the true value 99% of the time?

1.10. To test the performance of a pH meter, a large number of readings was taken with a buffer. The standard deviation, σ, was ± 0.025 pH unit. What is the minimum number of determinations that should be made in order that their mean would fall within

a) ± 0.050 pH unit (of the true value) with 95% certainty?
b) ± 0.020 pH unit with 95% certainty?
c) ± 0.010 pH unit with 95% certainty?
d) ± 0.050 pH unit with 99% certainty?
e) ± 0.020 pH unit with 99% certainty?
f) ± 0.010 pH unit with 99% certainty?

1.11. If the true value for the iron content of a sample is 10.39%, and the mean and standard deviation of 4 determinations are 10.21% and 0.078%, respectively, determine if a determinate error is indicated at the 95% and 99% confidence levels.

Statistical tests

1.12. Two students analyze the same sample. Student A makes 6 determinations and obtains $s_A = 0.14$, whereas student B makes 5 determinations and obtains $s_B = 0.05$. Are s_A and s_B significantly different at the 95% confidence level?

1.13. The following results were obtained during the testing of the validity of a method by analyzing an NBS sample that contains 21.20% Ca (true value, μ): $\bar{x} = 21.24\%$, $s = 0.12$, $N = 10$. Is there a significant difference between $\bar{x}$ and μ at the 90% and 99% confidence levels?

1.14. The only information for bottles A and B found in a storehouse was that they contained calcium carbonate. The following results were obtained for their calcium content. Are these results significantly different at the 95% confidence level?

% Ca

Bottle A	*Bottle B*
40.07	39.78
39.99	39.70
40.00	39.78
	39.73

1.15. The following results were obtained in the determination calcium in serum by two methods, fluorimetry and atomic absorption spectroscopy (AAS). Is there a significant difference in the precision of the two methods?

Ca, mg/100 mL	
AAS	*Fluorimetry*
11.4	9.7
10.6	10.2
11.1	12.0
11.7	12.1
10.2	9.8
10.5	10.6
	11.7

1.16. The results for the uranium content of two samples are:

Sample 1: 10.32, 10.39, 10.19, 10.21
Sample 2: 10.45, 10.41, 10.51, 10.36, 10.34, 10.39

Are the two samples significantly different at the 95% confidence level?

Rejection of a result

1.17. Use the Q test to test the following results: 6.02, 7.28, 7.32, 7.02, 7.92, 7.12, 7.32, 7.22.

Propagation of errors in computation

1.18. A 0.2125 ± 0.0001 g impure sample containing $CaCO_3$ and inert materials is dissolved in 50.00 ± 0.05 mL of 0.2250 ± 0.0002 $\underline{M}$ HCl and the excess HCl is titrated with 26.47 ± 0.02 mL of 0.2872 ± 0.0003 $\underline{M}$ NaOH. Calculate the percentage of $CaCO_3$ present in the sample and estimate the accuracy we should expect in its determination.

Significant figures

1.19. The meniscus of the liquid contained in a 1.000-mL microburet can be read with an accuracy of 0.001 mL. What is the smallest volume, in milliliters, that should be used in a titration in order to insure that the volumetric error is not to exceed 0.5%?

1.20. Add the following concentrations of H^+: 6.0×10^{-4} $\underline{M}$ from $HClO_4$, 4.1×10^{-6} $\underline{M}$ from HNO_3 and 1.7×10^{-11} $\underline{M}$ from ionization of water. Express the result using the correct number of significant figures.

Statistics of linear relationships

1.21. The following data were obtained in the spectrophotometric determination of iron. a) Determine the linear least squares line. b) Compute the concentration of iron in a solution, in ppm Fe, if its absorbance is 0.452.

[Fe], ppm	A
1.0	0.240
2.0	0.460
3.0	0.662
4.0	0.876

1.22. The following results were obtained in the determination of glucose in 8 serum samples by two methods, a standard spectrophotometric method and a tested enzymatic method. Determine whether there is a significant difference between the two methods and calculate the correlation coefficient. Assume the two methods have similar precisions.

Glucose, mg/100 mL

Sample number	*Standard method*	*Enzymatic method*
1	90	75
2	182	155
3	175	143
4	142	122
5	110	90
6	101	85
7	122	98
8	100	86

Chapter 2

Chemical Reactions and Chemical Equations

2.1 Introduction

Chemical reactions taking place in aqueous solutions involve molecules and/or ions. Purely molecular reactions are most common in organic chemistry and frequently are slow, whereas ionic reactions are encountered mainly in inorganic chemistry and are usually extremely fast.

Chemical reactions are represented by chemical equations, molecular or ionic. In ionic chemical equations, one includes only those species which are reactants or products in the reaction. A chemical equation describes not only the chemical changes occurring in a chemical reaction, but it also gives the relative numbers of molecules or ions of reactants and products. From this, the proportions by weight of reactants and products and the proportions by volume for gases, can also be obtained. In this way, the chemical equations can be the basis for various stoichiometric calculations.

In writing ionic equations, the strong electrolytes are usually written as ions, because they dissociate completely, whereas solid substances, gases and soluble weak electrolytes are written as molecules. One uses a single arrow for a reaction that takes place completely toward the indicated direction, and two arrows for a reversible reaction. Two unequal arrows are often written if the equilibrium of the reversible reaction is generally displaced in one direction, e.g., $Ag^+ + Cl^- \rightleftharpoons \mathbf{AgCl}$. The position of the equilibrium may, of course, be altered by changing the concentrations (Le Châtelier's Principle). In this text, the formulas of solid substances are written in boldface. For a chemical equation to be correct, it must agree with the law of mass preservation, and the law of electrical charge preservation.

2.2 Classes of Chemical Reactions

Chemical reactions are classified in two general categories, *displacement reactions*, in which no change in oxidation number occurs, and *oxidation-reduction (redox) reactions*, in which a change in oxidation number does occur.

Displacement reactions. The displacement reactions take place because of the removal of one or more products from the reaction, which causes the equilibrium to shift to the right. The removal involves the formation of a precipitate or a gas or a slightly ionized compound. For example:

$$[Ag(NH_3)_2]^+ + Cl^- + 2H^+ \rightleftharpoons \mathbf{AgCl} + 2NH_4^+$$

$$S^{2-} + 2H^+ \rightleftharpoons H_2S \uparrow$$

$$\mathbf{Cu(OH)_2} + 4NH_3 \rightleftharpoons [Cu(NH_3)_4]^{2+} + 2\,OH^-$$

$$Hg^{2+} + 2Cl^- \rightleftharpoons HgCl_2$$

Oxidation-reduction reactions. In order to elucidate more easily the transfer of electrons in redox reactions, the term *oxidation number* (oxidation state) is used. This is defined as the charge that an atom *appears* to have when the electrons are counted according to some arbitrary rules:

1. The oxidation number of a *free element* is zero.
2. In a *simple ion*, which contains only one atom, the oxidation number is equal to the charge on the atom.
3. The oxidation number of oxygen is -2, except in peroxides and the compound F_2O, where it is -1 and $+2$, respectively.
4. The oxidation number of hydrogen is $+1$, except in certain hydrides (e.g., those of alkali and alkaline earth metals, such as NaH and CaH_2), where it is -1.
5. The algebraic sum of the oxidation numbers of the atoms in a neutral molecule is zero, while in a complex ion it is equal to the charge of the ion.

In general, the oxidation number of an atom is positive if the atom is combined with a more electronegative element, negative if the atom is combined with a less electronegative element, and zero if it is combined with an identical atom. In the case of carbon, C is zero with respect to C, negative with respect to H (i.e., -1), and positive with respect to O and N containing groups. The actual number with respect to O and N depends on the oxidation state of the O or N; O is generally -2, except in the case of peroxides, where it is -1, while N is often -3 in amines (RNH_2) and $+3$ in nitro compounds (RNO_2). The relative electronegativities are $H < C < N < O$. In the case in which an atom (e.g., carbon) is combined with two or more atoms of different electronegativity, the oxidation number is calculated from the algebraic sum of the oxidation numbers. Consider, for example, the carbon in the alcoholic group of a primary alcohol: $-\overset{|}{\underset{|}{C^0}}-\overset{H^{+1}}{\underset{H^{+1}}{C^n}}-O^{-2}H^{+1}$. The oxidation states of each atom, except the one in question, are indicated on the structure. The value n for the alcoholic group carbon is equal to 2(−1) from the 2 Hs, plus 0 from the C, plus (+1) from the -OH group (the − 2 on the O is half neutralized by the H, i.e., the carbon is +1 with respect to the -OH *group*). So the alcoholic carbon has n = −1. Similarly, for the secondary alcohol

$$\begin{array}{c} \mathrm{H} \\ | \\ -\mathrm{C}-\mathrm{C}-\mathrm{C}- \\ | \\ \mathrm{OH} \end{array}$$

, the alcoholic group has an oxidation number given by (−1) from the H, plus (+1) from the -OH group, plus 2(0) from the 2 Cs, or a net oxidation number of zero. The aldehyde carbon in the group

$$\begin{array}{c} \mathrm{H} \\ | \\ -\mathrm{C}-\mathrm{C}=\mathrm{O} \end{array}$$

has an oxidation number given by (−1) from the H, plus (+2) from the O, plus zero from the C, or a net of +1.

2.3 Chemical Equations of Redox Reactions

The basic presumption for balancing chemical equations of redox reactions is that the oxidation and reduction take place simultaneously, and that the oxidant and the reductant react in equivalent amounts (the total electron change of each must be the same, although in opposite sign). The *balancing* of the equation, i.e., the calculation of the numerical coefficients of each chemical species, is usually done by the ***half-reaction method*** or by the *oxidation number method.* In both methods, the products of the reaction must be known.

The half-reaction or half-cell method. In this method, the reaction is considered to be the summation of two half-reactions. A separate chemical equation is written for each half-reaction: one a reduction half-reaction and one an oxidation half-reaction. The method is based on the fact that the number of electrons lost by the reductant is equal to the number of electrons gained by the oxidant. The half-reaction method is mainly used in ionic reactions, and involves the following six steps:

1. The partial (unbalanced) equations of the oxidation and reduction half-reactions are written; these include the species being oxidized or reduced, plus any other known species in the reaction. The hydrogen or oxygen will be balanced in the following steps with H^+, OH^- or H_2O.
2. Each partial equation is balanced with respect to mass. This is accomplished in the following steps:
   (a) Balance the oxygens in the half-reaction by adding the correct number of H_2O molecules on the appropriate side of the half reaction.
   (b) Balance the hydrogen in the half-reaction by adding the correct number of protons (H^+) on the appropriate side of the half reaction. *If the solution is alkaline*, then add the proper number of hydroxyl ions (OH^-) on each side of the half-reaction to neutralize the protons previously added (each H^+ will

then be converted to a water molecule, H_2O; add or cancel with previous H_20 molecules where possible).

3. Each partial equation is balanced with respect to charge, by adding electrons so as to balance the ionic charges.
4. The balanced equation for the reduction half-reaction is multiplied by the number of electrons in the oxidation half-reaction, and the balanced oxidation half-reaction is multiplied by the number of electrons in the reduction reaction. (For simplification, the two multipliers are divided by a common denominator, if there is one.)
5. The resulting equations are summed and similar terms on opposite sides of the equation are cancelled.
6. The final equation is checked. The number of atoms of each element, as well as the algebraic sum of the charges, must be the same on both sides.

Example 2.1. Write an equation for the reduction of nitrate ion to ammonia by aluminum in alkaline medium.

Solution.

1. Partial equations,

$$NO_3^- \rightleftharpoons NH_3$$
$$\mathbf{Al} \rightleftharpoons [Al(OH)_4]^- \text{ (or } AlO_2^- + 2H_2O)$$

2. Mass balance,

(a) $NO_3^- \rightleftharpoons NH_3 + 3H_2O$
(b) $NO_3^- + 9H^+ \rightleftharpoons NH_3 + 3H_2O$
$NO_3^- + 9H^+ + 9\,OH^- \rightleftharpoons NH_3 + 3H_2O + 9\,OH^-$
$NO_3^- + 6H_2O \rightleftharpoons NH_3 + 9\,OH^-$

(a) $\mathbf{Al} + 4H_2O \rightleftharpoons [Al(OH)_4]^-$
(b) $\mathbf{Al} + 4H_2O \rightleftharpoons [Al(OH)_4]^- + 4H^+$
$\mathbf{Al} + 4H_2O + 4\,OH^- \rightleftharpoons [Al(OH)_4]^- + 4H^+ + 4\,OH^-$
$\mathbf{Al} + 4\,OH^- \rightleftharpoons [Al(OH)_4]^-$

3. Electrical charge balance,

$$NO_3^- + 6H_2O + 8e \rightleftharpoons NH_3 + 9\,OH^- \text{ (reduction)}$$

$$\mathbf{Al} + 4\,OH^- \rightleftharpoons [Al(OH)_4]^- + 3e \text{ (oxidation)}$$

4. The equation for the reduction half-reduction is multiplied by 3 and the one for the oxidation half-reaction by 8, so that the electrons are equal, i.e.,

$$3NO_3^- + 18H_2O + 24e \rightleftharpoons 3NH_3 + 27\,OH^-$$

$$8\mathbf{Al} + 32\,OH^- \rightleftharpoons 8[Al(OH)_4]^- + 24e$$

5. Summing up and cancellation of similar terms,

$$8\mathbf{Al} + 3NO_3^- + 18H_2O + 5\,OH^- \rightleftharpoons 3NH_3 + 8[Al(OH)_4]^-$$

6. Checking of the final equation. The equation is correct (atoms and charges are equal on each side).

The oxidation number method. This is the only method for balancing the equations for molecular reactions, but it is also used equally well for ionic reactions. An oxidant takes on electrons to be reduced, while a reductant loses electrons to be oxidized. The method includes the following five steps:

1. The partial equation and the oxidation numbers of the atoms being oxidized or reduced are written for the reaction.
2. The number of electrons gained by each molecule or ion of the oxidant is determined. The number of electrons lost by each molecule or ion of the reductant are determined.
3. The coefficients for mass balance of the oxidant and the reductant in the equation are determined. If the coefficients have common denominators, both are divided by the maximum common denominator, in order to obtain smaller coefficients.
4. The equation is balanced with respect to mass and electrical charges, using H^+ and H_2O in acidic solutions, or OH^- and H_2O in alkaline solutions.
5. The final equation is checked, exactly as in the half-reaction method.

Example 2.2. Balance the equation

$$\mathbf{FeCr_2O_4} + K_2CO_3 + O_2 \rightarrow \mathbf{Fe_2O_3} + K_2CrO_4 + CO_2$$

The species being oxidized are Fe and Cr, and reduced is O (in O_2). See Note below.

Solution.

1. $\overset{0}{\mathbf{Fe}}\overset{+4}{\mathbf{Cr_2}}\mathbf{O_4} + K_2CO_3 + \overset{0}{O_2} \rightarrow \overset{+3}{\mathbf{Fe_2}}\mathbf{O_3} + K_2\overset{+6}{Cr}O_4 + CO_2.$

   The number of electrons lost or gained by the molecules upon reaction to the indicated products is:

2. $\overset{0}{\mathbf{Fe}}\overset{+4}{\mathbf{Cr_2}}\mathbf{O_4} + K_2CO_3 + \overset{0}{O_2} \rightarrow \overset{+3}{\mathbf{Fe_2}}\mathbf{O_3} + K_2\overset{+6}{Cr}O_4 + CO_2$

   $$\underbrace{\downarrow\ 3e \quad \downarrow\ (2\times 2)e}_{-7e/\text{molecule}} \qquad \overbrace{\uparrow}^{}\ +4e/\text{molecule}$$

   $\therefore$ 7 molecules of O_2 must react with 4 molecules of $FeCr_2O_4$:

3. $\mathbf{4FeCr_2O_4} + K_2CO_3 + 7O_2 \rightarrow \mathbf{2Fe_2O_3} + 8K_2CrO_4 + CO_2$
4. $\mathbf{4FeCr_2O_4} + 8K_2CO_3 + 7O_2 \rightarrow \mathbf{2Fe_2O_3} + 8K_2CrO_4 + 8CO_2$ (final).
5. Checking of the final equation; it is correct.

Note. The oxidation numbers 0 and +4 are assigned to Fe and Cr, respectively, in $FeCr_2O_4$. Other possible (fictitious) oxidations numbers may be assigned if the actual ones are not known, so long as the molecule is neutral. Examples are +2 for Fe and +3 for Cr or +3 for Fe and +2.5 for Cr (if we assume that only Cr is oxidized), and so forth. In fact, we could assume that all but one of the atoms in a molecule have the same oxidation state as in the products following reaction, and a fictitious oxidation state can be assigned to that one atom, i.e., all electron transfer is assumed to occur at that atom. However, in all cases the coefficients of the equation remain the same, because there is always a loss of 7 electrons from each $FeCr_2O_4$ molecule.

Example 2.3. Write the equation for the oxidation of glycerin to CO_2 by $Cr_2O_7^{2-}$ ion in acidic medium.

Solution. Using the rules given in Section 2.2 for oxidation-reduction reactions, the oxidation states of the carbons in glycerin and the Cr species are as follows:

1. $\overset{-1}{C}H_2OH\overset{0}{C}HOH\overset{-1}{C}H_2OH + \overset{+6}{Cr_2}O_7^{2-} \rightleftharpoons \overset{+4}{C}O_2 + \overset{+3}{Cr^{3+}}$

   The number of electrons lost or gained by the two molecules is:

2. $\overset{-1}{C}H_2OH\overset{0}{C}HOH\overset{-1}{C}H_2OH + \overset{+6}{Cr_2}O_7^{2-} \rightleftharpoons \overset{+4}{C}O_2 + \overset{+3}{Cr^{3+}}$

   $\downarrow$ $\downarrow$ $\downarrow$ $\uparrow$

   $\underbrace{5e \quad 4e \quad 5e}_{-\,14e/\text{molecule}} \qquad \overbrace{2 \times 3 = +6e/\text{molecule}}$

   $\therefore$ They react in a 6:14 ratio or 3:7:

3. $3CH_2OHCHOHCH_2OH + 7Cr_2O_7^{2-} \rightleftharpoons CO_2 + Cr^{3+}$

   Balancing H,O, and the charge with H^+ and H_2O:

4. $3CH_2OHCHOHCH_2OH + 7Cr_2O_7^{2-} + 56H^+ \rightleftharpoons 9CO_2 + 14Cr^{3+} + 40H_2O$
5. Checking of the final equation; it is correct.

Example 2.4. Balance the following equation, where M and A are a metal and a nonmetal, respectively.

$$aM_3A + bHNO_3 \rightleftharpoons xM(NO_3)_3 + yH_3AO_4 + zNO + wH_2O$$

Solution.

1. Since the identities of M and A are unknown, we arbitrarily assign an oxidation number of +1 for M, whereupon the oxidation number of A is −3.

$$a\overset{+1}{M_3}\overset{-3}{A} + bH\overset{+5}{N}O_3 \rightleftharpoons x\overset{+3}{M}(NO_3)_3 + yH_3\overset{+5}{A}O_4 + z\overset{+2}{N}O + wH_2O$$

2. $\overset{+1\;-3}{aM_3A} + b\overset{+5}{HNO_3} \rightleftharpoons x\overset{+3}{M}(NO_3)_3 + yH_3\overset{+5}{A}O_4 + z\overset{+2}{N}O + wH_2O$

$$\underbrace{(3\times2)e8e}_{-14e/\text{molecule}} \quad \overbrace{+3e}/\text{molecule}$$

3. $3M_3A + 14HNO_3 \rightleftharpoons xM(NO_3)_3 + yH_3AO_4 + zNO + wH_2O$
4. For a = 3, x must be equal to 9, whereupon we have 27 nitrate ions on the right side of the equation, and only 14 N atoms on the left side. These 27 nitrate ions, in which the oxidation number of N remains unchanged, should be added to the 14 nitrate ions in which the oxidation number is changed, whereupon b = 41. Subsequently, the other coefficients are calculated, i.e., y = 3, z = 14, and w = 16. Hence,

$$3M_3A + 41HNO_3 \rightleftharpoons 9M(NO_3)_3 + 3H_3AO_4 + 14NO + 16H_2O$$

5. The checking of the final equation proves that is correct.

Note. The problem of the excess nitrate ions would not have existed, had the reaction been written in ionic form, that is,

$$3M_3A + 14NO_3^- + 41H^+ \rightleftharpoons 9M^{3+} + 3H_3AO_4 + 14NO + 16H_2O$$

Algebraic method for calculating the coefficients of a redox reaction. In molecular equations of redox reactions, the coefficients can also be found algebraically, on the basis of the principle of mass balance, i.e., that the number of atoms of each element must be the same on both sides of the equation. In ionic equations, besides the mass balance, we also take into consideration the principle that the algebraic sum of charges must be the same on both sides of the equation. As illustrated below, a system of N equations with N+1 unknowns is obtained for each chemical equation. But we can arbitrarily set one of the coefficients to unity and solve for the others relative to this. If fractional coefficients result, we multiply all coefficients by the smallest common multiplier of the denominators, thus obtaining integer coefficients. Thus, for Example (2.4) we have the following system:

$$aM_3A + bHNO_3 \rightleftharpoons xM(NO_3)_3 + yH_3AO_4 + zNO + wH_2O$$

The M atoms must be balanced on each side, and so 3a = x. Likewise, for A: a = y, for H: b = 3y + 2w, for N: b = 3x + z, and for O: 3b = 9x + 4y + z + w. We have six unknowns but only five equations. We can arbitrarily set one coefficient to unity and solve for the other five relative to this. We set a = 1, whereupon by solving the system of simultaneous equations, we find the following values: b = 41/3, x = 3, y = 1, z = 14/3, and w = 16/3. Multiplying these values by 3, we find that a = 3, b = 41, x = 9, y = 3, z = 14, and w = 16, i.e., the same coefficients we found by the oxidation number method.

Example 2.5. Calculate algebraically the coefficients of the ionic equation

$$aCr_2O_7^{2-} + bH^+ + cI^- \rightleftharpoons xCr^{3+} + y[I_3]^- + zH_2O$$

Solution. We have the system: For Cr: 2a = z, for O: 7a = z, for H: b = 2z, for I: c = 3y. For the electrical charges on each side to be equal, we have: b − 2a − c = 3x − y. Thus, we have a system of 5 equations with 6 unknowns. We set a = 1, whereupon by solving the system we find the following values: b = 14, c = 9, x = 2, y = 3, and z = 7. Hence, the equation becomes:

$$Cr_2O_7^{2-} + 14H^+ + 9I^- \rightleftharpoons 2Cr^{3+} + 3[I_3]^- + 7H_2O$$

Practice Equations. Complete and balance the following equations:

$$[Ag(NH_3)_2]^+ + HCHO + OH^- \rightleftharpoons \mathbf{Ag} + CO_3^{2-}$$

$$\mathbf{As_2S_3} + H_2O_2 + OH^- \rightleftharpoons SO_4^{2-} + AsO_4^{3-}$$

$$NO_2^- + \mathbf{Al} + OH^- \rightleftharpoons NH_3 \uparrow$$

$$CrI_3 + Cl_2 + KOH \rightleftharpoons K_2CrO_4 + KIO_4 + KCl$$

$$\mathbf{As_2S_3} + ClO_3^- \rightleftharpoons HAsO_4^{2-} + \mathbf{S} + Cl^-$$

$$VO^{2+} + MnO_4^- \rightleftharpoons HVO_3 + Mn^{2+}$$

$$Na_3H_2IO_6 + MnSO_4 + H_2SO_4 \rightleftharpoons HMnO_4 + Na_2SO_4 + NaIO_3$$

$$\mathbf{CuI} + HNO_3 \rightleftharpoons \mathbf{I_2} + NO \uparrow$$

$$\mathbf{HgNH_2Cl} + NO_3^- + Cl^- \rightleftharpoons N_2 \uparrow + NO \uparrow$$

$$\mathbf{Fe_3P} + NO_3^- + H^+ \rightleftharpoons Fe^{3+} + H_2PO_4^- + NO \uparrow$$

$$\mathbf{Hg_2Cl_2} + IO_3^- + Cl^- \rightleftharpoons HgCl_2 + ICl_2^-$$

$$K_2Na[Co(NO_2)_6] + MnO_4^- + H^+ \rightleftharpoons NO_3^- + Co^{2+} + Mn^{2+}$$

$$\mathbf{Cr(OH)_3} + IO_3^- + OH^- \rightleftharpoons CrO_4^{2-} + I^-$$

Answers. By working as in Examples 2.1 to 2.5, we have

$$4[Ag(NH_3)_2]^+ + HCHO + 6\,OH^- \rightleftharpoons 4\mathbf{Ag} + CO_3^{2-} + 8NH_3 + 4H_2O$$

$$\mathbf{As_2S_3} + 14H_2O_2 + 12\,OH^- \rightleftharpoons 3SO_4^{2-} + 2AsO_4^{3-} + 20H_2O$$

$$NO_2^- + 2\mathbf{Al} + OH^- + 5H_2O \rightleftharpoons NH_3 \uparrow + 2[Al(OH)_4]^-$$

$$2CrI_3 + 27Cl_2 + 64KOH \rightleftharpoons 2K_2CrO_4 + 6KIO_4 + 54KCl + 32H_2O$$

$$3\mathbf{As_2S_3} + 5ClO_3^- + 9H_2O \rightleftharpoons 6HAsO_4^{2-} + 9\mathbf{S} + 5Cl^- + 12H^+$$

$$5VO^{2+} + MnO_4^- + 6H_2O \rightleftharpoons 5HVO_3 + Mn^{2+} + 7H^+$$

$$5Na_3H_2IO_6 + 2MnSO_4 + 3H_2SO_4 \rightleftharpoons 2HMnO_4 + 5Na_2SO_4 + 5NaIO_3 + 7H_2O$$

$$\mathbf{6CuI} + 4HNO_3 + 16H^+ \rightleftharpoons 3I_2 + 4NO\uparrow + 6Cu^{2+} + 8H_2O$$

$$\mathbf{2HgNH_2Cl} + 2NO_3^- + 6Cl^- + 4H^+ \rightleftharpoons N_2\uparrow + 2NO\uparrow + 2[HgCl_4]^{2-} + 4H_2O$$

$$\mathbf{3Fe_3P} + 14NO_3^- + 38H^+ \rightleftharpoons 9Fe^{3+} + 3H_2PO_4^- + 14NO\uparrow + 16H_2O$$

$$\mathbf{2Hg_2Cl_2} + IO_3^- + 6Cl^- + 6H^+ \rightleftharpoons 4HgCl_2 + ICl_2^- + 3H_2O$$

$$5K_2Na[Co(NO_2)_6] + 11MnO_4^- + 28H^+ \rightleftharpoons 30NO_3^- + 5Co^{2+} + 11Mn^{2+} + 10K^+ + 5Na^+ + 14H_2O$$

$$\mathbf{2Cr(OH)_3} + IO_3^- + 4\,OH^- \rightleftharpoons 2CrO_4^{2-} + I^- + 5H_2O$$

Practice Equations. Complete and balance the following equations involving organic molecules:

$$CH_3OH + MnO_4^- \rightleftharpoons CO_2 + MnO_4^{2-}$$

$$CH_2OHCHOHCH_2OH + MnO_4^- + OH^- \rightleftharpoons CO_2 + MnO_4^{2-}$$
(glycerin)

$$CH_2OHCH_2OH + MnO_4^- + OH^- \rightleftharpoons CO_2 + MnO_4^{2-}$$

$$C_6H_5OH + MnO_4^- + OH^- \rightleftharpoons CO_2 + MnO_4^{2-}$$

$$CH_2OHCHOHCH_2OH + Ce^{4+} + H_2O \rightleftharpoons HCOOH + Ce^{3+} + H^+$$

$$HOOCCH(OH)CH(OH)COOH + Ce^{4+} \rightleftharpoons CO_2 + HCOOH + Ce^{3+}$$
(tartaric acid)

$$HOOCCH_2COOH + Ce^{4+} \rightleftharpoons CO_2 + HCOOH + Ce^{3+}$$
(malonic acid)

$$CH_2OHCH_2OH + IO_4^- \rightleftharpoons HCHO + IO_3^-$$

$$HOCH_2CH(NH_2)COOH + IO_4^- \rightleftharpoons HCHO + OHCCOOH + IO_3^-$$
(serine)

Answers. By working as in Example 2.3, we have

$$CH_3OH + 6MnO_4^- + 6\,OH^- \rightleftharpoons CO_2^+ + 6MnO_4^{2-} + 5H_2O$$

$$CH_2OHCHOHCH_2OH + 14MnO_4^- + 14\,OH^- \rightleftharpoons 3CO_2 + 14MnO_4^{2-} + 11H_2O$$

$$CH_2OHCH_2OH + 10MnO_4^- + 10\,OH^- \rightleftharpoons 2CO_2 + 10MnO_4^{2-} + 8H_2O$$

$$C_6H_5OH + 28MnO_4^- + 28\,OH^- \rightleftharpoons 6CO_2 + 28MnO_4^{2-} + 17H_2O$$

$$CH_2OHCHOHCH_2OH + 8Ce^{4+} + 3H_2O \rightleftharpoons 3HCOOH + 8Ce^{3+} + 8H^+$$

$$HOOCCH(OH)CH(OH)COOH + 6Ce^{4+} + 2H_2O \rightleftharpoons 2HCOOH + 2CO_2 + 6Ce^{3+} + 6H^+$$

$$HOOCCH_2COOH + 6Ce^{4+} + 2H_2O \rightleftharpoons 2CO_2 + HCOOH + 6Ce^{3+} + 6H^+$$

$$CH_2OHCH_2OH + IO_4^- \rightleftharpoons 2HCHO + IO_3^- + H_2O$$

$$HOCH_2CH(NH_2)COOH + IO_4^- \rightleftharpoons HCHO + OHCCOOH + IO_3^- + NH_3$$

2.4 Problems

Stoichiometry

2.1. Depending on the conditions, elements X and Z react by a weight ratio of either 1 part of X to 1.982 parts of Z, or 1 part of X to 3.964 parts of Z, and form two different compounds. Which couple of the following substances does not agree with the above data? a) $X_2Z - X_2Z_2$, b) $X_3Z_2 - X_3Z_4$, c) $XZ - XZ_2$, and d) $XZ - X_2Z$.

2.2. When copper reacts with concentrated HNO_3, under certain conditions a mixture of NO_2 and NO at a volume ratio of 5:2 is evolved. How many grams of HNO_3 reacted with 5.0 g of copper?

Chemical equations

2.3. Complete and balance the following reactions of nitrogen compounds:

(a) $MnO_4^- + N_2H_5^+ \rightleftharpoons Mn^{2+} + N_2\uparrow$

(b) $[I_3]^- + N_3^- \rightleftharpoons N_2\uparrow + I^-$

(c) $Fe^{3+} + NH_2OH \rightleftharpoons N_2\uparrow$

(d) $OBr^- + NH_3 \rightleftharpoons N_2\uparrow + Br^-$

(e) $NH_4^+ + NO_2^- \rightleftharpoons N_2\uparrow$

(f) $NH_4^+ + NO_2^- \rightleftharpoons NO\uparrow$

(g) $HNO_2 + SO_2 \rightleftharpoons H_3^+NOH$

(h) $N_2H_5^+ + NO_3^- \rightleftharpoons NO_2\uparrow$

2.4. Write the reactions of nitric acid with metallic zinc, in which the nitrate ion is reduced to all possible oxidation states of nitrogen.

2.5. Complete and balance the following equations:

(a) $CH_3COCOCH_3 + Ce(IV) \rightleftharpoons CH_3COOH + Ce(III)$

(b) $CH_2OHCHOHCH_2OH + IO_4^- \rightleftharpoons 2HCHO + HCOOH + IO_3^-$

(c) $CH_3CHOHCH_2OH + IO_4^- \rightleftharpoons CH_3CHO + HCHO + IO_3^-$

(d) $C_6H_5OH + Br_2 \rightleftharpoons C_6H_2Br_3OH + HBr$

(e) $BrO_3^- + Cr^{3+} \rightleftharpoons Br^- + Cr_2O_7^{2-} + H^+$

2.6. Write at least two reactions in which the following species participate as oxidants: a) MnO_4^-, b) $Cr_2O_7^{2-}$, c) Ce(IV), d) $[I_3]^-$, e) IO_4^-, and the following ones as reductants: a) $C_2O_4^{2-}$, b) $S_2O_3^{2-}$, c) I^-, d) Br^-.

2.7. Complete and balance the following equations:

(a) $AsH_3 + AgNO_3 \rightleftharpoons \mathbf{As_2O_3} + HNO_3 + \mathbf{Ag}$

(b) $Mn^{2+} + MnO_4^- + H_2O \rightleftharpoons \mathbf{MnO_2}$

(c) $\mathbf{Pb} + \mathbf{PbO_2} + H_2SO_4 \rightleftharpoons \mathbf{PbSO_4}$

(d) $\mathbf{Cd} + \mathbf{Ni_2O_3} \rightleftharpoons \mathbf{Cd(OH)_2} + \mathbf{Ni(OH)_2}$

2.8. Complete and balance the following equations:

(a) $AlCl_3 + Na_2S_2O_3 \rightleftharpoons \mathbf{Al(OH)_3} + \mathbf{S} + SO_2 \uparrow$

(b) $\mathbf{Na_2B_4O_7} + CH_3OH + H_2SO_4 \rightleftharpoons (CH_3)_3BO_3 + Na_2SO_4$

(c) $Fe^{3+} + CO_3^{2-} \rightleftharpoons \mathbf{Fe(OH)_3}$

(d) $\mathbf{Na_2B_4O_7} + \mathbf{CuO} \rightleftharpoons \mathbf{Cu(BO_2)_2} + NaBO_2$

(e) $[SbCl_4]^- + H_2O \rightleftharpoons \mathbf{SbOCl}$

(f) $\mathbf{Zn} + SCN^- \rightleftharpoons Zn^{2+} + H_2S \uparrow + HCN \uparrow$

(g) $\mathbf{S} \rightleftharpoons SO_3^{2-} + S^{2-}$

(h) $C_2H_5OH + Cr_2O_7^{2-} + H^+ \rightleftharpoons CH_3CHO + Cr^{3+}$

(i) $HMnO_4 + AsH_3 + H_2SO_4 \rightleftharpoons H_3AsO_4 + MnSO_4 + H_2O$

(j) $\mathbf{P_2I_4} + \mathbf{P_4} + H_2O \rightleftharpoons PH_4I + H_3PO_4$

Chapter 3

Concentration of Solutions-Computations

3.1 Ways of Expressing the Concentrations of Solutions

Concentration is defined as the amount of dissolved substance (solute) per unit amount of solution or solvent. The concentration is usually expressed either in *physical units*, i.e., weight and volume units, or in *chemical units*, i.e., in gram-formula weights, gram-molecular weights, and chemical equivalents (e.g., equivalent in number of protons donated or accepted in reacting).

The most commonly used units for expressing the concentration of a solution are given in Table 3.1.

A. Expressing concentration in physical units.

1. *Weight per cent*, %w/w, i.e., g of solute in 100 g of solution. The weight (mass) w of the solute in g, the volume v of the solution in mL, the density d in g/mL, and the per cent (%) content by weight, are interrelated by

$$w = v \times d \times \frac{\%}{100} \tag{3.1}$$

2. *Volume per cent*, %v/v, i.e., mL of solute in 100 mL of solution. This way of expressing the concentration is used only in the case of mixing of liquids.

3. *Weight-volume per cent*, %w/v, i.e., g of solute in 100 mL of solution.

4. *Grams of solute in one liter of solution.*

5. *Milligrams of solute in one milliliter of solution.*

6. *Parts per million*, ppm, i.e., number of weight parts of solute in million weight parts (or volume parts, if density = 1) of solvent. For example, mg/kg or μg/g, or mg/L or μg/L (μ represents micro, 10^{-6}).

Table 3.1. *Common units of concentration of solutions.*

Units of solute	Quantity of solution or solvent	Concentration units
		Physical units
g	100 g solution	g/100 g solution (weight per cent, w/w)
mL	100 mL solvent	mL/100 mL solution (volume per cent, v/v)
g	100 mL solution or solvent	g/100 mL solution or solvent (weight-volume per cent, w/v)
g	L solution	g/L solution
mg	mL solution	mg/mL solution
mg	L solution	mg/L solution (parts per million, ppm)
μg	mL solution	μg/mL solution (parts per million, ppm)
		Chemical units
mol	L solution	mol/L solution (molarity, M)
mmol	mL solution	mmol/mL solution (molarity, M)
gfw	L solution	gfw/L or fw/L solution (formality, F)
mfw	mL solution	mfw/mL solution (formality, F)
eq	L solution	eq/L solution (normality, N)
meq	mL solution	meq/mL solution (normality, N)
mol	1000 g solvent	mol/1000 g solvent (molality, m)

7. *Parts per billion*, ppb, is used for even smaller concentrations. This is the number of weight parts of solute in 1 billion weight parts (or volume parts, if d = 1) of solvent. For example, μg/kg or ng/g, or μg/L or ng/mL (*n* represents nano, 10^{-9}).

B. Expressing concentration in chemical units.

1. *Gram-molecular or molecular concentration or molarity*, M, i.e., gram-moles of solute in one liter of solution, mol/L. The gram-mole or mole of a substance is a quantity in grams numerically equal to its molecular weight [or to 6.023×10^{23} (Avagadro's number) molecules of the substance].

In the case of ionic compounds and ionic solutions, we sometimes use the terms *gram-formula weight*, and *gram-formula* or *formal concentration* or *formality*, F. These are identical in concept to gram-molecular weight and gram-molecular (molar) concentration. The latter are often used exclusively in texts, even for ionic substances. The *gram-formula weight* (gfw) of a substance is a quantity in grams numerically equal to the formal weight, i.e., the summation of the atomic weights of all the atoms in the chemical formula of the substance. *Formality* is defined as the number of gram-formula weights of solute in one liter of solution, gfw/L or fw/L (or mfw/mL). The use of formality is diminishing, in favor of *analytical concentration.*

The molarity and the formality of a given solution are usually the same in the case of molecular solutions, but they differ in solutions of strong and weak electrolytes. Thus, a solution prepared by dissolving 1 gram-formula weight of ethanol (46.07 grams) in water and diluting to exactly 1 liter is by definition, 1.000 $\underline{F}$; it is also 1.000 $\underline{M}$. Similarly, a solution prepared by dissolving 1 gram-formula weight of monochloroacetic acid (94.50 grams) in water and diluting to exactly 1 liter is 1.000 $\underline{F}$. The molarity, however, will be less than one because monochloroacetic acid is dissociated to the extent of 3.68%. Consequently, a 1.000 $\underline{F}$ solution of $ClCH_2COOH$ is actually only 0.963 $\underline{M}$ with respect to the undissociated $ClCH_2COOH$ species. An example of a strong electrolyte would be a 1.000 $\underline{F}$ solution of potassium chloride (74.56 g KCl/L). Such a solution will also be 1.000 $\underline{F}$ with respect to K^+ or Cl^-. The molarity of the species KCl, however, will be zero, because the potassium chloride undergoes complete dissociation. From these examples it can be seen that the formality of a solute can *always* be calculated from its formula weight and the amount of solute dissolved in a given volume. On the other hand, the molarity of a solute cannot be specified, unless we are fully aware of the chemistry of the dissolved compound in the solution.

The following relations are valid for molarity and formality:

$$\text{mol} = \frac{\text{g}}{\text{gmw}}; \ \text{mmol} = \frac{\text{mg}}{\text{gmw}}$$

$$\underline{M} = \frac{\text{mol}}{\text{L}} = \frac{\text{g}}{\text{gmw} \times \text{L}} = \frac{\text{mmol}}{\text{mL}} = \frac{\text{mg}}{\text{gmw} \times \text{mL}} \tag{3.2}$$

$$\text{fw} = \frac{\text{g}}{\text{gfw}}; \ \text{mfw} = \frac{\text{mg}}{\text{gfw}}$$

$$\underline{F} = \frac{\text{fw}}{\text{L}} = \frac{\text{g}}{\text{gfw} \times \text{L}} = \frac{\text{mfw}}{\text{mL}} = \frac{\text{mg}}{\text{mfw} \times \text{mL}} \tag{3.3}$$

where

L, mL = volume of the solution in liters and milliliters, respectively

g, mg = weight of solute in grams and milligrams, respectively

gmw = grams of solute per mole or milligrams per millimole

gfw = grams of solute per gram-formula weight and

mfw = milligrams of solute per gram-milliformula weight

2. *Normality*, $\underline{N}$, i.e., gram-equivalents of solute in one liter of solution, eq/L, or gram-milliequivalents in one milliliter, meq/mL. The *gram-equivalent weight* (gew) of a substance is that quantity of the substance which gives or accepts or substitutes for or is *chemically equivalent* to one gram-equivalent of hydrogen, i.e., 1.008 g of hydrogen. Practically speaking, the gram equivalent weight is equal to the gram formula weight divided by the number of reaction units per molecule. In displacement reactions (acid-base, precipitation and complex-formation reactions), it is equal to the quotient of the gram-formula weight of the substance and the total oxidation number of the reacting positive or negative ions, regardless of the algebraic sign. More specifically, in acid-base reactions, the gram-equivalent weight is the weight in grams of a substance required to furnish or react with 1.008 g of H^+. In precipitation and complex-forming reactions,

the gram-equivalent weight is the weight in grams of a substance required to furnish or react with one mole of a univalent cation, or 1/2 mole of a divalent cation, etc. In redox reactions, the gram-equivalent weight of the oxidant (reductant) is equal to the quotient of the gram-formula weight and the number of gram-electrons gained (lost) per mole. That is, the gram-equivalent weight is the weight in grams of a substance required to furnish or react with one gram electron (6.02×10^{23} electrons). Oftentimes, instead of gram-equivalent weight, the term *equivalent weight*, ew, is used (as molecular weight may be used in place of gram-molecular weight). In contrast to the gram-molecular weight (mole), which is always the same for a given substance, the equivalent weight of a substance refers to a particular chemical reaction of the substance, and therefore its value for a given chemical will depend on the reaction. Certain examples, explaining the calculation of the gram-equivalent weight, are given in Table 3.2.

It should be emphasized, that while the gram-equivalent weight is usually smaller than the mole, in certain cases it is larger. For example, in the reaction

$$Ag^+ + 2CN^- \rightleftharpoons [Ag(CN)_2]^-$$

we have 1 eq $CN^- \equiv$ 1 eq $Ag^+ \equiv$ 2 mol CN^-. Hence, for KCN salt, 1 $\underline{N}$ KCN $\equiv$ 2 $\underline{M}$ KCN.

The following relations are valid for the normality:

$$\text{eq} = \frac{\text{g}}{\text{ew (g/eq)}}; \ \text{meq} = \frac{\text{mg}}{\text{ew (mg/eq)}}$$

$$\underline{\text{N}} = \frac{\text{eq}}{\text{L}} = \frac{\text{meq}}{\text{mL}} = \frac{\text{g}}{\text{ew} \times \text{L}} = \frac{\text{mg}}{\text{ew} \times \text{mL}} \tag{3.4}$$

$$\text{ew} = \frac{\text{gmw}}{\text{n}} = \frac{\text{gfw}}{\text{n}} \tag{3.5}$$

where L, mL, g, mg, gmw, and gfw have the same meaning as in Relations (3.2) and (3.3), ew is the number of grams of solute per gram-equivalent weight or of milligrams per gram-milliequivalent weight, and n is the number of equivalents per mole or gram-formula weight. Hence, for H_2SO_4 and $Ba(OH)_2$, in which both protons or hydroxide ions normally react,

$$\underline{\text{N}} = n\underline{\text{M}}; \ \underline{\text{M}} = \frac{\text{N}}{n} \ [\text{e.g., } 1 \ \underline{\text{M}} \ H_2SO_4 \equiv 2 \ \underline{\text{N}} \ H_2SO_4] \tag{3.6}$$

and

$$\underline{\text{N}} = n\underline{\text{F}}; \ \text{F} = \frac{\text{N}}{n} \ [\text{e.g., } 1 \ \underline{\text{F}} \ Ba(OH)_2 \equiv 2 \ \underline{\text{N}} \ Ba(OH)_2] \tag{3.7}$$

Note. For the reaction between solutions of substances A and B, the following relations are always valid:

$$\begin{array}{c} \text{equivalents of A} = \text{equivalents of B} \\ \text{milliequivalents of A} = \text{milliequivalents of B} \end{array} \tag{3.8}$$

or

$$V_A \times \underline{N}_A = V_B \times \underline{N}_B \tag{3.8a}$$

Table 3.2. *Gram-equivalent weights of electrolytes.*

Reaction Type	Reactant	Gram-equivalent weight: Method of calculation	Gram-equivalent weight: Examples
1. Displacement	Acid	$\frac{\text{(Gram-formula weight)}}{\text{(Number of released } H^+)}$	$\frac{HCl}{1} = 36.46$ g $\frac{H_2SO_4}{2} = 49.04$ g
	Base	$\frac{\text{(Gram-formula weight)}}{\text{(Number of } H^+ \text{ taken)}}$	$\frac{NH_3}{1} = 17.03$ g $\frac{Ba(OH)_2}{2} = 85.68$ g
	Salt	$\frac{\text{(Gram-formula weight)}}{\text{(Total oxidation number of positive or negative ions)}}$	$\frac{Al_2(SO_4)_3}{6} = 57.02$ g for $2Al^{3+}$ or $3SO_4{}^{2-}$ $\frac{Na_2HPO_4}{1} = 141.96$ g for H^+ $\frac{Na_2HPO_4}{2} = 70.98$ g for Na^+ $\frac{Na_2HPO_4}{3} = 47.32$ g for $PO_4{}^{3-}$ $\frac{FeSO_4}{2} = 75.95$ g for Fe^{2+} or $SO_4{}^{2-}$ $\frac{KHC_2O_4}{1} =$ 128.13 g for $HC_2O_4{}^- \rightleftharpoons H^+ + C_2O_4{}^{2-}$
2. Redox	Oxidant	$\frac{\text{(Gram-formula weight)}}{\text{(Number of gram-electrons gained)}}$	$\frac{FeSO_4}{1} = 151.91$ g for $Fe^{2+} \rightarrow Fe^{3+}$ $\frac{KMnO_4}{5} = 31.61$ g for $MnO_4{}^- \rightarrow Mn^{2+}$ $\frac{KMnO_4}{3} = 52.68$ g for $MnO_4{}^- \rightarrow MnO_2$ $\frac{KMnO_4}{1} = 158.04$ g for $MnO_4{}^- \rightarrow MnO_4{}^{2-}$
	Reductant	$\frac{\text{(Gram-formula weight)}}{\text{(Number of gram-electrons lost)}}$	$\frac{H_2S}{2} = 17.04$ g for $S^{2-} \rightarrow S$ $\frac{KHC_2O_4}{2} =$ 64.06 g for $HC_2O_4{}^- \rightarrow CO_2$ (for 2 C atoms)

where V_A and V_B are the volumes of the reacting solutions (expressed in the same units, usually in mL), and $\underline{N}_A$ and $\underline{N}_B$ the corresponding normalities, i.e., meq/mL. Thus, $(mL_A)(meq_A/mL_A) = (mL_B)(meq_B/mL_B)$.

3. *Molality, m,* i.e., moles of solute in 1000 g of solvent. Molality is independent of temperature, and is mainly used in physical chemistry, especially in the colligative properties of solutions, that is, those properties which depend on the number of particles, rather than on the kind.

4. *Concentration in mole fractions.* The mole fraction x_A of component A of a solution is defined as the ratio of moles of A to the total number of moles of all components of the solution, i.e.,

$$x_A = \frac{n_A}{n_A + n_B + \ldots + n_X} \tag{3.9}$$

5. *Titer.* This unit of concentration ($\underline{T}$) is often used for routine titrations, in order to shorten calculations. $\underline{T}$ represents the amount of analyte being determined that reacts with a unit volume (1 mL) of the titrant (reactant), and is usually expressed in mg/mL. The following relation holds:

$$\underline{T}\ (\text{mg/mL}) = \underline{N}\ (\text{meq/mL}) \times \text{ew}\ (\text{mg/meq})$$

or

$$\underline{T}\ (\text{mg/mL}) = \underline{M}\ (\text{mmol/mL}) \times \text{gfw}\ (\text{mg/mmol}) \tag{3.10}$$

where $\underline{N}$ ($\underline{M}$) is the normality (molarity) of the standard solution, and ew (gfw) is the equivalent weight (gram-formula weight) of the substance being determined.

Example 3.1. A 5.00-mL portion of a vinegar, of density 1.008 g/mL, requires 43.03 mL of 0.1002 $\underline{N}$ NaOH solution for titration. Calculate the weight per cent (w/w) of acetic acid in the sample.

Solution. The titrated portion of the sample contains

$$43.03\ \text{mL} \times 0.1002\ \text{meq/mL} \times 0.06005\ \text{g}\ CH_3COOH/\text{meq} = 0.2589\ \text{g}\ CH_3COOH.$$

Hence, the weight per cent of acetic acid in the sample is equal to

$$\%CH_3COOH\ (\text{w/w}) = \frac{0.2589\ \text{g}\ CH_3COOH}{5.00\ \text{mL} \times 1.008\ \text{g solution/mL}} \times 100 = \mathbf{5.14\%}.$$

Example 3.2. Express ppm and ppb concentrations in weight/weight (w/w), volume/volume (v/v), and weight/volume (w/v) (assume that d = 1.000 g/mL).

Solution. We have

$$\text{ppm} = \mu\text{g/g}(10^{-6}\text{g/g});\ \text{or mg/kg}\ (10^{-3}\text{g}/10^{3}\text{g})\ (\text{w/w})$$
$$\text{ppm} = \text{nL/mL}\ (10^{-9}\text{L}/10^{-3}\text{L});\ \text{or}\ \mu\text{L/L}\ (10^{-6}\ \text{L/L})\ (\text{v/v})$$

ppm = μg/mL (10^{-6}g/mL); or mg/L (10^{-3}g/10^{-3}mL) (w/v)
(Note: 1 mL $\approx$ 1 g)

ppb = ng/g (10^{-9}g/g); or μg/kg (10^{-6}g/10^{3}g) (w/w)

ppb = pL/mL (10^{-12}L/10^{-3}L); or nL/L (10^{-9}L/L) (v/v)

ppb = ng/mL (10^{-9}g/mL); or μg/L (10^{-6}/10^{3}mL) (w/v)

Example 3.3. Potassium acid iodate, $KH(IO_3)_2$, is used as a primary standard in various types of titrations. For each of the following cases, calculate the gram-equivalent weight of $KH(IO_3)_2$:

(a) $KH(IO_3)_2 + NaOH \rightleftharpoons KIO_3 + NaIO_3 + H_2O$

(b) $KH(IO_3)_2 + 2AgNO_3 \rightleftharpoons \mathbf{2AgIO_3} + KNO_3 + HNO_3$

(c) $KH(IO_3)_2 + 4KMnO_4 + 5KOH \rightleftharpoons 2KIO_4 + 4K_2MnO_4 + 3H_2O$

Solution. We have

(a) $ew_{KH(IO_3)_2} = \dfrac{389.92\text{ g}}{1} = \mathbf{389.92}$ g (neutralization reaction, $1H^+$/molecule to react with 1 OH^-)

(b) $ew_{KH(IO_3)_2} = \dfrac{389.92\text{ g}}{2} = \mathbf{194.96}$ g (displacement reaction, −2 charges/molecule to react with $2Ag^+$)

(c) $ew_{KH(IO_3)_2} = \dfrac{389.92\text{ g}}{4} = \mathbf{97.48}$ g (redox reaction, $-4e^-$/molecule: each

$$\overset{+5}{IO_3^-} \rightarrow \overset{+7}{IO_4^-})$$

Example 3.4. A solution of $KHC_2O_4 \cdot H_2C_2O_4 \cdot 2H_2O$ is 0.2000 N as a reductant (product is CO_2). How many milliliters of water should be added to 30.00 mL of the solution in order to make it 0.1000 N as an acid (3 reacting protons)?

Solution. The solution is 0.2000 (eq/L) / 4 (eq/mol) = 0.05000 M (mol/L).Thus, as an acid it is 3 (eq/mol) × 0.05000 (mol/L) = 0.1500 N (eq/L). Let y mL be the required volume of water, whereupon to form 0.1000 N acid (since milliequivalents before and after dilution will be the same), we have:

$$(30.00 + y)\text{ mL} \times 0.1000\text{ (meq/mL)} = 30.00\text{ mL} \times 0.1500\text{ (meq/mL)}$$

or

$$y = \mathbf{15.00}\text{ mL.}$$

Example 3.5. A 50.00-mL portion of a $AgNO_3$ solution, containing 2.0000 g Ag/L, reacts with 15.00 mL of a KCN solution, according to the reaction

$$Ag^+ + 2CN^- \rightleftharpoons [Ag(CN)_2]^-$$

Calculate the molarity and the normality of the KCN solution.

Solution. We have

$$\underline{M}_{AgNO_3} = \frac{2.0000 \text{ g/L}}{107.87 \text{ g/mol}} = 0.01854.$$

$$\underline{M}_{KCN} \text{ (mmol/mL)} \times 15.00 \text{ mL} = 0.01854 \text{ (mmol } Ag^+\text{/mL)} \times 50.00 \text{ mL} \times 2 \text{ (mmol } CN^-\text{/mmol } Ag^+)$$

Hence, $\underline{M}_{KCN} = 0.01854 \times 2 \times (50.00/15.00) =$ **0.1236**. We have 1 eq KCN $\equiv$ 1 eq $Ag^+ \equiv$ 2 mol KCN. Hence, 1 $\underline{N}$ KCN $\equiv$ 2 $\underline{M}$ KCN, whereupon $\underline{N}_{KCN} = \underline{M}_{KCN}/2 = 0.1236/2 =$ **0.0618**.

Example 3.6. A sulfuric acid solution has a density of 1.28 g/mL and contains 37.0% H_2SO_4 by weight. Calculate a) the formality of the solution, b) the normality, c) the molality, and d) the mole fraction of H_2SO_4. e) How many milliliters of water should be added to 100 mL of the solution to make it 1.00 $\underline{N}$ in H_2SO_4? f) How many milligrams of H_2SO_4 are there in 15.0 mL of 1.00 $\underline{N}$ H_2SO_4 solution? (It is assumed that the volume of a mixture equals the sum of the volumes of the solutions mixed, and that both protons of H_2SO_4 react).

Solution. a) The weight of one liter of solution is equal to

$$\frac{1280 \text{ g solution}}{\text{mL}} \times \frac{1000 \text{ mL}}{\text{L}} = 1280 \text{ g solution/L}$$

37.0% of this weight is H_2SO_4, i.e.,

$$\frac{1280 \text{ g solution}}{\text{L}} \times \frac{37.0 \text{ g } H_2SO_4}{100 \text{ g solution}} = 473.6 \text{ g } H_2SO_4\text{/L}.$$

The gram-formula weight of H_2SO_4 is 98.08. Hence,

$$\underline{F} = \frac{473.6 \text{ g } H_2SO_4\text{/L}}{98.08 \text{ g } H_2SO_4\text{/gfw}} = \mathbf{4.83} \text{ gfw/L}.$$

b) The equivalent weight of H_2SO_4 is 98.08/2 = 49.04 g. Hence,

$$\underline{N} = \frac{473.6 \text{ g } H_2SO_4\text{/L}}{49.04 \text{ g } H_2SO_4\text{/eq}} = \mathbf{9.66} \text{ eq/L}.$$

The normality can also be calculated as follows:

$$\underline{N} = 4.83 \text{ gfw/L} \times 2 \text{ eq/gfw} = \mathbf{9.66} \text{ eq/L}.$$

c) One liter of solution weighs 1280 g and contains 473.6 g H_2SO_4 and 1280 − 473.6 = 806.4 g H_2O. The solution contains 473.6 g H_2SO_4/806.4 g H_2O $\equiv$ 587.3 g H_2SO_4/1000 g H_2O. Hence,

$$m = \frac{587.3 \text{ g } H_2SO_4\text{/1000 g } H_2O}{98.08 \text{ g } H_2SO_4\text{/mol } H_2SO_4} = \mathbf{5.99} \frac{\text{mol } H_2SO_4}{1000 \text{ g } H_2O}.$$

d) Let x_2 be the mole fraction of H_2SO_4. We have

$$x_2 = \frac{\text{mol } H_2SO_4}{\text{mol } H_2SO_4 + \text{mol } H_2O} = \frac{4.83 \text{ mol}}{4.83 \text{ mol} + \dfrac{806.4 \text{ g } H_2O}{18.02 \text{ g } H_2O/\text{mol}}} = \mathbf{0.097}.$$

e) Let V mL be the volume of the 1.00 $\underline{N}$ H_2SO_4 obtained by dilution. The number of gram-equivalents of H_2SO_4 is the same before and after dilution. Therefore, we have

$$100 \text{ mL} \times 9.66 \text{ meq/mL} = \text{V mL} \times 1.00 \text{ meq/mL}$$

or V = 966 mL. Hence, 966 − 100 = **866** mL of water are required.

f) mg H_2SO_4 = 15.0 mL × 1.00 meq/mL × 49.04 mg/meq = **736** mg.

Example 3.7. a) How many milliliters of a sulfuric acid solution, that has a density of 1.84 g/mL and contains 96% H_2SO_4 by weight, should be diluted with water to obtain one liter of solution containing 50% H_2SO_4 by weight? b) What is the normality of the initial and final solutions?

Solution. a) Let y mL be the required volume of H_2SO_4, and consequently (1000−y) mL be the volume of water. The weight of H_2SO_4 in the initial and the final solutions is the same. Hence, we have

$$\frac{1.84 \text{ g soln}}{\text{mL}} \times \frac{96 \text{ g } H_2SO_4}{100 \text{ g soln}} \times \text{y mL}$$

$$= \left[\frac{1.84 \text{ g soln}}{\text{mL}} \times \text{y mL} + \frac{1.00 \text{ g } H_2O}{\text{mL}} \times (1000 - \text{y})\text{mL}\right] \times \frac{50 \text{ g } H_2SO_4}{100 \text{ g soln}}$$

or y = **371.4** mL.

b) The normality of the initial solution is equal to

$$\frac{1.84 \text{ g soln}}{\text{mL}} \times \frac{96 \text{ g } H_2SO_4}{100 \text{ g soln}} \times \frac{1 \text{ meq } H_2SO_4}{0.04904 \text{ g } H_2SO_4} = \mathbf{36.0} \text{ meq/mL}$$

and of the final solution is equal to

$$\underline{N}_f = 36.0 \text{ meq/mL} \times (371.4/1000) = \mathbf{13.4} \text{ meq/mL}.$$

Example 3.8. A 2.00 $\underline{M}$ $CaCl_2$ solution has a density of 1.08 g/mL. Calculate: a) the per cent chlorine in the solution, b) the molality of $CaCl_2$, and c) the mole fraction of $CaCl_2$.

Solution. a) We have

$$\%\text{Cl} = \frac{\left(\dfrac{2.00 \text{ mol } CaCl_2}{\text{L}}\right)\left(\dfrac{2 \text{ mol } Cl^-}{\text{mol } CaCl_2}\right)\left(\dfrac{35.45 \text{ g } Cl^-}{\text{mol } Cl^-}\right)}{(1000 \text{ mL/L})(1.08 \text{ g/mL})} \times 100 = \mathbf{13.1\%} \text{ (w/w)}.$$

b) Molality, m =

$$\frac{2.00 \text{ mol } CaCl_2/\text{L solution}}{[(1000 \text{ mL} \times 1.08 \text{ g/mL}) - (2.00 \text{ mol} \times 111 \text{ g/mol } CaCl_2)] \text{ g } H_2O/\text{L solution}}$$

$$= \frac{2.00 \text{ mol } CaCl_2}{858 \text{ g } H_2O} \equiv \frac{2.00 \text{ mol } CaCl_2}{858 \text{ g } H_2O} \times 1000 = \mathbf{2.33} \text{ mol } CaCl_2/1000 \text{ g } H_2O.$$

c) Mole fraction of $CaCl_2$ =

$$x_{CaCl_2} = \frac{\frac{\text{mol } CaCl_2}{\text{L solution}}}{\frac{\text{mol } CaCl_2}{\text{L solution}} + \frac{\text{mol } H_2O}{\text{L solution}}} = \frac{2.00 \text{ mol}}{2.00 \text{ mol} + \frac{858 \text{ g } H_2O}{18.02 \text{ g } H_2O/\text{mol}}} = \mathbf{0.0403}.$$

Example 3.9. The normality of acids A and B is 1.50 and 2.10, respectively. How many milliliters should be taken from each acid to prepare 250 mL of 1.65 N acid?

Solution. The sum of the milliequivalents of each acid will equal the milliequivalents of the final acid (250 mL × 1.65 N). Let y mL be the required volume of acid A, and consequently (250 −y) mL be the volume of acid B. We have

$$1.50 \text{ (meq/mL)} \times \text{y mL} + (250 - \text{y}) \text{ mL} \times 2.10 \text{ (meq/mL)} = 250 \text{ mL} \times 1.65 \text{ (meq/mL)}$$

or y = 187.5. Hence, we should take **187.5** mL of acid A and **62.5** mL of acid B.

Example 3.10. Express in all possible ways the concentration of ethanol in a solution obtained by mixing 23.00 g of ethanol with 500 mL of water. The final solution has a density of 0.992 g/mL. The density of water is equal to 1.000 g/mL.

Solution. The volume of the solution is equal to

$$\frac{23.00 \text{ g } C_2H_5OH + 500 \text{ g } H_2O}{0.992 \text{ g/mL}} = 527\text{mL} = 0.527\text{L}.$$

Hence, we have

Molarity:

$$\underline{M} = \frac{23.00 \text{ g } C_2H_5OH}{(46.07 \text{ g/mol})(0.527 \text{ L})} = \mathbf{0.947} \text{ mol/L}.$$

Formality:

$$\underline{F} = \frac{23.00 \text{ g } C_2H_5OH}{(46.07 \text{ g/fw})(0.527 \text{ L})} = \mathbf{0.947} \text{ fw/L}.$$

Molality:

$$m = \frac{\frac{23.00 \text{ g } C_2H_5OH}{500 \text{ g } H_2O}}{46.07 \text{ g/mol}} \times 1000 = \mathbf{0.998} \text{ mol/1000 g } H_2O.$$

Weight per cent:

$$\%w/w = \frac{23.00 \text{ g}}{523 \text{ g}} \times 100 = \mathbf{4.40\%} \text{ w/w}.$$

Weight-volume per cent:

$$\%w/v = \frac{23.00 \text{ g}}{527 \text{ mL}} \times 100 = \mathbf{4.36\%} \text{ w/v}.$$

Example 3.11. How many grams of $Na_2C_2O_4$ should be added to 200 mL of 0.100 $\underline{F}$ $KHC_2O_4 \cdot H_2C_2O_4$ solution so that the normality of the solution as a reducing agent is three times its normality as an acid (it is assumed that the addition of $Na_2C_2O_4$ causes no change in solution volume)?

Solution. Let y mfw be the required amount of $Na_2C_2O_4$, whereupon the solution contains

$$\text{meq } C_2O_4^{2-} \text{ (redox)} = \text{meq from } KHC_2O_4 \cdot H_2C_2O_4 + \text{meq from } Na_2C_2O_4$$
$$= (200 \text{ mL})(0.100 \text{ mfw/mL})(4 \text{ meq } C_2O_4^{2-}/\text{mfw } KHC_2O_4 \cdot H_2C_2O_4) +$$
$$\text{y mfw } (2 \text{ meq } C_2O_4^{2-}/\text{mfw } Na_2C_2O_4) = (80 + 2y) \text{ meq } C_2O_4^{2-}$$

Also, meq $C_2O_4^{2-}$ (acid) = meq from $KHC_2O_4 \cdot H_2C_2O_4$

$$(200 \text{ mL})(0.100 \text{ mfw/mL})(3 \text{ meq } H^+/\text{mfw } KHC_2O_4 \cdot H_2C_2O_4) = 60 \text{ meq } H^+.$$

Since for the final solution, meq(redox) = 3 × meq(acid), we have (80 + 2y) meq(redox) = 3 × 60 meq (H^+) or y = 50 mfw $Na_2C_2O_4$. Consequently, we should add

$$(50 \text{ mfw})(134.0 \text{ mg } Na_2C_2O_4/\text{mfw})(0.001 \text{ g/mg}) = \mathbf{6.70} \text{ g } Na_2C_2O_4.$$

Example 3.12. If 1.00 mL of 4.0 $\underline{M}$ HNO_3 is added to a precipitate consisting of 12.1 mg of $Pb(OH)_2$ and 78 mg of $Bi(OH)_3$, what percentage of the acid will be consumed in the dissolution of the precipitate?

Solution. 12.1 mg of $Pb(OH)_2$ corresponds to

$$\frac{12.1 \text{ mg}}{241.21 \text{ mg/mmol}} = 0.05016 \text{ mmol } Pb(OH)_2 \equiv 2 \times 0.05016$$
$$= 0.1003 \text{ mmol } H^+ \; [\mathbf{Pb(OH)_2} + 2H^+ \rightleftharpoons Pb^{2+} + 2H_2O],$$

whereas 78 mg of $Bi(OH)_3$ correspond to $\frac{78 \text{ mg}}{260 \text{ mg/mmol}} = 0.300$ mmol $Bi(OH)_3 \equiv 0.300 \times 3 = 0.900$ mmol H^+ [$\mathbf{Bi(OH)_3} + 3H^+ \rightleftharpoons Bi^{3+} + 3H_2O$]. Hence, a total of 0.1003 + 0.900 = 1.000 mmol H^+ is required to dissolve the precipitate. 1.00 mL of 4.0 $\underline{M}$ HNO_3 contains 4.0 mmol HNO_3. Hence, (1.000/4.0)100 = **25.0%** of the acid is consumed for the dissolution of the precipitate.

Example 3.13. X grams of the hydroxide $M(OH)_2$ is neutralized by 20.0 mL of 1.000 $\underline{M}$ HCl solution. After the evaporation of the solution, the salt residue weighs 1.11 g. Calculate X and the molecular weight of the hydroxide.

Solution. We have

$$M(OH)_2 + 2HCl \rightleftharpoons MCl_2 + 2H_2O$$

and

$$(20.0 \text{ mL HCl})(1.000 \text{ mmol/mL}) = 20.0 \text{ mmol HCl} \equiv 10.0 \text{ mmol } M(OH)_2 \equiv 10.0 \text{ mmol } MCl_2 \equiv 1.11 \text{ g } MCl_2 \equiv 0.111 \text{ g } MCl_2\text{/mmol}.$$

Hence, the molecular weight of MCl is 111, the atomic weight of M is equal to $111 - (2 \times 35.45) = 40.1$ and the molecular weight of $M(OH)_2$ is equal to $40.1 + (2 \times 17.01) = \mathbf{74.1}$. Hence,

$$X = (74.1 \text{ g/mol})(0.010 \text{ mol}) = \mathbf{0.741} \text{ g } M(OH)_2.$$

Example 3.14. A HNO_3 solution is 1.00 N as an acid. How many milliliters of water should be added to 50.0 mL of the acid to make it 1.00 N as an oxidant, given that HNO_3 is reduced to NO?

Solution. The solution is $1.00 \times 3 = 3.00$ N as an oxidant ($\overset{+5}{H N O_3} \rightarrow \overset{+2}{NO}$). Let y mL be the required volume of water, whereupon we have $50.0 \times 3.00 = (50.0 + y)$ or $y = \mathbf{100}$ mL of water.

Example 3.15. Twenty eight milliliters of 0.500 M HCl were required to dissolve 0.132 g of an Al-Mg alloy. Calculate the per cent composition of the alloy.

Solution. The dissolution reactions are

$$2Al + 6HCl \rightleftharpoons 2AlCl_3 + 3H_2 \uparrow$$

$$Mg + 2HCl \rightleftharpoons MgCl_2 + H_2 \uparrow$$

Let y mg of Al and z mg of Mg represent the content of 0.132 g of alloy. The sum of milliequivalents of Al and Mg is equal to the milliequivalents of HCl. Therefore, we have the equations

$$y + z = 132 \tag{3.11}$$

$$\frac{y \text{ mg Al}}{26.98/3 \text{ mg Al/meq}} + \frac{z \text{ mg Mg}}{24.30/2 \text{ mg Mg/meq}} = 28.0 \text{ mL} \times 0.500 \text{ meq/mL} \tag{3.12}$$

Solving the system of Equations (3.11) and (3.12), we find that $y = 108.5$ mg Al and $z = 23.5$ mg Mg. Hence, the alloy contains

$$\frac{108.5 \text{ mg}}{132 \text{ mg}} \times 100 = \mathbf{82.2\%} \text{ Al}$$

and

$$\frac{23.5 \text{ mg}}{132 \text{ mg}} \times 100 = \mathbf{17.8\%} \text{ Mg}.$$

3.2 Effective Concentration or Activity

The ionic atmosphere surrounding each ion reduces its velocity and reactivity, causing the solution to behave as if its concentration is smaller than the stoichiometric (real or analytical) concentration C. (At very high ionic concentrations, the reverse may be true due to a decrease in the effective concentration of water.) Therefore, in order to achieve agreement between calculated and experimental data involving ionic or molecular reactions, we have to multiply the concentration C by a correction factor f, called the *activity coefficient*. Thus, we obtain the effective concentration a, called the *activity*. The following relation is valid:

$$a = fC \qquad (3.13)$$

The value of the coefficient f depends on the charge z of the ion and the ionic strength μ of the solution, according to the equation

$$-\log f = Az^2\sqrt{\mu} \qquad (3.14)$$

where A = constant = 0.51 for the water at 25°C, and μ = ionic strength of the solution, which is defined by the equation

$$\mu = \frac{1}{2}(C_1{z_1}^2 + C_2{z_2}^2 + \ldots + C_n{z_n}^2) = \frac{1}{2}\sum C_i{z_i}^2 \qquad (3.15)$$

where C_1, $C_2 \ldots C_n$ are the molarities of the various ions of the solution and $z_1, z_2 \ldots z_n$ are their charges. For better accuracy, the following relation is used

$$-\log f = Az^2\sqrt{\mu}/(1+\sqrt{\mu}) \text{ (extended Debye – Hückel equation)} \qquad (3.16)$$

For very dilute ionic solutions ($\mu \to 0$), the influence of the ionic atmosphere on the ion of interest becomes increasingly small and we have $f \to 1$ and $a \to C$.

The relation between the ionic strength μ and the concentration C for salts of various charge types is:

M^+A^-, $\mu = C$; $M^{2+}{A_2}^-$ or ${M_2}^+A^{2-}$, $\mu = 3C$; $M^{2+}A^{2-}$, $\mu = 4C$;
$M^{3+}{A_3}^-$ or ${M_3}^+A^{3-}$, $\mu = 6C$.

Example 3.16. Calculate the concentration and the activity of each ion in a mixture obtained by mixing 25.0 mL of 0.120 M $MnCl_2$ solution and 35.0 mL of 0.060 M KCl solution.

Solution. The volume of the mixture is 60.0 mL. Assuming that the dissociation of the salts is complete, we have

$$\text{Molarity of } Mn^{2+}, \underline{M} = \frac{25.0 \text{ mL} \times 0.120 \text{ mmol/mL}}{60.0 \text{ mL}} = \mathbf{0.0500} \text{ mmol/ml}$$

$$\text{Molarity of } K^{+}, \underline{M} = \frac{35.0 \text{ mL} \times 0.060 \text{ mmol/mL}}{60.0 \text{ mL}} = \mathbf{0.035} \text{ mmol/mL}$$

$$\text{mmol } Cl^- = 2 \times \text{mmol } MnCl_2 + 1 \times \text{mmol KCl}$$

Molarity of Cl^-, $\underline{M}$ =

$$\frac{(25.0 \text{ mL})\left(\dfrac{0.120 \text{ mmol}}{\text{mL}}\right)\left(2\dfrac{\text{mmol } Cl^-}{\text{mmol } MnCl_2}\right) + (35.0 \text{ mL})\left(\dfrac{0.0060 \text{ mmol } Cl^-}{\text{mL}}\right)}{60.0 \text{ mL}}$$

$$= \mathbf{0.135} \text{ mmol/ml.}$$

The ionic strength of the solution is equal to

$$\mu = \frac{1}{2}(C_{Mn^{2+}}z^2_{Mn^{2+}} + C_{K^+}z^2_{K^+} + C_{Cl^-}z^2_{Cl^-})$$

$$\mu = \frac{1}{2}[(0.0500)(2)^2 + (0.035)(1)^2 + (0.135)(-1)^2] = 0.185.$$

Using Equation (3.16), we have

$$-\log f_{Mn^{2+}} = \frac{0.51 \times (2)^2 \times \sqrt{0.185}}{1 + \sqrt{0.185}} \text{ or } f_{Mn^{2+}} = 0.243$$

$$-\log f_{K^+} = -\log f_{Cl^-} = \frac{0.51 \times 1^2 \times \sqrt{0.185}}{1 + \sqrt{0.185}} \text{ or } f_{K^+} = f_{Cl^-} = 0.702.$$

Hence,

$$a_{Mn^{2+}} = f_{Mn^{2+}}C_{Mn^{2+}} = 0.243 \times 0.0500 = \mathbf{0.0122}\ \underline{M}$$
$$a_{K^+} = f_{K^+}C_{K^+} = 0.702 \times 0.035 = \mathbf{0.0246}\ \underline{M}$$
$$a_{Cl^-} = f_{Cl^-}C_{Cl^-} = 0.702 \times 0.135 = \mathbf{0.095}\ \underline{M}.$$

3.3 Charge Balance

According to the *principle of electroneutrality*, all solutions are electrically neutral, i.e., there is no solution containing a detectable excess of positive or negative charge, because the sum of the positive charges equals the sum of negative charges. The use of charge balance and mass balance (Section 3.4 below) is useful in solving for *equilibrium* concentrations of species (as opposed to total or analytical concentrations). The example below illustrates the method to arrive at charge balance.

Example 3.17. Are the following statements true or false? Justify your answer.

a) In an H_2S solution, $[H^+]$ is two times $[S^{2-}]$.

b) In a saturated CdS solution, the following relation is valid: $[Cd^{2+}] = [S^{2-}]$.

c) In a saturated solution of Bi_2S_3, the following relation is valid: $[Bi^{3+}] = [S^{2-}] + [OH^-]$.

d) In a saturated 0.1 M Na_2S solution, the relation $[Na^+] = 2[S^{2-}]$ is false.

Solution. a) The statement is false. There are two sources of H^+, from the dissociation of both H_2S and H_2O. H_2S dissociates stepwise. The equilibria are:

$$H_2S \rightleftharpoons HS^- + H^+$$
$$HS^- \rightleftharpoons S^{2-} + H^+$$
$$H_2O \rightleftharpoons H^+ + OH^-$$

Dissociation of H_2S gives H^+ and two anionic species, HS^- and S^{2-}, and that of water gives H^+ and OH^-. The amount of H^+ from that portion of *completely* dissociated H_2S is equal to twice the amount of S^{2-} formed, and from *partial* (first step) dissociation is equal to the amount of HS^- formed. That is, for each S^{2-} formed, there are 2 H^+, for each HS^- formed, there is 1 H^+, and for each OH^- formed, there is 1 H^+. Now, for the singly charged species, the *charge* concentration is identical to the concentration of the *species*. But for S^{2-}, the *charge* concentration is twice that of the *species*, so we must multiply the S^{2-} concentration by two to arrive at the charge concentration from it. According to the principle of electroneutrality, positive charge concentration must equal the negative charge concentration. Hence,

$$[H^+] = 2[S^{2-}] + [HS^-] + [OH^-]$$

Note that while there may be more than one source for a given species (H^+ in this case), the total charge from all sources is always equal to the net *equilibrium* concentration of the species multiplied by its charge.

b) The statement is false. The equilibria are:

$$CdS \rightleftharpoons Cd^{2+} + S^{2-}$$
$$S^{2-} + H_2O \rightleftharpoons HS^- + OH^-$$
$$HS^- + H_2O \rightleftharpoons H_2S + OH^-$$
$$H_2O \rightleftharpoons H^+ + OH^-$$

Again, the charge concentration for the singly charged species (H^+, OH^-, HS^-) will be equal to the concentrations of the species. But for Cd^{2+} and S^{2-}, the charge concentration will be twice their concentrations. We must again equate the positive and negative charge concentrations:

$$2[Cd^{2+}] + [H^+] = 2[S^{2-}] + [HS^-] + [OH^-]$$

c) The statement is false. The cations present are Bi^{3+} and H^+, while the anions present are S^{2-}, HS^-, and OH^-. As before, we must multiply the concentrations of the multiply charged species by their charges to arrive at the equivalent concentration of charge. Hence, we have

$$3[Bi^{3+}] + [H^+] = 2[S^{2-}] + [HS^-] + [OH^-]$$

d) The statement is correct, because for charge balance, we have the relation

$$[Na^+] + [H^+] = 2[S^{2-}] + [HS^-] + [OH^-]$$

3.4 Mass Balance

The ***principle of mass balance*** is based on the law of mass conservation, and states that the number of atoms of an element remains constant in usual chemical reactions (between stable isotopes), because then no atoms are produced or destroyed. The principle is expressed mathematically by equating the concentrations, usually the molarities.

Example 3.18. Write the equations of mass balance for a 1.00×10^{-5} $\underline{M}$ $[Ag(NH_3)_2]Cl$ solution.

Solution. The equilibria are:

$$[Ag(NH_3)_2]Cl \rightarrow [Ag(NH_3)_2]^+ + Cl^-$$
$$[Ag(NH_3)_2]^+ \rightleftharpoons [Ag(NH_3)]^+ + NH_3$$
$$[Ag(NH_3)]^+ \rightleftharpoons Ag^+ + NH_3$$
$$NH_3 + H^+ \text{ (from } H_2O) \rightleftharpoons NH_4^+$$

The Cl^- concentration is equal to the concentration of the salt that dissociated:

$$[Cl^-] = 1.00 \times 10^{-5} \ \underline{M}.$$

Likewise, the sum of the concentrations of *all* silver species is equal to the concentration of Ag in the original salt that dissociated:

$$[Ag^+] + [Ag(NH_3)^+] + [Ag(NH_3)_2{}^+] = 1.00 \times 10^{-5} \ \underline{M}.$$

We have the following nitrogen-containing species:

$$NH_4{}^+, \ NH_3, \ [Ag(NH_3)]^+, \ [Ag(NH_3)_2]^+.$$

The concentration of N from the last species is twice the concentration of the $[Ag(NH_3)_2]^+$. The total concentration of the nitrogen is twice the concentration of the original salt, since there are two NH_3 per molecule. Hence, we can write:

$$[NH_4{}^+] + [NH_3] + [Ag(NH_3)^+] + 2[Ag(NH_3)_2{}^+] = 2.00 \times 10^{-5} \ \underline{M}.$$

3.5 Problems

For the solutions of all the problems in this book, it is assumed, unless stated otherwise, that no volume change occurs when adding a solid or a gas in a solution, and that the volume of a mixture obtained by mixing various solutions equals the sum of the volumes of the solutions mixed. The density of water is taken as equal to 1.000 g/mL. The activity is taken as equal to concentration, except if stated otherwise.

Equivalent weights

3.1. For each of the following reactions, write the equivalent weight of the substance being titrated (titrand) as a function of the molecular weight:

Titrand

(a) $PO_4^{3-} + 3Ag^+ \rightleftharpoons \mathbf{Ag_3PO_4}$
(b) $2CN^- + Ag^+ \rightleftharpoons [Ag(CN)_2]^-$
(c) $ClO_3^- + 6Fe^{2+} + 6H^+ \rightleftharpoons Cl^- + 6Fe^{3+} + 3H_2O$
(d) $Ba^{2+} + SO_4^{2-} \rightleftharpoons \mathbf{BaSO_4}$
(e) $B_4O_7^{2-} + 2H^+ + 5H_2O \rightleftharpoons 4H_3BO_3$
(f) $\mathbf{I_2} + 2S_2O_3^{2-} \rightleftharpoons 2I^- + S_4O_6^{2-}$
(g) $5NO_2^- + 2MnO_4^- + 6H^+ \rightleftharpoons 5NO_3^- + 2Mn^{2+} + 3H_2O$
(h) $2RSH + \mathbf{I_2} \rightleftharpoons RSSR + 2HI$
(i) $3N_2H_4 + 2BrO_3^- \rightleftharpoons 3N_2 + 2Br^- + 6H_2O$
(j) $4CN^- + Ni^{2+} \rightleftharpoons [Ni(CN)_4]^{2-}$

Concentrations

3.2. The molar solubility of $PbSO_4$ is 1.14×10^{-4} mol/L. How many micrograms of Pb^{2+} are there in 1 mL of a saturated (solid in equilibrium with ions) $PbSO_4$ solution?

3.3. What weight of each of the following salts is required for the preparation of 500 mL of a solution containing a) 10.0 mg of cation/mL (series A), b) 10.0 mg of anion/mL (series B)?

A : 1) $Hg_2(NO_3)_2 \cdot 2H_2O$, 2) $SbCl_3$, 3) $Fe(NO_3)_3 \cdot 6H_2O$, 4) NH_4NO_3.
B : 1) $Na_2C_2O_4$, 2) $KNaC_4H_4O_6 \cdot 4H_2O$, 3) Na_2SO_4, 4) $K_3[Fe(CN)_6]$.

3.4. A hundred milliliters of a $Ni(NO_3)_2$ solution containing 2.00 g Ni/L reacts with 30.0 mL of KCN solution, according to the reaction

$$Ni^{2+} + 4CN^- \rightleftharpoons [Ni(CN)_4]^{2-}$$

Calculate the molarity and the normality of the KCN solution.

3.5. How many milliliters of water should be added to each of the following solutions, in order to obtain solutions having the concentrations indicated in parentheses?

(a) 24.5 mL of 4.0 M H_2SO_4 (0.500 N)

(b) 120 mL of 0.120 N $Ba(OH)_2$ (0.0500 M)

(c) 50.0 mL HCl, of density 1.095 g/mL, containing 20.0% HCl by weight (1.00 M)

3.6. Quantities of 25.0 mL of 0.100 M H_2SO_4, 10.0 mL of 0.200 N HCl, and 45.0 mL of 0.120 N KOH are mixed in a flask. Is the mixture acidic, neutral or alkaline?

3.7. How much NaF must be added to 1000 L of water so that the concentration of F^- ion is equal to 1.00 ppm?

3.8. Calculate the formality, the normality and the molality of the following solutions:

(a) Acetic acid, density 1.05 g/mL, 99.5% CH_3COOH by weight

(b) Ammonia, density 0.898 g/mL, 28.0% NH_3 by weight

(c) Hydrochloric acid, density 1.18 g/mL, 35.6% HCl by weight

(d) Nitric acid, density 1.41 g/mL, 67.5% HNO_3 by weight

3.9. A sulfuric acid solution has a density of 1.28 g/mL and contains 37.0% H_2SO_4 by weight. In what proportion should this solution be mixed with a 0.500 N H_2SO_4 solution to obtain 10.0 L of 1.00 M H_2SO_4 solution?

3.10. How many milliliters of a sulfuric acid solution that has a density of 1.85 g/mL and contains 98.0% H_2SO_4 by weight should be diluted with water to obtain one liter of a solution containing 66.0% H_2SO_4 by weight? What is the normality of the initial solution (both protons react)?

3.11. A concentrated nitric acid solution has a density of 1.42 g/mL and contains 70.0% HNO_3 by weight. a) What is the molarity of the solution? b) What volume of this acid contains 9.94 g of HNO_3? c) How will 500 mL of 0.1580 M HNO_3 be prepared from the concentrated HNO_3 solution?

3.12. A sulfuric acid solution has a density of 1.28 g/mL and contains 37.0% H_2SO_4 by weight. An ammonia solution has a density of 0.898 g/mL and contains 28.0% NH_3 by weight. In which of the two solutions is the mole fraction of the solute larger?

3.13. How many grams of $KHC_2O_4 \cdot H_2C_2O_4$ should be mixed with 1.000 g of $Na_2C_2O_4$ so that when dissolving the mixture in water, the normality of the resulting solution as a reductant (product is CO_2) is 3.10 times its normality as an acid (3 reacting protons)?

3.14. How many milliliters of a sulfuric acid solution that has a density of 1.835 g/mL and contains 93.1% H_2SO_4 by weight are required to prepare one liter of 1.000 N H_2SO_4 solution (both protons react)?

3.15. How many milliliters of water should be added to 990 mL of 0.1038 N HCl solution to make it exactly 1.000 N?

3.16. A 32.40-mL portion of 0.0980 M HCl solution is mixed with 34.40 mL of 0.1080 N HCl solution. Calculate the normality of the solution obtained by mixing these two solutions, and its normality when it is diluted with water to 100.0 mL.

3.17. How many grams of solute are contained in each of the following solutions?

(a) 20.00 meq NaCN ($Ag^+ + 2CN^- \rightleftharpoons [Ag(CN)_2]^-$)

(b) 50.00 mL 0.1000 N $Na_2C_2O_4$ ($C_2O_4^{2-} \rightarrow 2CO_2 + 2e$)

(c) 5.00 meq $[Fe(CN)_6]^{4-}$ ($[Fe(CN)_6]^{4-} \rightleftharpoons [Fe(CN)_6]^{3-} + e$)

(d) 50.00 mL 0.2000 N Na_2CO_3 ($CO_3^{2-} + 2H^+ \rightleftharpoons H_2CO_3$)

3.18. Calculate the molality, the molarity, and the formality of methanol in a solution that has a density of 1.4633 g/mL, and is obtained by dissolving 50.00 mL of methanol (d = 0.7961 g/mL) in 950.0 mL of chloroform.

Activity

3.19. Calculate the concentration and the activity coefficient of each ion in a mixture obtained by mixing 20.0 mL of 0.100 M $CaCl_2$ solution and 30.0 mL of 0.200 M NaCl solution.

Stoichiometry

3.20. How many milliliters of 0.200 M $KMnO_4$solution are required to react (product Mn^{2+}) with 10.00 mL of 0.2500 N $Na_2C_2O_4$ solution (product CO_2) in strongly acidic medium?

3.21. A 0.1000 M HCl solution is ionized quantitatively, but a 0.1000 M CH_3COOH solution is ionized by only 1.34%. How many milliliters of 0.2500 N NaOH solution are required to neutralize 100 mL of each of these solutions?

3.22. Fill the vacancies below so that in each case the quantities of the substances will be equivalent:

(a) 0.654 g $KHC_2O_4 \cdot H_2C_2O_4 \overset{H^+}{\equiv}$ __________ mL 0.2500 N $KMnO_4$ solution

(b) 32.7 mg $KHC_2O_4 \cdot H_2C_2O_4 \equiv$ __________ mL 0.1000 N NaOH solution

(c) 5.88 g $K_2Cr_2O_7 \overset{H^+}{\equiv}$ __________ mL 0.5000 M KI solution

3.23. Calculate the formality and the normality of a solution containing 0.604 g of $Cu(NO_3)_2 \cdot 3H_2O$ in 20.0 mL. If the copper is precipitated quantitatively as CuS, how many milligrams of precipitate will result from 2.0 mL of solution?

3.24. a) What is the molarity of a sulfuric acid solution, 50.00 mL of which gives 2.1015 g of $BaSO_4$?

b) How many milliliters of a concentrated sulfuric acid solution that has a density of 1.84 g/mL and contains 95.9% H_2SO_4 by weight are required for the preparation of 1 liter of the above solution in (a)?

3.25. a) How many milliliters of water should be added to 50.0 mL of a hydrochloric acid solution that has a density of 1.10 g/mL and contains 20.0% HCl by weight to obtain a solution that has a density of 1.04 g/mL and contains 8.16% HCl by weight?

b) How many milliliters of 0.1250 $\underline{M}$ $Ba(OH)_2$ solution are required to neutralize 8.00 mL of the resulting HCl solution?

3.26. If 1.0 mL of 4 $\underline{M}$ HNO_3 is added to a precipitate consisting of 8.67 mg of CdS and 51.4 mg of Bi_2S_3, what percentage of the acid will be consumed for the dissolution of the precipitate?

3.27. If an excess of NaClO oxidizes I_2 to HIO_3, calculate the quantity of iodine remaining in the I_2 form, when 0.10 mL of 5.0% w/v NaClO solution is mixed with a solution containing 4.0 mg of I^-.

Indirect analysis

3.28. Forty milliliters of 0.4000 N HCl solution were required to dissolve 0.7690 g of a $CaCO_3$ - $MgCO_3$ mixture. Calculate the per cent $CaCO_3$ and $MgCO_3$ in the mixture.

Mass balance

3.29. Write the equations of mass balance and electroneutrality for a 0.100 $\underline{M}$ $[Cd(NH_3)_4]Cl_2$ solution.

3.30. Prove the following relations using the principles of electroneutrality and mass balance:

(a) $[NO_2^-] = [H^+] - [OH^-]$ for 0.2 $\underline{M}$ HNO_2 solution
(b) $[CH_3COOH] = 0.2 - [H^+] + [OH^-]$ for 0.2 $\underline{M}$ CH_3COOH solution
(c) $[H_2C_2O_4] = 0.1 - [H^+] + [OH^-] + [C_2O_4^{2-}]$ for 0.1 $\underline{M}$ $H_2C_2O_4$ solution
(d) $[HCN] = [OH^-] - [H^+]$ for 0.1 $\underline{M}$ KCN solution
(e) $[H_2PO_4^-] = \dfrac{[OH^-] - [H^+] - [HPO_4^{2-}] - 3[H_3PO_4]}{2}$ for 0.1 $\underline{M}$ Na_3PO_4 solution
(f) $[HSO_4^-] = 0.2 - [H^+] + [OH^-]$ for 0.1 $\underline{M}$ H_2SO_4 solution (it is assumed that the dissociation of H_2SO_4 to H^+ and HSO_4^- is quantitative).

3.31. Write equations of mass balance for an aqueous saturated solution of BaF_2 containing the species F^-, HF, HF_2^-, and Ba^{2+}.

3.32. Write an equation of mass balance for an aqueous solution of $Ba_3(PO_4)_2$.

Chapter 4

Reaction Rate and Chemical Equilibrium

4.1 Reaction Rate and Factors Affecting It

The *reaction rate* is defined as the change in concentration of one of the components of the chemical system per unit *time*. For the reaction

$$A + B \rightleftharpoons C + D \tag{4.1}$$

the reaction rate or velocity v is given mathematically by the relations

$$v = \frac{d[C]}{dt} = \frac{d[D]}{dt} = -\frac{d[A]}{dt} = -\frac{d[B]}{dt} \tag{4.2}$$

where $d[C], (d[D], d[A], d[B])$ is an infinitesimal change in the concentration of C (D, A, B) during the infinitesimal time dt; the negative signs denote that A and B are disappearing. (If a particles of A react with each B, then A is consumed a times as fast as B. Therefore,

$$-\frac{1}{a} \cdot \frac{d[A]}{dt} = -\frac{d[B]}{dt}.)$$

The concentration is usually expressed in moles per liter, and therefore, the reaction rate represents the change in molarity (M) per unit time, e.g., $M{\cdot}s^{-1}$.

The reaction rate is affected by many factors, such as the nature of reactants (usually, ionic reactions are fast, whereas molecular reactions are relatively slow), the temperature (the reaction rate increases with increasing temperature), the presence of catalysts (they usually accelerate the reaction), the concentration (the reaction rate increases with increasing concentration), the solvent, etc.

4.2 The Law of Mass Action. Chemical Equilibrium and Factors Affecting It

The reaction rate for the reaction

$$aA + bB + cC + \ldots \rightarrow dD + \ldots \quad (4.3)$$

is often given by the relation

$$v = k[A]^x[B]^y[C]^z \quad (4.4)$$

where k is a proportionality constant, called the *reaction rate constant* or *specific reaction rate*, which is equal to the reaction rate when the molar concentrations of each of the reactants is unity. The exponents x, y, z represent the *order* of the reaction with respect to each reactant. We should note that many rate expressions (for reversible reactions) do not fall in this simple form and products can enter into the rate expression.

From Equation (4.4) it can be seen that when all other factors (temperature, solvent, etc.) are constant, the rate of a chemical reaction is proportional to the molar concentration of the reactants, each concentration being raised to an appropriate power. Equation (4.4) expresses the *law of mass action* (law of Guldberg and Waage, 1867), and it is also known as the *rate law*. The sum $x + y + z + \ldots$ gives the total reaction order. The rate law of a reaction is determined experimentally and it can not be derived from the chemical equation of the reaction. The exponents of the various terms in the rate law usually do not coincide with the corresponding coefficients in the chemical equation, and the rate law may include terms with negative or fractional exponents.

Under proper conditions, many chemical reactions are *reversible*, that is, the products react with each other and reproduce the initial substances. We can write an *equilibrium constant* K for such reactions. When equilibrium is reached, each step in the reaction mechanism must also have reached equilibrium, and the equilibrium constant is independent of the reaction mechanism. Hence, we can write the correct expression for K from the balanced equations for the reactions by applying the rate law accordingly. (The overall equilibrium constant is actually the product of the corresponding constants for each mechanistic step, and their rate coefficients add to give the overall reaction coefficients.) Applying the law of mass action to the reversible reaction

$$A + B \underset{k_2}{\overset{k_1}{\rightleftharpoons}} C + D \quad (4.5)$$

we have

$$v_1 = k_1[A][B] \quad (4.6)$$

$$v_2 = k_2[C][D] \quad (4.7)$$

where v_1 is the reaction rate for the forward reaction, and v_2 is the rate of the opposite reaction. At the equilibrium state we have

$$k_1[A][B] = k_2[C][D] \quad (4.8)$$

or

$$\frac{[C][D]}{[A][B]} = \frac{k_1}{k_2} = K \quad (4.9)$$

or, for the general reaction

$$aA + bB \rightleftharpoons cC + dD \quad (4.5a)$$

we have

$$\frac{[C]^c[D]^d}{[A]^a[B]^b} = K \tag{4.9a}$$

The constant K is called the *equilibrium constant.* K varies with temperature and is almost independent of concentrations. K is constant when activities are used instead of concentrations, whereupon it is known as the *thermodynamic equilibrium constant.* The value of K can be found by determining the concentrations of the components of the system at the equilibrium state (Example 4.1). It should be noted that the equilibrium is dynamic, not static.

The main factors affecting the equilibrium state are the nature of the solvent (when the system contains ions), the temperature, the pressure (when gases are involved), and the concentration. By varying these factors, one can shift the equilibrium to the desired direction. The direction of the shift is predicted by Le Châtelier's principle, whereas the extent of the shift and the composition of the system at the new equilibrium state is calculated from the value of K. According to Le Châtelier's principle, if one of the factors that affect a system at the equilibrium state is changed, then the system readjusts to reduce the imposed change (stress).

Typical examples of equilibria and equilibrium constants are given in Table 4.1, for 25°C. These equilibria are studied in Chapters 5–8.

Example 4.1. The chemical reaction among A, B, and C is represented by the equation

$$A + B + C \rightleftharpoons D + E$$

The initial rate of the reaction was measured for various initial concentrations of A, B, and C at constant temperature. The results are given below:

Experiment	Initial concentrations, M			Initial rate,
	A	B	C	M·s^{-1}
1	0.200	0.50	0.40	8.0×10^{-5}
2	0.40	0.50	0.40	3.2×10^{-4}
3	0.40	2.00	0.40	1.28×10^{-3}
4	0.100	0.250	1.60	4.0×10^{-5}

Find the rate law and the reaction order and calculate the initial rate of the reaction, when the initial concentration is 0.250 M for A, 0.160 M for B, and 0.124 M for C.

Solution. The sought-for rate law is

$$v = k[A]^a[B]^b[C]^c \tag{4.10}$$

From a comparison of the data for experiments 1 and 2, it can be seen that when the initial concentration of A is doubled and for the same initial concentrations of B and

Table 4.1. *Examples of Chemical Equilibria and Equilibrium Constants at 25° C.*

Type of equilibrium	Name and symbol of equilibrium constant	Example	Expression and values of the equilibrium constant
Dissociation of water	Ion-product of water, K_w	$H_2O \rightleftharpoons H^+ + OH^-$	$K^* = [H^+][OH^-] = 1.00 \times 10^{-14}$
Dissociation of a weak acid or base	Dissociation constant, K_a or K_b	$CH_3COOH \rightleftharpoons H^+ + CH_3COO^-$	$K = \frac{[H^+][CH_3COO^-]}{[CH_3COOH]} = 1.80 \times 10^{-5}$
Heterogeneous equilibrium between a slightly soluble ionic solid and its ions in a saturated solution	Solubility product, K_{sp}	$\mathbf{Ag_2S} \rightleftharpoons 2Ag^+ + S^{2-}$	$K^* = [Ag^+]^2[S^{2-}] = 1 \times 10^{-50}$
Dissociation of a complex ion	Instability constant, K_{inst}	$[Cu(NH_3)_4]^{2+} \rightleftharpoons Cu^{2+} + 4NH_3$	$K = \frac{[Cu^{2+}][NH_3]^4}{[Cu(NH_3)_4{}^{2+}]} = 1 \times 10^{-12}$
Redox equilibria	K	$2Ag^+ + \mathbf{Zn} \rightleftharpoons \mathbf{2Ag} + Zn^{2+}$	$K^* = \frac{[Zn^{2+}]}{[Ag^+]^2} = 6.7 \times 10^{37}$

* Water and solid (pure) substances are omitted because their concentrations are constant (their activities are unity).

C, the initial rate v is quadrupled. Hence, the initial rate is proportional to $[A]^2$, that is, a = 2. Similarly, from experiments 2 and 3 it is deduced that the initial rate v is proportional to [B], that is, b = 1. Substituting the values of a and b in Equation (4.10), we find for experiments 1 and 4 the relations

$$v_1 = k(0.200)^2(0.50)(0.40)^c = 8.0 \times 10^{-5} \tag{4.11}$$

$$v_2 = k(0.100)^2(0.250)(1.60)^c = 4.0 \times 10^{-5} \tag{4.12}$$

Dividing Equations (4.11) and (4.12), we have $(0.25)^c = 0.25$; hence, c = 1 (we find the same result if we use the data for experiments 2 and 4 or 3 and 4 instead of the data for experiments 1 and 4). Therefore, the rate law is

$$v = k[A]^2[B][C] \tag{4.13}$$

Hence, the reaction is fourth order. Substituting the data for any experiment in Equation (4.13), we find the value of the reaction rate constant k. For example, from experiment 1 we have

$$k = (8.0 \times 10^{-5})/(0.200)^2(0.50)(0.40) = 1.00 \times 10^{-2}$$

Hence, the sought-for initial rate is equal to

$$v = (1.00 \times 10^{-2})(0.250)^2(0.160)(0.124) = \mathbf{1.24 \times 10^{-5}}\ \underline{M} \cdot s^{-1}.$$

Example 4.2. It is found experimentally that an equimolar mixture of anhydrous CH_3COOH and absolute C_2H_5OH is 66.7% esterified (ester formed by reaction below). How many moles of acetic acid should be mixed with 5.00 moles of absolute ethanol, in order to esterify ethanol by 90.0%.

Solution. We have

$$\begin{array}{ccccccc} CH_3COOH & + & C_2H_5OH & \rightleftharpoons & CH_3COOC_2H_5 & + & H_2O \\ 1-0.667 & & 1-0.667 & & 0.667 & & 0.667 \end{array}$$

$$K = \frac{(0.667)^2}{(1-0.667)^2} = 4.01.$$

Assume that y moles of CH_3COOH are required, whereupon at the equilibrium state we have 4.50 moles each of ester and water, 0.500 mole of C_2H_5OH and (y−4.50) moles of CH_3COOH. Hence,

$$\frac{(4.50)^2}{(y-4.50)(0.500)} = 4.01$$

and

$$y = \mathbf{14.60}\ \text{mol}\ CH_3COOH.$$

Example 4.3. At the equilibrium state of the system

$$H_2(g) + I_2(g) \rightleftharpoons 2HI(g),$$

the concentrations of the various substances at 400°C are:

$$[H_2] = [I_2] = 2.0 \times 10^{-3}\ \underline{M},\ [HI] = 1.6 \times 10^{-2}\ \underline{M}.$$

a) Calculate the equilibrium constant. b) Calculate the concentrations of the various substances at the equilibrium state, when 8.0 moles H_2 and 8.0 moles I_2 are enclosed in a 10-liter flask, at 400°C.

Solution. a) We have

$$K = \frac{[HI]^2}{[H_2][I_2]} = \frac{(1.6 \times 10^{-2})^2}{(2.0 \times 10^{-3})^2} = \mathbf{64}.$$

b) Assume that y moles H_2/L and y moles I_2/L were consumed. Hence, (0.80 − y) mole H_2/L and (0.80 − y) mole I_2/L remained, and 2y moles HI/L were produced. We have

$$\frac{(2y)^2}{(0.80 - y)^2} = 64,$$

from which y = 0.64 mol/L. Hence, the concentrations of the various substances at the equilibrium state are: $[H_2] = [I_2] = 0.80 - 0.64 =$ **0.16** mol/L, $[HI] = 2 \times 0.64 =$ **1.28** mol/L.

The concentrations of gases are expressed in mol/L, but instead of concentrations we can use the partial pressures of the gases. In both cases, the equilibrium constant is a constant, K_c and K_p, respectively, but its relative values and units will differ in cases where the number of moles of gases changes during the reaction. The relation between K_p and K_c is found as follows: Assume that we have the system $N_2O_4(g) \rightleftharpoons 2NO_2(g)$, which is at the equilibrium state. The pressure P and the molar concentration C (C = n/V, where n is the number of moles in a volume of V liters) are interrelated by the equation

$$P = CRT \text{ (law of ideal gases)} \tag{4.14}$$

Hence, we have

$$K_c = \frac{C^2_{NO_2}}{C_{N_2O_4}} \tag{4.15}$$

and

$$K_p = \frac{P^2_{NO_2}}{P_{N_2O_4}} = \frac{C^2_{NO_2}(RT)^2}{C_{N_2O_4}(RT)} = K_cRT \tag{4.16}$$

or, in general,

$$K_p = K_c(RT)^{\Delta n} \tag{4.17}$$

where Δn is the change in the total number of moles of gases during the reaction. For $\Delta n = 0$, $K_p = K_c$.

Example 4.4. A 5.0-liter flask containing 0.50 mole of PCl_5 is heated at 250°C, whereupon PCl_5 decomposes partially, according to the equation $PCl_5(g) \rightleftharpoons PCl_3(g) +$

$Cl_2(g)$. At the equilibrium state, there is 0.25 mole chlorine. How much chlorine would have been produced, if the volume of the flask was 0.050 L?

Solution. Since at the equilibrium state there is 0.25 mole of chlorine, there is also 0.25 mole of PCl_3 and $0.50 - 0.25 = 0.25$ mole of PCl_5. Hence, $[Cl_2] = [PCl_3] = [PCl_5] = 0.25\ mol/5.0\ L = 0.050\ mol/L$. The equilibrium constant, expressed as a function of the concentrations of the gases, is equal to

$$K_c = \frac{[PCl_3][Cl_2]}{[PCl_5]} = \frac{(0.050)^2}{0.050} = \mathbf{0.050}.$$

If the volume of the flask was 0.050 L, before the start of the decomposition we would have had $[PCl_5] = 0.50\ mol/0.050\ L = 10.0\ mol/L$. Let y be the concentration of chlorine at the equilibrium state, whereupon $[Cl_2] = [PCl_3] = y$, and $[PCl_5] = 10.0 - y$. Substituting these values at the expression for the equilibrium constant K_c, we have

$$\frac{y^2}{10.0 - y} = 0.050,$$

from which $y = 0.682\ mol\ Cl_2/L$. Hence, if the volume of the flask was 0.050 L, then $(0.682\ mol\ Cl_2/L)(0.050\ L) =$ **0.034** mole of Cl_2 would have been produced.

Note. According to Le Châtelier's principle, an increase in pressure results in a decrease in the decomposition of PCl_5. Indeed, the decomposition decreased from 50% $[(0.25/0.50) \times 100 = 50]$ to 6.8% $[(0.034/0.50) \times 100 = 6.8]$, when the volume of the flask decreased from 5.0 to 0.050 L, that is, when the pressure increased 100 times.

Example 4.5. When gaseous PCl_5 is heated, it decomposes to gaseous PCl_3 and Cl_2. At 273°C and a pressure of 1.00 atm, the density of the mixture, containing PCl_5, PCl_3, and Cl_2, is 0.0025 g/mL. Calculate the equilibrium constants K_c and K_p of the decomposition reaction.

Solution. We have the reaction

$$PCl_5(g) \overset{\Delta}{\rightleftharpoons} PCl_3(g) + Cl_2(g).$$

Assume that initially we have 1 mole $PCl_5 \equiv 208.3$ g PCl_5, and that after equilibrium has been established we have y moles each of PCl_3 and Cl_2, and consequently $(1 - y)$ moles of PCl_5. Then we have

$$\text{density d} = \frac{W(g)}{V_{tot}(L)} = 2.5\ g/L = \frac{208.3}{V_{PCl_3} + V_{Cl_2} + V_{PCl_5}}$$

$$= \frac{208.3}{y \cdot 22.4 \left(\frac{273 + 273}{273}\right) + y \cdot 22.4 \cdot \frac{546}{273} + (1 - y) \cdot 22.4 \cdot \frac{546}{273}}$$

from which $y = 0.86$. Therefore, $V_{tot} = (0.86 \times 22.4 \times 2) + (0.86 \times 22.4 \times 2) + (1 - 0.86)(22.4)2 = 83.3$ L. Hence,

$$K_c = \frac{[PCl_3][Cl_2]}{[PCl_5]} = \frac{(0.86/83.3)^2}{(0.14/83.3)} = \mathbf{0.063}$$

and

$$K_p = K_cRT = 0.063 \times 0.082055 \times 546 = \mathbf{2.82}.$$

4.3 Distribution Equilibria Between Immiscible Solvents

A simple type of equilibrium is the distribution of a substance, e.g., Br_2, between two immiscible liquids, e.g., H_2O and CCl_4. In this case, the equilibrium constant K_D is called the *distribution coefficient* and is given by the relation

$$K_D = \frac{[Br_2]_{CCl_4}}{[Br_2]_{H_2O}} \tag{4.18}$$

If dissociation or coupling or polymerization or complexation of the solute takes place, the equilibrium constant D is referred to as the *distribution ratio* and is given by the relation

$$D = \frac{\text{analytical concentration of A in the organic phase}}{\text{analytical concentration of A in the aqueous phase}} \tag{4.19}$$

The distribution principle is extensively used in analytical chemistry during the extraction of various substances, mainly to increase the sensitivity or selectivity of certain tests.

Example 4.6. Forty milliliters of an aqueous solution of substance A are shaken with 40 mL of an organic solvent that is used either in one extraction, or in two or four extractions of 20 or 10 mL each. Prepare a table of the nonextracted fractions of substance A in each case, if the distribution coefficient is a) 1.00, b) 10.0, and c) 100 (assume that activities equal concentrations).

Solution. Assume that W_0 is the initial weight of substance A in volume V_w of water, W_1 is the weight of A remaining in the aqueous phase after its extraction with volume V_o of an organic solvent, and K_D is the distribution coefficient. We have

$$K_D = \frac{[A]_o}{[A]_w} = \frac{\dfrac{W_0 - W_1}{V_o}}{\dfrac{W_1}{V_w}} = \frac{(W_0 - W_1)V_w}{V_oW_1}$$

or

$$K_dV_oW_1 + W_1V_w = W_0V_w \text{ or } W_1 = \left(\frac{V_w}{V_oK_D + V_w}\right) W_0.$$

Similarly, we have

$$W_2 = \left(\frac{V_w}{V_oK_D + V_w}\right) W_1 = \left(\frac{V_w}{V_oK_D + V_w}\right)^2 W_0,$$

Table 4.2. *Fractions of substance A not extracted in successive extractions.*

K_D	V_o, mL	f_n
a) 1.00	40	0.500
	20	0.444
	10	0.410
b) 10.0	40	0.0909
	20	0.0278
	10	0.00666
c) 100	40	0.00990
	20	0.000384
	10	0.00000219

and in general,

$$W_n = \left(\frac{V_w}{V_o K_D + V_w}\right)^n W_0 \tag{4.20}$$

where W_n is the weight of substance A in the aqueous phase after n extractions with organic solvent, each time with volume V_o of solvent. If f_n represents the fraction of substance A remaining in the aqueous layer after n extractions, on the basis of Equation (4.20), we have

$$f_n = \frac{W_n}{W_0} = \left(\frac{V_w}{V_o K_D + V_w}\right)^n \tag{4.21}$$

The fractions of substance A that were not extracted, calculated by Equation (4.21) for various values of K_D, are given in Table 4.2.

Example 4.7. A solution containing 0.1340 g of iodine in 10.0 mL of CCl_4 is shaken with 100 mL of water. At equilibrium, the aqueous layer is found to contain 0.0140 g of iodine. a) Calculate the distribution coefficient, K_D. b) How many grams of iodine remain in the aqueous layer, if the initial concentration $[I_2]$ of iodine in the aqueous solution A is 1.00×10^{-3} $\underline{M}$, and 100 mL of solution A are extracted with two successive 50-mL portions of CCl_4? c) If 100 mL of a 1.00×10^{-3} $\underline{M}$ aqueous solution of iodine are extracted with n successive 50-mL portions of CCl_4, what is the minimum value of n, so that 0.0066 mg of iodine remains in the aqueous layer? d) How many milliliters of CCl_4 would be required to extract 99.9% of iodine from 100 mL of 1.00×10^{-3} $\underline{M}$ iodine solution, in one extraction?

Solution. a) The distribution coefficient is equal to

$$K_D = \frac{[I_2]_{CCl_4}}{[I_2]_{H_2O}} = \frac{[(0.1340 - 0.0140)\text{ g}/253.8\text{ g } I_2/\text{mol}]/0.0100\text{ L}}{(0.0140\text{ g}/253.8\text{ g } I_2/\text{mol})/0.100\text{ L}} = \mathbf{85.7}.$$

b) Assume that 100 mL of aqueous solution A contains W_0 g of iodine initially, and W_2 g after two extractions. We have

$$W_0 = (1.00 \times 10^{-3}\text{ mmol/mL}) \times (100\text{ mL}) \times (0.2538\text{ g } I_2/\text{mmol}) = 0.02538\text{ g}.$$

Substituting these values in Equation (4.20), we have

$$W_2 = \left(\frac{100}{(85.7 \times 50) + 100}\right)^2 0.02538 = \mathbf{1.32 \times 10^{-5}} \text{ g of iodine.}$$

c) Substituting in Equation (4.20), we have

$$W_n = 0.0000066 = \left(\frac{100}{(85.7 \times 50) + 100}\right)^n 0.02538$$

or

$$\frac{0.0000066}{0.02538} = 0.00026 = (0.02281)^n$$

or

$$n = \log 0.00026 / \log 0.02281 = 2.2.$$

Hence, three extractions are required.

d) 100 mL of aqueous solution contain initially $100 \times 0.00100 = 0.100$ mmol I_2. After the extraction, 0.1% of the initial amount of iodine remains in the aqueous layer, that is, $(0.100 \times 0.1)/100 = 0.000100$ mmol I_2, and we have $[I_2]_{H_2O} = 0.000100/100 = 1.00 \times 10^{-6}$ M. The CCl_4 layer contains $0.100 - 0.000100 = 0.0999$ mmol I_2. Hence, we have

$$K_D = 85.7 = \frac{[I_2]_{CCl_4}}{1.00 \times 10^{-6}}$$

or

$$[I_2]_{CCl_4} = 8.57 \times 10^{-5} = 0.0999 \text{ mmol } I_2/V_{CCl_4}$$

or

$$V_{CCl_4} = 0.0999/8.57 \times 10^{-5} = \mathbf{1166} \text{ mL.}$$

4.4 Problems

Reaction rate and chemical equilibrium

4.1. The chemical reaction between A and B is represented by the equation

$$A(g) + 2B(g) \rightleftharpoons C(g)$$

The initial rate of the reaction was measured for various initial concentrations of A and B, at constant temperature. The results are given below:

Experiment	Initial concentration, M		Initial rate, M·min^{-1}
	A	B	
1	0.25	0.25	4.0×10^{-4}
2	0.25	0.50	1.6×10^{-3}
3	0.50	0.50	3.2×10^{-3}

a) Find the rate law and the reaction order.

b) If the initial concentration of A is 2.0 M and of B is 1.0 M, which of the following values is the most probable for the initial reaction rate? 1) 6.4×10^{-3}, 2) 3.2×10^{-3}, 3) 7.2×10^{-3}, and 4) 5.12×10^{-2} M·min^{-1}.

c) If the volume of the flask in which experiment 3 is performed is doubled, which of the following values will be the most probable initial reaction rate? 1) 1.6×10^{-3}, 2) 3.6×10^{-3}, 3) 6.4×10^{-3}, and 4) 4.0×10^{-4} M · min^{-1}.

4.2. It is found experimentally that a mixture obtained by mixing 1 mole of CH_3COOH with 1 mole of C_2H_5OH contains 2/3 mole of ester and 2/3 mole of water at the equilibrium state. If 1 more mole of C_2H_5OH is added to this mixture, what will be the composition of the mixture at the new equilibrium state?

4.3. What is the effect of each of the following changes on the equilibrium position of the reaction

$$H_2(g) + I_2(g) \rightleftharpoons 2HI(g) + 3Kcal?$$

a) increase in pressure, b) doubling of the hydrogen molecules, c) decrease in temperature, d) addition of a catalyst.

4.4. Substances A, B, and C react a 150°C, in a 10-liter flask, according to the equation

$$A(g) + 2B(g) + 3C(g) \rightleftharpoons 4D(g)$$

At the equilibrium state, there are in the flask 2.0 moles of substance A, 4.0 moles of B, 6.0 moles of C, and 8.0 moles of D. How many moles of substance D will there be in the flask, under the same conditions, if at the equilibrium state there are 4.0 moles of each of the substances A, B, and C?

4.5. The following data were obtained for the reaction among A, B, and C:

Experiment	Initial concentration, M			Initial rate, M·min^{-1}
	A	B	C	
1	0.20	0.20	0.20	4.8×10^{-4}
2	0.60	0.40	0.20	5.76×10^{-3}
3	0.60	0.20	0.20	1.44×10^{-3}
4	0.60	0.40	0.40	1.152×10^{-2}

Find the rate law, the reaction order, and the value of the reaction rate constant k.

4.6. A sample containing 5.0 moles of HI is enclosed in a 5.0-liter flask and heated to 628°C, whereupon equilibrium is established, according to the equation $2HI(g) \rightleftharpoons H_2(g) + I_2(g)$. Given that the equilibrium constant K_C of this reaction is 0.0380, calculate a) the degree of dissociation of HI, and b) the concentration of each species at the equilibrium state. c) Is the degree of dissociation independent of the initial concentration of HI?

4.7. A sample containing 0.50 mole of Br_2 gas is enclosed in a 5.0-liter flask and heated to 1480°C. Under these conditions, the sample is found to be 3.0% dissociated. Calculate the equilibrium constant K_C.

4.8. What should be the value of the equilibrium constant K of the reaction $A + 2B \rightleftharpoons C + 2D$, so that when A and B react in 1:2 ratio of initial concentrations, A is 99.9% transformed to products at the equilibrium state?

Distribution equilibria

4.9. How many extractions are required to remove 99.9% of substance A from 10.0 mL of an aqueous solution, if each extraction is carried out with 5.00 mL of CCl_4 and the distribution coefficient is 10.0?

4.10. Phenacetin is a component of the common analgesic preparation APC (Aspirin-Phenacetin-Caffeine), and is much less soluble in water than the other two components. What must be the smallest value of the distribution ratio D for the extraction of phenacetin with chloroform from an aqueous solution, so that by extraction with four successive 10.0-mL portions of chloroform, 99.9% of the phenacetin present in one APC tablet, dissolved in 50.0 mL of water, is extracted?

Chapter 5

Equilibria Involving Weak Acids and Bases

5.1 Dissociation of Weak Monoprotic Acids and Weak Monoacidic Bases

The general equation

$$HA + H_2O \rightleftharpoons H_3O^+ + A^-, \tag{5.1}$$

which describes the reaction that takes place when a weak molecular acid is dissolved in water, is characterized by an equilibrium constant, i.e.,

$$K_a = \frac{[H_3O^+][A^-]}{[HA]} \tag{5.2}$$

or, if we disregard the hydration of the proton,

$$K_a = \frac{[H^+][A^-]}{[HA]} \tag{5.2a}$$

The expressions (5.2) and (5.2*a*), which give the *dissociation constant* of the weak acid HA, are equivalent. However, the expression (5.2*a*) is used more often, because it is simpler.

We should note that for a strong acid like HCl, the dissociation is complete (i.e., Reaction 5.1 is quantitatively to the right), and so an equilibrium constant is not written for these (it would be infinite). This complete dissociation of the acid is known as the solvent leveling effect.

The dissociation of weak cationic acids is described in a similar way. For example, the dissociation of NH_4^+ and its dissociation constant are given by the equations

$$NH_4{}^+ + H_2O \rightleftharpoons NH_3 + H_3O^+ \tag{5.3}$$

or

$$NH_4{}^+ \rightleftharpoons NH_3 + H^+ \tag{5.3a}$$

$$K_a = \frac{[NH_3][H_3O^+]}{[NH_4^+]} \tag{5.4}$$

or

$$K_a = \frac{[NH_3][H^+]}{[NH_4^+]} \tag{5.5}$$

The constant K_a for a cationic acid is also written as K_h, whereupon it is called the *hydrolysis constant* (Section 5.8). K_h and K_a are completely identical.

Analogous equations are valid for weak monoacidic bases. For example, in the case of ammonia we have the equations

$$NH_3 + H_2O \rightleftharpoons NH_4^+ + OH^- \tag{5.6}$$

and

$$K_b = \frac{[NH_4^+][OH^-]}{[NH_3]} \tag{5.7}$$

where K_b is the dissociation constant of the base NH_3.

Dissociation of water. Applying the law of chemical equilibrium to the dissociation of water, we have

$$K = \frac{[H^+][OH^-]}{[H_2O]} \tag{5.8}$$

Since the product $K[H_2O]$ is essentially constant, Equation (5.8) is simplified to

$$[H^+][OH^-] = [H_2O]K = K_w = 1.00 \times 10^{-14} \text{ (at 25°C)} \tag{5.9}$$

The constant K_w is known as the *ion-product of water*, or as the *water constant.*

The dissociation constant. The strength of an electrolyte is expressed quantitatively by the dissociation constant or by the *degree of dissociation or ionization*, α. The degree of dissociation denotes the fraction of the electrolyte molecules which dissociate to ions in water, i.e., α is equal to the ratio of dissociated molecules over the sum of the electrolyte molecules. Therefore , the terms K_a and K_b and α should be interrelated. For example, for a solution of the weak acid HA of total concentration, C, we have

$$\alpha = \frac{[H^+]}{C} = \frac{[A^-]}{C} \tag{5.10}$$

or

$$[H^+] = [A^-] = \alpha C \tag{5.10a}$$

(The exact relation is $[H^+] = [OH^-]+[A^-]$, according to the principle of electroneutrality (Section 3.3), but $[OH^-]$ is considered negligible—see below, Equation (5.19).)

$$HA = C - \alpha C = C(1 - \alpha) \tag{5.11}$$

and

$$K_a = \frac{\alpha C \cdot \alpha C}{C(1-\alpha)} = \frac{\alpha^2 C}{1-\alpha} \tag{5.12}$$

Relation (5.12), or the corresponding relation $K_a = \alpha^2/(1-\alpha)V$ (where V is the solution volume that contains 1 mole of the substance, i.e., $V = 1/C$), is known as the *Ostwald dilution law.* For $\alpha < 0.1$, Equation (5.12) is simplified to

$$K_a \simeq \alpha^2 C \text{ or } \alpha \simeq \sqrt{K_a/C} \tag{5.13}$$

Concentration of H^+ in a solution of a weak monoprotic acid. Suppose we have a solution of the weak monoprotic acid HA, of analytical concentration C. We have

$$HA \rightleftharpoons H^+ + A^-, \; K_a = \frac{[H^+][A^-]}{[HA]} \tag{5.14}$$

Mass balance:

$$C = [A^-] + [HA] \tag{5.15}$$

Electroneutrality:

$$[H^+] = [A^-] + [OH^-], \tag{5.16}$$

from which

$$[A^-] = [H^+] - [OH^-] \tag{5.16a}$$

Combining Equations (5.15) and (5.16*a*), we have

$$[HA] = C - [A^-] = C - [H^+] + [OH^-] \tag{5.17}$$

Equations (5.9), (5.14), (5.16*a*), and (5.17) constitute a system of 4 equations with 4 unknowns, i.e., $[H^+]$, $[OH^-]$, $[HA]$, and $[A^-]$. Substituting the values of $[A^-]$ and $[HA]$ from Equations (5.16*a*) and (5.17) into (5.14) and taking into account Equation (5.9), we have

$$K_a = \frac{[H^+]([H^+] - K_w/[H^+])}{C - [H^+] + K_w/[H^+]} = \frac{[H^+]^3 - K_w[H^+]}{C[H^+] - [H^+]^2 + K_w}$$

or

$$[H^+]^3 + K_a[H^+]^2 - (CK_a + K_w)[H^+] - K_wK_a = 0 \tag{5.18}$$

Equation (5.18) gives the exact $[H^+]$ in a solution of the acid HA. In usual concentrations of solutions of weak acids, $[OH^-]$ (which $= K_w/[H^+]$ in the equation) is negligible in comparison to C or $[H^+]^2$, since the dissociation of water to give $[OH^-]$ is suppressed in the presence of the acid, whereupon we have

$$K_a \simeq \frac{[H^+]^2}{C - [H^+]} \tag{5.19}$$

or

$$[H^+]^2 + K_a[H^+] - CK_a \simeq 0 \tag{5.19a}$$

If the degree of dissociation of the acid is small, then $[H^+] << C$, and Equation (5.19) is simplified to

$$K_a \simeq [H^+]^2/C \text{ or } [H^+] \simeq \sqrt{CK_a} \tag{5.20}$$

In approximate calculations, it is sufficient to have $[H^+] < C/10$, in order to consider the approximation of Equation (5.20) as satisfactory. This simplification generally holds if the initial concentration, C, is $>> K_a$.

Example 5.1. Calculate the concentrations of ions and molecules in a 0.00100 M HF solution.

Solution. We have the relations

$$HF \rightleftharpoons H^+ + F^-, \quad K_a = \frac{[H^+][F^-]}{[HF]} = 6.9 \times 10^{-4} \text{ (at 25°C)}.$$

If $[OH^-]$ (and $[H^+]$) from the dissociation of water is negligible (1st assumption), then

$$[H^+] \simeq [F^-] \text{ and } [HF] = 0.00100 - [F^-] \simeq 0.00100 - [H^+].$$

That is, Equation (5.19) applies. Hence,

$$\frac{[H^+]^2}{0.00100 - [H^+]} = 6.9 \times 10^{-4} \tag{5.21}$$

We assume that $[HF] >> [F^-]$ (2nd assumption), whereupon $[HF] \simeq 0.00100$ M, and Equation (5.21) is simplified to $[H^+]^2/0.00100 = 6.9 \times 10^{-4}$, from which we have $[H^+] = 8.3 \times 10^{-4}$ M. At this point we check the correctness of the above assumptions. From the relation $[H^+][OH^-] = 1.00 \times 10^{-14}$, we have $[OH^-] = 1.00 \times 10^{-14}/8.3 \times 10^{-4} = 1.2 \times 10^{-11}$ M. Hence, the first assumption is correct, and Equation (5.19) applies. On the other hand, the second assumption is not valid, because 0.00100, [HF], is not much greater than 8.3×10^{-4}, $[F^-]$; the difference is only about 20%. Therefore, we solve the quadratic Equation (5.21), whereupon we find that $[H^+] = 5.6 \times 10^{-4}$ M. Consequently, the concentrations of the various species in the solution are

$$[H^+] = \mathbf{5.6 \times 10^{-4}} \text{ M} \simeq [F^-]$$

$$[HF] = 0.00100 - 5.6 \times 10^{-4} = \mathbf{4.4 \times 10^{-4}} \text{ M}$$

$$[OH^-] = 1.0 \times 10^{-14}/5.6 \times 10^{-4} = \mathbf{1.8 \times 10^{-11}} \text{ M}.$$

Example 5.2. Solution D of the weak acid HX is obtained by mixing solutions A and B of the same acid in a proportion of 1:1. The degree of dissociation of the acid in solutions A and B is 0.0134 and 0.00424, respectively. Calculate the degree of dissociation in solution D (we neglect the dissociation of water).

Solution. *1st method.* Let C_1, C_2, and C_3 be the concentrations of the acid in solutions A, B, and D, respectively, and α_1, α_2, and α_3 be the corresponding degrees of dissociation. We have

$$K_a = \frac{\alpha_1^2 C_1}{1-\alpha_1} = \frac{\alpha_2^2 C_2}{1-\alpha_2} = \frac{\alpha_3^2 C_3}{1-\alpha_3} \tag{5.22}$$

or

$$\frac{C_1}{\frac{1-\alpha_1}{\alpha_1^2}} = \frac{C_2}{\frac{1-\alpha_2}{\alpha_2^2}} = \frac{C_3}{\frac{1-\alpha_3}{\alpha_3^2}} = \frac{C_1 + C_2}{\frac{2(1-\alpha_3)}{\alpha_3^2}} = \frac{C_1 + C_2}{\frac{1-\alpha_1}{\alpha_1^2} + \frac{1-\alpha_2}{\alpha_2^2}}$$

or

$$\frac{2(1-\alpha_3)}{\alpha_3^2} = \frac{1-\alpha_1}{\alpha_1^2} + \frac{1-\alpha_2}{\alpha_2^2} \tag{5.23}$$

Substituting the values of α_1 and α_2 in Equation (5.23), we have

$$\frac{2(1-\alpha_3)}{\alpha_3^2} = \frac{1-0.0134}{(0.0134)^2} + \frac{1-0.00424}{(0.00424)^2},$$

from which $\alpha_3 = \mathbf{0.00572}$.

2nd method. Substituting the values of α_1 and α_2 in Equation (5.22), we find that $C_1 = 0.0992C_2$. Substituting this value and α_2 in the equation

$$\frac{\alpha_2^2 C_2}{1-\alpha_2} = \frac{\alpha_3^2(C_1+C_2)}{2(1-\alpha_3)}$$

we find that $\alpha_3 = \mathbf{0.00572}$.

Concentration of OH^- in a solution of a weak monoacidic base. Equations analogous to the above ones are also valid for a weak monoacidic base, in which $[OH^-]$ substitutes for $[H^+]$, and K_b replaces K_a. For example, for a solution of the weak monoacidic molecular base B, of total concentration C, we have

$$[OH^-]^3 + K_b[OH^-]^2 - (CK_b + K_w)[OH^-] - K_wK_b = 0 \tag{5.24}$$

$$K_b \simeq \frac{[OH^-]^2}{C-[OH^-]} \tag{5.25}$$

or

$$[OH^-]^2 + K_b[OH^-] - CK_b \simeq 0 \tag{5.25a}$$

$$K_b \simeq [OH^-]^2/C \text{ or } [OH^-] \simeq \sqrt{CK_b} \tag{5.26}$$

5.2 Effect of Ionic Strength on the Dissociation Constant

The value of the dissociation constant K_a of a weak acid (or of K_b of a weak base) is a function of the ionic strength μ of the solution, whereas the *thermodynamic dissociation constant*, K_a° (or K_b°), which is calculated using activities, is independent of μ. The two constants are interrelated by the relation

$$K_a^\circ = K_a f' \tag{5.27}$$

where f′ is the *activity factor* which is calculated from the activity coefficients of the ions and the undissociated molecules of the acids. For example, for CH_3COOH we have

$$K_a^\circ = \frac{a_{H^+}a_{CH_3COO^-}}{a_{CH_3COOH}} = \frac{[H^+][CH_3COO^-]}{[CH_3COOH]} \times \frac{f_{H^+}f_{CH_3COO^-}}{f_{CH_3COOH}} = K_a f' \tag{5.28}$$

The relation between K_a and μ depends on the type of the acid.

Example 5.3. Derive equations relating the dissociation constants K_a, K_a° and the ionic strength μ for the following acids: a) CH_3COOH, b) HPO_4^{2-}, c) NH_4^+.

Solution. a) On the basis of Equation (3.14), we have

$$\log f_{H^+} \simeq \log f_{CH_3COO^-} = -A\sqrt{\mu} \tag{5.29}$$

The undissociated molecules have no charge, and therefore, we can assume that $f_{CH_3COOH} \simeq 1$. Taking the logarithm of Equation of (5.28), in combination with (5.29), we have

$$\log K_a^\circ = \log K_a - A\sqrt{\mu} - A\sqrt{\mu}$$

or

$$\log K_a = \log K_a^\circ + 2A\sqrt{\mu} \tag{5.30}$$

b) For the anionic acid HPO_4^{2-}, we have

$$\log K_a^\circ = \log K_a + \log f_{H^+} + \log f_{PO_4^{3-}} - \log f_{HPO_4^{2-}} = \log K_a - A\sqrt{\mu} - 9A\sqrt{\mu} + 4A\sqrt{\mu}$$

or

$$\log K_a = \log K_a^\circ + 6A\sqrt{\mu} \tag{5.31}$$

c) For the cationic acid NH_4^+, we have

$$\log K_a^\circ = \log K_a + \log f_{H^+} + \log f_{NH_3} - \log f_{NH_4^+}$$

or, since $f_{NH_3} \simeq 1$ and $f_{H^+} \simeq f_{NH_4^+}$,

$$\log K_a = \log K_a^\circ \tag{5.32}$$

Note. Similar relations are valid for the various types of weak bases.

5.3 Acidity, Neutrality, and Alkalinity in Aqueous Solutions

The ion product of water, K_w, permits the calculation of the H^+ and OH^- ion concentrations, not only in water, but also in any aqueous solution, because Equation (5.9) is always valid. For example, from Equation (5.9) we find that in water we have $[H^+] = [OH^-] = 1.00 \times 10^{-7}$ $\underline{M}$ (at 25°C). In a 0.200 $\underline{M}$ HCl solution, we have $[H^+] = 0.200$ $\underline{M}$; hence, $[OH^-] = 1.00 \times 10^{-14}/0.200 = 5.00 \times 10^{-14}$ $\underline{M}$. Similarly, in a 0.500 $\underline{M}$ NaOH solution, we have $[OH^-] = 0.500$ $\underline{M}$; hence, $[H^+] = 1.00 \times 10^{-14}/0.500 = 2.00 \times 10^{-14}$

M. By definition, an aqueous solution having $[H^+] = 1.00 \times 10^{-7}$ M at 25°C is said to be ***neutral***, whereas if $[H^+] > 1.00 \times 10^{-7}$ M, i.e., if $[H^+] > [OH^-]$, the solution is said to be ***acidic***, and if $[H^+] < 1.00 \times 10^{-7}$ M, i.e., if $[H^+] < [OH^-]$, the solution is said to be *alkaline*.

In order to simplify calculations involving H^+ ion concentrations, Sõrenson defined the term pH by the equation

$$pH = -\log a_{H+} = \log \frac{1}{a_{H+}} \simeq -\log[H^+] \tag{5.33}$$

Since we generally deal with concentrations less than 1 molar, the negative sign in the definition generally avoids negative units or negative powers of 10 in expressing the acidity of solutions. Similarly, we have the relations.

$$pOH = -\log a_{OH-} \simeq -\log[OH^-]$$

and

$$pK = -\log K.$$

Taking the logarithm of Equation (5.9), and combining it with (5.33), we have at 25°C

$$(-\log[H^+]) + (-\log[OH^-]) = -\log(1.00 \times 10^{-14})$$

or

$$pH + pOH = 14.00 \tag{5.34}$$

By using Equation (5.34), we can calculate pH in water when pOH is known, and vice versa. Usually, we use the term pH, whereupon a solution having pH = 7.00 is neutral, pH < 7.00 is acidic, and pH > 7.00 is alkaline (basic).

Example 5.4. Calculate the terms $[H^+]$, $[OH^-]$, pH, and pOH for a 0.025 M $Ca(OH)_2$ solution (assume complete dissociation of the base).

Solution. We have $[OH^-] = 2 \times 0.025 =$ **0.050 M**. Hence, $[H^+] = 1.00 \times 10^{-14}/0.050 = \mathbf{2.00 \times 10^{-13}}$ M, pH $= -\log(2.00 \times 10^{-13}) = 13.00 - 0.30 =$ **12.70**, and pOH $= 14.00 - 12.70 =$ **1.30**.

Example 5.5. Calculate the pH of a 1.00×10^{-7} M HCl solution.

Solution. We have pH $= -\log(1.00 \times 10^{-7}) = 7.00$. This value is incorrect, because it is impossible for a solution of strong acid to be neutral. We reached this improper conclusion because we neglected the dissociation of water, which cannot be neglected in the case of very dilute solutions. The correct value of pH is calculated as follows: According to the principle of electroneutrality, we have

$$[H^+] = [Cl^-] + [OH^-] = 1.00 \times 10^{-7} + [OH^-] \text{ or } [OH^-] = [H^+] - 1.00 \times 10^{-7}.$$

Substituting this value in the ion product of water, we have

$$[H^+]([H^+] - 1.00 \times 10^{-7}) = 1.00 \times 10^{-14}.$$

Solving the quadratic equation, we find that

$$[H^+] = 1.62 \times 10^{-7} \text{ \underline{M}}.$$

Hence, pH $= -\log(1.62 \times 10^{-7}) = \mathbf{6.79}$.

Example 5.6. Calculate the pH of a solution obtained by mixing equal volumes of strong acid and strong base solutions, having pH values of 2.00 and 13.00, respectively.

Solution. We have $[H^+]_{acid} = 0.0100$ M and $[OH^-]_{base} = 0.100$ M. Hence, after mixing the solutions, we have $[OH^-]_{excess} = (0.100 - 0.0100)/2 = 0.045$ M, and pH $= -\log(1.00 \times 10^{-14}/0.045) = \mathbf{12.65}$.

Example 5.7. Calculate the concentrations of the various ions in a 0.0200 M H_2SO_4 solution.

Solution. The H_2SO_4 is dissociated completely to H^+ and HSO_4^-; hence, we have $[H^+] = [HSO_4^-] = 0.0200$ M. The acid HSO_4^- is further partially dissociated to H^+ and SO_4^{2-}. Let $[SO_4^{2-}] = y$, whereupon $[HSO_4^-] = 0.0200 - y$, and $[H^+] = 0.0200 + y$. Substituting these values in the expression for the equilibrium constant K_2, we have

$$0.012 = \frac{(0.0200 + y)y}{0.0200 - y},$$

from which y = 0.0063. Hence, the concentrations of the ions in the solution are

$$[SO_4^{2-}] = \mathbf{0.0063} \text{ \underline{M}},$$

$$[HSO_4^-] = 0.0200 - 0.0063 = \mathbf{0.0137} \text{ \underline{M}},$$

$$[H^+] = 0.0200 + 0.0063 = \mathbf{0.0263} \text{ \underline{M}}, \text{ and}$$

$$[OH^-] = 1.00 \times 10^{-14}/0.0263 = \mathbf{3.80 \times 10^{-13}} \text{ \underline{M}}.$$

Acid-base (protolytic) indicators, which undergo definite color changes in well-defined pH ranges are used for the experimental determination of pH. The acid-base indicators are weak organic acids or weak organic bases, the undissociated molecules of which have a different color from that of the dissociated form. For a weak acid HIn, with an acidity constant, K_{HIn}, which acts as an acid-base indicator, the following relations are valid:

$$\underset{\text{(acid color)}}{HIn} \rightleftharpoons H^+ + \underset{\text{(base color)}}{In^-} \tag{5.35}$$

$$K_{HIn} = \frac{[H^+][In^-]}{[HIn]} \tag{5.36}$$

$$pH = pK_{HIn} - \log \frac{[HIn]}{[In^-]} \tag{5.37}$$

The color of the indicator solution is determined by K_{HIn} and the ratio $[HIn]/[In^-]$, and hence by the pH. Generally, only one color of the indicator is seen at a ratio of 1:10 or 10:1, corresponding to a pH range of $pK_{HIn} \pm 1$.

Example 5.8. The acid-base indicator HIn undergoes a color change when it is converted by 1/5 to its ionic form. Calculate the dissociation constant of the indictor, K_{HIn}, given that the color change takes place at pH 6.40.

Solution. We have $[H^+] = 10^{-6.40}\ \underline{M} = 3.98 \times 10^{-7}\underline{M}$. Hence,

$$K_{HIn} = \frac{[H^+][In^-]}{[HIn]} = \frac{(3.98 \times 10^{-7})(1)}{4} = \mathbf{9.95 \times 10^{-8}}$$

Example 5.9. The dissociation constant of the indicator HIn is 4.00×10^{-9}. What percentage of the indicator is present in the acid form at pH a) 7.12, b) 8.12, c) 9.40?

Solution. *1st method.* a) Let C be the total (analytical) concentration of the indicator, the acid form of which (acid color) is HIn. We have

$$\frac{[HIn]}{[In^-]} = \frac{[H^+]}{K_a} = \frac{10^{-7.12}}{4.00 \times 10^{-9}} = 18.96$$

and

$$[HIn] + [In^-] = [HIn^-] + \frac{[HIn]}{18.96} = 1.053[HIn] = C,$$

or $[HIn] = 0.950C$. That is, at pH 7.12, **95.0%** of the indicator is present in the HIn form.

b) We have

$$\frac{[HIn]}{[In^-]} = \frac{10^{-8.12}}{4.00 \times 10^{-9}} = 1.896$$

and $[HIn] + [In^-] = [HIn] + ([HIn]/1.896) = 1.527[HIn] = C$, or $[HIn] = 0.655C$. That is, at pH 8.12, **65.5%** of the indicator is present in the HIn form.

c) Similarly, we find that at pH 9.40, we have $[HIn] = 0.0905C$, that is, **9.05%** of the indicator is present in the HIn form.

2nd method. We have from Equation (5.57) below,

$$\alpha_0 = \frac{[HIn]}{C} = \frac{[H^+]}{[H^+] + K_a} = \frac{7.59 \times 10^{-8}}{(7.59 \times 10^{-8}) + (4.00 \times 10^{-9})} = 0.950,$$

i.e., **95.0%** of the indicator is present in the acid form HIn. Likewise, we find that the values of α_0 is cases b and c are 0.655 and 0.0905, respectively, i.e., **65.5%** and **9.05%** of the indicator is present in the HIn form.

5.4 Dissociation of Polyprotic Acids

Polyprotic (polybasic) acids dissociate in multiple steps which are characterized by the consecutive dissociation constants, K_1, $K_2 \ldots K_n$ ($K_1 > K_2 \ldots > K_n$). The calculation of the concentrations of ions and molecules in a solution of a weak polyprotic acid is made as in the case of a weak monoprotic acid. Suppose, for example, that the concentrations of ions and molecules in a H_2S solution of concentration C are sought for. We have the relations

$$H_2S \rightleftharpoons H^+ + HS^-, \ K_1 = \frac{[H^+][HS^-]}{[H_2S]} = 1.0 \times 10^{-7} \quad (5.38)$$

$$HS^- \rightleftharpoons H^+ + S^{2-}, \ K_2 = \frac{[H^+][S^{2-}]}{[HS^-]} = 1.0 \times 10^{-14} \quad (5.39)$$

$$H_2O \rightleftharpoons H^+ + OH^-, \ K_w = [H^+][OH^-] = 1.00 \times 10^{-14} \quad (5.9)$$

$$\text{Mass balance (for S)}: \ C = [H_2S] + [HS^-] + [S^{2-}] \quad (5.40)$$

$$\text{Electroneutrality}: \ [H^+] = [HS^-] + [OH^-] + 2[S^{2-}] \quad (5.41)$$

Hence, we have a system of five equations with five unknowns, i.e., $[H^+]$, $[OH^-]$, $[S^{2-}]$, $[SH^-]$, and $[H_2S]$. Solving this system, we find the sought-for concentrations. Thus, in a manner similar to the one use in the derivation of Equations (5.18) and (5.24), we find that $[H^+]$ is given by the equation

$$[H^+]^4 + K_1[H^+]^3 + (K_1K_2 - CK_1 - K_w)[H^+]^2 - (K_1K_w + 2CK_1K_2)[H^+] - K_1K_2K_w = 0 \quad (5.42)$$

The exact solution of Equation (5.42) and of the system of five equations is very laborious, but is seldom required (only in the case of very dilute solutions). In solutions commonly used, $[OH^-]$ is negligible, and since often $K_2 << K_1$, we can neglect the second dissociation step, whereupon $[S^{2-}]$ is considered negligible, and we have $[H^+] \simeq [HS^-]$, the equivalent of a monoprotic acid. Hence, Equations (5.40) and (5.41) are simplified to

$$C = [H_2S] + [HS^-] \quad (5.43)$$

$$[H^+] \simeq [HS^-] \quad (5.44)$$

Combining Equations (5.38), (5.43), and (5.44), we have

$$K_1 = 1.0 \times 10^{-7} = \frac{[H^+]^2}{C - [H^+]} \quad (5.45)$$

from which we calculate $[H^+]$. (This equation compares with Equations (5.21) and (5.25) for monoprotic species since the second ionization of H_2S is negligible.) For a saturated solution of H_2S (such solutions are used in semimicro qualitative analysis), we have $C = 0.10$ $\underline{M}$ at 25°C and 1 atm, and $[H^+] << C$. Hence,

$$[H^+] = \sqrt{1.0 \times 10^{-7} \times 0.10} = \mathbf{1.0 \times 10^{-4}} \ \underline{M} \simeq [HS^-]$$

$$[H_2S] = 0.10 - 0.0001 \simeq \mathbf{0.10} \ \underline{M}$$

$$[OH^-] = 1.00 \times 10^{-14}/1.0 \times 10^{-4} = \mathbf{1.0 \times 10^{-10}} \ \underline{M}$$

$[S^{2-}]$ is calculated by Equation (5.39), i.e.,

$$K_2 = 1.0 \times 10^{-14} = \frac{[1.0 \times 10^{-4}][S^{2-}]}{[1.00 \times 10^{-4}]}.$$

Hence, $[S^{2-}] = \mathbf{1.0 \times 10^{-14}}$ $\underline{M}$.

Generally, for any diprotic acid, if $K_2 << K_1$, we have $[A^{2-}] \simeq K_2$. We solve for the cases of solutions of polyacidic bases in a similar way.

Calculation of concentrations of species present in solutions of polyprotic acids at a given pH. A solution of a polyprotic acid contains the undissociated molecules of the acid and the anions resulting from its dissociation. The concentrations of these species can be calculated as a function of the $[H^+]$ of the solution, the consecutive dissociation constants of the polyprotic acid, and its analytical concentration C.

Example 5.10. Show that the following relations are valid for an H_2S solution of total concentration C:

$$[H_2S] = \frac{[H^+]^2 C}{[H^+]^2 + K_1[H^+] + K_1K_2}$$

$$[HS^-] = \frac{K_1[H^+]C}{[H^+]^2 + K_1[H^+] + K_1K_2}$$

$$[S^{2-}] = \frac{K_1K_2C}{[H^+]^2 + K_1[H^+] + K_1K_2}$$

Solution. If α_0, α_1, and α_2 are the fractions of the total concentrations in the form of H_2S, HS^-, and S^{2-}, respectively (the α-functions—α_0, α_1, α_2—should not be confused with the degree of dissociation, α), we have the relations

$$H_2S \rightleftharpoons H^+ + HS^-, \quad K_1 = \frac{[H^+][HS^-]}{[H_2S]} \tag{5.38}$$

$$HS^- \rightleftharpoons H^+ + S^{2-}, \quad K_2 = \frac{[H^+][S^{2-}]}{[HS^-]} \tag{5.39}$$

$$C = [H_2S] + [HS^-] + [S^{2-}] \tag{5.46}$$

$$\alpha_0 = [H_2S]/C \tag{5.47}$$

$$\alpha_1 = [HS^-]/C \tag{5.48}$$

$$\alpha_2 = [S^{2-}]/C \tag{5.49}$$

and

$$\alpha_0 + \alpha_1 + \alpha_2 = 1 \tag{5.50}$$

From Equations (5.38) and (5.39) we have

$$[HS^-] = \frac{K_1[H_2S]}{[H^+]} \tag{5.38a}$$

$$[S^{2-}] = \frac{K_1K_2[H_2S]}{[H^+]^2} \tag{5.51}$$

Combining Equations (5.46), (5.38*a*), and (5.51), we have

$$C = [H_2S] + \frac{K_1[H_2S]}{[H^+]} + \frac{K_1K_2[H_2S]}{[H^+]^2} \tag{5.52}$$

Solving Equation (5.52) for $[H_2S]$, we have

$$[H_2S] = \frac{C[H^+]^2}{[H^+]^2 + K_1[H^+] + K_1K_2} \tag{5.53}$$

Substituting the value of $[H_2S]$ from Equation (5.53) into (5.47), we have

$$\alpha_0 = \frac{[H^+]^2}{[H^+]^2 + K_1[H^+] + K_1K_2} \tag{5.54}$$

In a similar way, by substituting in Equation (5.46) to obtain an expression containing only the species of interest (i.e., HS^- or S^{2-}) analogous to Equation (5.52), we find that

$$\alpha_1 = \frac{K_1[H^+]}{[H^+]^2 + K_1[H^+] + K_1K_2} \tag{5.55}$$

$$\alpha_2 = \frac{K_1K_2}{[H^+]^2 + K_1[H^+] + K_1K_2} \tag{5.56}$$

From Equations (5.48) and (5.55) we have

$$[HS^-] = \frac{K_1[H^+]C}{[H^+]^2 + K_1[H^+] + K_1K_2},$$

whereas from Equations (5.49) and (5.56) we have

$$[S^{2-}] = \frac{K_1K_2C}{[H^+]^2 + K_1[H^+] + K_1K_2}$$

Note. The following general formulas are valid for the case of the polyprotic acid H_nA:

$$\alpha_0 = \frac{[H^+]^n}{[H^+]^n + K_1[H^+]^{n-1} + K_1K_2[H^+]^{n-2} + \ldots + K_2K_2 \ldots K_n} \tag{5.57}$$

$$\alpha_1 = \frac{K_1[H^+]^{n-1}}{[H^+]^n + K_1[H^+]^{n-1} + K_1K_2[H^+]^{n-2} + \ldots + K_2K_2 \ldots K_n} \tag{5.57a}$$

$$\alpha_2 = \frac{K_1K_2[H^+]^{n-2}}{[H^+]^n + K_1[H^+]^{n-1} + K_1K_2[H^+]^{n-2} + \ldots + K_1K_2 \ldots K_n} \tag{5.57b}$$

$$\vdots$$

$$\alpha_n = \frac{K_1K_2 \ldots K_n}{[H^+]^n + K_1[H^+]^{n-1} + K_1K_2[H^+]^{n-2} + \ldots + K_1K_2 \ldots K_n} \tag{5.57n}$$

Equations (5.57), (5.57*a*), (5.57*b*) ... (5.57*n*) allow the calculation of the concentration of any species at a given pH (these equations are also valid for monoprotic acids). Using these equations we can construct *distribution diagrams* which give the values of α_i for the various species as a function of pH (such a diagram in logarithmic form is shown in Figure 5.1). The values of α_1, $\alpha_2 \ldots \alpha_n$ can also be used to calculate the amount of strong base required to react with the weak acid, H_nA, to prepare a certain volume of buffer solution (Example 5.23). Also, by using α-functions, we calculate the solubility of a sparingly soluble salt derived from a weak acid or weak base at a given pH, and we construct logarithmic diagrams that are useful for readily identifying the equilibrium concentrations of various species (Section 5.6 below).

Example 5.11. A 250-mL portion of a 1.00 $\underline{M}$ NH_3 – 1.80 $\underline{M}$ NH_4Cl buffer solution contains 0.750 g of the diprotic acid H_2A (MW 150). For this solution, we have the relations $\alpha_0 = 0.150$, and $[H_2A] = 3[A^{2-}]$. Calculate the pH of the solution, the dissociation constants K_1 and K_2, and the concentrations of the species H_2A, HA^- and A^{2-} (assume that the pH of the buffer is not affected by the presence of the (small) amount of H_2A).

Solution. *1st method.* We have

$$[OH^-] = \frac{1.80 \times 10^{-5} \times 1.00}{1.80} = 1.00 \times 10^{-5}\ \underline{M},\ [H^+] = 1.00 \times 10^{-14}/1.00 \times 10^{-5} =$$

$$1.00 \times 10^{-9}\ \underline{M} \text{ and } pH = -\log(1.00 \times 10^{-9}) = \mathbf{9.00}.$$

The molarity of the solution is equal to

$$\frac{0.750 \text{ g } H_2A}{150 \text{ g/mol} \times 0.250 \text{ L}} = 0.0200\ \underline{M}.$$

We have

$$[H_2A] = \alpha_0 C = 0.150 \times 0.0200 = \mathbf{0.00300}\ \underline{M}$$

$$[A^{2-}] = [H_2A]/3 = 0.00300/3 = \mathbf{0.00100}\ \underline{M}$$

and

$$[HA^-] = C - [H_2A] - [A^{2-}] = 0.0200 - 0.00300 - 0.00100 = \mathbf{0.0160}\ \underline{M},$$

whereupon

$$K_1 = \frac{[H^+][HA^-]}{[H_2A]} = \frac{(1.00 \times 10^{-9})(0.0160)}{0.00300} = \mathbf{5.33 \times 10^{-9}}$$

$$K_2 = \frac{[H^+][A^{2-}]}{[HA^-]} = \frac{(1.00 \times 10^{-9})(0.00100)}{0.0160} = \mathbf{6.25 \times 10^{-11}}.$$

2nd method. We calculate, as above, that pH = 9.00 and $C_{H_2A} = 0.0200\ \underline{M}$. We have $\alpha_2 = \alpha_0/3 = 0.150/3 = 0.050$; consequently, $\alpha_1 = 1.00 - 0.150 - 0.050 = 0.80$. Hence, we have

$$0.150 = \frac{[H^+]^2}{[H^+]^2 + K_1[H^+] + K_1K_2} \qquad (5.58)$$

$$0.80 = \frac{K_1[H^+]}{[H^+]^2 + K_1[H^+] + K_1K_2} \tag{5.59}$$

$$0.050 = \frac{K_1K_2}{[H^+]^2 + K_1[H^+] + K_1K_2} \tag{5.60}$$

Dividing Equations (5.59) and (5.58), we have

$$K_1/[H^+] = 5.33;\ \text{hence,}\ K_1 = \mathbf{5.33 \times 10^{-9}}.$$

Dividing Equations (5.58) and (5.60), we have

$$[H^+]^2 = 3K_1K_2;\ \text{hence,}\ K_2 = (1.00 \times 10^{-9})^2/(3 \times 5.33 \times 10^{-9}) = \mathbf{6.25 \times 10^{-11}}.$$

Consequently, we have

$$[H_2A] = 0.150 \times 0.0200 = \mathbf{0.00300}\ \underline{M}$$

$$[HA^-] = 0.80 \times 0.0200 = \mathbf{0.0160}\ \underline{M}$$

$$[A^{2-}] = 0.050 \times 0.0200 = \mathbf{0.00100}\ \underline{M}.$$

5.5 Proton Balance

The proton balance convention is useful for the derivation of relations between the concentrations of the various species in a solution. It states that the sum of the molarities of the species formed by the release of protons must equal the sum of the molarities of the species formed by taking protons. The molarity of each species is multiplied by a coefficient which is equal to the number of protons released or consumed in the formation of the species.

Note. The proton balance coincides with the charge balance for solutions of molecular acids (HCl), but not for solutions of cationic acids (NH_4^+) or anionic acids (HPO_4^{2-}).

Example 5.12. Write the proton balance for a) HCl, b) NH_4Cl, c) Na_2HPO_4 solutions.

Solution. a) The proton balance for an HCl solution gives the relation

$$[H^+] = [OH^-] + [Cl^-] \tag{5.61}$$

Equation (5.61) is valid, because for each OH^- ion formed ($H_2O \rightleftharpoons H^+ + OH^-$), one H^+ is also formed, and for each Cl^- ion formed, one H^+ ion is also formed ($HCl \rightarrow H^+ + Cl^-$).

b) The proton balance for an NH_4Cl solution gives the relation

$$[H^+] = [NH_3] + [OH^-] \tag{5.62}$$

That is, one NH_3 is formed from the release of a proton from NH_4^+ (or one NH_4OH from its hydrolysis: $NH_4^+ + H_2O \rightleftharpoons NH_4OH + H^+$).

c) The proton balance for a Na_2HPO_4 solution gives the relation

$$[H^+] + [H_2PO_4^-] + 2[H_3PO_4] = [OH^-] + [PO_4^{3-}] \tag{5.63}$$

That is, one PO_4^{3-} is formed in the release of a proton by HPO_4^{2-}, while one proton is consumed in forming $H_2PO_4^-$ and two protons are consumed in forming H_3PO_4 (during the hydrolysis of HPO_4^-). Relation (5.63) can also be derived algebraically on the basis of mass balance and charge balance, as follows: Let C be the total concentration of the Na_2HPO_4 solution, whereupon

$$\text{Mass balance (Na)}: \ 2C = [Na^+] \tag{5.64}$$

$$\text{Mass balance (P)}: \ C = [H_3PO_4] + [H_2PO_4^-] + [HPO_4^{2-}] + [PO_4^{3-}] \tag{5.65}$$

$$\text{Electroneutrality}: \ [H^+] + [Na^+] = [OH^-] + [H_2PO_4^-] + 2[HPO_4^{2-}] + 3[PO_4^{3-}] \tag{5.66}$$

Combining Equations (5.64) and (5.65), we have

$$[Na^+] = 2[H_3PO_4] + 2[H_2PO_4^-] + 2[HPO_4^{2-}] + 2[PO_4^{3-}] \tag{5.67}$$

From Equations (5.66) and (5.67), we have

$$[H^+] + [H_2PO_4^-] + 2[H_3PO_4] = [OH^-] + [PO_4^{3-}] \tag{5.67a}$$

5.6 Logarithmic Acid-Base Concentration Diagrams

The concentrations and the ratios of the various species in ionic acid-base equilibria as a function of pH can be represented diagrammatically by logarithmic concentration diagrams, $[\log C_i = f(pH)]$. Suppose we have a solution of a monoprotic acid of concentration C and dissociation constant K_a. We have the relations

$$\alpha_{HA} = \alpha_0 = \frac{[HA]}{C} = \frac{[H^+]}{[H^+] + K_a} \text{ or } [HA] = \frac{C[H^+]}{[H^+] + K_a} \tag{5.68}$$

$$\alpha_{A^-} = \alpha_1 = \frac{[A^-]}{C} = \frac{K_a}{[H^+] + K_a} \text{ or } [A^-] = \frac{CK_a}{[H^+] + K_a} \tag{5.69}$$

Equations (5.68) and (5.69) allow the determination of the concentrations [HA] and $[A^-]$ at any $[H^+]$. A plot of the logarithm of the concentration of the species present in the solution as a function of pH is known as a *logarithmic concentration diagram* or simply as a *logarithmic diagram*.

Example 5.13. Construct a logarithmic concentration diagram for 1.00 M HCN solution.

Solution. The logarithmic diagram contains the curves represented by the functions

$$\log[\mathrm{HCN}] = f(\mathrm{pH}) \tag{5.70}$$

$$\log[\mathrm{CN^-}] = g(\mathrm{pH}) \tag{5.71}$$

Suppose that y is the intersection point of the two curves, where we have $[\mathrm{HCN}] = [\mathrm{CN^-}]$ and $K_a = [\mathrm{H^+}]$. For point y, by taking the logarithm of the equation of the dissociation constant, we have $\log K_a = \log[\mathrm{H^+}]$, or $\mathrm{pH} = \mathrm{p}K_a$, and $[\mathrm{HCN}] = [\mathrm{CN^-}] = 0.500$ M. Thus, the point y is defined with coordinates $[\mathrm{p}K_a, \log(C/2)]$ or $(9.4 - 0.3)$.

For $[\mathrm{H^+}] >> 4 \times 10^{-10}$ M, i.e., at a pH significantly smaller than $\mathrm{p}K_a$, Equations (5.68) and (5.69) are simplified to $(\mathrm{HCN} \equiv \mathrm{HA})$

$$[\mathrm{HCN}] = C = 1.00 \text{ M} \tag{5.72}$$

$$[\mathrm{CN^-}] = CK_a/[\mathrm{H^+}] \tag{5.73}$$

Taking the logarithms, we have

$$\log[\mathrm{HCN}] = \log C = 0 (\log[\mathrm{HCN}] \text{ independent of pH}) \tag{5.74}$$

$$\log[\mathrm{CN^-}] = \log C - \mathrm{p}K_a + \mathrm{pH} = -9.4 + \mathrm{pH} \text{ (a straight line with a slope of } +1) \tag{5.75}$$

For $[\mathrm{H^+}] << 4 \times 10^{-10}$ M, i.e., at a pH significantly larger than $\mathrm{p}K_a$, Equations (5.68) and (5.69) are simplified to

$$[\mathrm{HCN}] = C[\mathrm{H^+}]/K_a \tag{5.76}$$

$$[\mathrm{CN^-}] = C = 1.00 \text{ M}. \tag{5.77}$$

Taking the logarithms, we have

$$\log[\mathrm{HCN}] = \log C + \mathrm{p}K_a - \mathrm{pH} = 9.4 - \mathrm{pH} \quad \text{(a straight line with a slope of } -1) \tag{5.78}$$

$$\log[\mathrm{CN^-}] = \log C = 0 (\log[\mathrm{CN^-}] \text{ independent of pH}) \tag{5.79}$$

For any aqueous solution we have

$$\log[\mathrm{H^+}] = -\mathrm{pH} \text{(a straight line with a slope of } -1) \tag{5.80}$$

$$\log[\mathrm{OH^-}] = \log(K_w/[\mathrm{H^+}]) = -14 + \mathrm{pH} \quad \text{(at 25°C, a straight line with a slope of } +1) \tag{5.81}$$

On the basis of Equations (5.74) to (5.81), we construct the logarithmic diagram of Figure 5.1. From this diagram, the equilibrium concentrations of any of the species are readily available for a given pH at the specified analytical concentration. In a similar way, logarithmic diagrams are constructed for solutions of polyprotic acids.

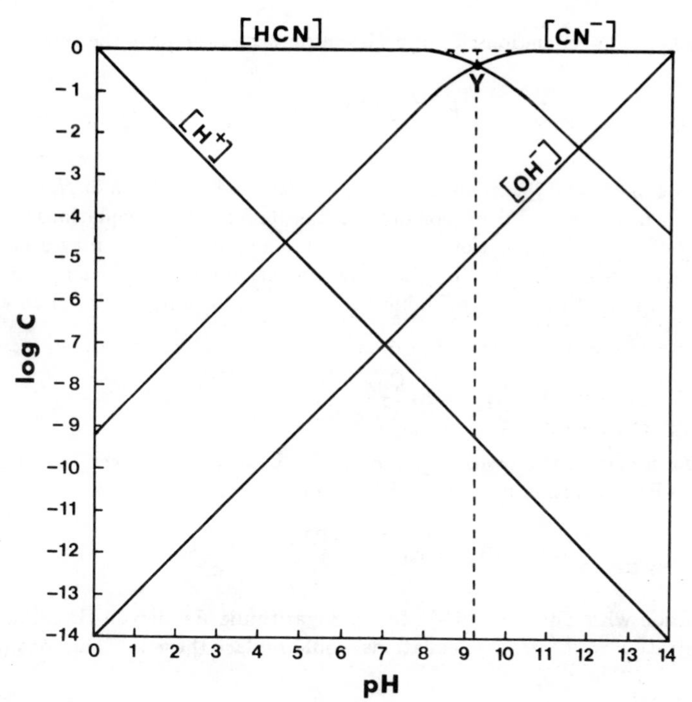

Figure 5.1. *Logarithmic concentration diagram for 1.00 M HCN solution.*

5.7 Buffers

A *buffer solution* is one that maintains its pH approximately constant when small amounts of acid or base are added or when it is diluted. Buffer solutions, also known as buffers, usually consist of a mixture of a weak acid and its salt, e.g., CH_3COOH–CH_3COONa, or of a weak base and its salt, e.g., NH_3–NH_4Cl. Their action is related to the *common ion* effect, i.e., the fact that the addition of a common ion in a solution of a weak acid or base results in shifting of the equilibrium towards the undissociated molecule of the acid or the base. When an acid or base is added, it is consumed in shifting the equilibrium, and the pH is determined by the ratio of the buffer components (Example 5.15). For a buffer of the weak monoprotic acid HA and its salt MA, of concentrations C_{HA} and C_A, respectively, the following equation is valid

$$pH = pK_a + \log \frac{C_A}{C_{HA}} \tag{5.82}$$

It is derived by solving for $[H^+]$ in the K_a expression and taking the negative logarithm of both sides of the equation. Likewise, for a buffer of the weak monoacidic base B and its salt BHX, of concentrations C_B and C_A, respectively, we have the relation

$$pOH = pK_b + \log \frac{C_A}{C_B} \tag{5.83}$$

Equations (5.82) and (5.83) are known as *Henderson-Hasselbalch* or *Henderson* equations, and are used for the calculation of the pH of buffers. These equations are not valid in the cases of very dilute solutions, where the dissociation of water may become appreciable, nor when the acid or the base is a strong electrolyte. Rigorous treatment of the equilibria in a HA-MA buffer (including the H_2O equilibrium) results in an expression for the exact value of $[H^+]$

$$[H^+] = K_a \frac{C_{HA} - [H^+] + [OH^-]}{C_A + [H^+] - [OH^-]}, \tag{5.84}$$

This is used in combination with Equation (5.9). Likewise, the exact value of $[OH^-]$ in the above B-BHX solution is calculated by the equation

$$[OH^-] = K_b \frac{C_B - [OH^-] + [H^+]}{C_A + [OH^-] - [OH^-]} \tag{5.85}$$

in combination with Equation (5.9). In the logarithmic Henderson-Hasselbalch forms, the concentration ratios in these equations would replace those in Equations (5.82) and (5.83).

Example 5.14. The pH of a 0.025 M $NaHCO_3$-0.025 M Na_2CO_3 subsidiary buffer used for the calibration of pH meters is 10.00. Examine whether the pH calculated using activities agrees with the experimental value.

Solution. The ionic strength of the solution is equal to

$$\mu = [0.025 \times (1)^2 + (0.075) \times (1)^2 + (0.025) \times (-2)^2]/2 = 0.100.$$

Using Equation (3.16), we find that for $\mu = 0.100$, the activity coefficients for the HCO_3^- and CO_3^{2-} ions are 0.754 and 0.324, respectively. Substituting these values in the relation

$$a_{H+} = K_2 \frac{[HCO_3^-]f_{HCO_3^-}}{[CO_3^{2-}]f_{CO_3^{2-}}} \tag{5.86}$$

we have

$$a_{H+} = 4.8 \times 10^{-11} \frac{(0.025)(0.754)}{(0.025)(0.324)} = 1.12 \times 10^{-10}$$

and $pH = -\log(1.12 \times 10^{-10}) = 9.95$. The agreement between the calculated value of 9.95 and the experimental value of 10.00 is relatively good.

Note. For better accuracy, μ is calculated by using the Davies equation

$$\log f_z = z^2 \left(-\frac{A\sqrt{\mu}}{1 + \sqrt{\mu}} + 0.15\mu \right) \tag{5.87}$$

Use of this equation gives pH = 10.00.

Using Equations (5.82) and (5.83) we can calculate changes in pH occurring upon the addition of a strong acid or a strong base to a buffer, as shown in the following example.

Example 5.15. Calculate the changes in $[H^+]$ and pH in a 0.500 $\underline{M}$ CH_3COOH-0.500 $\underline{M}$ CH_3COONa buffer that result when a) solid NaOH is added until the solution (before reaction) becomes 0.010 $\underline{M}$ in NaOH, b) HCl is added until the solution (before reaction) becomes 0.010 $\underline{M}$ in HCl (the change in solution volume is considered negligible).

Solution. a) In the initial solution we have $[H^+] = 1.8 \times 10^{-5}(0.500/0.500) = 1.8 \times 10^{-5}$ $\underline{M}$, and $pH = -\log(1.8 \times 10^{-5}) = 4.74_5$. When NaOH is added to the buffer, the following reaction takes place (assume 1 L of buffer)

$$CH_3COOH + OH^- \rightleftharpoons CH_3COO^- + H_2O$$

$$0.010 \text{ mol} \quad 0.010 \text{ mol} \quad 0.010 \text{ mol}$$

Let $[H^+] = y$, after the addition of NaOH, whereupon $[CH_3COOH] = 0.490 - y$, and $[CH_3COO^-] = 0.510 + y$. We have from Equation (5.84), and neglecting $[OH^-]$ in this acid solution

$$[H^+] = 1.8 \times 10^{-5} \times \frac{0.490 - y}{0.510 + y} \simeq 1.8 \times 10^{-5} \times \frac{0.490}{0.510} \simeq 1.73 \times 10^{-5}\underline{M}$$

and $pH = -\log(1.73 \times 10^{-5}) = 4.76_2$. Hence, the addition of NaOH results in a decrease in $[H^+]$ from 1.8×10^{-5} to 1.73×10^{-5} $\underline{M}$, i.e., by about 4%, whereas the pH is increased by $4.76_2 - 4.74_5 = 0.017$ units.

b) When HCl is added to the buffer, the following reaction takes place

$$CH_3COO^- + H^+ \rightleftharpoons CH_3COOH$$

$$0.010 \text{ mol} \quad 0.010 \text{ mol} \quad 0.010 \text{ mol}$$

Let $[H^+] = z$, after the addition of HCl, whereupon $[CH_3COOH] = 0.510 - z$, and $[CH_3COO^-] = 0.490 + z$. We have

$$[H^+] = 1.8 \times 10^{-5} \times \frac{0.510 - z}{0.490 + z} \simeq 1.8 \times 10^{-5} \times \frac{0.510}{0.490} \simeq 1.87 \times 10^{-5}\ \underline{M}$$

and

$$pH = -\log(1.87 \times 10^{-5}) = 4.72_8.$$

Hence, the addition of HCl results in an increase in $[H^+]$ from 1.8×10^{-5} to 1.87×10^{-5} $\underline{M}$, i.e., by about 4%, whereas the pH is decreased by $4.74_5 - 4.72_8 = 0.017$ units.

Note. The changes in $[H^+]$ and pH resulting from the addition of a strong acid or a strong base to an NH_3-NH_4Cl buffer are calculated in a similar manner, using Equation (5.85).

The *buffer capacity*, β (also called *buffer index* or *buffer intensity*), is a measure of the buffer's ability to maintain its pH constant upon the addition of strong acid or strong base, and is defined by the equation

$$\beta = \frac{dC_B}{dpH} = -\frac{dC_A}{dpH} \tag{5.88}$$

where dC_B and dC_A represent the number of moles of strong base and acid, respectively, added to one liter of buffer, and dpH is the resulting pH change. In all cases, β is a positive number. It can be shown that β is given by the relation

$$\beta = 2.303\left[\frac{K_w}{[H^+]} + [H^+] + \frac{CK_a[H^+]}{([H^+] + K_a)^2}\right] \tag{5.89}$$

where C is the sum of the analytical concentrations C_{HA} and C_A of the acid and the salt, respectively.

For buffer solutions having $C > 10^{-3}$ M, Equation (5.89) is simplified to

$$\beta = 2.303\frac{C_{HA} \cdot C_A}{C_{HA} + C_A} \tag{5.90}$$

For example, using Equations (5.90) and (5.88), we have for Example (5.15):

$$\beta = 2.303\frac{(0.500)^2}{0.500 + 0.500} = 0.576,$$

whereupon

$$\text{a) } dpH = \frac{dC_B}{\beta} = \frac{0.010}{0.576} = 0.017 = \Delta pH,$$

$$\text{b) } dpH = -\frac{dC_A}{\beta} = \frac{0.010}{0.576} = -0.017 = \Delta pH.$$

Preparation of buffer solutions. Buffers are prepared in three ways: 1) by mixing a weak acid and its conjugate base or a weak base and its conjugate acid, 2) by mixing an excess of weak acid with a limited amount of strong base, and 3) by mixing an excess of weak base with a limited amount of strong acid.

Example 5.16. Calculate the pH and the buffer capacity of a solution obtained by mixing 112 mL of 0.1325 M H_3PO_4 and 136 mL of 0.1450 M Na_2HPO_4.

Solution. Upon mixing, the following neutralization reaction takes place:

$$H_3PO_4 + Na_2HPO_4 \rightleftharpoons 2NaH_2PO_4$$

Hence, after the mixing we have

$$C_{Na_2HPO_4} = \frac{(136 \text{ mL})(0.1450 \text{ mmol/mL}) - (112 \text{ mL})(0.1325 \text{ mmol/mL})}{(136 + 112) \text{ mL}} = 0.01968 \text{ \underline{M}},$$

$$C_{NaH_2PO_4} = \frac{(112 \text{ mL})(0.1325 \text{ mmol/mL})2}{248 \text{ ml}} = 0.1197 \text{ \underline{M}}$$

and

$$\text{pH} = \text{pK}_2 + \log\frac{[\text{HPO}_4{}^{2-}]}{[\text{H}_2\text{PO}_4{}^-]} = 7.21 + \log\frac{0.01968}{0.1197} = \mathbf{6.43}.$$

The buffer capacity is equal to

$$\beta = 2.303\frac{(0.1197 \times 0.01968)}{(0.1197 + 0.01968)} = \mathbf{0.0389}.$$

Example 5.17. Two buffers, X and Z, of pH 5.00 and 6.00, respectively, were prepared from the acid HA and the salt NaA. Both solutions are 0.500 M in HA. If equal volumes of the two buffers are mixed, what is the pH of the resulting solution ($K_{HA} = 1.00 \times 10^{-5}$)?

Solution. In buffer X we have from K_a

$$[\text{A}^-] = (1.00 \times 10^{-5})(0.500)/1.0 \times 10^{-5} = 0.500 \text{ \underline{M}},$$

whereas in buffer Z we have

$$[\text{A}^-] = (1.00 \times 10^{-5})(0.500)/1.0 \times 10^{-6} = 5.00 \text{ \underline{M}}.$$

If equal volumes of X and Z are mixed, we have $[\text{HA}] = 0.500\underline{\text{M}}$, and $[\text{A}^-] = (0.500 + 5.00)/2 = 2.75$ M. Hence, the pH of the resulting solution is equal to

$$\text{pH} = 5.00 + \log\frac{2.75}{0.500} = \mathbf{5.74}.$$

Example 5.18. Calculate the volumes of 0.500 M NH_4Cl and 0.500 M NH_3 required for the preparation of 200 mL of a buffer of pH 8.00.

Solution. We have $[H^+] = 1.0 \times 10^{-8}$ M. The following relations are valid:

$$\text{NH}_4{}^+ \rightleftharpoons \text{NH}_3 + \text{H}^+,\ K_a = \frac{[\text{NH}_3][\text{H}^+]}{[\text{NH}_4{}^+]} = \frac{K_w}{K_b} = \frac{1.00 \times 10^{-14}}{1.8 \times 10^{-5}} = 5.6 \times 10^{-10}$$

Suppose that y mL of NH_4Cl and consequently (200 – y) mL of NH_3 are required, whereupon we have

$$(\text{y mL})(0.500 \text{ mmol/mL}) = 0.500\text{y mmol NH}_4{}^+$$

and

$$(200 - \text{y}) \text{ mL } (0.500 \text{ mmol/mL}) = (100 - 0.500\text{y}) \text{ mmol NH}_3$$

Substituting these values in the expression for K_a, we have

$$5.6 \times 10^{-10} = \frac{(1.0 \times 10^{-8})[NH_3]}{[NH_4^+]} = \frac{(1.0 \times 10^{-8})(100 - 0.500y)\ \text{mmol}}{(0.500y)\ \text{mmol}}$$

Solving the equation, we have y = 189. Hence, **189** mL of NH_4Cl and 200 − 189 = **11** mL of NH_3 are required. It should be noted that the ratio $[NH_3]/[NH_4^+]$ was given directly in millimoles and not in molarities or mmol/mL. The ratio of molarities equals the ratio of millimoles because the volumes cancel.

Example 5.19. Calculate the pH of a buffer that is prepared by mixing 100 mL of 0.200 M NaOH and 150 mL of 0.400 M CH_3COOH.

Solution. Initially, we have

$$(100\ \text{mL})(0.200\ \text{mmol/mL}) = 20.0\ \text{mmol NaOH}$$

and

$$(150\ \text{mL})(0.400\ \text{mmol/mL}) = 60.0\ \text{mmol } CH_3COOH$$

During the neutralization, 20.0 mmol of CH_3COO^- are produced. Hence, 40.0 mmol of CH_3COOH remain, and we have

$$\text{pH} = 4.74 + \log\frac{20.0/250}{40.0/250} = \mathbf{4.44}.$$

It should be noted that it is possible to cancel the volumes from the beginning, as was done in Example (5.18).

Example 5.20. Forty millimoles of NaOH are added to 100 mL of a 1.20 M HA – y M NaA buffer, whereupon a solution of pH 5.30 is produced. Calculate the pH of the buffer. Assume that the volume of the buffer remains unchanged ($K_{HA} = 1.00 \times 10^{-5}$).

Solution. After the addition of NaOH, we have

$$[HA] = \frac{(100 \times 1.20) - 40.0}{100}$$

$$[A^-] = \frac{(100y + 40.0)}{100}$$

Substituting the values of [HA] and $[A^-]$ in Equation (5.82), we have

$$5.30 = 5.00 + \log\frac{(100y + 40.0)/100}{(120 - 40.0)/100},$$

from which y = 1.20 M. Hence, the pH of the buffer is equal to

$$\text{pH} = 5.00 + \log\frac{1.20}{1.20} = \mathbf{5.00}.$$

Example 5.21. How many grams of CH_3COONa should be added to one liter of 0.100 M HCl solution so that the pH becomes 4.44?

Solution. *1st method.* At pH 4.44 we have

$$pH = 4.44 = 4.74 + \log \frac{[CH_3COO^-]}{[CH_3COOH]} \text{ or } [CH_3COOH]/[CH_3COO^-] = 2.00.$$

$1000 \times 0.100 = 100$ mmol of CH_3COONa are required to neutralize HCl ($HCl + CH_3COONa \rightleftharpoons CH_3COOH + NaCl$), whereupon 100 mmol of CH_3COOH are formed. Let y mmol of CH_3COONa be the additional amount required so that $[CH_3COOH]/[CH_3COO^-] = 2.00$. We have $[CH_3COOH]/[CH_3COO^-] = 100/y$ or $y = 50$. Therefore, a total of $100 + 50 = 150$ mmol of CH_3COONa is required; hence, we should add

$$(150 \text{ mmol})(0.08203 \text{ g } CH_3COONa/\text{mmol}) = \mathbf{12.3} \text{ g } CH_3COONa.$$

2nd method. Suppose that C mol CH_3COONa/L are required. We have

$$\alpha_0 = \frac{[CH_3COOH]}{C} = \frac{[H^+]}{[H^+] + K_a} = \frac{10^{-4.44}}{10^{-4.44} + 1.8 \times 10^{-5}} = 0.6686$$

$$= \frac{0.100}{C}(HCl + CH_3COO^- \rightleftharpoons CH_3COOH + Cl^-),$$

from which C = 0.150 M. Hence, we should add

$$(0.150 \text{ mol})(82.03 \text{ g } CH_3COONa/\text{mol}) = \mathbf{12.3} \text{ g } CH_3COONa/L.$$

Example 5.22. Twenty five milliliters of 0.120 M HA solution are titrated with 0.100 M NaOH solution. After the addition of 17.5 mL NaOH, the pH = 5.80. Calculate the dissociation constant K_a of the weak monoprotic acid HA.

Solution. After the addition of 17.5 mL NaOH,

$$(17.5 \text{ mL})(0.100 \text{ mmol/mL}) = 1.75 \text{ mmol NaA}$$

are formed, leaving

$$(25.0 \text{ mL})(0.120 \text{ mmol/mL}) - 1.75 \text{ mmol} = 1.25 \text{ mmol HA}.$$

Hence, we have

$$pH = pK_a + \log \frac{[NaA]}{[HA]} = 5.80 = pK_a + \log \frac{(1.75\text{mmol}/42.5 \text{ mL})}{(1.25 \text{ mmol}/42.5 \text{ mL})}$$

or $pK_a = 5.65_4$. Hence, $K_a = 10^{-5.65_4} = \mathbf{2.22 \times 10^{-6}}$.

Example 5.23. How many milliliters of 0.500 M NaOH solution should be added to 40.0 mL of 0.100 M H_3PO_4 to prepare a buffer of pH 7.00?

Solution. *1st method.* Let V mL be the volume of the buffer. Since the pH is 7.00, $pK_1 = 2.12$, and $pK_2 = 7.21$, it follows that the buffer consists of NaH_2PO_4 and Na_2HPO_4. When NaOH is added, the following reactions take place:

$$H_3PO_4 + NaOH \rightleftharpoons NaH_2PO_4 + H_2O \tag{5.91}$$

$$NaH_2PO_4 + NaOH \rightleftharpoons Na_2HPO_4 + H_2O \tag{5.92}$$

$40.0 \times 0.100 = 4.00$ mmol of NaOH are added for Reaction (5.91), and an additional y mmol for Reaction (5.92), i.e., a total of (4.00 + y) mmol of NaOH are added. y mmol of Na_2HPO_4 are formed in Reaction (5.92), leaving (4.00 − y) mmol of NaH_2PO_4. We have

$$pH = pK_2 + \log\frac{[HPO_4^{2-}]}{[H_2PO_4^-]} = 7.00 = 7.21 + \log\frac{[HPO_4^{2-}]}{[H_2PO_4^-]}$$

or

$$\frac{[HPO_4^{2-}]}{[H_2PO_4^-]} = 10^{7.00-7.21} = 0.617 = \frac{y/V}{(4.00-y)/V}$$

or y = 1.53. Hence, 4.00 + 1.53 = 5.53 mmol of NaOH should be added, i.e., 5.53/0.500 = **11.06** mL of 0.500 M NaOH solution.

2nd method. Let C be the analytical concentration of phosphate. We have

$$\alpha_0 = \frac{[H_3PO_4]}{C} = \frac{(1.00\times10^{-7})^3}{(1.00\times10^{-7})^3 + 7.5\times10^{-17} + 4.65\times10^{-17} + 4.65\times10^{-22}}$$

$$= \frac{10^{-21}}{1.215\times10^{-16}} = 8.2\times10^{-6}$$

$$\alpha_1 = \frac{[H_2PO_4^-]}{C} = \frac{7.5\times10^{-17}}{1.215\times10^{-16}} = 0.62$$

$$\alpha_2 = \frac{[HPO_4^{2-}]}{C} = \frac{4.65\times10^{-17}}{1.215\times10^{-16}} = 0.38$$

$$\alpha_3 = \frac{[PO_4^{3-}]}{C} = \frac{4.65\times10^{-22}}{1.215\times10^{-16}} = 3.8\times10^{-6}$$

From the values of α_1 and α_2, it can be seen that the solution essentially contains only NaH_2PO_4 and Na_2HPO_4 (the quantities of H_3PO_4 and Na_3PO_4 are negligible). The solution contains initially

$$(40.0\ \text{mL})(0.100\ \text{mmol/mL}) = 4.00\ \text{mmol}\ H_3PO_4.$$

The buffer contains $4.00 \times 0.62 = 2.48$ mmol of NaH_2PO_4 and $4.00 \times 0.38 = 1.52$ mmol of Na_2HPO_4, and for their preparation $2.48 + (2 \times 1.52) = 5.52$ mmol of NaOH were added, i.e., 5.52/0.500 = **11.04** mL of 0.500 M NaOH solution.

Example 5.24. How many milliliters of 0.100 M H_3PO_4 and 0.100 M NaOH solutions are required to prepare one liter of a buffer of pH 6.90?

Solution. From the values of pH and the dissociation constants of H_3PO_4 it is concluded that the main phosphate species in the buffer are $H_2PO_4^-$ and HPO_4^{2-}. Suppose that we take x mmol of H_3PO_4, which react with x mmol of NaOH, forming x mmol of NaH_2PO_4. Suppose that an additional y mmol of NaOH are added, whereupon y mmol of Na_2HPO_4 are formed and (x−y) mmol of NaH_2PO_4 are left. Therefore, a total of x mmol of H_3PO_4 are required, i.e., 10.0x mL of 0.100 M H_3PO_4, and (x+y) mmol of NaOH, i.e., 10.0(x+y) mL of 0.100 M NaOH. We have

$$\frac{[H_2PO_4^-]}{[HPO_4^{2-}]} = \frac{[H^+]}{K_2} = \frac{1.26\times10^{-7}}{6.2\times10^{-8}} = 2.032 = \frac{x-y}{y} \tag{5.93}$$

and

$$10.0x + 10.0(x + y) = 1000. \tag{5.94}$$

Solving the system of Equations (5.93) and (5.94), we find that y = 14.2, and x = 42.9. Hence, **429** mL of 0.100 $\underline{M}$ H_3PO_4 and **571** mL of 0.100 $\underline{M}$ NaOH are required.

Example 5.25. Solutions corresponding to 0.600 mol of HCl, 0.0500 mol of H_3PO_4, 0.0800 mol of NaH_2PO_4, 0.350 mol of Na_3PO_4, and 0.110 mol of NaOH are mixed in a flask. Calculate the pH of the resulting solution.

Solution. Let V liters be the volume of the resulting solution. First, we assume complete dissociation of all the electrolytes and we calculate the total quantity of each ion in the solution. Then, we assume that the H^+ ions react with the bases of the solution, successively, in order of decreasing strength. We have

$$\text{mol ionizable } H^+ : \; 0.600 + (0.0500 \times 3) + (0.0800 \times 2) = 0.910$$

$$\text{mol } OH^- : \; 0.110$$

$$\text{mol } PO_4{}^{3-} : \; 0.0500 + 0.0800 + 0.350 = 0.480$$

First, 0.110 mol of H^+ reacts with 0.110 mol of OH^-, leaving 0.910−0.110 = 0.800 mol of H^+. Then 0.480 mol of H^+ reacts with 0.480 mol of $PO_4{}^{3-}$ forming 0.480 mol of $HPO_4{}^{2-}$ and leaving 0.800−0.480 = 0.320 mol of H^+, which subsequently reacts with 0.320 mol of $HPO_4{}^{2-}$, forming 0.320 mol of $H_2PO_4{}^-$ and leaving 0.480−0.320 = 0.160 mol of $HPO_4{}^{2-}$. The solution contains 0.320 mol of NaH_2PO_4, 0.160 mol of Na_2HPO_4, and 0.110 mol of NaCl; therefore, it is a buffer with a pH equal to

$$pH = pK_2 + \log \frac{[Na_2HPO_4]}{[NaH_2PO_4]} = 7.21 + \log \frac{0.160/V}{0.320/V} = \mathbf{6.91}.$$

Example 5.26. How many milliliters of 0.100 $\underline{M}$ NaOH solution and how many grams of NaCl should be added to 500 mL of 0.100 $\underline{M}$ CH_3COOH solution, to prepare a buffer of pH 4.50 and ionic strength 0.250 (use activities)?

Solution. We have

$$a_{H^+} = 10^{-4.50} = 3.16 \times 10^{-5} \; \underline{M} \text{ and } -\log f_{CH_3COO^-} =$$

$$(0.51(1)^2\sqrt{0.250})/(1 + \sqrt{0.250}) = 0.170 \text{ or } f_{CH_3COO^-} = 0.676 \text{ and } f_{CH_3COOH} \simeq 1.$$

Substituting these values in the relation

$$K_a^\circ = 1.8 \times 10^{-5} = \frac{a_{H^+} f_{CH_3COO^-}[CH_3COO^-]}{f_{CH_3COO^-}[CH_3COOH]}$$

we have

$$1.8 \times 10^{-5} = \frac{3.16 \times 10^{-5} \times 0.676[CH_3COO^-]}{[CH_3COOH]},$$

from which $[CH_3COO^-]/[CH_3COOH] = 0.843$. Let y mL of 0.100 $\underline{M}$ NaOH be the required volume, whereupon we have

$$\frac{[CH_3COO^-]}{[CH_3COOH]} = \frac{0.100y}{(500 \times 0.100) - 0.100y} = 0.843,$$

from which y = 229 mL. We have 500 + 229 = 729 mL of buffer of pH 4.50, but with an ionic strength equal to $\mu \simeq C_{CH_3COONa} = [CH_3COO^-] = (229 \times 0.100)/729 = 0.0314$. In order to have $\mu = 0.250$, we should add 0.250−0.0314 = 0.2186 mol NaCl/L ≡ (0.2186 mol NaCl/L)(58.44 g/mol)(0.729 L) = **9.3** g of NaCl.

5.8 Hydrolysis

Salts derived from weak acids or bases will react with water to associate with a proton or hydroxide ion ($A^- + H_2O \rightleftharpoons HA + OH^-$ or $B^+ + H_2O \rightleftharpoons BOH + H^+$). Hence, they act as weak bases or weak acids, respectively. The reactions are called *hydrolysis reactions*, and are essentially acid-base reactions. An equilibrium exists between the nonhydrolyzed substances and the hydrolysis products, and the equilibrium may be called the *hydrolysis constant*, designated as K_h. It actually represents either a basicity constant, K_b, or an acidity constant, K_a. In analogy to the degree of dissociation of a weak electrolyte, Equation (5.10), we have the *degree of hydrolysis*, α, which represents the fraction of the hydrolyzed ions or molecules of an electrolyte.

Hydrolysis of monovalent anions. These are treated like any other weak base. In a solution of the salt MA of the weak monoprotic acid HA, of concentration C, which is hydrolyzed according to the reaction

$$A^- + H_2O \rightleftharpoons HA + OH^-, \quad K_h = \frac{[HA][OH^-]}{[A^-]} \equiv K_b \tag{5.95}$$

the following relations are valid:

$$[OH^-]^3 + K_h[OH^-]^2 - (CK_h + K_w)[OH^-] - K_wK_h = 0 \tag{5.96}$$

$$K_h \simeq \frac{[OH^-]^2}{C - [OH^-]} \tag{5.97}$$

$$K_h \simeq [OH^-]^2/C \text{ or } [OH^-] \simeq \sqrt{CK_h} \tag{5.98}$$

Equation (5.96) gives the *exact* value of $[OH^-]$, but for convenience it is used only in the case of very dilute solutions. In solutions of ordinary concentrations, Equation (5.97) or (5.98) (when the degree of hydrolysis of the salt is small) is used. Multiplying the numerator and denominator of Equation (5.95) by $[H^+]$, we have

$$K_h = \frac{[HA][OH^-][H^+]}{[A^-][H^+]} = \frac{K_w}{K_a} \tag{5.99}$$

Combining Equations (5.98) and (5.99), we have

$$[OH^-] \simeq \sqrt{CK_w/K_a} \tag{5.100}$$

If we substitute K_h by K_b in Equations (5.96), (5.97), and (5.98), they become identical to Equations (5.24), (5.25), and (5.26), which are valid for a solution of a weak monoacidic base. This is so because there is no real difference between the hydrolysis of an anion and the dissociation of a weak base. Therefore, in this case the following relation is valid:

$$K_h = \frac{\alpha^2 C}{1-\alpha} \simeq \alpha^2 C, \tag{5.101}$$

which is identical to Equation (5.12).

Example 5.27. Calculate the pH and degree of hydrolysis of 0.180 M CH_3COONa solution.

Solution. We have $K_h = \frac{[CH_3COOH][OH^-]}{[CH_3COO^-]} = \frac{[OH^-]^2}{C-[OH^-]} = \frac{K_w}{K_a} = \frac{1.00 \times 10^{-14}}{1.8 \times 10^{-5}} =$ 5.56×10^{-10}. We have $[OH^-] << C$; hence, $[OH^-] = \sqrt{(5.56 \times 10^{-10})(0.180)} = 1.00 \times 10^{-5}$ M, $pOH = -\log(1.00 \times 10^{-5}) = 5.00$, and $pH = 14.00 - 5.00 = \mathbf{9.00}$.

The degree of dissociation $\alpha = \frac{[OH^-]}{C} = \frac{1.00 \times 10^{-5}}{0.180} = \mathbf{5.56 \times 10^{-5}}$ or $\mathbf{5.56 \times 10^{-3}}$%.

Hydrolysis of ammonium ions. Ammonium ions and derivatives of the ion NH_4^+, upon hydrolysis give acidic solutions, and they are treated like any other weak acid. Suppose, for example, that we have an NH_4Cl solution of concentration C. We have

$$NH_4^+ + H_2O \rightleftharpoons NH_3 + H_3O^+, \; K_h = \frac{[NH_3][H_3O^+]}{[NH_4^+]} = \frac{[NH_3^+][H^+]}{[NH_4^+]} = K_{NH_4^+} \equiv K_a \tag{5.102}$$

$$[H^+]^3 + K_h[H^+]^2 - (CK_h + K_w)[H^+] - K_wK_h = 0 \tag{5.103}$$

(This becomes identical to Equation (5.18), if we substitute K_h by K_a.)

$$K_h = \frac{K_w}{K_b} \simeq \frac{[H^+]^2}{C-[H^+]} \tag{5.104}$$

$$[H^+] \simeq \sqrt{CK_h} \simeq \sqrt{CK_w/K_b} \tag{5.105}$$

Hydrolysis of metal ions. Hydrated metal ions, of the general formula $[M(H_2O)_x]^{y+}$, act as Brönsted acids (proton donors) and are hydrolyzed (dissociate) according to the equation

$$[M(H_2O)_x]^{y+} + H_2O \rightleftharpoons [M(H_2O)_{x-1}(OH)]^{(y-1)+} + H_3O^+ \tag{5.106}$$

for which we have the relation

$$K_h = \frac{[M(H_2O)_{x-1}(OH)^{(y-1)+}][H_3O^+]}{[M(H_2O)_x{}^{y+}]} = K_a = \frac{K_w}{K_b} \tag{5.107}$$

Example 5.28. Calculate the pH and the degree of hydrolysis of 0.200 M $AlCl_3$ solution.

Solution. We have

$$K_h = \frac{[Al(H_2O)_5(OH)^{2+}][H^+]}{[Al(H_2O)_6{}^{3+}]} \simeq \frac{[H^+]^2}{0.200 - [H^+]} = K_a = 1.12 \times 10^{-5}.$$

Solving the quadratic equation, we have $[H^+] = 1.49 \times 10^{-3}$ M. Hence, $pH = -\log(1.49 \times 10^{-3}) = \mathbf{2.83}$.

$$\text{The degree of hydrolysis } \alpha = \frac{1.49 \times 10^{-3}}{0.200} = \mathbf{0.00746} \text{ or } \mathbf{0.746\%}.$$

Hydrolysis of salts of weak monoprotic acids and weak monoacidic bases. Suppose we have a solution of the salt BHA, of the weak acid HA and the weak base B, of concentration C. We have

$$BH^+ + A^- + 2H_2O \rightleftharpoons B + HA + H_3O^+ + OH^-$$

$$(\text{or } BH^+ + A^- \rightleftharpoons B + HA),$$

$$K_h = \frac{[B][HA]}{[BH^+][A^-]} \tag{5.108}$$

The following equation is valid ($K_{HA} = K_a$ of HA, $K_{BH^+} = K_a$ of BH^+)

$$\begin{gathered}[H^+]^4 + (C + K_{HA} + K_{BH^+})[H^+]^3 + (K_{HA}K_{BH^+} - K_w)[H^+]^2 - \\ (CK_{HA}K_{BH^+} + K_wK_{HA} + K_wK_{BH^+})[H^+] - K_wK_{HA}K_{BH^+} = 0\end{gathered} \tag{5.109}$$

$$K_h = \frac{[B][HA][H^+][OH^-]}{[BH^+][A^-][H^+][OH^-]} = \frac{K_w}{K_aK_b} \tag{5.110}$$

$$K_h = \frac{\alpha^2}{(1-\alpha)^2} \tag{5.111}$$

It can be shown that

$$[H^+] = \sqrt{\frac{K_aK_w}{K_b}} \tag{5.112}$$

The exact value of $[H^+]$ for any solution of the salt BHA can be found by Equation (5.109). The equation is usually solved by the method of approximations.

If $[H^+]$ is known, the degrees of hydrolysis of the BH^+ and A^- ions are calculated by the equations

$$\alpha_{BH^+} = \frac{K_{BH^+}}{K_{BH^+} + [H^+]} \tag{5.113}$$

$$\alpha_{A^-} = \frac{[H^+]}{K_{HA} + [H^+]} \tag{5.114}$$

Example 5.29. Calculate the pH of 1.00×10^{-4} M NH_4CN solution and the degree of hydrolysis of the NH_4^+ and CN^- ions.

Solution. Using Equation (5.112), we find the approximate value of $[H^+]$, i.e., $[H^+] = \sqrt{(4.0 \times 10^{-10})(1.00 \times 10^{-14})/1.8 \times 10^{-5}} = 4.7 \times 10^{-10}$ M. Hence, pH = 9.33. Solving Equation (5.109) by the method of approximations, we find the exact value of $[H^+] = 5.5 \times 10^{-10}$ M. Hence, pH $= -\log(5.5 \times 10^{-10}) = \mathbf{9.26}$. Using Equations (5.113) and (5.114), we have

$$\alpha_{NH_4^+} = \frac{5.6 \times 10^{-10}}{5.6 \times 10^{-10} + 5.5 \times 10^{-10}} = \mathbf{0.50} \text{ or } \mathbf{50\%}$$

$$\alpha_{CN^-} = \frac{5.5 \times 10^{-10}}{4.0 \times 10^{-10} + 5.5 \times 10^{-10}} = \mathbf{0.58} \text{ or } \mathbf{58\%}.$$

Hydrolysis of acid salts of weak acids. Suppose we have a solution of the salt NaHA, of total concentration C. The anion of the salt is an ampholyte, that is, it acts as an acid and as a base,

$$HA^- + H_2O \rightleftharpoons H_2A + OH^-, \; K_h = \frac{[H_2A][OH^-]}{[HA^-]} = \frac{K_w}{K_1} \tag{5.115}$$

$$HA^- + H_2O \rightleftharpoons A^{2-} + H_3O^+, \; K_2 = \frac{[H^+][A^{2-}]}{[HA^-]} \tag{5.116}$$

The solution is alkaline if $K_h > K_2$, and acidic if $K_h < K_2$. In such a solution, $[H^+]$ is calculated by the relation

$$[H^+] \simeq \sqrt{\frac{K_1K_2C + K_1K_w}{C + K_1}} \tag{5.117}$$

If $K_1 << C$ and $K_1K_w << K_1K_2C$, Equation (5.117) is simplified to

$$[H^+] \simeq \sqrt{K_1K_2} \tag{5.118}$$

Example 5.30. Calculate the pH and the concentrations of the various species in a 0.010 M $NaHCO_3$ solution.

Solution. Using Equation (5.117), we have

$$[H^+] = \sqrt{\frac{(4.2 \times 10^{-7})(4.8 \times 10^{-11})(0.010) + (4.2 \times 10^{-7})(1.00 \times 10^{-14})}{4.2 \times 10^{-7} + 0.010}} = \mathbf{4.5_4 \times 10^{-9}} \text{ M}.$$

Hence, pH $= -\log(4.5_4 \times 10^{-9}) = \mathbf{8.34}$. Substituting the above value of $[H^+]$ in the expression for the dissociation constant K_1, we calculate $[H_2CO_3]$, i.e.,

$$4.2 \times 10^{-7} = \frac{(4.5 \times 10^{-9})(0.010)}{[H_2CO_3]} \text{ or } [H_2CO_3] = \mathbf{1.1 \times 10^{-4}} \text{ M}.$$

Similarly, $[CO_3^{2-}$ is calculated from K_2, i.e.,

$$4.8 \times 10^{-11} = \frac{(4.5 \times 10^{-9})[CO_3^{2-}]}{0.010} \text{ or } [CO_3^{2-}] = \mathbf{1.1 \times 10^{-4}}\ \underline{M}.$$

From the relation

$$[HCO_3^-] = C - [H_2CO_3] - [CO_3^{2-}] \tag{5.119}$$

we have

$$[HCO_3^-] = 0.010 - 1.1 \times 10^{-4} \simeq \mathbf{0.0098}\ \underline{M}.$$

We also have

$$[Na^+] = C = \mathbf{0.010}\ \underline{M}.$$

From the calculated values, it is concluded that $[H_2CO_3] + [CO_3^{2-}] \simeq 0.02[HCO_3^-]$. Hence, the use of Equation (5.117) was valid. However, if for some other value of C we find that $[H_2CO_3]+[CO_3^{2-}] \geq 0.1$ C, then more accurate values should be calculated for the concentrations of the various species, as follows: The calculated value of $[HCO_3^-]$ is substituted in the equation

$$[H^+] = \sqrt{\frac{K_1K_2[HCO_3^-] + K_1K_w}{[HCO_3^-] + K_1}} \tag{5.120}$$

to obtain a more accurate value for $[H^+]$, which in combination with the values of K_1 and K_2 gives more accurate values for the concentrations of $[H_2CO_3]$ and $[CO_3^{2-}]$. Substituting these values in Equation (5.119), we find an even more accurate value of $[HCO_3^-]$, and the round of calculations is repeated until two successive values of $[H^+]$ agree with each other at a predetermined level. Usually, it is sufficient to repeat the round of calculations (reiterations) once to obtain a sufficiently accurate value of $[H^+]$.

Example 5.31. Calculate a) the pH of 0.100 M glycine, H_2NCH_2COOH, and b) the isoelectric point of glycine ($pK_1 = 2.35$, $pK_2 = 9.78$).

Solution. a) Let HGly be the formula of glycine. We have

$$H_2Gly^+ \rightleftharpoons H^+ + HGly,\ K_1 = \frac{[H^+][HGly]}{[H_2Gly^+]} = 10^{-2.35} = 4.47 \times 10^{-3}$$

$$HGly \rightleftharpoons H^+ + Gly^-,\ K_2 = \frac{[H^+][Gly^-]}{[HGly]} = 10^{-9.78} = 1.66 \times 10^{-10}.$$

Since glycine is an ampholyte, we have

$$\begin{aligned}[H^+] &= \sqrt{\frac{K_1K_2[HGly] + K_1K_w}{[HGly] + K_1}} \\ &= \sqrt{\frac{10^{-2.35} \times 10^{-9.78} \times (0.100) + 10^{-2.35} \times 10^{-14}}{0.100 + 10^{-2.35}}} = 8.43 \times 10^{-7}\ \underline{M}\end{aligned}$$

(we assume that $[HGly] \simeq 0.100$ M). Hence, $pH = -\log(8.43 \times 10^{-7}) = \mathbf{6.07}$.

b) At the isoelectric point we have $[H_2Gly^+] = [Gly^-]$. Hence,

$$K_1K_2 = \frac{[H^+]^2[Gly^-]}{[H_2Gly^+]} = [H^+]^2 \text{ and } [H^+] = \sqrt{K_1K_2}$$

or

$$pH = \frac{pK_1 + pK_2}{2} = \mathbf{6.07}.$$

5.9 Problems

Dissociation of weak monoprotic acids and weak monoacidic bases

5.1. What is the concentration of a CH_3COOH solution which is 10% dissociated?

5.2. Show that the *exact* value of $[OH^-]$ in a solution of the weak monoacidic base B, of total concentration C, is given by the equation

$$[OH^-]^3 + K_b[OH^-]^2 - (CK_b + K_w)[OH^-] - K_wK_b = 0.$$

5.3. Show that the *exact* value of $[H^+]$ in an NH_4Cl solution of concentration C is given by the equation

$$[H^+]^3 + K_h[H^+]^2 - (CK_h + K_w)[H^+] - K_wK_h = 0.$$

Ion Product of water

5.4. The degree of hydrolysis of 0.100 M NaA solution is $7.5 \times 10^{-3}\%$ at 25°C. If the dissociation constant of the acid HA is 1.80×10^{-5}, calculate the ion product of water at 25°C.

5.5. Calculate the ion product of water at 25°C, given that the degree of hydrolysis of 0.100 M NH_4Cl solution is $7.5 \times 10^{-3}\%$ at 25°C.

Calculation of the dissociation constant

5.6. A sample of weak acid HA is titrated with NaOH solution. A total of 40.0 mL of NaOH solution is required to neutralize the HA sample. The pH of the titrated sample after the addition of 8.0 mL of base was 5.14. Calculate the dissociation constant K_a of HA.

5.7. If the pH of a 0.25 M solution of the weak acid HA is 3.20, what is the dissociation constant of the acid?

5.8. A sample of the weak monoprotic acid HA is titrated with NaOH solution. After 10.00 mL of NaOH has been added, the pH is 5.70. If a total of 30.00 mL of NaOH is required until the equivalence point, calculate the dissociation constant of the acid.

5.9. A 0.3600-g sample of benzoic acid is dissolved in 180 mL of water, and 10.00 mL of 0.0869 M NaOH is added to the solution, which is then diluted with water to 200.0 mL. The pH of the final solution is found equal to 3.82 at 25°C. Calculate the dissociation constant of benzoic acid.

5.10. A sample of the weak monoprotic acid HA is titrated with NaOH solution, 37.20 mL of base being consumed until the equivalence point. At this point, 18.60 mL of HCl solution of the same normality as the NaOH solution is added, and the pH is found equal to 4.30. Calculate K_a for HA.

5.11. A 50.0-mL portion of a 0.120 M solution of the weak monoprotic acid HA is titrated with 0.200 M NaOH solution. After the addition of 17.5 mL of the base, the pH of the solution was 4.80. Calculate K_a for HA.

Degree of dissociation

5.12. If the acid HA is 4.0% dissociated in a 0.20 M solution, what is the degree of dissociation in a 0.50 M HA solution?

5.13. Calculate the maximum degree of dissociation a) of HCN, b) of CH_3COOH, and c) of the NH_4^+ ion in aqueous solutions, at 25°C.

5.14. To what volume should 200 mL of 0.100 M HA solution be diluted so that the degree of dissociation is tripled, given that K_a for HA is 1.00×10^{-5}?

5.15. At what concentration of monochloroacetic acid is the degree of dissociation equal to the dissociation degree of 0.0200 M CH_3COOH? (Neglect the dissociation of water.)

Concentrations of ions and molecules

5.16. Calculate the concentrations of ions and molecules in a 1.00 M HNO_2 solution, as well as its degree of dissociation.

Effect of ionic strength on the dissociation constant

5.17. Derive relations combining the dissociation constants K_a, K_a°, and the ionic strength μ for the following acids: a) HCOOH, b) $H_2PO_4^-$, c) $CH_3NH_3^+$. For the same change in μ, in which of the three acids is the largest change in K_a observed and in which one the minimum?

Acidity, neutrality, and alkalinity

5.18. How many grams of gaseous HCl should be added to one liter of 0.100 $\underline{M}$ CH_3COOH solution so that the concentration of H^+ ions is doubled?

5.19. How many milliliters of water should be added to 200 mL of 0.100 $\underline{M}$ CH_3COOH so that the $[H^+]$ of the solution becomes 0.00100 $\underline{M}$?

5.20. To what volume should 10.0 mL of 0.300 $\underline{M}$ NH_3 solution be diluted to give a solution having $[OH^-] = 4.8 \times 10^{-5}$ $\underline{M}$?

Calculation of $[H^+]$ and/or $[OH^-]$

5.21. Calculate $[H^+]$ for each of the following solutions:

a) Five milliliters of hydrochloric acid, of density 1.095 g/mL, containing 20.0% HCl by weight, and 40.0 mL of 1.00 $\underline{M}$ NH_3 are diluted with water to 200 mL.

b) Twenty five milliliters of ammonia, of density 0.898 g/mL, containing 28.0% NH_3 by weight, are diluted with water to one liter.

5.22. Calculate a) the $[H^+]$ of an HCl solution of pH 1.40, b) the $[OH^-]$ of an NaOH solution of pH 12.60.

5.23. How many grams of NH_4Cl should be added to 250 mL of 0.100 $\underline{M}$ NH_3 in order to increase $[H^+]$ a hundred-fold?

5.24. Calculate the terms $[H^+]$, $[OH^-]$, pH, and pOH for a 0.050 $\underline{M}$ $Ba(OH)_2$ solution (assume complete dissociation of the base).

pH

5.25. How many grams of NH_4Cl are required to prepare 250 mL of a solution having pH 4.45?

5.26. How many grams of $CH_3COONa \cdot 3H_2O$ should be added to 200 mL of 0.100 $\underline{M}$ CH_3COOH so that its pH increases by two units?

5.27. If in a solution we have $[H^+] = 10[OH^-]$, what is the pH of the solution?

5.28. What is the change in $[H^+]$ of a solution, when its pH is increased by 1.60 units?

5.29. Calculate the pH of 1.00×10^{-2} $\underline{M}$ oxalic acid solution.

5.30. Calculate the pH and the concentrations of the various ions and molecules in the following solutions: a) 0.225 $\underline{M}$ Na_2SO_3, b) 0.0500 $\underline{M}$ $(NH_4)_2CO_3$, c) 0.070 $\underline{M}$ $Al(NO_3)_3$, d) 0.200 $\underline{M}$ NH_4NO_3.

5.31. Calculate the pH of a solution obtained by mixing equal volumes of 0.200 $\underline{M}$ NaOH, 0.160 $\underline{M}$ HCl, 0.120 $\underline{M}$ Na_2HPO_4, and 0.120 $\underline{M}$ NaH_2PO_4.

5.32. Which of the following results in the largest increase in the pH of 50 mL of 0.100 $\underline{M}$ CH_3COOH solution? a) dilution with water to one liter, b) addition of 20.00 mL of 0.050 $\underline{M}$ NaOH, or c) addition of 0.082 g of CH_3COONa?

5.33. An aqueous solution containing 2.00 mmol of hydrazine and 16.00 mmol of $NaHCO_3$ is titrated with a standard iodine solution ($N_2H_4 + 2[I_3]^- + 4HCO_3^- \rightleftharpoons N_2 + 6I^- + 4H_2CO_3$). If the solution volume at the end point is 50.00 mL, what is the pH of the solution at the end point?

5.34. Derive the pH of a solution of the salt NaA of the weak acid HA, as a function of the salt concentration C and the constants K_a of HA and K_w of water (when $C >> [OH^-]$).

5.35. What is the molarity of an H_2SO_4 solution having pH 2.00?

5.36. Calculate the pH of a solution obtained by mixing equal volumes of HCl solutions having pH 2.00 and 3.00.

5.37. Calculate the pH of the following solutions: a) 1.00×10^{-7} $\underline{M}$ NaOH, b) 1.00×10^{-8} $\underline{M}$ NaOH, c) 1.00×10^{-8} $\underline{M}$ HCl.

5.38. Calculate the pH of a solution obtained by mixing 50.0 mL of 0.0500 $\underline{M}$ HCl solution and 50.0 mL of a) H_2O, b) 0.0400 $\underline{M}$ NaOH, c) 0.0250 $\underline{M}$ $Ba(OH)_2$, d) 0.0100 $\underline{M}$ $AgNO_3$, e) 0.0500 $\underline{M}$ NH_3.

5.39. Calculate the pH of a solution obtained by mixing 25.0 mL of 0.0840 $\underline{M}$ NH_3 solution and 20.0 mL of a) H_2O, b) 0.0600 $\underline{M}$ HBr, c) 0.0600 $\underline{M}$ NH_4Cl, d) 0.1050 $\underline{M}$ CH_3COOH.

5.40. Calculate the pH of a solution obtained by mixing 75.0 mL of each of the following solutions: 0.500 $\underline{M}$ NH_3, 0.0500 $\underline{M}$ HCl, 0.100 $\underline{M}$ NH_4Cl.

5.41. Calculate the pH of a solution obtained by mixing equal volumes of three HNO_3 solutions, having pH 1.20, 2.00, and 4.00.

Protolytic indicators

5.42. The dissociation constant of the indicator HIn is 4.00×10^{-8}. What percentage of the indicator is present in the acid form at pH a) 7.16, b) 8.16, c) 9.60?

5.43. How many milliliters of 1.00 M NaOH should be added to 200 mL of 0.100 M CH_3COOH-1.00 × 10^{-4} M bromcresol blue indicator solution (pK_{HIn} = 4.9), to obtain a solution having a) [HIn]/[In^-] = 10/1, b) [HIn][In^-] = 1/10?

5.44. A tiny amount of the indicator HIn is added to a 0.050 M CH_3COONa solution. What percentage of the indicator will be present in the alkaline form (base color) in this solution (dissociation constant $K_{HIn} = 3.16 \times 10^{-10}$)?

Dissociation of polyprotic acids

5.45. Show that for an $H_2C_2O_4$ solution of total concentration C:

$$[HC_2O_4^-] = \frac{K_1[H^+]C}{[H^+]^2 + K_1[H^+] + K_1K_2}$$

5.46. Calculate the concentrations of ions and molecules in 0.0100 M H_2SO_4 solution.

Dissociation of polyacidic bases

5.47. Show that in an aqueous solution of the diacidic base B of concentration C, the following relation is valid:

$$[OH^-]^4 + K_1[OH^-]^3 + (K_1K_2 - K_w - CK_1)[OH^-]^2 - (K_1K_w + 2CK_1K_2)[OH^-] - K_1K_2K_w = 0.$$

Calculation of concentrations of species at a given pH

5.48. Which arsenic species (ions or molecules) exist in a larger percentage in an arsenic acid solution ($C_{H_3AsO_4}$ = 1.00 M) having a pH of a) 1.4, b) 6.0, c) 10.5, d) 12.5?

5.49. For integer values of pH between 0 and 14, calculate the concentrations of the species resulting from the following acids: a) H_2S, b) H_2CO_3, c) H_3AsO_4. Construct the corresponding distribution diagrams of the species as a function of pH (acid concentration = C).

Proton balance

5.50. Write the proton balance for the following solutions: a) Na_2HAsO_4, b) $(NH_4)_2HPO_4$, c) $NH_4H_2PO_4$, d) CH_3COONa, e) H_3PO_4, f) $NaHCO_3$, g) HNO_3.

Logarithmic acid-base concentration diagrams

5.51. Construct a logarithmic concentration diagram for a 1.00 M H_2A solution (pK_1 = 2.96, pK_2 = 4.16).

Buffers

5.52. A CH_3COOH solution is 1.00% dissociated. What will $[H^+]$ be, if 0.82 g of CH_3COONa is added to 100 mL of this solution?

5.53. Calculate $[H^+]$ in a solution obtained by adding 20.0 mL of 0.20 $\underline{M}$ NaOH to 160 mL of 0.30 $\underline{M}$ CH_3COOH.

5.54. How many milliliters of hydrochloric acid solution, of density 1.18 g/mL, containing 35.6% HCl by weight, should be added to 100 mL of 1.00 $\underline{M}$ CH_3COONa solution so that $[H^+]$ of the resulting solution is 1.80×10^{-5} $\underline{M}$?

5.55. A 200-mL portion of a 1.00 $\underline{M}$ NH_3-1.80 $\underline{M}$ NH_4Cl buffer solution contains 0.3000 g of the diprotic acid H_2A (MW 150). For this solution, we have the relations $\alpha_0 = 0.100$, and $[H_2A] = 2[A^{2-}]$. Calculate the pH of the solution, the concentrations of the species H_2A, HA^- and A^{2-}, and the dissociation constants K_1 and K_2. (Assume that the pH of the buffer is not affected by the presence of the (small) quantity of H_2A.)

5.56. A CH_3COOH-CH_3COONa buffer has a pH of 4.85 and contains 0.500 mol of acetic species per liter. Calculate the molarity of the CH_3COO^- ions in the solution.

5.57. Forty millimoles of HCl are added to 100 mL of a 1.20 $\underline{M}$ NH_3-x $\underline{M}$ NH_4Cl buffer, and the resulting solution has a pH of 8.96. Calculate the pH of the buffer.

5.58. When an $(NH_4)_2SO_4$ solution is treated with NaOH, the NH_4^+ ion is converted to NH_3. What should the pH of the solution be so that 99.95% of NH_4 is converted to NH_3?

5.59. Calculate the change in pH of a 0.500 $\underline{M}$ $NaHSO_4$-0.500 $\underline{M}$ Na_2SO_4 buffer, when a) 20.0 mL of 0.100 $\underline{M}$ NaOH, b) 20.0 mL of 0.100 $\underline{M}$ H_2SO_4, are added to 180 mL of buffer. c) What is the buffer capacity of the buffer?

5.60. Calculate the change in pH of a 0.170 $\underline{M}$ NH_3-0.187 $\underline{M}$ NH_4Cl buffer, when a) 20.0 mL of 0.100 $\underline{M}$ HCl, b) 20.0 mL of 0.100 $\underline{M}$ NaOH, are added to 200 mL of buffer.

5.61. Calculate the changes in $[H^+]$ and pH of a 0.100 $\underline{M}$ CH_3COOH-0.100 $\underline{M}$ CH_3COONa buffer that results when a) solid NaOH is added until the solution (before reaction) becomes 0.010 $\underline{M}$ in NaOH, b) HCl is added until the solution (before reaction) becomes 0.010 $\underline{M}$ in HCl (the solution volume remains unchanged).

5.62. How many milliliters of 0.200 $\underline{M}$ NH_3 solution should be added to 200 mL of 0.200 $\underline{M}$ NH_4Cl solution to obtain a solution of pH 9.00?

5.63. How many grams of NH_4Cl and how many milliliters of concentrated ammonia solution (15.0 $\underline{M}$) are required to prepare one liter of 0.500 $\underline{M}$ NH_4Cl-NH_3 buffer of pH 10.00?

5.64. How would you prepare one liter of an NH_3-NH_4Cl buffer having $[NH_3] + [NH_4^+] = 0.200$ $\underline{M}$ and pH 9.00, from 1.50 $\underline{M}$ NH_3 solution, 1.50 $\underline{M}$ HCl solution, and water?

5.65. Calculate the volumes of 0.500 M CH_3COOH and 0.500 M CH_3COONa solutions which are required to prepare 400 mL of a buffer having pH 4.00.

5.66. How many milliliters from each of the following solutions, 0.50 M H_3PO_4, 0.50 M NaOH, and 2.00 M $NaClO_4$, are required to prepare one liter of a buffer having pH 7.00, total phosphate concentration 0.100 M, and ionic strength 0.500?

5.67. How many grams of NaOH should be added to 250 mL of 0.100 M tartaric acid solution to prepare a buffer of pH 4.00? (**Hint.** Use the alpha-functions α_0, α_1, and α_2, which represent the fractions of total tartrate present in the form of H_2A, HA^-, and A^{2-}, respectively, where H_2A represents the tartaric acid.)

5.68. How many milliliters of 1.00 M NaOH solution should be mixed with 60.0 mL of 0.100 M $NaHCO_3$ solution to prepare a buffer containing 2.00 mol HCO_3^-/mol CO_3^{2-}? What is the pH of the buffer?

5.69. How many grams of CH_3COONa should be added to 50.0 mL of 0.240 M HCl solution to prepare a buffer of pH 5.04?

5.70. How many milliliters of 0.100 M NaOH solution should be added to 25.0 mL of 0.100 M CH_3COOH solution to prepare a buffer of pH 5.00?

5.71. How many milliliters of 1.00 M NaOH solution should be added to 100 mL of 1.00 M HA solution to prepare a buffer of pH 3.70? The dissociation constant of HA is 1.00×10^{-4}.

5.72. How many grams of NH_4Cl are there in one liter of a buffer having a pH of 10.00 and containing 21.4 mg NH_3/mL of solution?

5.73. Calculate the pH of a solution which is prepared by adding 5.00 mL of 0.100 M NaOH solution to 204 mg of KH_2PO_4 and diluting with water to 200 mL.

5.74. Buffer X contains one mole of CH_3COOH and one mole of CH_3COONa per liter. Calculate the weight of NaOH that should be added to 500 mL of solution X so that its pH changes by one unit.

5.75. A 10.70-g portion of NH_4Cl is added to 200 mL of 0.250 M NaOH solution. Calculate the pH of the resulting buffer.

5.76. The H^+ ion concentration in a CH_3COOH-CH_3COONa buffer is 9.0×10^{-6} M. Calculate the molarities of CH_3COOH and CH_3COONa, given that $[H^+]$ becomes 1.00×10^{-5} M, when 10.0 mmol of HCl is added to one liter of buffer.

5.77. Calculate the pH of a buffer which is prepared by dissolving 4.92 g of CH_3COONa in 200 mL of 0.200 M HCl solution ($V_{final} = 200$ mL).

5.78. Calculate the pH of a buffer which is prepared by mixing 200 mL of 1.00 M NaOH solution with 300 mL of 2.00 M CH_3COOH solution.

5.79. How many milliliters of 2.00 M HCl solution should be added to one liter of 0.100 M NH_3 solution to prepare a buffer of pH 9.00?

5.80. How many grams of NH_4Cl should be added to 200 mL of 0.600 $\underline{M}$ NH_3 to prepare a buffer of pH 9.30?

5.81. How many milliliters of 1.00 $\underline{M}$ HCl solution should be added to 200 mL of 0.600 $\underline{M}$ NH_3 to prepare a buffer of pH 9.30?

5.82. How many grams of CH_3COOH and CH_3COONa are there in one liter of a buffer having a pH of 5.00 and an ionic strength of 0.100?

5.83. How many grams of NaCl should be added to 100 mL of a HCl acid solution to prepare a buffer having a pH of 1.83 and an ionic strength of 0.100?

5.84. A 0.100 $\underline{M}$ HA-0.100 $\underline{M}$ NaA buffer has a pH of 5.04. How many grams of NaOH should be added to one liter of buffer to obtain a solution of pH 5.10?

5.85. At what ratio should solutions of 0.100 $\underline{M}$ of weak base B and 0.0500 $\underline{M}$ HCl be mixed to prepare one liter of buffer of pH 9.00? The dissociation constant of B, K_b, is 5.0×10^{-5}.

5.86. What is the pH of the following buffers? a) glycine (H_2NCH_2COOH)-HCl at a mole ratio of 2:1, b) glycine-NaOH, at a mole ratio of 2:1 ($pK_1 = 2.35$, $pK_2 = 9.78$).

5.87. How many grams of $NaHSO_4$ and Na_2SO_4 are required to prepare 500 mL of a buffer having a pH of 1.62 and $C_S = 1.00$ $\underline{M}$?

5.88. How many milliliters of 1.00 $\underline{M}$ NaOH solution should be added to 0.9004 g of oxalic acid to give a buffer of pH 4.43 when diluted to 100 mL?

5.89. Two CH_3COOH-CH_3COONa buffers, each having $[CH_3COOH] + [CH_3COO^-] = 0.100$ $\underline{M}$, have a pH of 4.00 and 5.00. If equal volumes of the two buffers are mixed, what is the pH of the resulting solution?

5.90. How many milliliters of NH_3-NH_4Cl buffer of pH 10.00 and $C_N = 0.200$ $\underline{M}$ should be mixed with 100 mL of NH_3-NH_4Cl buffer of pH 9.00 and $C_N = 0.100$ $\underline{M}$ to obtain a buffer of pH 9.50?

5.91. A buffer solution of pH 5.00 is prepared from the weak acid HX ($pK_{HX} = 5.30$) and its salt NaX. Ten millimoles of HCl are added to 100 mL of buffer. What should the initial concentrations of HX and NaX be in the buffer so that the pH changes by 0.20 upon the addition of the acid?

5.92. At what ratio should a 0.100 $\underline{M}$ solution of the weak acid HA and a 0.0500 $\underline{M}$ NaOH solution be mixed to prepare 100 mL of buffer of pH 5.00? (Dissociation constant $K_{HA} = 5.0 \times 10^{-5}$.)

5.93. a) How many grams of $NaHCO_3$ should be added to 3.18 g of Na_2CO_3 to prepare with water 500 mL of buffer of pH 10.68?

b) What will be the pH of the above buffer, if 1.00 mL of 1.00 $\underline{M}$ HCl solution is added to it?

c) How many milliliters of 1.00 M NaOH solution should be added to 2.52 g of $NaHCO_3$ to prepare with water 500 mL of buffer of pH 10.20?

Buffer capacity

5.94. Compare the buffer capacities of the following three buffers: a) 0.010 M CH_3COOH-0.10 M CH_3COONa, b) 0.010 M CH_3COOH-0.0040 M CH_3COONa, c) 0.010 M CH_3COOH-0.0010 M CH_3COONa.

Degree of hydrolysis

5.95. The pH of 0.100 M HCOONa is 8.34. Calculate the degree of hydrolysis of the salt and the dissociation constant of HCOOH.

5.96. Calculate the pH and the degree of hydrolysis of a 0.100 M $AlCl_3$ solution.

5.97. Calculate the pH of a 1.00×10^{-5} M NH_4CN solution and the degree of hydrolysis of the NH_4^+ and CN^- ions.

5.98. Calculate the concentrations of the various species in a 0.100 M Na_2S solution, as well as its pH and its degree of hydrolysis.

Isoelectric point

5.99. Calculate a) the pH of a 0.100 M solution of glutamic acid, $HOOC(CH_2)_2CH(NH_2)COOH$, b) the isoelectric point of glutamic acid ($pK_1 = 2.23$, $pK_2 = 4.42$, $pK_3 = 9.95$).

Chapter 6

Equilibria Involving Precipitates and Their Ions

6.1 Solubility Product — Solubility

The general equation

$$M_mX_x \rightleftharpoons mM^{a+} + xX^{b-}, \tag{6.1}$$

which describes the ionization of the slightly soluble ionic solid M_mX_x in its saturated solution (solid in contact with the solution), is characterized by an equilibrium constant, i.e.,

$$K = \frac{[M^{a+}]^m[X^{b-}]^x}{[M_mX_x]} \tag{6.2}$$

or, since the concentration of M_mX_x is constant,

$$K[M_mX_x] = K_{sp} = [M^{a+}]^m[X^{b-}]^x \tag{6.3}$$

The constant, K_{sp}, is called the *solubility product constant* or simply the *solubility product.*

If S is the molar solubility (i.e., in moles per liter) of the substance M_mX_x, we have $[M^{a+}] = mS$ and $[X^{b-}] = xS$. Substituting these values in Equation (6.3), we have

$$K_{sp} = (mS)^m(xS)^x = m^m x^x S^{m+x} \tag{6.4}$$

For the calculation of S from K_{sp}, Equation (6.4) is used in the form

$$S = \sqrt[m+x]{K_{sp}/m^m x^x} \tag{6.4a}$$

For example, for a saturated $Ca_3(PO_4)_2$ solution we have $[Ca^{2+}] = 3S$, and $[PO_4^{3-}] = 2S$. Hence, $K_{sp} = [Ca^{2+}]^3[PO_4^{3-}]^2 = (3S)^3(2S)^2 = 108S^5$, or directly, using Equation (6.4),

$$K_{sp} = 3^3 \cdot 2^2 \cdot S^{3+2} = 108S^5,$$

from which $S = \sqrt[5]{K_{sp}/108}$.

When the *ion product*, I.P., of the dissolved substance exceeds the constant K_{sp}, a precipitate is formed. In the general case of the slightly soluble substance M_mX_x, precipitation takes place only if

$$\text{I.P.} = [M^{a+}]^m[X^{b-}]^x > K_{sp} \tag{6.5}$$

If I.P. = K_{sp}, the solution is saturated, but no precipitation takes place. If I.P. < K_{sp}, the solution is unsaturated, and if there is a precipitate, it dissolves, until I.P. = K_{sp}.

Example 6.1. The concentration of Ag^+ ion in a saturated aqueous solution of Ag_2CrO_4 was found potentiometrically equal to 1.56×10^{-4} M. Calculate the solubility of Ag_2CrO_4 in g/L, and its solubility product.

Solution. Each mole of Ag_2CrO_4 gives 2 moles of Ag^+ ions and 1 mole of CrO_4^{2-} ions. Consequently, the molarity of the dissolved salt and the CrO_4^{2-} ion is equal to one-half the concentration of the Ag^+ ion, i.e., 7.8×10^{-5} M. The molecular weight of Ag_2CrO_4 is 331.7. Hence, the solubility of Ag_2CrO_4 in g/L is equal to

$$(7.8 \times 10^{-5} \text{mol/L})(331.7 \text{ g/mol}) = \mathbf{0.0259} \text{ g/L}.$$

The solubility product of Ag_2CrO_4 is equal to

$$K_{sp} = [Ag^+]^2[CrO_4{}^{2-}] = (1.56 \times 10^{-4})^2(7.80 \times 10^{-5}) = \mathbf{1.90 \times 10^{-12}}.$$

Example 6.2. The solubility product of the compound M_2X was found equal to 3.58×10^{-13}, on the basis of solubility measurements and the assumption of quantitative ionization to M^+ and X^{2-} ions. Later, it was found that the compound is quantitatively ionized to $M_2{}^{2+}$ and X^{2-} ions. Calculate the real solubility product of M_2X, and identify M_2X on the basis of data given in Appendix A (Table A.3).

Solution. If S is the molar solubility of M_2X, and we assume an ionization according to the equation

$$\mathbf{M_2X} \rightleftharpoons 2M^+ + X^{2-}$$

we have $[M^+] = 2S$ and $[X^{2-}] = S$. Hence,

$$K_{sp(M_2X)} = [M^+]^2[X^{2-}] = (2S)^2S = 4S^3 = 3.58 \times 10^{-13}$$

or

$$S = \sqrt[3]{3.58 \times 10^{-13}/4} = 4.47 \times 10^{-5} \text{ M}.$$

However, since the real ionization of the salt is given by the equation

$$\mathbf{M_2X} \rightleftharpoons M_2^{2+} + X^{2-}$$

it follows that

$$K_{sp(M_2X)_{real}} = [M_2{}^{2+}][X^{2-}] = S^2 = (4.47 \times 10^{-5})^2 = \mathbf{2.00 \times 10^{-9}} \ (M_2X \equiv Hg_2CrO_4).$$

Example 6.3. How many grams of NH_4Cl should be added to 50 mL of 0.20 $\underline{M}$ NH_3 solution so that after mixing of the resulting solution with 50 mL of 0.020 $\underline{M}$ $MnCl_2$ solution, no precipitate of $Mn(OH)_2$ will be formed?

Solution. After the mixing of the solutions, the concentrations of NH_3 and Mn^{2+} are 0.10 $\underline{M}$ and 0.010 $\underline{M}$, respectively. In order to avoid the formation of a $Mn(OH)_2$ precipitate, we should have $[Mn^{2+}][OH^-]^2 \leq K_{sp(Mn(OH)_2)}$, i.e., $(0.010)[OH^-]^2 \leq 2 \times 10^{-13}$ or $[OH^-] \leq 4.47 \times 10^{-6}$ $\underline{M}$. The $[NH_4^+]$ required in order to have $[OH^-] = 4.47 \times 10^{-6}$ $\underline{M}$, is calculated from K_b for ammonia, i.e.,

$$\frac{[NH_4^+][OH^-]}{[NH_3]} = \frac{[NH_4^+](4.47 \times 10^{-6})}{0.10} = 1.8 \times 10^{-5},$$

from which we find that $[NH_4^+] = 0.403$ $\underline{M}$. The final volume of the solution is 100 mL. Hence, we should add at least

$$(0.403 \text{ mol/L})(0.100 \text{ L})(53.5 \text{ g } NH_4Cl/\text{mol}) = \mathbf{2.16} \text{ g } NH_4Cl.$$

Example 6.4. The pH of a saturated aqueous solution of zinc ammonium phosphate ($ZnNH_4PO_4$) is 7.20, whereas the concentration of Zn^{2+} ion in the solution is 8.4×10^{-5} $\underline{M}$. Calculate the solubility product of $ZnNH_4PO_4$.

Solution. The solubility product of $ZnNH_4PO_4$ is given by the equation

$$K_{sp(ZnNH_4PO_4)} = [Zn^{2+}][NH_4^+][PO_4^{3-}] \tag{6.6}$$

We can calculate $[NH_4^+]$ and $[PO_4^{3-}]$ from the concentration of H^+ ion ($[H^+] = 10^{-7.20} = 6.3 \times 10^{-8}$ $\underline{M}$) and the total (analytical) concentration C of the ammoniacal and phosphate species ($C = [Zn^{2+}] = 8.4 \times 10^{-5}$ $\underline{M}$). We have

$$[NH_4^+] = \alpha_0 C = \frac{[H^+]C}{[H^+] + K_a} = \frac{(6.3 \times 10^{-8})(8.4 \times 10^{-5})}{(6.3 \times 10^{-8}) + (1.00 \times 10^{-14}/1.8 \times 10^{-5})} =$$

$$= 8.3 \times 10^{-5} \ \underline{M}$$

$$[PO_4^{3-}] = \alpha_3 C = \frac{K_1K_2K_3C}{[H^+]^3 + K_1[H^+]^2 + K_1K_2[H^+] + K_1K_2K_3} =$$

$$\frac{(7.5 \times 10^{-3})(6.2 \times 10^{-8})(1 \times 10^{-12})(8.4 \times 10^{-5})}{(6.3 \times 10^{-8})^3 + (7.5 \times 10^{-3})(6.3 \times 10^{-8})^2 + (7.5 \times 10^{-3})(6.2 \times 10^{-8})(6.3 \times 10^{-8}) + (7.5 \times 10^{-3})(6.2 \times 10^{-8})(1 \times 10^{-12})}$$

$$= 6.6 \times 10^{-10} \ \underline{M}.$$

Substituting the values of $[Zn^{2+}]$, $[NH_4^+]$, and $[PO_4^{3-}]$ in Equation (6.6), we have

$$K_{sp(ZnNH_4PO_4)} = (8.4 \times 10^{-5})(8.3 \times 10^{-5})(6.6 \times 10^{-10}) = \mathbf{4.6 \times 10^{-18}}.$$

Example 6.5. What is the minimum concentration of excess Ag^+ ion required for the quantitative precipitation (at least by 99.9%) of oxalate ions from a 0.0303 M $Na_2C_2O_4$ solution, buffered at pH 4.00?

Solution. We have

$$[C_2O_4{}^{2-}] = \alpha_2 C = \frac{(3.8\times10^{-2})(5.0\times10^{-5})(0.0303)}{(1.00\times10^{-4})^2 + (3.8\times10^{-2})(1.00\times10^{-4}) + (3.8\times10^{-2})(5.0\times10^{-5})} = 0.0101\ \underline{M}.$$

If at least 99.9% of oxalate is precipitated, the concentration of $C_2O_4{}^{2-}$ remaining in the solution is $(0.1/100)(0.0101) = 1.01 \times 10^{-5}$ M. Hence, the minimum concentration of excess Ag^+ required for the quantitative precipitation of $C_2O_4{}^{2-}$ is equal to

$$[Ag^+] = \sqrt{K_{sp}/[C_2O_4{}^{2-}]} = \sqrt{1\times10^{-11}/1.01\times10^{-5}} = \mathbf{1.00\times10^{-3}}\ \underline{M}.$$

That is, sufficient $AgNO_3$ should be added so that the oxalate is precipitated and an excess of 1.00×10^{-3} M Ag^+ remains in the solution.

6.2 Factors that Affect the Solubility and/or the Solubility Product

1. **Common ion effect.** If an electrolyte is added to a saturated solution of a slightly soluble electrolyte and a common ion is present in both electrolytes, the equilibrium is shifted in favor of the solid phase (according to Le Châtelier's principle). In this way, the solubility product remains essentially constant (actually K_{sp} increases because of increasing ionic strength, but the increase may be considered negligible, unless a very large excess of the common ion is added). We make use of the common ion effect in the case of precipitation reactions, where a small excess of reagent is usually added to decrease the solubility of the slightly soluble substance and assure quantitative precipitation, as in gravimetric analysis.

Example 6.6. Calculate the molar solubility of Ag_2CrO_4 a) in water, b) in a 0.050 M Na_2CrO_4 solution, c) in a 0.050 M $AgNO_3$ solution (disregard the hydrolysis of chromate ions).

Solution. a) Let x be the molar solubility of Ag_2CrO_4 in water. According to Equation (6.4a), we have

$$x = \sqrt[3]{1.9\times10^{-12}/4} = \mathbf{7.8\times10^{-5}}\ \text{mol/L}.$$

b) Let y be the molar solubility of Ag_2CrO_4, whereupon $[Ag^+] = 2y$. The $CrO_4{}^{2-}$ ions result from two sources, i.e., from Ag_2CrO_4 which gives y mol/L, and from Na_2CrO_4

which gives 0.050 mol/L. Hence, $[CrO_4^{2-}] = y + 0.050$. Substituting these values in the K_{sp} expression, we have

$$K_{sp} = [Ag^+]^2[CrO_4^{2-}] = (2y)^2(y + 0.050) = 1.9 \times 10^{-12}.$$

By solving the cubic equation, we find the exact value of y, but such a procedure is tedious. From the relatively small value of K_{sp}, it can be seen that the solubility of Ag_2CrO_4 is small (especially in the presence of excess CrO_4^{2-}), and we assume that $y << 0.050$, whereupon $y + 0.050 \simeq 0.050$. Hence, the equation is simplified to

$$(2y)^2(0.050) = 1.9 \times 10^{-12}$$

from which $y = \mathbf{3.1 \times 10^{-6}}$ mol/L. From the value of y it is concluded that the assumption made is correct, because indeed $0.050 + 3.1 \times 10^{-6} \simeq 0.050$.

c) Let z be the molar solubility of Ag_2CrO_4, whereupon $[Ag^+] = 0.050 + 2z$, and $[CrO_4^{2-}] = z$. Hence, we have the equation

$$(0.050 + 2z)^2 z = 1.9 \times 10^{-12}$$

or, since $z << 0.050$,

$$(0.050)^2 z = 1.9 \times 10^{-12}$$

from which $z = \mathbf{7.6 \times 10^{-9}}$ mol/L.

Note. From the results, it is concluded that the solubility of Ag_2CrO_4 is decreased in the presence of a common ion, much more in the presence of Ag^+ ion (a decrease in solubility of $7.8 \times 10^{-5}/7.6 \times 10^{-9} = 10300$ times) than in the presence of CrO_4^{2-} ion of the same concentration (a decrease in solubility of $7.8 \times 10^{-5}/3.1 \times 10^{-6} = 25$ times). This is due to the presence of 2 Ag^+ ions and the squared dependence, and only 1 CrO_4^{2-} ion in the Ag_2CrO_4 molecule.

Example 6.7. Twenty five milliliters of 0.200 $\underline{M}$ $AgNO_3$ solution are mixed with 50.0 mL of 0.076 $\underline{M}$ K_2CrO_4 solution. a) Find the composition of the solution at equilibrium. b) Is the precipitation of silver quantitative (assume $[CrO_4^{2-}] = C_{K_2CrO_4}$)?

Solution. a) We mix

$$25.0 \text{ mL } \times 0.200 \text{ mmol/mL} = 5.00 \text{ mmol } AgNO_3$$

and

$$50.0 \text{ mL } \times 0.076 \text{ mmol/mL} = 3.80 \text{ mmol } K_2CrO_4.$$

Hence, 2.50 mmol of CrO_4^{2-} will react with 5.00 mmol of Ag^+ ($2Ag^+ + CrO_4^{2-} \rightleftharpoons \mathbf{Ag_2CrO_4}$), leaving an excess of 1.30 mmol of K_2CrO_4, whereupon $[CrO_4^{2-}] = 1.30/75.0 = 0.0173$ $\underline{M}$. Let y be the molar solubility of Ag_2CrO_4, whereupon $[Ag^+] = 2y$ and $[CrO_4^{2-}] = y + 0.0173$. Substituting the values of $[Ag^+]$ and $[CrO_4^{2-}]$ in the K_{sp} expression, we have

$$K_{sp(Ag_2CrO_4)} = 1.9 \times 10^{-12} = (2y)^2(y + 0.0173)$$

or, since $y << 0.0173$ (as concluded from the small value of K_{sp}), $(2y)^2(0.0173) = 1.9 \times 10^{-12}$, from which we find that $y = 5.23 \times 10^{-6}$. Hence, we have

$[Ag^+] = 2 \times 5.23 \times 10^{-6} = \mathbf{1.05 \times 10^{-5}}\ \underline{M}$
$[CrO_4{}^{2-}] = 5.23 \times 10^{-6} + 0.0173 \simeq \mathbf{0.0173}\ \underline{M}$
$[K^+] = (50.0 \times 2 \times 0.076)/75.0 = \mathbf{0.101}\ \underline{M}$
and $[NO_3{}^-] = (25.0 \times 0.200)/75.0 = \mathbf{0.0667}\ \underline{M}$.

b) The percentage of silver that was precipitated is equal to

$$\frac{5.00 \text{ mmol} - (75.0 \text{ mL} \times 1.05 \times 10^{-5} \text{ mmol/mL})}{5.00 \text{ mmol}} \times 100 = \mathbf{99.98\%}.$$

Hence, the precipitation of silver is quantitative.

Example 6.8. How many grams of K_2CrO_4 should be added to one liter of a saturated Ag_2CrO_4 solution, so that $[CrO_4{}^{2-}] = 2[Ag^+]$ (assume that Cr exists only as $CrO_4{}^{2-}$)?

Solution. *1st Method.* If S is the molar solubility of Ag_2CrO_4, we have $S = \sqrt[3]{1.9 \times 10^{-12}/4} = 7.8 \times 10^{-5}\ \underline{M} = [CrO_4{}^{2-}]$, whereas $[Ag^+] = 2S = 1.56 \times 10^{-4}\ \underline{M}$. Suppose that we should add y mol K_2CrO_4/L so that $[CrO_4{}^{2-}] = 2[Ag^+]$. If $[Ag^+] = z$, we have $[CrO_4{}^{2-}] = 2z$, and $z^2 \cdot 2z = 1.9 \times 10^{-12}$, or

$$z = \sqrt[3]{1.9 \times 10^{-12}/2} = 9.83 \times 10^{-5}\ \underline{M} = [Ag^+] \text{ and } [CrO_4{}^{2-}] = 2z = 1.97 \times 10^{-4}\ \underline{M}.$$

Suppose that upon the addition of y mol K_2CrO_4/L, w mol of Ag_2CrO_4 will be precipitated (common ion effect), whereupon $[CrO_4{}^{2-}]$ will be decreased by w mol/L and $[Ag^+]$ by 2w mol/L. Finally, we have

$$[CrO_4{}^{2-}] = 7.8 \times 10^{-5} + y - w = 1.97 \times 10^{-4} \quad (6.7)$$

$$[Ag^+] = 1.56 \times 10^{-4} - 2w = 9.83 \times 10^{-5} \quad (6.8)$$

From Equation (6.8) we have $w = 2.89 \times 10^{-5}$, whereupon from (6.7) we have

$$y = 1.97 \times 10^{-4} + 2.89 \times 10^{-5} - 7.8 \times 10^{-5} = 1.48 \times 10^{-4} \text{ mol } K_2CrO_4/L,$$

that is,

$$(1.48 \times 10^{-4} \text{ mol/L})(194.2 \text{ g } K_2CrO_4/\text{mol}) = \mathbf{0.0287} \text{ g } K_2CrO_4/L.$$

2nd Method. Setting $[CrO_4{}^{2-}] = 2[Ag^+]$ in the solubility product expression, we have $[Ag^+]^2(2[Ag^+]) = 2[Ag^+]^3 = 1.9 \times 10^{-12}$, or $[Ag^+] = \sqrt[3]{1.9 \times 10^{-12}/2} = 9.83 \times 10^{-5}\ \underline{M}$. We consider $[H^+] = [OH^-]$, whereupon for $[CrO_4{}^{2-}] = 2[Ag^+]$ and $[Ag^+] = 9.83 \times 10^{-5}$ $\underline{M}$, the electroneutrality principle becomes

$$2[CrO_4{}^{2-}] = [Ag^+] + [K^+] = 2(2[Ag^+]) = 4[Ag^+]$$

or

$$[K^+] = 3[Ag^+] = 3(9.83 \times 10^{-5}) = 2.95 \times 10^{-4}\ \underline{M} \equiv (2.95 \times 10^{-4})/2\ \underline{M}\ K_2CrO_4.$$

Hence, we should add

$$(2.95 \times 10^{-4}/2 \text{ mol/L})(194.2 \text{ g } K_2CrO_4/\text{mol}) = \mathbf{0.0286} \text{ g } K_2CrO_4/\text{L}.$$

Example 6.9. Calculate the molar solubilities of CH_3COOAg and $Ag_2C_2O_4$ in a solution saturated by both salts. By what percentage is the solubility of $Ag_2C_2O_4$ decreased by the presence of CH_3COOAg (ignore the hydrolysis of CH_3COO^- and $C_2O_4^{2-}$)?

Solution. We have the equilibria

$$\begin{array}{cccc} \mathbf{CH_3COOAg} & \rightleftharpoons & Ag^+ & + \; CH_3COO^- \\ y & & y & y \end{array}$$

$$\begin{array}{cccc} \mathbf{Ag_2C_2O_4} & \rightleftharpoons & 2Ag^+ & + \; C_2O_4^{2-} \\ z & & 2z & z \end{array}$$

and the relations

$$[Ag^+][CH_3COO^-] = (y + 2z)y = 4 \times 10^{-3}$$
$$[Ag^+]^2[C_2O_4^{2-}] = (y + 2z)^2 z = 1 \times 10^{-11}$$

We assume that $y + 2z \simeq y$ (because of the small value of $K_{sp(Ag_2C_2O_4)}$), whereupon $y = \sqrt{4 \times 10^{-3}} = 6.32 \times 10^{-2}$ and $z = 1 \times 10^{-11}/4 \times 10^{-3} = 2.5 \times 10^{-9}$ (indeed, $2z = 5.0 \times 10^{-9} << y = 6.32 \times 10^{-2}$). Hence, in a saturated solution of both salts, we have $S_{CH_3COOAg} = [CH_3COO^-] = \mathbf{6.32 \times 10^{-2}}$ mol/L and $S_{Ag_2C_2O_4} = [C_2O_4^{2-}] = \mathbf{2.5 \times 10^{-9}}$ mol/L.

In a saturated solution of $Ag_2C_2O_4$, we have $S_{Ag_2C_2O_4} = \sqrt[3]{1 \times 10^{-11}/4} = 1.36 \times 10^{-4}$ mol/L. Hence the solubility of $Ag_2Cr_2O_4$ is decreased by

$$\left(\frac{1.36 \times 10^{-4} - 2.5 \times 10^{-9}}{1.36 \times 10^{-4}}\right) 100 = \mathbf{99.998\%}.$$

---

2. **Diverse ion effect (effect of ionic strength).** The addition of a strong electrolyte to a saturated solution of a slightly soluble electrolyte with which there is no common ion increases the ionic strength of the solution, and consequently the solubility and the solubility product of the slightly soluble electrolyte as well. Because of this, the decrease in solubility in the presence of a common ion is smaller than the one expected on the basis of theoretical calculations using concentrations. For a given concentration of diverse ions, the increase in solubility increases with the charge of the diverse ions and of the ions of the slightly soluble electrolyte.

The value of the solubility product K_{sp} is a function of the ionic strength μ of the solution, whereas the *thermodynamic solubility product*, K°_{sp}, which is calculated using activities, is independent of μ. The two constants are interrelated by the equation

$$K^{\circ}_{sp} = K_{sp}f',$$

where f′ is the *activity factor*, calculated from the activity coefficients of the ions. For example, for $BaSO_4$ we have

$$K^\circ_{sp} = a_{Ba^{2+}} a_{SO_4^{2-}} = f_{Ba^{2+}}[Ba^{2+}] f_{SO_4^{2-}}[SO_4^{2-}] = K_{sp} f_{Ba^{2+}} f_{SO_4^{2-}} = K_{sp} f' \qquad (6.9)$$

Example 6.10. Calculate the solubility product K_{sp} and the molar solubility of $BaSO_4$ in 0.010 M KNO_3 solution. Use activities (Equation 3.14).

Solution. The ionic strength of the solution is equal to

$$\mu = \frac{1}{2}[0.0100(1)^2 + 0.010(-1)^2] = 0.010.$$

Since $K^\circ_{sp} = K_{sp} f_{Ba^{2+}} f_{SO_4^{2-}}$, we have $\log K_{sp} = \log K^\circ_{sp} - \log f_{Ba^{2+}} - \log f_{SO_4^{2-}} = -8.82 + 0.51(2)^2\sqrt{0.010} + 0.51(-2)^2\sqrt{0.010} = -8.41$. Hence,

$$K_{sp} = 10^{-8.41} = \mathbf{3.9 \times 10^{-9}}, \text{ and } S_{BaSO_4} = \sqrt{3.9 \times 10^{-9}} = \mathbf{6.2} \times \mathbf{10^{-5}} \text{ mol/L}.$$

Note. The solubility product increased from 1.5×10^{-9} in a saturated solution of $BaSO_4$ ($\mu \simeq 0$) to 3.9×10^{-9} in a saturated $BaSO_4$ solution in 0.010 M KNO_3 ($\mu = 0.010$).

---

3. **Effect of pH.** The solubility of a salt of a weak acid depends on the pH of the solution. The H^+ ion combines with the anion of the salt to form the weak acid, thus increasing the solubility of the salt.

Example 6.11. a) Derive a relation giving the molar solubility of CaC_2O_4 as a function of its solubility product, the concentration of H^+ ion, and the dissociation constants of oxalic acid. b) Calculate the molar solubility of CaC_2O_4 in a solution, the $[H^+]$ of which is maintained at 1.2×10^{-4} M.

Solution. a) Let S be the molar solubility of CaC_2O_4 in a solution in which the H^+ concentration is maintained at a known value, $[H^+]$. We have the relations

$$\mathbf{CaC_2O_4} \rightleftharpoons Ca^{2+} + C_2O_4^{2-}, \; K_{sp} = [Ca^{2+}][C_2O_4^{2-}] \qquad (6.10)$$

$$H_2C_2O_4 \rightleftharpoons H^+ + HC_2O_4^-, \; K_1 = \frac{[H^+][HC_2O_4^-]}{[H_2C_2O_4]} \qquad (6.11)$$

$$HC_2O_4^- \rightleftharpoons H^+ + C_2O_4^{2-}, \; K_2 = \frac{[H^+][C_2O_4^{2-}]}{[HC_2O_4^-]} \qquad (6.12)$$

$$S = [Ca^{2+}] = [H_2C_2O_4] + [HC_2O_4^-] + [C_2O_4^{2-}] \qquad (6.13)$$

Equations (6.10), (6.11), (6.12), and (6.13) constitute a system of four equations with four unknowns, i.e., $[Ca^{2+}]$, $[H_2C_2O_4]$, $[HC_2O_4^-]$, and $[C_2O_4^{2-}]$. Solving this system we find $[Ca^{2+}]$, and consequently S as well, for a known $[H^+]$. Combining the above equations we have

$$\begin{aligned} S &= [Ca^{2+}] = \frac{[H^+]^2K_{sp}}{K_1K_2[Ca^{2+}]} + \frac{[H^+]K_{sp}}{K_2[Ca^{2+}]} + \frac{K_{sp}}{[Ca^{2+}]} = \frac{K_{sp}}{[Ca^{2+}]}\left(\frac{[H^+]^2}{K_1K_2} + \frac{[H^+]}{K_2} + 1\right) \\ &= \frac{K_{sp}}{S}\left(\frac{[H^+]^2 + K_1[H^+] + K_1K_2}{K_1K_2}\right) \end{aligned}$$

or

$$S = \sqrt{K_{sp}([H^+]^2 + K_1[H^+] + K_1K_2)/K_1K_2}. \tag{6.14}$$

Equation (6.14) gives the sought-for relation which combines the molar solubility S with the equilibrium constants K_{sp}, K_1, K_2, and $[H^+]$.

b) *1st method.* Substituting $[H^+] = 1.2 \times 10^{-4}$ M in Equation (6.14), we have

$$\begin{aligned} S &= \sqrt{\begin{array}{l} 1.3 \times 10^{-9}[(1.2 \times 10^{-4})^2 + (3.8 \times 10^{-2})(1.2 \times 10^{-4}) + \\ (3.8 \times 10^{-2})(5.0 \times 10^{-5})]/(3.8 \times 10^{-2})(5.0 \times 10^{-5}) \end{array}} \\ &= \mathbf{6.7 \times 10^{-5}}\ \text{mol/L}. \end{aligned}$$

2nd method. It is known that we can calculate $[C_2O_4{}^{2-}]$ as a function of the total concentration C of the oxalate species ($C = [Ca^{2+}] = S$). On the basis of Equations (5.49) and (5.56), we have

$$\begin{aligned} \alpha_2 &= \frac{[C_2O_4{}^{2-}]}{C} = \frac{[C_2O_4{}^{2-}]}{[Ca^{2+}]} \\ &= \frac{(3.8 \times 10^{-2})(5.0 \times 10^{-5})}{(1.2 \times 10^{-4})^2 + (3.8 \times 10^{-2})(1.2 \times 10^{-4}) + (3.8 \times 10^{-2})(5.0 \times 10^{-5})} = 0.293 \end{aligned}$$

from which

$$[C_2O_4{}^{2-}] = 0.293[Ca^{2+}] \tag{6.15}$$

Combining Equations (6.10) and (6.15), we have

$$0.293[Ca^{2+}]^2 = 1.3 \times 10^{-9},$$

from which $[Ca^{2+}] = \sqrt{1.3 \times 10^{-9}/0.293} = 6.7 \times 10^{-5}$ M. Hence, $S = [Ca^{2+}] =$ $\mathbf{6.7 \times 10^{-5}}$ mol/L.

---

4. Effect of hydrolysis. In the determination of the solubility of a slightly soluble salt in water, the hydrolysis of the basic anion should also be taken into consideration, because if it is ignored, the calculated solubility values often are several thousand times smaller than the real ones (Example 6.12). (The cation can also be hydrolyzed, e.g., $Fe^{3+} + H_2O \rightleftharpoons [FeOH]^{2+} + H^+$, or other reactions can take place between the ions of the salt and other constituents of the solution, such as complex formation. In these cases we have a system of multiple equilibria, and all the equilibria should be taken into account when calculating the solubility.) The magnitude of the error depends on the values of the solubility product and hydrolysis constants. The accurate calculation of the solubility is tedious, but it can usually be simplified by appropriate assumptions. For example, in the case of a highly insoluble salt, such as HgS, it is assumed that the dissolution of the salt does not affect appreciably the H^+ and OH^- concentrations of the solution, and that essentially we have $[H^+] = [OH^-] \simeq 10^{-7}$ $\underline{M}$ (Example 6.12). Conversely, in the case of a not too sparingly soluble salt, such as MnS, it is assumed that enough OH^- ions are produced from the hydrolysis of the salt (i.e., S^{2-}), so the contribution of OH^- ions resulting from the dissociation of water can be ignored as comparatively insignificant (Example 6.13).

Example 6.12. Calculate the molar solubility of HgS in water (take into account the hydrolysis of S^{2-} ions).

Solution. Since the molar solubility S_{HgS} of HgS is extremely small, essentially we have $[H^+] = 1 \times 10^{-7}$ $\underline{M}$. We have

$$\begin{aligned}\alpha_2 &= \frac{[S^{2-}]}{S_{HgS}} = \frac{K_1K_2}{[H^+]^2 + K_1[H^+] + K_1K_2} \\ &= \frac{(1.0 \times 10^{-7})(1.0 \times 10^{-14})}{(1.0 \times 10^{-7})^2 + (1.0 \times 10^{-7})(1.0 \times 10^{-7}) + (1.0 \times 10^{-7})(1.0 \times 10^{-14})} \\ &= 5.0 \times 10^{-8} = \frac{[S^{2-}]}{[Hg^{2+}]},\end{aligned}$$

from which

$$[S^{2-}] = 5.0 \times 10^{-8}[Hg^{2+}].$$

Hence,

$$K_{sp(HgS)} = 1 \times 10^{-50} = [Hg^{2+}][S^{2-}] = 5.0 \times 10^{-8}[Hg^{2+}]^2.$$

Consequently, $[Hg^{2+}] = S_{HgS} = \sqrt{1 \times 10^{-50}/5.0 \times 10^{-8}} = \mathbf{4.5 \times 10^{-22}}$ mol/L.

Note. Had we ignored the hydrolysis of S^{2-} ion, the solubility of HgS would have been found equal to $y = \sqrt{1 \times 10^{-50}} = 1 \times 10^{-25}$, i.e., about 4500 times smaller than the real one.

Example 6.13. Calculate the molar solubility of MnS in water (take into account the hydrolysis of S^{2-} ions).

Solution. Let S be the molar solubility of MnS. Since the hydrolysis of HS^- ions is negligible ($HS^- + H_2O \rightleftharpoons H_2S + OH^-$, $K_{h_2} = K_w/K_1 = 1.00 \times 10^{-14}/1.0 \times 10^{-7} = 1.0 \times 10^{-7}$) compared to the hydrolysis of S^{2-} ion ($S^{2-} + H_2O \rightleftharpoons HS^- + OH^-$, $K_{h_1} = K_w/K_2 = 1.00 \times 10^{-14}/1.0 \times 10^{-14} = 1.0$), we have

$$\underset{}{\mathbf{MnS}} + H_2O \rightleftharpoons \underset{S}{Mn^{2+}} + \underset{S}{HS^-} + \underset{S}{OH^-}$$

The equilibrium constant of the above reaction is equal to

$$\begin{aligned}K &= [Mn^{2+}][HS^-][OH^-] = S^3 = \frac{[Mn^{2+}][HS^-][OH^-][H^+][S^{2-}]}{[H^+][S^{2-}]} \\ &= \frac{K_{sp(MnS)}K_w}{K_2} = \frac{8 \times 10^{-14} \times 1.00 \times 10^{-14}}{1.0 \times 10^{-14}} = 8 \times 10^{-14},\end{aligned}$$

whereupon $S = \sqrt[3]{8 \times 10^{-14}} = \mathbf{4.3 \times 10^{-5}}$ mol/L.

Note. Had we ignored the hydrolysis of S^{2-} ion, the solubility of MnS would have been found equal to $y = \sqrt{8 \times 10^{-14}} = 2.8 \times 10^{-7}$, i.e., about 154 times smaller than the real one.

---

5. Effect of complexation. The solubility of a slightly soluble salt in water increases with complex formation, and it can be calculated on the basis of the concentration of the complexing agent, the solubility product of the salt, and the formation constants of the complex ions (Section 7.3).

6. Effect of other factors. The solubility of a slightly soluble salt also depends on the nature of the solvent, the temperature, and the particle size. Freshly precipitated colloidal particles have a higher solubility than larger particles formed after standing. (This is why in gravimetry small particles tend to dissolve and reprecipitate on the larger ones when allowed to stand, giving a more filterable precipitate.)

6.3 Fractional Precipitation and its Applications in Analytical Chemistry

The selective precipitation of a single ion or a group of ions from a mixture of ions is carried out by the technique of *fractional (controlled) precipitation*. This is achieved by adjusting the conditions so that the solubility product constant is exceeded for only one or for certain ions of the mixture, although all ions can react with the common reagent. Generally, during the gradual increase of the reagent concentration, the substance requiring the smallest concentration (quantity) of reagent is precipitated first (not necessarily the substance with the smallest solubility product—remember, the relative solubilities depend on the relative charges at a given K_{sp}). Further addition of reagent may result in the precipitation of other ions, simultaneously or after the completion of the precipitation of the first ion, depending on solubility differences of the precipitated substances and the concentrations of the ions in the mixture. A classic example of fractional precipitation is the separation of the cations of the second group from the cations of the third group by S^{2-} ion in acid medium ($[H^+] = 0.1 - 0.3$ $\underline{M}$), which is the basis of the systematic analysis of cations. Fractional precipitation is also used for the separation of various ions in the form of hydroxides (Problem 6.39), carbonates, chlorides, and so forth.

Example 6.14. To a solution 0.010 $\underline{M}$ in each of Tl^+, Pb^{2+}, and Ag^+ ions, chloride ions are added. a) What is the order of precipitation of the chlorides? b) What percentage of each ion has been precipitated when the next ion starts to precipitate?

Solution. a) We calculate the $[Cl^-]$ required to start the precipitation of each ion, using the solubility product constants of the chlorides, i.e.,

$$\text{for TlCl}: \quad [Cl^-] = 3.5 \times 10^{-4}/0.010 = 3.5 \times 10^{-2}\ \underline{M}$$

$$\text{for } PbCl_2: \quad [Cl^-] = \sqrt{1.6 \times 10^{-5}/0.010} = 4.0 \times 10^{-2}\ \underline{M},$$

and for

$$AgCl: \ [Cl^-] = 1.8 \times 10^{-10}/0.010 = 1.8 \times 10^{-8}\ \underline{M}.$$

Hence, AgCl is precipitated first, because it requires the smallest $[Cl^-]$, subsequently TlCl, and finally $PbCl_2$.

b) When $[Cl^-]$ exceeds 1.8×10^{-8} $\underline{M}$, AgCl starts to precipitate, and consequently the $[Ag^+]$ in the solution decreases and a larger $[Cl^-]$ is required to continue the precipitation. When $[Cl^-]$ exceeds 3.5×10^{-2} $\underline{M}$, TlCl starts to precipitate. At the start of the TlCl precipitation, the $[Ag^+]$ in the solution is equal to $1.8 \times 10^{-10}/3.5 \times 10^{-2} = 5.1 \times 10^{-9}$ $\underline{M}$, i.e., equal to $(5.1 \times 10^{-9}/0.010)100 = 5.1 \times 10^{-5}\%$ of the initial concentration. Hence, 99.99995% of Ag^+ ions have been precipitated, i.e., the precipitation of AgCl is essentially complete before Tl^+ ions start to precipitate. Similarly, we find that when $PbCl_2$ starts to precipitate, the $[Tl^+]$ in the solution is equal to $3.5 \times 10^{-4}/0.040 = 8.75 \times 10^{-3}$ $\underline{M}$. Hence, the percentage of Tl^+ ions that have not been precipitated when $PbCl_2$ starts to precipitate is equal to $(8.75 \times 10^{-3}/0.010)100 = 87.5\%$. Consequently, the separation of Tl^+ ions from Pb^{2+} ions in the given solution by fractional precipitation in the form of chlorides is not possible.

Example 6.15. A solution 0.0050 $\underline{M}$ in Pb^{2+}, 0.0050 $\underline{M}$ in Mn^{2+}, and 0.30 $\underline{M}$ in H^+ is saturated with H_2S (0.1 $\underline{M}$). a) Which metal sulfide precipitates first? b) What should the hydrogen ion concentration be for MnS to start to precipitate?

Solution. a) We have

$$[S^{2-}] = K_1K_2[H_2S]/[H^+]^2 = (1.0 \times 10^{-7})(1.0 \times 10^{-14})(0.1)/(0.30)^2 = 1.11 \times 10^{-21}\ \underline{M}.$$

Substituting this value in the K_{sp} expression for PbS, we have

$$(0.0050)(1.11 \times 10^{-21}) = 5.6 \times 10^{-24} > 8 \times 10^{-28} = K_{sp(PbS)}.$$

Hence, PbS precipitates. Similarly, for the Mn^{2+} ions we have

$$(0.0050)(1.11 \times 10^{-21}) = 5.6 \times 10^{-24} < 8 \times 10^{-14} = K_{sp(MnS)}.$$

Hence, MnS does not precipitate.

b) To precipitate MnS, we should have $[Mn^{2+}][S^{2-}] > K_{sp(MnS)}$, i.e.,

$$(0.0050)[S^{2-}] > 8 \times 10^{-14} \text{ or } [S^{2-}] \geq 1.6 \times 10^{-11}\underline{M}.$$

The required $[H^+]$, so that $[S^{2-}] = 1.6 \times 10^{-11}$ $\underline{M}$, is equal to

$$\sqrt{K_1K_2[H_2S]/[S^{2-}]}, \text{ or } [H^+] = \sqrt{1.0 \times 10^{-22}/1.6 \times 10^{-11}} = \mathbf{2.5 \times 10^{-6}}\ \underline{M}.$$

Example 6.16. A solution contains Cd^{2+} and Fe^{2+} ions, each at a concentration of 10.0 mg/mL. Calculate the permissible H^+ ion concentration range for the quantitative separation of Cd^{2+} (precipitation by 99.99%) from Fe^{2+}, in the form of CdS, by saturating the solution with H_2S.

Solution. We have initially

$$[Cd^{2+}] = (10.0 \text{ mg Cd/mL})/(112.4 \text{ mg/mmol}) = 0.089 \ \underline{M}$$

and

$$[Fe^{2+}] = (10.0 \text{ mg Fe/mL})/(55.85 \text{ mg/mmol}) = 0.179 \ \underline{M}.$$

In a saturated solution of H_2S we have

$$K_{H_2S} = [H^+]^2[S^{2-}] = 1.0 \times 10^{-22} \tag{6.16}$$

We also have

$$[S^{2-}] = K_{sp}/[M^{2+}] \quad (M^{2+} = Cd^{2+} \text{ or } Fe^{2+}) \tag{6.17}$$

Combining Equations (6.16) and (6.17), we have

$$[H^+] = \sqrt{\frac{1.0 \times 10^{-22}[M^{2+}]}{K_{sp}}} \tag{6.18}$$

For the quantitative precipitation of Cd^{2+}, we should have $[Cd^{2+}] = (0.089 \times 0.01)/100 = 8.9 \times 10^{-6}$ $\underline{M}$. Substituting this value in Equation (6.18), we have

$$[H^+] = \sqrt{\frac{1.0 \times 10^{-22} \times 8.9 \times 10^{-6}}{6 \times 10^{-27}}} = 0.385 \ \underline{M}.$$

Similarly, to prevent precipitation of Fe^{2+}, we should have

$$[H^+] = \sqrt{\frac{1.0 \times 10^{-22} \times 0.179}{5 \times 10^{-18}}} = 1.9 \times 10^{-3} \ \underline{M}.$$

Hence, we should have $0.0019 \ \underline{M} < [H^+] < 0.385 \ \underline{M}$.

Example 6.17. Calculate the concentrations of the ions in a solution obtained by mixing 25 mL of 0.30 $\underline{M}$ NaCl, 50 mL of 0.40 $\underline{M}$ NaI, and 50 mL of 0.50 $\underline{M}$ $AgNO_3$, after equilibration.

Solution. We mix

$$25 \text{ mL} \times 0.30 \text{ mmol NaCl/mL} = 7.5 \text{ mmol NaCl},$$

$$50 \text{ mL} \times 0.40 \text{ mmol NaI/mL} = 20 \text{ mmol NaI},$$

and

$$50 \text{ mL} \times 0.50 \text{ mmol } AgNO_3\text{/mL} = 25 \text{ mmol } AgNO_3.$$

Consequently, essentially all I^- ions and 5.0 mmol Cl^- will be precipitated, and 2.5 mmol Cl^- will be left in a total volume of 125 mL. Hence, we have

$$[Cl^-] = 2.5 \text{ mmol}/125 \text{ mL} = \mathbf{0.020} \ \underline{M},$$
$$[Ag^+] = K_{sp(AgCl)}/[Cl^-] = 1.8 \times 10^{-10}/0.020 = \mathbf{9.0 \times 10^{-9}} \ \underline{M},$$
$$[I^-] = K_{sp(AgI)}/[Ag^+] = 8.5 \times 10^{-17}/9.0 \times 10^{-9} = \mathbf{9.4 \times 10^{-9}} \ \underline{M},$$
$$[Na^+] = (7.5 + 20) \text{ mmol}/125 \text{ mL} = \mathbf{0.22} \ \underline{M},$$
$$\text{and} \quad [NO_3^-] = 25 \text{ mmol}/125 \text{ mL} = \mathbf{0.20} \ \underline{M}.$$

6.4 Combination of Equilibrium Constants

The equilibrium constants for many reactions are not found in tables. However, they can often be calculated by combining other known equilibrium constants, such as dissociation constants, instability constants, solubility product constants, the ion product of water, etc. The calculation can be made with either one of the following two methods.

Method A. If a chemical equation can be represented as the sum or the difference of two equations, the equilibrium constant of the respective reaction is equal to the product or the quotient, respectively, of the equilibrium constants corresponding to these two equations.

Method B. We write the expression for the equilibrium constant for the given reaction, and by appropriate mathematical transformation, e.g., by multiplying the numerator and the denominator of a quotient by the same factor, we correlate it with known equilibrium constants.

Example 6.18. Calculate the equilibrium constant for the reaction between silver chromate and Cl^- ions.

Solution. The equation of the reaction is

$$\mathbf{Ag_2CrO_4} + 2Cl^- \rightleftharpoons 2\mathbf{AgCl} + CrO_4^{2-}, \ K = \frac{[CrO_4^{2-}]}{[Cl^-]^2} \tag{6.19}$$

According to method A, the above reaction is split into two reactions with known equilibrium constants, i.e.,

$$\mathbf{Ag_2CrO_4} \rightleftharpoons 2Ag^+ + CrO_4^{2-}, \ K_{sp} = [Ag^+]^2[CrO_4^{2-}] = 1.9 \times 10^{-12} \tag{6.20}$$

$$2\mathbf{AgCl} \rightleftharpoons 2Ag^+ + 2Cl^-, K_{sp}^2 = [Ag^+]^2[Cl^-]^2 = (1.8 \times 10^{-10})^2 \tag{6.21}$$

Equation (6.19) is equal to the difference between (6.20) and (6.21), whereas the equilibrium constant K is equal to the quotient of the equilibrium constants for (6.20) and (6.21), i.e.,

$$K = \frac{1.9 \times 10^{-12}}{(1.8 \times 10^{-10})^2} = \mathbf{5.9 \times 10^7}.$$

According to method B, we multiply the numerator and the denominator of Equation (6.19) by $[Ag^+]^2$, whereupon we have

$$K = \frac{[CrO_4^{2-}][Ag^+]^2}{[Cl^-]^2[Ag^+]^2} = \frac{K_{sp(Ag_2CrO_4)}}{K^2_{sp(AgCl)}} = \frac{1.9 \times 10^{-12}}{(1.8 \times 10^{-10})^2} = \mathbf{5.9 \times 10^{-7}}.$$

Note. Method B has been used for the calculation of hydrolysis constants (Section 6.2, Example 6.13). This method is also used for the calculation of the equilibrium constants for dissolution reactions of precipitates by complex formation (Chapter 7).

Example 6.19. After treating 0.1433 g of AgCl with 5.00 mL of 1.5 M Na_2CO_3 solution, it was found that the solution contained 0.0026 g Cl^-/L. a) Calculate the solubility product of AgCl. b) What percentage of AgCl was converted to Ag_2CO_3?

Solution. a) The equation for the reaction is

$$\mathbf{2AgCl} + CO_3^{2-} \rightleftharpoons \mathbf{Ag_2CO_3} + 2Cl^- \tag{6.22}$$

for which we have

$$K = \frac{[Cl^-]^2}{[CO_3^{2-}]} \tag{6.23}$$

We have $[Cl^-] = (0.0026 \text{ g } Cl^-/L)/(35.45 \text{ g/mol}) = 7.33 \times 10^{-5}$ M, and $[CO_3^{2-}] = 1.5 - (7.33 \times 10^{-5}/2) \simeq 1.5$ M. Multiplying the numerator and the denominator of Equation (6.23) by $[Ag^+]^2$, we have

$$K = \frac{[Cl^-]^2[Ag^+]^2}{[CO_3^{2-}][Ag^+]^2} = \frac{K^2_{sp(AgCl)}}{8.2 \times 10^{-12}} = \frac{(7.33 \times 10^{-5})^2}{1.5},$$

from which

$$K_{sp(AgCl)} = \mathbf{1.7 \times 10^{-10}}.$$

b) Initially, we have 0.1433 g AgCl/(0.1433 g AgCl/mmol) = 1.00 mmol AgCl. After the completion of the reaction, the solution contains

$$(5.00 \text{ mL})(7.33 \times 10^{-5} \text{ mmol } Cl^-/\text{mL}) = 3.67 \times 10^{-4} \text{ mmol } Cl^- \equiv 3.67 \times 10^{-4} \text{ mmol AgCl}.$$

Hence, AgCl was converted to Ag_2CO_3 by

$$\frac{3.67 \times 10^{-4}}{1.00} \times 100 = \mathbf{0.0367\%}.$$

6.5 Problems

Solubility product

6.1. The solubility of Ag_2CrO_4 in water is 0.00259 g/100mL. Calculate the solubility product of Ag_2CrO_4 .

6.2. Calculate the solubility product of PbS, on the basis of the following data:

1. $\mathbf{PbSO_4} \rightleftharpoons Pb^{2+} + SO_4^{2-}$, $K_{sp(PbSO_4)} = 1.3 \times 10^{-8}$
2. $\mathbf{PbSO_4} + 2I^- \rightleftharpoons \mathbf{PbI_2} + SO_4^{2-}$, $K_2 = 1.57$
3. $\mathbf{PbI_2} + CrO_4^{2-} \rightleftharpoons \mathbf{PbCrO_4} + 2I^-$, $K_3 = 4.15 \times 10^{-5}$
4. $\mathbf{PbS} + CrO_4^{2-} \rightleftharpoons \mathbf{PbCrO_4} + S^{2-}$, $K_4 = 4.0 \times 10^{-14}$

6.3. Solid $SrSO_4$ is shaken with a 0.0010 M K_2SO_4 solution until equilibration, whereupon it is found that the solution contains 0.042 g $SrSO_4$/L. Calculate the solubility product of $SrSO_4$.

6.4. A 25.0-mL portion of 0.0100 M MCl_2 solution is mixed with 25.0 mL of a 0.100 M NH_3-NH_4Cl buffer. Calculate the solubility product of $M(OH)_2$, given that 25.0 mL of the buffer should contain at least 0.75 g of NH_4Cl to prevent precipitation of $M(OH)_2$ upon mixing of the two solutions.

6.5. Express the solubility product of the following slightly soluble salts as a function of their molar solubility S: a) Ag_2S, b) $PbCl_2$, c) $PbSO_4$, d) PbClF, e) $Ca_3(PO_4)_2$ (disregard hydrolysis).

Solubility

6.6. Calculate the molar solubility S for each of the following slightly soluble salts, as a function of its solubility product K_{sp}: a) MX_2, b) MX_4, c) MY, d) M_2Y_3, e) M_3Z_2, where M = metal ion and X, Y, and Z = monovalent, divalent, and trivalent anion, respectively.

6.7. Calculate the molar solubility of $BaSO_4$ in a 1.00 M HNO_3 solution.

6.8. How many milligrams of $PbCl_2$ will dissolve in 500 mL of water at 25°C?

6.9. Complete the following table:

Substance	Solubility, g/100 mL	K_{sp}
AgBr		5×10^{-13}
$Ag_2C_2O_4$		1×10^{-11}
$Ca_3(PO_4)_2$	1.22×10^{-4}	
$PbBr_2$	0.384	
$MgNH_4PO_4$		2.0×10^{-13}

6.10. In passing H_2S through 100 mL of saturated $PbBr_2$ solution (at 25°C), 0.250 g of PbS precipitate was formed. Calculate the molar solubility and the solubility product of $PbBr_2$.

6.11. How many grams of ZnS are dissolved in one liter of a solution having $[H^+]$ = 1.00 M?

6.12. Derive equations giving the solubility of the following salts as a function of $[H^+]$, the solubility product of the salt, and of the dissociation constants of the respective acids: a) CH_3COOAg, b) Ag_2CO_3, c) Ag_3PO_4, d) CaF_2, e) $BiPO_4$.

6.13. Calculate the pH at which the solubility of CaF_2 is equal to 100 ppm (disregard the formation of H_2F_2).

6.14. What is the solubility, in mg/100 mL, of the oxalate salts of Ca, Sr, and Ba in a 0.25 M CH_3COOH solution?

Precipitation

6.15. Solid NaOH is added to a 0.0100 M solution of the salt XCl_3 until the pH becomes 6.50. a) Will a precipitate be formed (atomic weight of X = 52.0, solubility of $X(OH)_3$ = 0.13 μg/100 mL). b) Identify element X.

6.16. Sodium sulfide is added to a Mn^{2+} solution, until C_{Na_2S} = 0.100 M. Which compound will be precipitated first, MnS or $Mn(OH)_2$?

6.17. An aqueous solution which is 0.010 M in $ZnCl_2$, 0.10 M in CH_3COOH, and 0.62 M in CH_3COONa is saturated with H_2S and maintained as such. What percentage of zinc is not precipitated?

6.18. The molecular weight of the hydroxide $M(OH)_2$ is 242 and its solubility is 2.42 mg/L. Solid NaOH is added to 100 mL of 0.100 M M^{2+} until the pH becomes 8.80. What percentage of M^{2+} is not precipitated?

6.19. How many grams of K_2SO_4 should be added to one liter of a saturated $SrSO_4$ solution, so that $[SO_4^{2-}] = 2[Sr^{2+}]$? (Assume that S exists only as SO_4^{2-}.)

6.20. How many grams of NH_4Cl should be added to a mixture of 20.0 mL of 0.050 M $Mg(NO_3)_2$ solution and 50.0 mL of 0.50 M NaOH solution so that the precipitate formed is dissolved?

6.21. A 0.100 M $ZnCl_2$ solution is saturated with H_2S (0.10 M). What percentage of zinc is left in the solution?

Common ion effect

6.22. Calculate the molar solubility of Ag_2CrO_4 a) in a 0.100 M Na_2CrO_4 solution, b) in a 0.100 M $AgNO_3$ solution (disregard the hydrolysis of chromate ions).

6.23. Eight grams of $AgNO_3$ and 12.0 g $Na_2C_2O_4$ are dissolved in water and the solution is diluted to 200 mL. Calculate the concentrations of the ions after equilibration (assume $[C_2O_4^{2-}] = C_{Na_2C_2O_4}$).

6.24. A 25.0-mL portion of 0.050 M K_2CrO_4 solution is mixed with 25.0 mL of 0.120 M $AgNO_3$ solution. Calculate a) the molar solubility of Ag_2CrO_4 in the solution, b) the molarities of Ag^+, CrO_4^{2-}, K^+, and NO_3^- ions.

6.25. A 25.0-mL portion of 0.040 M $MgCl_2$ solution is mixed with 25.0 mL of 0.050 M $AgNO_3$ solution. Calculate the molarities of Mg^{2+}, Cl^-, Ag^+, and NO_3^- ions in the resulting solution.

6.26. What are the concentrations of the ions in a solution obtained by mixing 50 mL of 0.060 $\underline{M}$ $Mg(NO_3)_2$ solution with 25 mL of 0.090 $\underline{M}$ NaOH solution, at the equilibrium state?

Diverse ion effect

6.27. Calculate the molar solubility of $BaSO_4$ in a 0.100 $\underline{M}$ KNO_3 solution. Use activities (Equation 3.14).

Effect of pH on solubility

6.28. Calculate the molar solubility of AgCN in a solution, the pH of which is maintained at 7.00.

6.29. Calculate the molar solubility of ZnC_2O_4 in a solution, the pH of which is maintained at a) 4.00, b) 5.00, c) 7.00. What conclusion can be drawn with respect to the effect of pH on the solubility of ZnC_2O_4?

6.30. Calculate the molar solubility of CaF_2 in a solution, the pH of which is maintained at 4.00 (disregard the formation of H_2F_2).

6.31. Calculate the solubility of zinc oxalate, in mg/100 mL, in a solution, the pH of which is maintained at 5.00.

6.32. Calculate the molar solubility of CaF_2 in a 0.0020 $\underline{M}$ HCl solution.

6.33. Calculate the molar solubility of PbS in a solution whose $[H^+]$ is maintained at 1.00×10^{-4} $\underline{M}$.

Effect of hydrolysis on solubility

6.34. Calculate the molar solubility of PbS in water (take into account the hydrolysis of S^{2-}).

6.35. Calculate the solubility of $CoCO_3$ in water, in mg/100 mL (take into account the hydrolysis of CO_3^{2-}).

6.36. Calculate the molar solubilities of the following compounds in water, taking also into account the hydrolysis of the anion: a) $BaCO_3$, b) $PbCrO_4$, c) Ag_2CO_3.

Fractional precipitation

6.37. Calculate the required concentration of NH_4^+ ions to prevent precipitation of $MgCO_3$ from a solution containing 0.30 mol of NH_3, 0.0125 mol of Mg^{2+}, and 0.080 mol of HCO_3^- in one liter.

6.38. A solution given for the analysis of the fourth group of cations contains also 20 ppm of Mg^{2+}. a) Assuming that the pH is adjusted with an NH_3-NH_4Cl buffer, what should the pH be so that Mg^{2+} does not start to precipitate? b) If by mistake the pH is adjusted to 10.50, what percentage of Mg^{2+} will be precipitated?

6.39. A solution is 0.010 $\underline{M}$ each in Fe^{3+} and Mg^{2+}. Calculate the permissible concentration range of OH^- for the quantitative separation (by 99.9%) of Fe^{3+} from Mg^{2+} in the form of $Fe(OH)_3$.

6.40. $CrO_4{}^{2-}$ ions are added to a solution containing 0.030 $\underline{M}$ Ba^{2+}, 0.00030 $\underline{M}$ Sr^{2+} and 0.010 $\underline{M}$ Ca^{2+}. a) What is the order of precipitation of the chromates? b) What percentage of each ion has been precipitated when the next ion starts to precipitate?

6.41. Cl^- ions are added to a solution that contains 0.030 $\underline{M}$ Ag^+ and 0.010 $\underline{M}$ Pb^{2+}. What is the concentration of the Ag^+ ion left in the solution when $PbCl_2$ starts to precipitate?

6.42. Solid $AgNO_3$ is added to a solution which is 0.0090 $\underline{M}$ in $CrO_4{}^{2-}$ and 0.090 $\underline{M}$ in Cl^-. a) Which precipitate will be precipitated first? b) What percentage of Cl^- has been precipitated when Ag_2CrO_4 starts to precipitate?

6.43. A solution contains $CrO_4{}^{2-}$ and Cl^- ions in a concentration ratio of 1250:1. Which salt will be precipitated first upon the addition of Ag^+ ions, if the initial $[CrO_4{}^{2-}]$ is a) 0.0010 $\underline{M}$, b) 0.10 $\underline{M}$?

6.44. A solution is 0.010 $\underline{M}$ in X^{2+} and 0.0010 $\underline{M}$ in Y^+. Calculate the minimum value of the ratio $K_{sp(XS)}/K_{sp(Y_2S)}$ so that the quantitative separation of Y^+ (decrease of its concentration to 1.0×10^{-6} $\underline{M}$) is feasible, by precipitation in the form of Y_2S without coprecipitation of X^{2+}.

6.45. What is the value of the ratio $[Ba^{2+}]/[Ca^{2+}]$ in a solution saturated with both $BaCrO_4$ and $CaCrO_4$?

6.46. A solution containing 0.050 $\underline{M}$ X^{2+} and 0.050 $\underline{M}$ Z^{2+} is acidified until $[H^+] =$ 0.30 M, and is saturated with H_2S (0.10 $\underline{M}$). What is the minimum value of the ratio $K_{sp(XS)}/K_{sp(ZS)}$ so that $[Z^{2+}]$ can be decreased to 1.00×10^{-6} $\underline{M}$ without coprecipitation of XS?

6.47. A solution 0.010 $\underline{M}$ each in Fe^{3+}, Ni^{2+}, and Mg^{2+} ions is acidic at pH 0.5. The pH is increased gradually by adding a base. Calculate the pH values at which each ion starts to precipitate as the hydroxide, and the pH at which the metal ion is 1.0×10^{-6} $\underline{M}$. Is it theoretically possible to separate these three cations by fractional precipitation of their hydroxides?

6.48. A solution is 0.10 $\underline{M}$ in NH_3, 0.50 $\underline{M}$ in $NH_4{}^+$, and 0.10 $\underline{M}$ in $HCO_3{}^-$. a) What is the concentration of $CO_3{}^{2-}$ ions in the solution? b) If a solid mixture of $BaCl_2$ and $MgCl_2$ is added to the solution so that $[Ba^{2+}] = [Mg^{2+}] = 0.020$ $\underline{M}$, will $BaCO_3$, $MgCO_3$, and $Mg(OH)_2$ be precipitated?

6.49. Calculate the concentrations of Ba^{2+}, Ca^{2+}, and $CrO_4{}^{2-}$ ions in a saturated solution which is at equilibrium with solid $BaCrO_4$ and $CaCrO_4$.

6.50. A solution is 0.010 M in Zn^{2+} and 0.0010 M in Mn^{2+}. a) Calculate the permissible pH range for the quantitative separation of Zn^{2+} (decrease of $[Zn^{2+}]$ to less than 1.0×10^{-6} M) in the form of ZnS by saturating the solution with H_2S. b) What is the minimum permissible pH for the quantitative precipitation of manganese in the form of MnS?

6.51. Calculate how many milligrams of Cu^{2+} and Zn^{2+} will be left in one liter of a solution that is 0.30 M in H^+, and is maintained saturated with H_2S (0.10 M) (disregard hydrolysis).

Combination of equilibrium constants

6.52. With how many milliliters of 0.50 M Na_2CO_3 solution should 0.100 g of CaC_2O_4 be treated so that it is converted quantitatively to $CaCO_3$?

6.53. Calculate the quantity of PbS converted to $PbCO_3$, when treating an excess of solid PbS with 10.0 mL of 1.5 M Na_2CO_3 solution.

6.54. Ten milliliters of 1.5 M $NaCO_3$ solution are shaken with 2.00 g of $BaSO_4$. Calculate the percentage of $BaSO_4$ converted to $BaCO_3$.

6.55. What is the minimum required concentration of Na_2CO_3 in solution A, so that by shaking 2.00 g of $BaSO_4$ with 20.0 mL of solution A, $BaSO_4$ is converted quantitatively to $BaCO_3$?

6.56. Calculate the equilibrium constant for the reaction

$$\mathbf{Ag_2CO_3} + 2I^- \rightleftharpoons \mathbf{2AgI} + CO_3^{2-}$$

Chapter 7

Equilibria Involving Complex Ions

7.1 Instability Constant of a Complex Ion

Complex ions are dissociated in a stepwise fashion, just as weak polyprotic acids are. Each step is characterized by an equilibrium constant, called the *instability constant* (K_{inst}), because it is a measure of the instability of the complex. The reciprocal of the instability constant is called the *formation constant* or *stability constant* (K_f), and represents the reverse equilibrium, i.e., the formation of the complex. Both formation constants and instability constants are used in this text, but the former prevails in the literature. The product of the *successive* (stepwise) instability (or stability) constants equals the *overall* instability (or stability) constant of the complex ion. For the general use of the dissociation of the complex ion $[ML_n]^{a-nb}$, we have

$$[ML_n]^{a-nb} \rightleftharpoons M^{a+} + nL^{b-}, \quad K_{inst} = \frac{[M^{a+}][L^{b-}]^n}{[ML_n^{a-nb}]} = \frac{1}{K_f} \tag{7.1}$$

As an example, for the dissociation of the complex ion $[Ag(NH_3)_2]^+$ we can write

$$[Ag(NH_3)_2]^+ \rightleftharpoons [Ag(NH_3)]^+ + NH_3, \quad K_{inst_1} = \frac{[Ag(NH_3)^+][NH_3]}{[Ag(NH_3)_2{}^+]} = 1.18 \times 10^{-4} = \frac{1}{K_{f_2}} \tag{7.2}$$

$$[Ag(NH_3)]^+ \rightleftharpoons Ag^+ + NH_3, \quad K_{inst_2} = \frac{[Ag^+][NH_3]}{[Ag(NH_3)^+]} = 5.0 \times 10^{-4} = \frac{1}{K_{f_1}} \tag{7.3}$$

Addition of chemical equations (7.2) and (7.3) and multiplication of the corresponding equilibrium constants yields the overall equations, that is

$$[Ag(NH_3)_2]^+ \rightleftharpoons Ag^+ + 2NH_3, \quad K_{inst} = \frac{[Ag^+][NH_3]^2}{[Ag(NH_3)_2{}^+]} = 5.9 \times 10^{-8} = \frac{1}{K_f} \tag{7.4}$$

It should be noticed that Equation (7.4) does not include all the ions present in the solution. The complex ion $[Ag(NH_3)]^+$ is present in a solution of $[Ag(NH_3)_2]X$ (X = anion), but it does not appear in Equation (7.4).

Since the successive dissociation constants often are similar in magnitude (as opposed to successive ionization constants of polyprotic acids), no instability constant can then be considered negligible in calculations. However, in actual practice a large excess of complexing agent is used in complexation reactions (for example, in forming a colored complex or for masking a metal ion by complexation to prevent its interference in a separate analytical reaction) so that the metal ion is present almost quantitatively in the form of the complex with the largest number of ligands. Therefore, in such calculations the overall instability constant is used. Of course, in titrations the complexing agent is not in excess until beyond the equivalence point.

Example 7.1. A solution 0.020 $\underline{M}$ in $[Ag(NH_3)_2]^+$ and 0.20 $\underline{M}$ in NH_3 is mixed with an equal volume of 2.0×10^{-6} $\underline{M}$ NaI. Will AgI precipitate?

Solution. In the mixing, the concentrations of the various substances are halved. Thus, before reaction, $[Ag(NH_3)_2{}^+] = 0.010$ $\underline{M}$, $[NH_3] = 0.10$ $\underline{M}$, and $[I^-] = 1.0 \times 10^{-6}$ $\underline{M}$. If $[Ag^+] = x$, then at equilibrium $[Ag(NH_3)_2{}^+] = 0.010 - x \simeq 0.010$ $\underline{M}$, and $[NH_3] = 0.10 + 2x \simeq 0.10$ $\underline{M}$. Substituting these concentrations in Equation (7.4), we have

$$x(0.10)^2/0.010 = 5.9 \times 10^{-8}$$

or

$$x = [Ag^+] = 5.9 \times 10^{-8} \text{ } \underline{M}.$$

Therefore, we have

$$[Ag^+][I^-] = (5.9 \times 10^{-8})(1.0 \times 10^{-6}) = 5.9 \times 10^{-14} > 8.5 \times 10^{-17} = K_{sp(AgI)}.$$

Hence, since the ion product exceeds K_{sp}, AgI will precipitate.

Example 7.2. Assuming that the salt $[Ag(NH_3)_2]Cl$ exists as a solid, is it possible to prepare a 0.010 $\underline{M}$ aqueous solution of this salt, without precipitating AgCl (assume $[Ag(NH_3)^+] \simeq 0$)?

Solution. During the dissolution of $[Ag(NH_3)_2]Cl$ in water we have

$$[Ag(NH_3)_2]Cl \rightarrow [Ag(NH_3)_2]^+ + Cl^- \qquad (a)$$

$$[Ag(NH_3)_2]^+ \rightleftharpoons Ag^+ + 2NH_3 \qquad (b)$$

and possibly

$$Ag^+ + Cl^- \rightleftharpoons \mathbf{AgCl}. \qquad (c)$$

From the salt, $[Cl^-] = 0.010$ $\underline{M}$. If $[Ag^+] = x$, then from (*b*), $[NH_3] = 2x$ and $[Ag(NH_3)_2{}^+] = 0.010 - x$. Substituting these concentrations in Equation (7.4), we have

$$\frac{x(2x)^2}{0.010-x} = 5.9 \times 10^{-8}.$$

From the relatively small value of K_{inst} it can be reasonably assumed that the dissociation of the complex $[Ag(NH_3)_2]^+$ is small. Therefore, we will assume that $x << 0.010$. Then $0.010-x \simeq 0.010$ and $x = [Ag^+] = \sqrt[3]{5.9 \times 10^{-10}/4} = 5.3 \times 10^{-4}$ $\underline{M}$ (our assumption was reasonable). Hence,

$$[Ag^+][Cl^-] = (5.3 \times 10^{-4})(0.010) = 5.3 \times 10^{-6} > 1.8 \times 10^{-10} = K_{sp(AgCl)}.$$

Thus, since the ion product exceeds K_{sp}, part of the silver will precipitate as AgCl. Therefore, it is impossible to prepare a 0.010 M $[Ag(NH_3)_2]Cl$ solution from stoichiometric amounts of $Ag(NH_3)_2^+$ and Cl^-.

Example 7.3. 25 mL of 0.0020 M $AgNO_3$ is mixed with 25 mL of xM KCN. The mixture contains an excess of CN^- ions. The minimum concentration of I^- ions needed to start the precipitation of AgI is 1.0 M. Calculate the value of x.

Solution. We have $[Ag^+] = K_{sp(AgI)}/[I^-] = 8.5 \times 10^{-17}/1.0 = 8.5 \times 10^{-17}$ M and $[Ag(CN)_2^-] = (0.0020/2) - 8.5 \times 10^{-17} \simeq 0.0010$ M. Hence,

$$K_{inst[Ag(CN)_2]^-} = 1 \times 10^{-20} = \frac{[Ag^+][CN^-]^2}{[Ag(CN)_2^-]} = \frac{8.5 \times 10^{-17}[CN^-]^2}{0.0010}$$

or $[CN^-] = 3.4 \times 10^{-4}$ M. The 50-mL mixture contains (50 mL)(3.4×10^{-4} mmol/mL) = 0.017 mmol of free CN^- and (50 mL)(0.0010 mmol/mL) = 0.050 mmol $[Ag(CN)_2]^-$, which is equivalent to 0.100 mmol CN^-. Hence, the initial 25 mL of KCN solution contains a total of 0.017 + 0.100 = 0.117 mmol KCN. Therefore,

$$x = 0.117/25 = \mathbf{0.0047}\ \text{M}.$$

Example 7.4. A solution is 0.010 M in HA (a weak organic acid with $K_a = 1.0 \times 10^{-5}$) and 1.0×10^{-6} M in MCl_3. Assume that only a single complex MA_3 is formed ($K_{inst} = 1.0 \times 10^{-10}$) according to the reaction

$$M^{3+} + 3HA \rightleftharpoons MA_3 + 3H^+.$$

Calculate the ratio of complexed to uncomplexed metal ion, $[MA_3]/[M^{3+}]$, in solutions buffered at a) pH 5.00, b) pH 1.00. What conclusion can be drawn from the results concerning the possibility of controlling the extent of complexing through control of pH?

Solution. *1st method.* a) Since the concentration of the salt MCl_3 is very small compared to that of HA, $[MA_3]$ can be omitted, as negligible in consuming HA. The fraction of HA that exists as A^- is denoted $\alpha_1 (= [A^-]/C_{HA})$. Then at pH 5.00 (see chapter 5),

$$\alpha_1 = \frac{[A^-]}{0.010} = \frac{K_a}{[H^+] + K_a} = \frac{1.0 \times 10^{-5}}{(1.0 \times 10^{-5}) + (1.0 \times 10^{-5})} = 0.50.$$

Hence, $[A^-] = 0.0050$ M, and

$$K_{inst} = 1.0 \times 10^{-10} = \frac{[M^{3+}][A^-]^3}{[MA_3]} = \frac{[M^{3+}](0.0050)^3}{[MA_3]}$$

or

$$\frac{[MA_3]}{[M^{3+}]} = \frac{(0.0050)^3}{1.0 \times 10^{-10}} = \mathbf{1.25 \times 10^3}.$$

b) Similarly, at pH 1.00

$$\alpha_1 = \frac{[A^-]}{0.010} = \frac{1.0 \times 10^{-5}}{0.10 + 1.0 \times 10^{-5}} = 1.0 \times 10^{-4}.$$

Hence,

$$[A^-] = 1.0 \times 10^{-6}\ \underline{M}$$

and

$$\frac{[MA_3]}{[M^{3+}]} = \frac{(1.0 \times 10^{-6})^3}{1.0 \times 10^{-10}} = \mathbf{1.0 \times 10^{-8}}.$$

From these results it is concluded that the extent of complexing can be controlled through control of pH. Thus, at pH 1.00 the extent of complexing is negligible, whereas at pH 5.00 it is almost quantitative.

2nd method. a) We have $[A^-] = 0.010 - [HA] - 3[MA_3]$ (mass balance for A), or, since $[MA_3]$ is negligible, $[A^-] = 0.10 - [HA]$. Hence, at pH 5.00

$$K_a = 1.0 \times 10^{-5} = \frac{[H^+][A^-]}{[HA]} = \frac{(1.0 \times 10^{-5})(0.010 - [HA])}{[HA]}.$$

Therefore, $[HA] = 0.0050\ \underline{M}$.

For the reaction

$$M^{3+} + 3HA \rightleftharpoons MA_3 + 3H^+$$

we have

$$K = \frac{[MA_3][H^+]^3}{[M^{3+}][HA]^3} = \frac{[MA_3][H^+]^3}{[M^{3+}][HA]^3} \cdot \frac{[A^-]^3}{[A^-]^3} = \frac{K_a^3}{K_{inst}} = \frac{(1.0 \times 10^{-5})^3}{1.0 \times 10^{-10}} = 1.0 \times 10^{-5}.$$

Then

$$\frac{[MA_3]}{[M^{3+}]} = 1.0 \times 10^{-5} \cdot \frac{[HA]^3}{[H^+]^3} = 1.0 \times 10^{-5} \frac{(0.0050)^3}{(1.0 \times 10^{-5})^3} = \mathbf{1.25 \times 10^{-3}}.$$

b) At pH 1.00 we have

$$1.0 \times 10^{-5} = \frac{(0.10)(0.010 - [HA])}{[HA]}.$$

Therefore, $[HA] \simeq 0.010\ \underline{M}$. Then

$$\frac{[MA_3]}{[M^{3+}]} = 1.0 \times 10^{-5} \cdot \frac{(0.010)^3}{(0.100)^3} = \mathbf{1.0 \times 10^{-8}}.$$

Example 7.5. A solution is 0.0200 $\underline{M}$ in HA (a weak acid with $K_a = 2.00 \times 10^{-4}$) and contains a trace of the salt MNO_3, which is totally ionized. When the solution is buffered at pH 4.00, 0.250% of M(I) is in the uncomplexed form. At pH 3.00, 1.375% of M(I) is uncomplexed. Assuming that only a single complex is formed, $[MA_n]^{1-n}$, calculate n and the instability constant of the complex.

Solution. Since the quantity of MNO_3 is extremely small, $[MA_n^{1-n}]$ is considered negligible in comparison with [HA] and $[A^-]$. Then at pH 4.00

$$\alpha_1 = \frac{[A^-]}{C_{HA}} = \frac{[A^-]}{0.0200} = \frac{K_a}{[H^+] + K_a} = \frac{1.00 \times 10^{-4}}{(1.0 \times 10^{-4}) + (1.00 \times 10^{-4})} = 0.50.$$

Therefore, $[A^-] = 0.0100$ M. Similarly, at pH 3.00

$$[A^-] = 0.0200 \cdot \frac{1.00 \times 10^{-4}}{(1.0 \times 10^{-3}) + (1.00 \times 10^{-4})} = 0.00182 \text{ M}.$$

Let C_M be the analytical concentration of the metal ion. Since only a single complex is formed, $C_M = [M^+] + [MA_n^{1-n}]$. Therefore,

$$\frac{[M^+]}{[MA_n^{1-n}]} = \frac{K_{inst}}{[A^-]^n} = \frac{[M^+]}{C_M - [M^+]}.$$

Then at pH 4.00, we have

$$\frac{0.250}{100 - 0.250} = \frac{K_{inst}}{(0.0100)^n} \tag{7.5}$$

and at pH 3.00

$$\frac{1.375}{100 - 1.375} = \frac{K_{inst}}{(0.00182)^n} \tag{7.6}$$

Dividing Equations (7.5) and (7.6), we have

$$\frac{0.250/99.750}{1.375/98.625} = (0.00182/0.0100)^n.$$

Therefore, $n = 1.007 \simeq \mathbf{1}$.

Substituting the value of n in Equation (7.5) or (7.6), we find the value of K_{inst}. For example, from Equation (7.5)

$$K_{inst} = \frac{0.250}{100 - 0.250}(0.0100)^1 = \mathbf{2.5 \times 10^{-5}}.$$

Example 7.6. A 25.0-mL portion of 0.0100 M $CaCl_2$ is mixed with 75.0 mL of 0.0086 M EDTA (ethylenediaminetetracetic acid, $(HO_2CCH_2)_2NCH_2CH_2N(CH_2CO_2H)_2$, represented by H_4Y—in fact, the disodium salt of the acid, Na_2H_2Y, is used), and the mixture is buffered at pH 10.00. Calculate $[Ca^{2+}]$ in the mixture.

Solution. 25.0 mL×0.0100 mmol/mL = 0.250 mmol $CaCl_2$ is mixed with 75.0 mL × 0.0086 mmol/mL = 0.645 mmol EDTA. Since there is a large excess of EDTA, we will assume that calcium is complexed quantitatively, and check this at the end of our calculation. Since $[H_2Y^{2-}]$, $[H_3Y^-]$ and $[H_4Y]$ are negligible at pH 10.00 (see Chapter 14), after mixing we have

$$\text{Mass balance (Ca)}: \ 0.00250 = [Ca^{2+}] + [CaY^{2-}] \simeq [CaY^{2-}]$$

$$\text{Mass balance (EDTA)}: \ 0.00645 = [HY^{3-}] + [Y^{4-}] + [CaY^{2-}]$$

Hence, from the fourth acidity constant for EDTA,

$$K_4 = 5.5 \times 10^{-11} = \frac{(1.0 \times 10^{-10})[Y^{4-}]}{[HY^{3-}]}.$$

Therefore, $[HY^{3-}] = 1.82\ [Y^{4-}]$. Then

$$0.00645 = 1.82\ [Y^{4-}] + [Y^{4-}] + 0.00250,$$

and

$$[Y^{4-}] = 1.40 \times 10^{-3}\ \underline{M}.$$

Substituting the values of $[CaY^{2-}]$ and $[Y^{4-}]$ in the instability constant expression of the complex $[CaY]^{2-}$, we have

$$2 \times 10^{-11} = \frac{[Ca^{2+}](1.40 \times 10^{-3})}{2.50 \times 10^{-3}}.$$

Therefore, $[Ca^{2+}] = \mathbf{3.6 \times 10^{-11}}\ \underline{M}$. Since $[Ca^{2+}] = 3.6 \times 10^{-11} << 0.00250 = [CaY^{2-}]$, the initial assumption that the calcium is complexed quantitatively is valid. The assumption would actually not be true in acid solution for calcium due to the shifting of the equilibrium $Ca^{2+} + H_2Y^{2-} \rightleftharpoons [CaY]^{2-} + 2H^+$ to the left by the acid. For more stable complexes, e.g., $[PbY]^{2-}$, the assumption would remain valid in acidic solution.

7.2 Distribution and Logarithmic Diagrams of Complex Ions

Distribution diagrams of complex ions. The free metal ion M and the complex ions ML, ML_2, ..., ML_n (charges have been omitted for convenience) coexist in a solution of the complex ion ML_n. The concentrations of these species can be calculated as a function of the equilibrium ligand concentration, [L], and the *formation constants* of the complex ions, using the expressions (derived in an approach similar to that used for Equation (5.57) in Chapter 5, see Example 5.10).

$$\beta_0 = \frac{[M]}{C_M} = \frac{1}{1 + K_1[L] + K_1K_2[L]^2 + \ldots + K_1K_2\ldots K_n[L]^n} \quad (7.7)$$

$$\beta_1 = \frac{[ML]}{C_M} = \frac{K_1[L]}{1 + K_1[L] + K_1K_2[L]^2 + \ldots + K_1K_2\ldots K_n[L]^n} = \beta_0 K_1[L] \quad (7.8)$$

$$\beta_2 = \frac{[ML_2]}{C_M} = \frac{K_1K_2[L]^2}{1 + K_1[L] + K_1K_2[L]^2 + \ldots + K_1K_2\ldots K_n[L]^n} = \beta_0 K_1K_2[L]^2 \quad (7.9)$$

$$\vdots$$

$$\beta_n = \frac{[ML_n]}{C_M} = \frac{K_1K_2\ldots K_n[L]^n}{1 + K_1[L] + K_1K_2[L]^2 + \ldots + K_1K_2\ldots K_n[L]^n} = \beta_0 K_1K_2\ldots K_n[L]^n \quad (7.10)$$

where C_M is the total (analytical) concentration of the metal, that is, the sum of the concentrations of all species containing the metal, K_1, $K_2 \ldots$ K_n are the successive *formation constants* and the fractions β_0, β_1, $\beta_2 \ldots$ β_n are the ratios of the concentrations of the species M, ML, ..., ML_n, to the analytical concentration, respectively.

Using the above equations, we construct *distribution diagrams*, that is, $\beta_i = f_i(pL)$, where f_i represents the fraction, analogous to the distribution diagrams for species present in a solution of an acid of known pH (see Chapter 5 after Equation (5.57) and Figure 5.1), and the *concise distribution diagrams*, that is $\beta_n = f(pL)$, $\beta_n + \beta_{n-1} = f(pL)$, $\beta_n + \beta_{n-1} + \ldots + \beta_0 = f(pL)$ (Problem 7.5).

Example 7.7. A solution is made by mixing 20.0 mL of 0.804 M NH_3 and 20.0 mL of 0.0080 M $AgNO_3$. Calculate the concentrations of Ag^+, $[Ag(NH_3)]^+$ and $[Ag(NH_3)_2]^+$ ions.

Solution. After mixing the solutions, $C_{NH_3} = 0.402$ M and $C_{Ag} = 0.0040$ M. Since NH_3 is present in large excess, we may assume that silver is present as $[Ag(NH_3)_2]^+$. Then the concentration of free ammonia is $0.402 - (2 \times 0.0040) = 0.394$ M. Therefore,

$$\beta_0 = \frac{[Ag^+]}{C_{Ag}} = \frac{[Ag^+]}{0.0040} = \frac{1}{1 + (2.0 \times 10^3)(0.394) + (2.0 \times 10^3)(8.5 \times 10^3)(0.394)^2} = 3.79 \times 10^{-7},$$

and

$$[Ag^+] = \mathbf{1.52 \times 10^{-9}}\ \text{M},$$

$$\beta_1 = \frac{[Ag(NH_3)^+]}{0.0040} = \frac{(2.0 \times 10^3)0.394}{1 + (2.0 \times 10^3)(0.394) + (2.0 \times 10^3)(8.5 \times 10^3)(0.394)^2} = 2.99 \times 10^{-4},$$

and

$$[Ag(NH_3)^+] = \mathbf{1.20 \times 10^{-6}}\ \text{M},$$

$$\beta_2 = \frac{[Ag(NH_3)_2{}^+]}{0.0040} = \frac{(2.0 \times 10^3)(8.5 \times 10^3)(0.394)^2}{1 + (2.0 \times 10^3)(0.394) + (2.0 \times 10^3)(8.5 \times 10^3)(0.394)^2} = 0.9997,$$

and

$$[Ag(NH_3)_2{}^+] \simeq \mathbf{0.0040}\ \text{M}.$$

Logarithmic diagrams of complex ions. The logarithmic concentration diagrams for equilibria involving complex ions are described by the Equation $\log C_i = f_i(pL)$ and resemble the logarithmic diagrams of acids and bases (Section 5.6 and Figure 5.1). A logarithmic concentration diagram has the advantage in that it permits the estimation of very small concentrations of the various species, and in cases of linearity it is simple and fast to plot. From logarithmic concentration diagrams, the concentration of the various ions present in solutions of complex ions can be estimated, when the concentration of the free (uncomplexed) ligand, [L], and the analytical concentration of the metal, C_i, are known.

Example 7.8. Construct the logarithmic concentration diagram for a solution containing 0.0100 M of silver and free (uncomplexed) ammonia in the range 10^{-8} to 1 M. The successive formation constants of the silver ammine complexes are $10^{3.30}$ and $10^{3.93}$.

Solution. The concentrations of all silver containing species are calculated by using Equations (7.7), (7.8) and (7.9), and are used to construct the diagram of Figure 7.1. However, simpler equations can be used for the calculations, as follows:

1. For $10^{-8} < [NH_3] << 10^{-3.93}$, we have from Equation (7.7)

$$[Ag^+] = \frac{C_{Ag}}{1 + K_1[NH_3] + K_1K_2[NH_3]^2} \simeq C_{Ag}$$

(i.e., the second two terms in the denominator are << 1), and

$$\log[Ag^+] = \log C_{Ag} = -2.00.$$

From Equation (7.8)

$$[Ag(NH_3)^+] = \frac{C_{Ag}K_1[NH_3]}{1 + K_1[NH_3] + K_1K_2[NH_3]^2} \simeq C_{Ag}K_1[NH_3],$$

and

$$\log[Ag(NH_3)^+] = \log C_{Ag} + \log K_1 + \log[NH_3],$$

that is, a line with a slope of +1. For $[NH_3] = 1.00 \times 10^{-8}$ M, we have

$$\log[Ag(NH_3)^+] = -2.00 + 3.30 - 8.00 = -6.70.$$

From Equation (7.9)

$$[Ag(NH_3)_2{}^+] = \frac{C_{Ag}K_1K_2[NH_3]^2}{1 + K_1[NH_3] + K_1K_2[NH_3]^2} \simeq C_{Ag}K_1K_2[NH_3]^2,$$

and

$$\log[Ag(NH_3)_2{}^+] = \log C_{Ag} + \log K_1 + \log K_2 + 2\log[NH_3],$$

that is, a line with a slope of +2. For $[NH_3] = 1.00 \times 10^{-8}$ M, we have

$$\log[Ag(NH_3)_2{}^+] = -2.00 + 3.30 + 3.93 - 16.00 = -10.77.$$

2. For $10^{-3.30} << [NH_3] < 1$, we have

$$[Ag^+] = \frac{C_{Ag}}{1 + K_1[NH_3] + K_1K_2[NH_3]^2} \simeq \frac{C_{Ag}}{K_1K_2[NH_3]^2},$$

and

$$\log[Ag^+] = \log C_{Ag} - \log K_1 - \log K_2 - 2\log[NH_3],$$

that is, a line with a slope of −2. For $[NH_3] = 1.00$ M, we have $\log[Ag^+] = -2.00 - 3.30 - 3.93 = -9.23$.

$$[Ag(NH_3)^+] = \frac{C_{Ag}K_1[NH_3]}{1 + K_1[NH_3] + K_1K_2[NH_3]^2} \simeq \frac{C_{Ag}}{K_2[NH_3]},$$

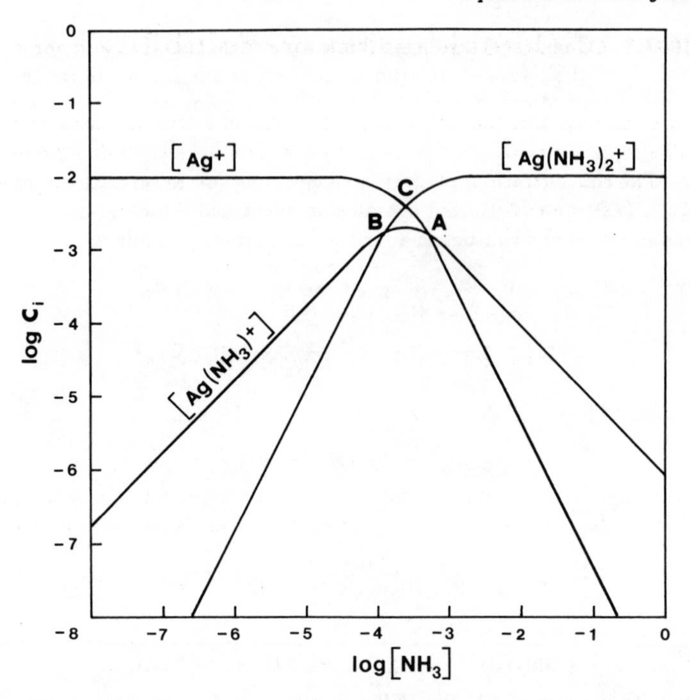

Figure 7.1. *Logarithmic concentration diagram of the silver-ammonia system.*

and

$$\log[Ag(NH_3)^+] = \log C_{Ag} - \log K_2 - \log[NH_3],$$

that is, a line with a slope of -1. For $[NH_3] = 1.00$ M, we have

$$\log[Ag(NH_3)^+] = -2.00 - 3.93 = -5.93.$$

$$[Ag(NH_3)_2{}^+] = \frac{C_{Ag}K_1K_2[NH_3]^2}{1 + K_1[NH_3] + K_1K_1[NH_3]^2} \simeq C_{Ag},$$

and $\log[Ag(NH_3)_2{}^+] = \log C_{Ag} = -2.00$.

Useful points for the completion of the logarithmic concentration diagram are the ones corresponding to the intersections of the ion lines. Their coordinates are determined easily, as follows: Point A (Figure 7.1), that is, the intersection of the $[Ag^+]$ and $[Ag(NH_3)^+]$ lines (where $[Ag^+] = [Ag(NH_3)^+]$), has an abscissa equal to $\log[NH_3] = pK_1$. Substituting this value in Equation (7.7), that is, $[NH_3] = 10^{-3.30}$, we have $\log[Ag^+] = -2.80 = \log[Ag(NH_3)^+]$. Similarly, point B, that is, the intersection of the $[Ag(NH_3)^+]$ and $[Ag(NH_3)_2{}^+]$ lines, has an abscissa equal to $\log[NH_3] = pK_2$. From Equation (7.8), for $[NH_3] = 20^{-3.93}$ we have $\log[Ag(NH_3)^+] = -2.80 = \log[Ag(NH_3)_2{}^+]$. Finally, point C, that is, the intersection of the $[Ag^+]$ and $[Ag(NH_3)_2{}^+]$ lines, has an ordinate equal to $\log[NH_3] = (pK_1 + pK_2)/2$. From Equation (7.9), for $[NH_3]^2 = 10^{-3.30} \times 10^{-3.93}$, that is, for $[NH_3] = 10^{-3.62}$, we have $\log[Ag(NH_3)_2{}^+] = -2.40 = \log[Ag^+]$.

7.3 Effect of Complexation on Solubility

The solubility of an insoluble (i.e., slightly soluble) salt in water increases with complex formation. The molar solubility S can be calculated from the solubility product of the insoluble salt and the formation constants of the complex ions. For example, if the insoluble salt MX is treated with the complexing agent L, we have:

$$\mathbf{MX} \rightleftharpoons M + X, \ K_{sp} = [M][X] \tag{7.11}$$

$$M + L \rightleftharpoons ML, \ K_1 = [ML]/[M][L] \tag{7.12}$$

$$ML + L \rightleftharpoons ML_2, \ K_2 = [ML_2]/[ML][L] \tag{7.13}$$

$$\vdots$$

$$ML_{n-1} + L \rightleftharpoons ML_n, \ K_n = [ML_n]/[ML_{n-1}][L] \tag{7.14}$$

where M and X may both be monovalent or polyvalent ions and L may be a molecule or ion. Therefore, the charges have been omitted for convenience. We also have

$$C_M = [M] + [ML] + [ML_2] + \ldots + [ML_n] \tag{7.15}$$

$$\beta_0 = [M]/C_M \tag{7.7}$$

Combining Equations (7.7) and (7.11) we have

$$K_{sp} = \beta_0 C_M [X] \tag{7.16}$$

The molar solubility S of MX is given by the expression

$$S = C_M = [X] \tag{7.17}$$

Combining Equations (7.16) and (7.17) we have

$$K_{sp} = \beta_0 S^2 \tag{7.18}$$

or

$$S = \sqrt{K_{sp}/\beta_0} \tag{7.18a}$$

In the absence of a complexing agent, β_0 is unity and

$$S = \sqrt{K_{sp}} \tag{7.19}$$

It can be seen from equations (7.18*a*) and (7.19) that the formation of complex ions results in an increase of the solubility of MX($\beta_0 < 1$). This increase is due to the decrease of the "free" metal ion concentration.

Example 7.9. Calculate the molar solubility of AgCl in a 0.10 M NH_3 solution.

Solution. *1st method.* Let S be the formal solubility of AgCl. From Equation (7.7), $1/\beta_0 = 1 + (2.0 \times 10^3)(0.10) + (2.0 \times 10^3)(8.5 \times 10^3)(0.10)^2 = 1.7 \times 10^5$. Substituting this value in Equation (7.18*a*), we have $S = \sqrt{(1.8 \times 10^{-10})(1.7 \times 10^5)} = 5.5 \times 10^{-3}$ mol/L. The value of 5.5×10^{-3} is only approximately accurate, because in the calculation we used the total concentration of ammonia and not the concentration of the free (uncomplexed) ammonia at the equilibrium state. Since NH_3 is present in excess, it may be assumed that all silver is present as $[Ag(NH_3)_2]^+$. Then, the concentration of free NH_3 is $0.10 - (2 \times 5.5 \times 10^{-3}) = 0.089$ M. Using the value 0.089 and the same procedure, we find that $\mathbf{S = 4.9 \times 10^{-3}}$ mol/L.

2nd method. Since NH_3 is present in excess, $[Ag(NH_3)^+]$ is considered negligible. In addition to Equation (7.4), we also have the following expressions:

$$\mathbf{AgCl} \rightleftharpoons Ag^+ + Cl^-, \; K_{sp(AgCl)} = [Ag^+][Cl^-] = 1.8 \times 10^{-10} \tag{7.20}$$

$$\text{Mass Balance (Cl)}: \; S = [Cl^-] \tag{7.21}$$

$$\text{Mass balance (Ag)}: \; S = [Ag^+] + [Ag(NH_3)_2{}^+] \simeq [Ag(NH_3)_2{}^+] \tag{7.22}$$

$$\text{Mass balance (N)}: \; 0.10 = [NH_3] + [NH_4{}^+] + 2[Ag(NH_3)_2{}^+] \simeq [NH_3] + 2[Ag(NH_3)_2{}^+]. \tag{7.23}$$

Combining Equations (7.22) and (7.23), we have

$$[NH_3] = 0.10 - 2S. \tag{7.24}$$

Substituting the values of $[Ag(NH_3)_2{}^+]$ and $[NH_3]$ from Equations (7.22) and (7.24) in the instability constant expression, Equation (7.4), we have

$$5.9 \times 10^{-8} = \frac{[Ag^+](0.10 - 2S)^2}{S} \tag{7.25}$$

Combining Equations (7.20), (7.21) and (7.25), we have

$$\frac{(5.9 \times 10^{-8})S}{(0.10 - 2S)^2} = \frac{1.8 \times 10^{-10}}{S} \tag{7.26}$$

Then $S = \mathbf{5.0 \times 10^{-3}}$ mol/L. The assumption made, $[Ag^+] << [Ag(NH_3)_2{}^+]$, is valid, because $[Ag^+] = 1.8 \times 10^{-10}/5.0 \times 10^{-3} = 3.6 \times 10^{-8} << 5.0 \times 10^{-3} = [Ag(NH_3)_2{}^+]$. The small difference in the values found by the two methods (about 2%) is due to different assumptions made in the two cases.

3rd method. The equilibrium constant of the dissolution reaction of AgCl,

$$\mathbf{AgCl} + 2NH_3 \rightleftharpoons [Ag(NH_3)_2]^+ + Cl^-,$$

is

$$K = \frac{[Ag(NH_3)_2{}^+][Cl^-]}{[NH_3]^2} \cdot \frac{[Ag^+]}{[Ag^+]} = \frac{K_{sp(AgCl)}}{K_{inst[Ag(NH_3)_2]^+}} = \frac{1.8 \times 10^{-10}}{5.9 \times 10^{-8}} \tag{7.27}$$

Let S = solubility of AgCl. Then $[NH_3] = 0.10 - 2S$. Substituting these values in Equation (7.27), we have

$$\frac{S^2}{(0.10 - 2S)^2} = \frac{1.8 \times 10^{-10}}{5.9 \times 10^{-8}} \tag{7.26a}$$

Then $S = \mathbf{5.0 \times 10^{-3}}$ mol/L.

Example 7.10. Calculate the molar solubility of Ag_2S in a 0.100 $\underline{M}$ NH_3 solution.

Solution. We have $[OH^-] = \sqrt{1.8 \times 10^{-6}} = 1.34 \times 10^{-3}$ $\underline{M}$ and $[H^+] = 1.00 \times 10^{-14}/1.34 \times 10^{-3} = 7.45 \times 10^{-12}$ $\underline{M}$. Let S = molar solubility of Ag_2S. Then

$$C_{Ag} = [Ag^+] + [Ag(NH_3)^+] + [Ag(NH_3)_2{}^+] = 2S$$

$$\beta_0 = [Ag^+]/C_{Ag}$$

or

$$[Ag^+] = \beta_0 C_{Ag} = 2\beta_0 S$$
$$S = [S^{2-}] + [HS^-] + [H_2S] = C_s$$
$$\alpha_2 = [S^{2-}]/C_s$$

or

$$[S^{2-}] = \alpha_2 C_s = \alpha_2 S.$$

Substituting the values of $[Ag^+]$ and $[S^{2-}]$ in the K_{sp} expression, we have

$$K_{sp(Ag_2S)} = [Ag^+]^2[S^{2-}] = (2\beta_0 S)^2(\alpha_2 S) = 4\beta_0^2\alpha_2 S^3$$

or

$$S = \sqrt[3]{\frac{K_{sp}}{4\beta_0^2\alpha_2}} = \sqrt[3]{\frac{K_{sp}}{4\alpha_2}\left(\frac{1}{\beta_0}\right)^2} \qquad (7.28)$$

For $[H^+] = 7.45 \times 10^{-12}$ $\underline{M}$, we have

$$\alpha_2 = \frac{(1.0 \times 10^{-7})(1.0 \times 10^{-14})}{(7.45 \times 10^{-12})^2 + (1.0 \times 10^{-7})(7.45 \times 10^{-12}) + (1.0 \times 10^{-7})(1.0 \times 10^{-14})} = 1.3 \times 10^{-3}.$$

For $[NH_3] = 0.100$ $\underline{M}$, we have $(1/\beta_0) = 1.7 \times 10^5$ (Example 7.9). Substituting the values of K_{sp}, α_2 and β_0 in Equation (7.28), we have

$$S = \sqrt[3]{\frac{1 \times 10^{-50}}{4 \times 1.3 \times 10^{-3}}(1.7 \times 10^5)^2} = \mathbf{4 \times 10^{-13}}\ \underline{M}.$$

Note. The actual concentration of free ammonia is $0.100 - 4(4 \times 10^{-13}) \simeq 0.100$ $\underline{M}$.

Example 7.11. How much NH_3 gas must be added to 100 mL of water that contains 0.94 g AgBr to dissolve all the AgBr (assume no volume change)?

Solution. *1st method.* We have 0.94 g AgBr/100 mL $\equiv$ 9.4 g AgBr/L $\equiv$ (9.4 g AgBr/L)/(187.8 g/mol) = 0.050 $\underline{M}$ = $[Br^-]$. Since an excess of NH_3 is required to dissolve AgBr (Example 7.9), $[Ag(NH_3)^+]$ is considered negligible. To prevent the formation of a AgBr precipitate, it must be that the ion product $[Ag^+][Br^-] < K_{sp(AgBr)}$ or $[Ag^+](0.050) < 5 \times 10^{-13}$ or $[Ag^+] < 1.0 \times 10^{-11}$ $\underline{M}$. Then $[Ag(NH_3)_2{}^+] \simeq 0.050$ $\underline{M}$ and

$$K_{inst} = 5.9 \times 10^{-8} = \frac{[Ag^+][NH_3]^2}{[Ag(NH_3)_2{}^+]} = \frac{(1.0 \times 10^{-11})[NH_3]^2}{0.050}$$

or

$$[NH_3] = \mathbf{17}\ \underline{M}.$$

An additional $2 \times 0.050 = 0.10$ mol NH_3 is present in 0.050 mol $[Ag(NH_3)_2]^+$. Therefore, needed is a total of $17 + 0.10 \simeq 17$ mol $NH_3/L \equiv (17 \text{ mol } NH_3/L)(17.0 \text{ g/mol}) = 289 \text{ g } NH_3/L \equiv$ **29** g NH_3/100 mL.

2nd method. The equilibrium constant of the dissolution reaction of AgBr,

$$\mathbf{AgBr} + 2NH_3 \rightleftharpoons [Ag(NH_3)_2]^+ + Br^-,$$

is

$$K = \frac{[Ag(NH_3)_2{}^+][Br^-]}{[NH_3]^2} \simeq \frac{(0.050)^2}{[NH_3]^2} = \frac{K_{sp(AgBr)}}{K_{inst[Ag(NH_3)_2]^+}} = \frac{5 \times 10^{-13}}{5.9 \times 10^{-8}}.$$

Then $[NH_3] = 17$ $\underline{M}$. Therefore, needed is a total of $17 + (2 \times 0.050) \simeq 17$ mol $NH_3/L \equiv (17 \text{ mol } NH_3/L)(17.0 \text{ g/mol}) = 289 \text{ g } NH_3/L \equiv$ **29** g NH_3/100 mL.

Example 7.12. To dissolve 0.010 mole of AgI, one liter of 0.36 $\underline{M}$ $Na_2S_2O_3$ is required. Calculate (a) the instability constant of the complex $[Ag(S_2O_3)_2]^{3-}$, and (b) the equilibrium constant of the dissolution reaction.

Solution. a) The dissolution reaction of AgI is

$$\mathbf{AgI} + 2S_2O_3{}^{2-} \rightleftharpoons [Ag(S_2O_3)_2]^{3-} + I^- \tag{7.29}$$

and 0.020 mol of $S_2O_3^{2-}$ is consumed for the dissolution of 0.010 mol of AgI. Hence, $[S_2O_3^{2-}]_{free} = 0.36 - 0.020 = 0.34$ $\underline{M}$. $[Ag^+]$ is calculated from $K_{sp(AgI)}$, that is, $[Ag^+] = 8.5 \times 10^{-17}/0.010 = 8.5 \times 10^{-15}$ $\underline{M}$. Substituting the values of $[Ag^+]$, $[S_2O_3^{2-}]$ and $[Ag(S_2O_3)_2^{3-}]$ in the instability constant expression we have

$$K_{inst[Ag(S_2O_3)_2]^{3-}} = \frac{[Ag^+][S_2O_3^{2-}]^2}{[Ag(S_2O_3)_2^{3-}]} = \frac{(8.5 \times 10^{-15})[0.34 + 2(8.5 \times 10^{-15})]^2}{0.010 - (8.5 \times 10^{-15})} = \mathbf{9.8 \times 10^{-14}}.$$

The equilibrium constant of Reaction (7.29) is

$$K = \frac{[Ag(S_2O_3)_2^{3-}][I^-]}{[S_2O_3{}^{2-}]^2} \cdot \frac{[Ag^+]}{[Ag^+]} = \frac{K_{sp(AgI)}}{K_{inst[Ag(S_2O_3)_2]^{3-}}} = \frac{8.5 \times 10^{-17}}{9.8 \times 10^{-14}} = \mathbf{8.7 \times 10^{-4}}.$$

Complex ion formation with excess of common (precipitating) ion. If a metal ion M forms a slightly soluble salt MX with the anion X (charges have been omitted for convenience), it often forms complex ions in the presence of excess X. The molar solubility of the salt is calculated from its solubility product and the formation constants of the complex ions. Given that S^0 is the *intrinsic solubility* of the undissociated salt in solution, [MX(sol)], that is, the concentration of the simplest possible complex ([M] : [X] = 1 : 1), we have

$$S = [M] + [MX(sol)] + [MX_2] + [MX_3] + \ldots + [MX_n] \tag{7.30}$$

$$\mathbf{MX} \rightleftharpoons M + X, \; K_{sp} = [M][X] \tag{7.11}$$

$$M + X \rightleftharpoons MX(sol), \; K_1 = [MX(sol)]/[M][X] \tag{7.31}$$

$$MX(sol) + X \rightleftharpoons MX_2, \; K_2 = [MX_2]/[MX(sol)][X] \tag{7.32}$$

$$MX_2 + X \rightleftharpoons MX_3, \ K_3 = [MX_3]/[MX_2][X] \tag{7.33}$$

$$\vdots$$

$$MX_{n-1} + X \rightleftharpoons MX_n, \ K_n = [MX_n]/[MX_{n-1}][X] \tag{7.34}$$

where K_1, $K_2 \ldots K_n$ are the successive formation constants. Combining Equations (7.11) and (7.31) we have

$$K_1 = [MX(sol)]/K_{sp} \tag{7.35}$$

Substituting the various terms of Equation (7.30) with their respective values from Equations (7.11) and (7.31) through (7.35) we have

$$S = K_{sp}/[X]+K_1K_{sp}+K_1K_2K_{sp}[X]+K_1K_2K_3K_{sp}[X]^2+\ldots+K_1K_2\ldots K_nK_{sp}[X]^{n-1} \tag{7.36}$$

If we designate K'_1, $K'_2 \ldots K'_n$ as the overall formation constants, then Equation (7.36) becomes

$$S = K_{sp}(1/[X] + K'_1 + K'_2[X] + K'_3[X]^2 + \ldots + K'_n[X]^{n-1}) \tag{7.37}$$

Example 7.13. Calculate the molar solubility of AgCl in solutions containing chloride at the following concentrations: 10^{-4}, 10^{-3}, 10^{-2}, 10^{-1}, 1 $\underline{M}$. Plot log S *vs.* log$[Cl^-]$. At what chloride concentration is the solubility minimal and what is the minimum solubility? The successive formation constants for the silver chloride complexes are $10^{3.04}$, $10^{2.00}$, 1, and $10^{0.26}$.

Solution. The complexes formed are [AgCl], $[AgCl_2]^-$, $[AgCl_3]^{2-}$ and $[AgCl_4]^{3-}$. The overall formation constants are $K'_1 = K_{f_1}$, $K'_2 = K_{f_1}K_{f_2}$, $K'_3 = K_{f_1}K_{f_2}K_{f_3}$, and $K'_4 = K_{f_1}K_{f_2}K_{f_3}K_{f_4}$. Accordingly, $K'_1 = 10^{3.04}$, $K'_2 = K'_3 = 10^{5.04}$, and $K'_4 = 10^{5.30}$. Substituting these values in equation (7.37), we have for $[Cl^-] = 10^{-4}$ $\underline{M}$

$$S = 1.8\times10^{-10}[1/10^{-4}+10^{3.04}+10^{5.04}(10^{-4})+10^{5.04}(10^{-4})^2+10^{5.30}(10^{-4})^3] = 2.0\times10^{-6}\ \underline{M}.$$

Similarly, for [Cl] $= 10^{-3}$, 10^{-2}, 10^{-1} and 1 $\underline{M}$, S $= 4.0 \times 10^{-7}$, 4.1×10^{-7}, 2.4×10^{-6} and 7.6×10^{-5} $\underline{M}$, respectively. From the calculated values (or from a plot of log S vs. log$[Cl^-]$) it is seen that the solubility S is minimal for $10^{-3} < [Cl^-] < 10^{-2}$ $\underline{M}$. For $[Cl^-] = \mathbf{3.2 \times 10^{-3}}$ $\underline{M}$, S has the minimum value of $\mathbf{3.2 \times 10^{-7}}$ $\underline{M}$.

7.4 Effect of pH on Equilibria Involving Complex Ions

Many complexing agents are Lewis and Brönsted bases. For example, NH_3, CN^-, [illegible] S^{2-} are electron pair donors and proton acceptors; hence, the fractions, α, of th[illegible] concentrations which are available for complexation vary with the pH of th[illegible] The calculation of the α fraction of the complexing agent is indispensable fo[illegible] involving complex ions and is done as previously described (Section 5.4). The pH also affects the values of the conditional formation constants which are used in calculations involving EDTA complexes.

Example 7.14. Calculate $[Cu^{2+}]$ in a solution containing 0.0020 M $Cu(NO_3)_2$ and 0.0030 M uncomplexed ammonia and having a pH of 10.00. The successive formation constants for copper-ammonia complexes are: $10^{4.31}$, $10^{3.67}$, $10^{3.04}$, and $10^{2.30}$.

Solution. We have (Equation (5.57*a*))

$$\alpha_1 = \frac{NH_3}{C} = \frac{K_1}{[H^+] + K_1} = \frac{10^{-9.26}}{10^{-10.00} + 10^{-9.26}} = 0.85,$$

where C is the concentration of free ammonia that is uncomplexed with copper and K_1 is the ionization constant of $NH_4{}^+$. So, 85% of the uncomplexed ammonia is present in the form of NH_3 and 15% in the form of NH_4^+. Hence, $[NH_3] = (0.0030)(0.85) = 0.00255$M. Then, from Equation (7.7)

$$\beta_0 = \frac{1}{1 + 10^{4.31}(0.00255) + 10^{4.31} \times 10^{3.67}(0.00255)^2 + 10^{4.31} \times 10^{3.67} \times 10^{3.04}(0.00255)^3 + 10^{4.31} \times 10^{3.67} \times 10^{3.04} \times 10^{2.30}(0.00255)^4}$$

or

$$\beta_0 = 1/3294 = 3.04 \times 10^{-4}.$$

From Equation (7.7),

$$[Cu^{2+}] = \beta_0 C_M = (3.04 \times 10^{-4})(0.0020) = \mathbf{6.1 \times 10^{-7}}\ \underline{M}.$$

7.5 Problems

Instability constant of a complex ion

7.1. How many grams of KCl must be added to 250 ml of 0.10 M $Na[Ag(CN)_2]$ in order for AgCl to start to precipitate?

7.2. A solution contains 0.0005 mol of Cl^- and 10 mol of ammonia per liter. Will AgCl precipitate, if 0.0010 mol of solid $AgNO_3$ is added to 1 L of this solution?

7.3. What is the equilibrium concentration of S^{2-} ions required to precipitate cadmium as CdS from a 0.600 M $[Cd(NH_3)_4]^{2+}$ solution?

7.4. The total (analytical) concentration of complexed silver in a solution is 0.00400 M. What must be the concentrations of ammonia and ammonium ions so that the pH of the solution is 10.00 and the average number of ammonia molecules per silver ion is 1.40?

Distribution and logarithmic diagrams of complex ions

7.5. Construct a) a diagram showing the fraction of cadmium present as each of the complexes as a function of pI, b) the distribution diagram for cadmium iodide complexes. The successive formation constants of the cadmium iodide complexes are: $10^{2.28}$, $10^{1.64}$, $10^{1.08}$, and $10^{1.10}$.

7.6. Construct the logarithmic concentration diagram for a solution containing 0.0100 M of cadmium and free (uncomplexed) ammonia in the range 10^{-5} to 1 M. The successive formation constants of the cadmium-ammonia complexes are $10^{2.51}$, $10^{1.96}$, $10^{1.30}$, and $10^{0.79}$.

7.7. Construct the logarithmic concentration diagram for a solution containing 0.0100 M of copper and free ethylenediamine (en) in the range 10^{-14} to 10^{-7} M. The successive formation constants of the copper ethylenediamine complexes are $10^{10.7}$ and $10^{8.9}$.

Effect of complexation on solubility

7.8. How many milliliters of an ammonia solution, which has a density of 0.898 g/mL and is 28.0% (w/w) in ammonia, is required to dissolve 14.33 g of AgCl? Assume that the only complex formed is $[Ag(NH_3)_2]^+$.

7.9. Calculate the molar solubility S of AgSCN in solutions containing thiocyanate at the following concentrations: 10^{-6}, 10^{-5}, 10^{-4}, 10^{-3}, 10^{-2}, and 10^{-1} M. Plot log S versus $\log[SCN^-]$ and find the $[SCN^-]$ at which the solubility is minimal. The successive formation constants for silver thiocyanate complexes are: $10^{4.75}$, $10^{3.48}$, $10^{1.22}$, and $10^{0.22}$.

7.10. What must be the molarity of an ammonia solution so that 500 mL dissolves 0.050 mol of AgBr? Assume that the only complex formed is $[Ag(NH_3)_2]^+$.

7.11. How many milliliters of a 0.100 M $Na_2S_2O_3$ solution is required to dissolve 0.0939 g of AgBr in a photographic film?

7.12. A series of solutions each 0.060 M in Cd^{2+} and with various concentrations of HI is saturated with H_2S. In the solution that is 0.3 M in HI a trace of CdS precipitates. Assume that at higher HI concentrations no CdS precipitate is formed because of the formation of the complex ion $[CdI_4]^{2-}$, and calculate the instability constant of the complex.

7.13. If 4.94 mmol of AgCl is dissolved in one liter of ammonia solution, the final concentration of free NH_3 being 0.089 M, calculate the concentration of uncomplexed Ag^+ in the solution.

7.14. Calculate the molar solubility of Ag_2S in a 1.00 M NH_3 solution.

7.15. Calculate the molar solubility of (a) AgBr, (b) AgI, in a 0.100 M NH_3 solution.

Chapter 8

Equilibria Involving Redox Systems

8.1 Electrode Potentials—Strength of Oxidizing and Reducing Agents

As we have seen (Chapter 2), every oxidation-reduction (redox) reaction can be considered as the sum of two half-reactions, that is, an oxidation half-reaction and a reduction half-reaction, which take place simultaneously. Each of the two half-reactions is represented by an ionic equation, which shows the transfer of electrons between the two oxidation states of the same element, that constitute a *conjugated redox couple.* The equation of a redox half-reaction can be written in the general form

$$A_{ox} + ne \rightleftharpoons A_{red} \tag{8.1}$$

The stronger an oxidizing agent A_{ox} is, i.e., the greater its tendency to gain electrons, the more the above equilibrium is displaced towards the reducing agent A_{red}.

The absolute tendencies of various substances to gain electrons cannot be determined from electrode potential measurements, but the *relative* tendencies can be determined by measurement of relative potentials using as a common reference point the potential of the half-reaction $2H^+ + 2e \rightleftharpoons H_2$, which is arbitrarily defined as zero. The results are expressed as standard potentials, which are measured with galvanic cells. A *galvanic cell* consists of two *half-cells,* that is, two electrodes immersed in two solutions, which are separated from one another but electrically connected. Upon connecting the electrodes, an oxidation-reduction half-reaction takes place at each, resulting in the development of a potential at the electrode, which is called the *electrode potential.* The difference in the two electrode *potentials* represents the cell *voltage* that is measured if a voltmeter is placed between the two electrodes. The measurement of the oxidizing and reducing powers in the galvanic cells for determining standard electrode potentials (see below) is performed under defined (standard) conditions, as follows: The activities of the various substances are equal to one, the gases are under a partial pressure of one atmosphere, the temperature is 25°C, and one of the half-cells consists of the standard (normal) hydrogen electrode (SHE or NHE). The cell voltage measured under standard conditions, expressed in volts, is the *standard electrode potential,* E^0, of the second half-cell redox couple relative to the SHE, which is also the *standard reduction potential* of the redox couple (it is called a *reduction potential*, since by convention, half-reactions for the potential scale are written as reductions). A positive value of standard potential means that the oxidized form of the redox couple is a stronger oxidizing agent than the H^+ ion, whereas a negative value means that the reduced form of the couple is a stronger reducing agent than H_2. The larger the value of E^0 for the reduction half-reaction (8.1), the stronger the oxidizing agent A_{ox} is, and consequently, the weaker the reducing agent A_{red} of the redox couple is. In appendix B the standard potentials, E^0, of many half-reactions are given.

8.2 Factors that Affect the Electrode Potential—The Nernst Equation

The electrode potential E of any redox couple depends upon the temperature plus the activities of the species involved in the couple, and therefore, it depends upon any factor that affects these activities. Factors include the ionic strength of the solution, the presence of complexing agents, the pH of the solution, the nature of the solvent, and so forth.

Effect of concentration on the electrode potential. The relation between the potential E of a redox half-reaction and the activities is given by the *Nernst equation*, which for the general half-reaction

$$A_{ox} + ne \rightleftharpoons A_{red} \tag{8.1}$$

is the following

$$E = E^0 - \frac{RT}{nF} \ln \frac{a_{A_{red}}}{a_{A_{ox}}} \tag{8.2}$$

or

$$E = E^0 - \frac{2.303RT}{nF} \log \frac{a_{A_{red}}}{a_{A_{ox}}} \tag{8.2a}$$

where

E^0 = standard electrode potential in volts

R = gas constant = 8.31441 volt coulomb K^{-1} mol^{-1}

T = absolute temperature (in K)

F = Faraday constant = 96484.56 coulombs/eq

n = number of electrons involved in the half-reaction (eq/mol)

$a_{A_{red}}$ = the activity of the reduced form of the redox couple

$a_{A_{ox}}$ = the activity of the oxidized form of the redox couple

Substituting the activities* in Equation (8.2) with the concentrations multiplied by the corresponding activity coefficients (Equation 3.13), we have

$$E = E^0 - \frac{RT}{nF} \ln \frac{f_{A_{red}}}{f_{A_{ox}}} - \frac{RT}{nF} \ln \frac{[A_{red}]}{[A_{ox}]} \tag{8.3}$$

and by combining the first two terms

$$E = E^{0\prime} - \frac{RT}{nF} \ln \frac{[A_{red}]}{[A_{ox}]} \tag{8.4}$$

where the potential $E^{0\prime}$ which is called the *formal potential* of the redox couple, is the value of E under defined solution conditons, usually at unit formal concentrations of the oxidant and the reductant and other species such as acids, rather than at unit activities.

We see that the concentration affects the electrode potential E in various ways, depending on the type of half-reaction. For example, in the half-reaction $Fe^{2+} + 2e \rightleftharpoons Fe$, E varies with the concentration of the Fe^{2+} ions only, and therefore, the absolute concentration is important (the activity of a pure substance such as Fe is unity). On the other hand, in the half-reaction $Fe^{3+} + e \rightleftharpoons Fe^{2+}$, E is independent of the absolute concentrations of the Fe^{3+} and Fe^{2+} ions, and depends only on the concentration ratio, $[Fe^{3+}]/[Fe^{2+}]$. Finally, in half-reactions in which the coefficients of the oxidant and the reductant in the half-reaction equation are not the same, as for example in the half-reaction $[I_3]^- + 2e \rightleftharpoons 3I^-$, E depends simultaneously on the absolute concentrations of the $[I_3]^-$ and I^- ions and on the concentration ratio, i.e., $[I^-]^3/[I_3{}^-]$.

For the general reaction

$$aA + bB \rightleftharpoons cC + dD \tag{8.5}$$

the Nernst equation is

$$E = E^0 - \frac{RT}{nF} \ln \frac{a_C^c\, a_D^d}{a_A^a\, a_B^b} \tag{8.6}$$

That is, all the products and reactants affect E, and their activities are raised to a power equal to the coefficients in the balanced reaction.

By substituting the values of the constants R and F in Equation (8.6), and dividing the natural logarithms by 2.303 in order to convert them to decimal ones, the Nernst equation at 25°C (298 K) becomes

$$E = E^0 - \frac{0.05916}{n} \log \frac{a_C^c\, a_D^d}{a_A^a\, a_B^b} \tag{8.7}$$

In the following calculations, assume concentrations can be used in place of activities unless otherwise specified.

Example 8.1. Calculate the electrode potentials of the following systems:

*See Chapter 3 for a discussion of activities. Concentrations are often substituted for activities, particularly for dilute solutions.

(a) **Pt** | MnO_4^-(0.200 M), Mn^{2+}(2.00×10^{-4} M), H^+(0.100 M)

(b) **Pt** | $Cr_2O_7^{2-}$(1.00×10^{-4} M), Cr^{3+}(1.00×10^{-3} M), H^+(0.100 M)

The vertical line indicates a phase boundary between the platinum electrode and the solution (See Section 8.3).

Solution. a) The equation for the reduction half-reaction (as mentioned before, half-reactions, by convention, are always written as reductions for the Nernst equation) is

$$MnO_4^- + 8H^+ + 5e \rightleftharpoons Mn^{2+} + 4H_2O$$

Hence, we have the equation

$$E = E^0_{MnO_4^-,\ Mn^{2+}} - \frac{0.05916}{5}\log\frac{[Mn^{2+}]}{[MnO_4^-][H^+]^8} = +1.51 - \frac{0.05916}{5}\log\frac{2.00 \times 10^{-4}}{(0.200)(0.100)^8} = \mathbf{+1.45}\ V.$$

b) The equation for the reduction half-reaction is

$$Cr_2O_7^{2-} + 14H^+ + 6e \rightleftharpoons 2Cr^{3+} + 7H_2O$$

Hence, we have the equation

$$\begin{aligned} E &= E^0_{Cr_2O_7^{2-},\ Cr^{3+}} - \frac{0.05916}{6}\log\frac{[Cr^{3+}]^2}{[Cr_2O_7^{2-}][H^+]^{14}} \\ &= +1.33 - \frac{0.05916}{6}\log\frac{(1.00 \times 10^{-3})^2}{(1.00 \times 10^{-4})(0.100)^{14}} = \mathbf{+1.21}\ V. \end{aligned}$$

Effect of complex or precipitate formation on the electrode potential. Whenever one of the species involved in a half-reaction forms complexes or a precipitate, the potential E and the oxidizing strength of the system change. In general, in the complexation of the two constituents of a redox couple, if the oxidized species forms more stable complexes than the reduced species, as is often the case, the potential and the oxidizing strength of the system decrease.

Example 8.2. Calculate the electrode potentials of the following systems:

(a) **Ag** | Ag^+(1.00×10^{-3} M)

(b) **Ag** | $[Ag(NH_3)_2]^+$(1.00×10^{-3} M), NH_3(1.00 M)

(c) **Ag** | **AgBr**, Br^-(1.00 M)

Solution. a) We have

$$\begin{aligned} E &= E^0_{Ag^+, Ag} - 0.05916 \log(1/[Ag^+]) = E^0_{Ag^+, Ag} + 0.05916 \log[Ag^+] \\ &= +0.7994 + 0.05916 \log(1.00 \times 10^{-3}) = \mathbf{+0.622} \text{ V}. \end{aligned}$$

b) Since there is a large excess of NH_3, the $[Ag(NH_3)^+]$ is considered negligible. We have

$$K_{inst} = 5.9 \times 10^{-8} = \frac{[Ag^+](1.00)^2}{1.00 \times 10^{-3}},$$

from which $[Ag^+] = 5.9 \times 10^{-11}$ $\underline{M}$. Hence, we have

$$E = +0.7994 + 0.05916 \log(5.9 \times 10^{-11}) = \mathbf{+0.194} \text{ V}.$$

We observe that because of Ag^+ complexation, the potential E, and consequently the oxidizing power of the Ag^+-Ag system as well, is decreased.

c) We have $[Ag^+] = K_{sp(AgBr)}/[Br^-] = 5 \times 10^{-13}/1.00 = 5 \times 10^{-13}$ $\underline{M}$, whereupon

$$E = +0.7994 + 0.05916 \log(5 \times 10^{-13}) = \mathbf{+0.072} \text{ V}.$$

We observe that because of Ag^+ precipitation, the potential E and the oxidizing strength of the Ag^+-Ag system are decreased.

Effect of pH on the electrode potential. Whenever H^+ or OH^- ions participate in a half-reaction, the electrode potential is changed with the pH. For example, for the half-reaction

$$A_{ox} + mH^+ + ne \rightleftharpoons A_{red} \tag{8.8}$$

the Nernst equation is

$$E = E^0 - \frac{RT}{nF} \ln \frac{a_{A_{red}}}{a_{A_{ox}} a^m_{H+}} \tag{8.9}$$

or

$$E = E^0 - \frac{0.05916}{n} \log \frac{a_{A_{red}}}{a_{A_{ox}}} - \frac{m}{n} \times 0.05916 \text{ pH at } 25°\text{C} \tag{8.10}$$

From Equation (8.10) it can be seen that the potential E is a function of the pH of the solution. Because of this dependence of E on pH, the degree of completeness of a reaction, and sometimes the direction, can be changed by changing the pH of the solution.

Example 8.3. If the pH is increased by 2, what will be the change in the electrode potential of the following redox couples? a) MnO_4^--$\mathbf{MnO_2}$, b) MnO_4^--Mn^{2+}, c) H_3AsO_4-H_3AsO_3, d) NO_3^--HNO_2.

Solution. a) From Equation (8.10) it is concluded that when the pH is increased by 2, the electrode potential is decreased by $(m/n) \times 0.05916 \times 2 = 0.1183\ m/n$ volts. In the case of the MnO_4^--$\mathbf{MnO_2}$ couple, we have the half-reaction

$$MnO_4^- + 4H^+ + 3e \rightleftharpoons \mathbf{MnO_2} + 2H_2O$$

Hence, the electrode potential decreases by $(0.1183 \times 4)/3 = \mathbf{0.1577}$ V.

b) We have the half-reaction

$$MnO_4^- + 8H^+ + 5e \rightleftharpoons Mn^{2+} + 4H_2O$$

Hence, the electrode potential decreases by $(0.1183 \times 8)/5 = \mathbf{0.1893}$ V.

c) We have the half-reaction

$$H_3AsO_4 + 2H^+ + 2e \rightleftharpoons H_3AsO_3 + 2H_2O$$

Hence, the electrode potential decreases by $(0.1183 \times 2)/2 = \mathbf{0.1183}$ V.

d) We have the half-reaction

$$NO_3^- + 3H^+ + 2e \rightleftharpoons HNO_2 + H_2O$$

Hence, the electrode potential decreases by $(0.1183 \times 3)/2 = \mathbf{0.1774}$ V.

Example 8.4. Given the half-reactions

$$MnO_4^- + 8H^+ + 5e \rightleftharpoons Mn^{2+} + 4H_2O$$

$$Cr_2O_7^{2-} + 14H^+ + 6e \rightleftharpoons 2Cr^{3+} + 7H_2O$$

If the concentration of every ion but H^+ is 1 M, calculate the hydrogen ion concentration at which the oxidizing strengths of the two systems would be equal.

Solution. The potential of the first half-cell is given by the equation

$$E_1 = E^0_{MnO_4^-,\ Mn^{2+}} - \frac{0.05916}{5} \log \frac{[Mn^{2+}]}{[MnO_4^-][H^+]^8},$$

and that of the second one by the equation

$$E_2 = E^0_{Cr_2O_7^{2-},\ Cr^{3+}} - \frac{0.05916}{6} \log \frac{[Cr^{3+}]^2}{[Cr_2O_7^{2-}][H^+]^{14}}.$$

In order for the oxidizing strengths to be equal, it is required that $E_1 = E_2$, or

$$E^0_{MnO_4^-,\ Mn^{2+}} - \frac{0.05916}{5} \log \frac{[Mn^{2+}]}{[MnO_4^-][H^+]^8} = E^0_{Cr_2O_7^{2-},\ Cr^{3+}} - \frac{0.05916}{6} \log \frac{[Cr^{3+}]^2}{[Cr_2O_7^{2-}][H^+]^{14}}.$$

Substituting the data in the above equation, we have

$$1.51 - \frac{0.05916}{5} \log \frac{1}{[H^+]^8} = 1.33 - \frac{0.05916}{6} \log \frac{1}{[H^+]^{14}},$$

from which $\log[H^+] = 4.149$. Hence, $[H^+] = \mathbf{1.41 \times 10^4}$ M.

Note. Obviously, the concentration $[H^+] = 1.41 \times 10^4$ M cannot be accomplished experimentally, and the calculation illustrates that we cannot make these two systems equal in oxidizing strength under the given conditions.

8.3 Galvanic Cells—Equilibrium Constant of Redox Systems

Electrochemical cells are classified as *galvanic cells* (voltaic cells) when they use spontaneous chemical reactions to generate electrical energy (a voltage), and as *electrolytic cells* when they require electrical energy from an external source to drive non-spontaneous chemical reactions. Galvanic cells are represented schematically, as in the following cell:

$$\mathbf{Zn} \mid Zn^{2+}(a_{Zn^{2+}=1}) \quad \| \quad Cu^{2+}(a_{Cu^{2+}=1}) \mid \mathbf{Cu}$$

left-hand electrode **right-hand electrode**

A vertical line indicates a phase boundary with a potential difference across it, whereas a double vertical line stands for a salt bridge. The electrode at which oxidation occurs is called the anode, whereas the one at which reduction occurs is called the cathode. In a galvanic cell (e.g., a battery), the electrode at which oxidation occurs to release electrons is the negative pole (the one where externally electrons are provided), so the anode in this case is the negative pole. Conversely, the cathode (where electrons are consumed in reduction) is the positive pole. (The opposite is true in electrolytic cells, where an external voltage is applied to force an electrolytic reaction to occur in a direction opposite to the spontaneous chemical reaction—i.e., the negative pole constitutes the cathode where reduction is forced to take place.)

Calculation of the electromotive force of a galvanic cell. The electromotive force (emf) or voltage of a galvanic cell is equal to the algebraic difference of the potentials of the two electrodes, when the external circuit is open, that is, when no current is flowing. To find the chemical reaction and the emf of the cell, we proceed as follows:

1. We write the equations for the reduction half-reactions of the half-cells and their potentials, as they are calculated by the Nernst equation.

2. If necessary, we multiply one or both of the equations by proper integers so that the number of electrons is the same in both equations and the electrons cancel in subtracting the equations. (We do not multiply the potentials, because they are independent of the number of species participating in each half-reaction, that is, they are independent of the coefficients of the half-reaction.)

3. We subtract the equation of the half-reaction of the left-hand electrode from that of the right-hand electrode, and also the corresponding potentials. The resulting equation is the equation of the cell reaction, whereas the calculated voltage represents the emf of the cell, E_{cell}. If the sign of E_{cell} is positive, the cell reaction proceeds spontaneously from left to right, according to the final equation, whereas if the sign is negative, the reaction goes from right to left. If it is zero, the reaction is at equilibrium. The polarity of the right-hand electrode is the same as the sign of E_{cell}.

By convention, the anode of the galvanic cell is written on the left side and the cathode on the right, whereupon $E_{cell} = E_c - E_a$, where E_c and E_a are the potentials of the cathode and the anode, respectively.

The equilibrium constant K of the reaction of the galvanic cell and the emf of the cell under standard conditions, E^0_{cell}, and the standard free energy change, ΔG^0, are related by the equation

$$\Delta G^0 = -nFE^0_{cell} = -RT\ln K = -2.303RT\log K \tag{8.11}$$

from which

$$\log K = \frac{nFE^0_{cell}}{2.303RT} = \frac{nE^0_{cell}}{0.05916} = 16.903nE^0_{cell}\ (\text{at } 25°\text{C}) \tag{8.11a}$$

or

$$K = 10^{16.903nE^0_{cell}}. \tag{8.11b}$$

Example 8.5. Calculate the emf of the galvanic cell

$$\mathbf{Fe} \mid Fe^{2+}(0.200\ \underline{M}) \quad \| \quad Ag^+(0.100\ \underline{M}) \mid \mathbf{Ag}$$

$$E_2 \qquad\qquad\qquad E_1$$

left-hand electrode **right-hand electrode**

and find the polarity of its electrodes.

Solution. The equations of the half-reactions of the half-cells and the standard potentials are

$$Ag^+ + e \rightleftharpoons \mathbf{Ag},\ E^0_1 = +0.7994\ V$$

$$Fe^{2+} + 2e \rightleftharpoons \mathbf{Fe},\ E^0_2 = -0.440\ V$$

For the silver and iron half-cells, we have the relations

$$E_1 = +0.7994 - 0.05916\log\frac{1}{0.100} = +0.740\ V$$

$$E_2 = -0.440 - \frac{0.05916}{2}\log\frac{1}{0.200} = -0.461\ V,$$

respectively. Hence, we have

$$2Ag^+ + 2e \rightleftharpoons \mathbf{2Ag},\ E_1 = +0.740\ V \tag{8.12}$$

$$Fe^{2+} + 2e \rightleftharpoons \mathbf{Fe}, E_2 = -0.461\ V \tag{8.13}$$

Subtracting Equations (8.12) and (8.13), we have

$$2Ag^+ + \mathbf{Fe} \rightleftharpoons \mathbf{2Ag} + Fe^{2+},\ E_{cell} = +0.740 - (-0.461) = +\mathbf{1.201}\ V \tag{8.14}$$

The positive sign denotes that Equation (8.14) represents the spontaneous cell reaction, and that the silver electrode is the positive electrode (cathode) of the galvanic cell. Consequently, the iron electrode is the negative electrode (anode).

The emf can also be calculated as follows:

$$E_{cell} = E_1 - E_2 = E_1^0 - E_2^0 - \frac{0.05916}{2}\log\frac{[Fe^{2+}]}{[Ag^+]^2} = +0.7994 - (-0.440)$$

$$-\frac{0.05916}{2}\log\frac{0.200}{(0.100)^2} = +\mathbf{1.201}\text{ V}.$$

Example 8.6. Calculate the emf for each of the following cells. Would the cell, as written, have to be galvanic or electrolytic in order for the reaction to occur in the direction as given?

(a) **Pt** | $Cr^{3+}(1.00 \times 10^{-3}\ \underline{M})$, $Cr^{2+}(0.100\ \underline{M})$‖$Cu^{2+}(0.0100\ \underline{M})$ | **Cu**

(b) **Pt** | $Sn^{4+}(2.56\times10^{-4}\ \underline{M})$, $Sn^{2+}(0.040\ \underline{M})$‖$Fe^{2+}(2.00\times10^{-6}\ \underline{M})$, $Fe^{3+}(0.040\ \underline{M})$ | **Pt**

(c) **Pt** | Sn^{4+} $(8.0 \times 10^{-4}\ \underline{M})$, Sn^{2+} $(0.040\ \underline{M})$ ‖ $[Ag(CN)_2]^-(9.0 \times 10^{-4}\ \underline{M})$, CN^- $(0.030\ \underline{M})$ | **Ag**

(d) **Pt**, $H_2(0.100$ atm) | $H^+(0.0100\ \underline{M})$‖**AgCl**, KCl$(1.00\ \underline{M})$ | **Ag**

(e) **Pb** | $\mathbf{PbSO_4}$, $SO_4^{2-}(0.0500\ \underline{M})$‖$Ag^+(0.0100\ \underline{M})$ | **Ag**

Solution. a) We have

$$Cu^{2+} + 2e \rightleftharpoons \mathbf{Cu}(1),\ E_1 = +0.337 + \frac{0.05916}{2}\log(0.0100) = +0.278\text{ V}$$

$$2Cr^{3+} + 2e \rightleftharpoons 2Cr^{2+}(2),\ E_2 = -0.410 - 0.05916\log\frac{0.100}{1.00 \times 10^{-3}} = -0.528\text{ V}$$

$$(1) - (2):\ Cu^{2+} + 2Cr^{2+} \rightleftharpoons \mathbf{Cu} + 2Cr^{3+}$$

and $E_{cell} = E_1 - E_2 = +0.278 - (-0.528) = +\mathbf{0.806}$ V. Hence, since the E_{cell} is positive, the reaction will occur spontaneously and the cell is galvanic.

b) We have

$$2Fe^{3+} + Sn^{2+} \rightleftharpoons 2Fe^{2+} + Sn^{4+}$$

and

$$E_{cell} = +0.771 - 0.15 - \frac{0.05916}{2}\log\frac{(2.00 \times 10^{-6})^2(2.56 \times 10^{-4})}{(0.040)^2(0.040)} = +\mathbf{0.94}\text{ V}.$$

Hence, the cell is galvanic.

c) We have

$$2[Ag(CN)_2]^- + Sn^{2+} \rightleftharpoons 2\mathbf{Ag} + Sn^{4+} + 4CN^-$$

and

$$E_{cell} = E_1^0 - E_2^0 - \frac{0.05916}{2}\log\frac{[Sn^{4+}][CN^-]^4}{[Ag(CN)_2{}^-]^2[Sn^{2+}]} = -0.31 - 0.15 -$$

$$\frac{0.05916}{2}\log\frac{(8.0 \times 10^{-4})(0.030)^4}{(9.0 \times 10^{-4})^2(0.040)} = -\mathbf{0.41}\text{ V}.$$

Hence, the cell is electrolytic because the sign of the E_{cell} is negative, i.e, an external emf of 0.41 V or greater must be applied in order for the reaction to occur as given.

d) $\mathbf{2AgCl} + H_2 \rightleftharpoons \mathbf{2Ag} + 2H^+ + 2Cl^-$ (cell reaction)

$$E_{cell} = +0.2224 - 0.000 - \frac{0.05916}{2}\log\frac{(0.0100)^2(1.00)^2}{(0.100)} = +\mathbf{0.3111}\ \text{V}.$$

Hence, the cell is galvanic.

e) $\mathbf{Pb} + 2Ag^+ + SO_4^{2-} \rightleftharpoons \mathbf{PbSO_4} + \mathbf{2Ag}$ (cell reaction)

$$E_{cell} = E_{Ag^+,\,Ag} - E_{PbSO_4,\,Pb} = [+0.7994 + 0.05916\log(0.0100)] -$$

$$\left[-0.356 - \frac{0.05916}{2}\log(0.0500)\right] = +0.6811 - (-0.318) = +\mathbf{0.999}\ \text{V}.$$

Hence, the cell is galvanic.

Example 8.7. For each of the following cells 1) determine the direction of the cell reaction, and 2) calculate the equilibrium constant for each cell reaction.

(a) $\mathbf{Cu} \mid Cu^{2+}(0.0100\ \underline{M}) \| Fe^{2+}(0.200\ \underline{M}),\ Fe^{3+}(0.040\ \underline{M}) \mid \mathbf{Pt}$

(b) $\mathbf{Pt},\ Cl_2(0.20\ \text{atm}) \mid HCl(4.0\ \underline{M}) \| HCl(0.200\ \underline{M}) \mid H_2(1.00\ \text{atm}),\ \mathbf{Pt}$

(c) $\mathbf{Pb} \mid \mathbf{PbSO_4},\ SO_4^{2-}(0.100\ \underline{M}) \| MnCl_2(0.100\ \underline{M}) \mid \mathbf{Mn}$

Solution. a) 1. We have

$$E_{cell} = \left(+0.771 - 0.05916\log\frac{0.200}{0.040}\right) - \left(+0.337 - \frac{0.05916}{2}\log\frac{1}{0.0100}\right) = +0.452\ \text{V}.$$

The positive sign of E_{cell} denotes that the cell reaction proceeds spontaneously from left to right, that is, we have

$$2Fe^{3+} + \mathbf{Cu} \rightleftharpoons 2Fe^{2+} + Cu^{2+}$$

2. We have

$$K = 10^{16.903\times 2\times(0.771-0.337)} = \mathbf{4.7 \times 10^{14}}.$$

b) 1. We have

$$E_{cell} = 0.000 - 1.359 - \frac{0.05916}{2}\log\frac{1.00 \times 0.20}{(0.200)^2(4.0)^2} = -1.344\ \text{V}.$$

The negative sign of E_{cell} denotes that the cell reaction proceeds spontaneously from right to left, that is, the following reaction occurs from right to left:

$$2H^+ + 2Cl^- \rightleftharpoons H_2 + Cl_2$$

2. We have

$$K = 10^{16.903\times 2\times(0.000-1.359)} = 10^{-45.94} = \mathbf{1.1 \times 10^{-46}},$$

for the forward reaction, i.e., negligible reaction. The reciprocal of K represents the equilibrium constant for the spontaneous (reverse) reaction.

c) 1. We have

$$E_{cell} = -1.18 - (-0.356) - \frac{0.05916}{2}\log\frac{1\times 1}{(0.100)(0.100)} = -0.88 \text{ V}.$$

Hence, the cell reaction proceeds from right to left, that is, we have

$$\mathbf{Pb} + Mn^{2+} + SO_4{}^{2-} \rightleftharpoons \mathbf{PbSO_4} + \mathbf{Mn}$$

2. We have $K = 10^{16.903\times 2\times[-1.18-(-0.0356)]} = 10^{-27.86} = \mathbf{1.4 \times 10^{-28}}$.

Example 8.8. In a solution, Fe^{3+} ions are reduced to Fe^{2+} ions by metallic Fe. If $[Fe^{2+}] = 0.100$ M at the equilibrium state, what is the Fe^{3+} ion concentration in equilibrium with Fe^{2+} ions and metallic Fe?

Solution. For the reaction $2Fe^{3+} + \mathbf{Fe} \rightleftharpoons 3Fe^{2+}$ we have

$$K = 10^{16.903\times 2\times[0.771-(-0.440)]} = 8.7\times 10^{40} = \frac{[Fe^{2+}]^3}{[Fe^{3+}]^2}$$

Hence,

$$[Fe^{3+}] = \sqrt{\frac{[Fe^{2+}]^3}{8.7\times 10^{40}}} = \sqrt{\frac{(0.100)^3}{8.7\times 10^{40}}} = \mathbf{1.07\times 10^{-22}}\ \underline{M}.$$

Example 8.9. What is the required $[Zn^{2+}]$ so that in a 1.00×10^{-3} M $CuSO_4$ solution Cu^{2+} ions will not be reduced to metallic Cu by Zn? Is it possible to achieve such a concentration?

Solution. For the reaction

$$Cu^{2+} + \mathbf{Zn} \rightleftharpoons \mathbf{Cu} + Zn^{2+}$$

we have

$$K = 10^{16.903\times 2\times[0.337-(-0.763)]} = 1.5\times 10^{37} = \frac{[Zn^{2+}]}{[Cu^{2+}]}.$$

Hence,

$$[Zn^{2+}] = (1.5\times 10^{37})(1.00\times 10^{-3}) = \mathbf{1.5\times 10^{34}}\ \underline{M}.$$

Such an enormous concentration of Zn^{2+} ions cannot be achieved experimentally, and therefore, it is impossible to avoid the reduction of Cu^{2+} ions to metallic Cu.

Example 8.10. The emf of the galvanic cell

$$(-)\mathbf{Pt},\ H_2\ (1\text{ atm}) \mid HCl(x\underline{M}) \| KCl(1.00\ \underline{M}),\ \mathbf{Hg_2Cl_2} \mid \mathbf{Hg}(+) \qquad (8.15)$$

is 0.339 V, whereas that of the galvanic cell

$$(-)\mathbf{Pt},\ H_2\ (1\ \text{atm})\ |\ NaOH(y\underline{M})||KCl(1.00\ \underline{M}),\ \mathbf{Hg_2Cl_2}\ |\ \mathbf{Hg}(+) \qquad (8.16)$$

is 1.067 V. In what proportion should the HCl and NaOH solutions be mixed to prepare a solution in which $[H^+] = 0.0400$ M?

Solution. For the galvanic cell (8.15), we have

$$E_{cell} = +0.339 = E_c - E_a = E_{Hg_2Cl_2,\ Hg} - E_{H^+,\ H_2} = +0.280 - E_{H^+,\ H_2},$$

from which $E_{H^+,\ H_2} = -0.059$ V. Hence,

$$E_{H^+,\ H_2} = -0.059 = E^0_{H^+,\ H_2} - \frac{0.05916}{2}\log\frac{1}{[H^+]^2} = 0.000 + 0.05916\log[H^+],$$

from which $\log[H^+] = -0.059/0.05916 \simeq -1.00$. Hence, $[H^+] = 0.100$ M and x = 0.100.

For the galvanic cell (8.16) we have $+1.067 = +0.280 - E_{H^+,\ H_2}$. Hence, $E_{H^+,\ H_2} = -0.787 = 0.000 + 0.05916\log[H^+]$, from which $[H^+] = 5.0 \times 10^{-14}$ M. Hence, $[OH^-] = 1.00 \times 10^{-14}/5.0 \times 10^{-14} = 0.200$ M and y = 0.200. Suppose that a mL of 0.100 M HCl should be mixed with b mL of 0.200 M NaOH. We have $(0.100a - 0.200b)/(a+b) = 0.0400$ or a:b = **4:1**.

Example 8.11. The internal resistance of the galvanic cell

$$\mathbf{Pt}\ |\ [Fe(CN_6)^{4-}(0.0100\ \underline{M}),\ [Fe(CN)_6]^{3-}(2.50 \times 10^{-4}\ \underline{M})||Ag^+(2.00 \times 10^{-3}\ \underline{M})\ |\ \mathbf{Ag}$$

is 4.0 ohms. Each half-cell contains 100 mL of solution. Calculate a) the initial voltage, when a current of 0.0100 A is drawn momentarily from the cell, b) the voltage after drawing from the cell a constant current of 0.0100 A for 19.2 min (the change in resistance is considered negligible).

Solution. a) The internal resistance introduces a voltage drop equal to the product of the current (amps) and the resistance (ohms) (Ohm's law). We have

$$Ag^+ + [Fe(CN)_6]^{4-} \rightleftharpoons \mathbf{Ag} + [Fe(CN)_6]^{3-}$$

and

$$E_{cell} = E_c - E_a - iR = [+0.7994 + 0.05916\log(2.00 \times 10^{-3})] -$$

$$\left(+0.356 - 0.05916\log\frac{0.0100}{2.50 \times 10^{-4}}\right) - (0.0100 \times 4.0) = +\mathbf{0.339}\ V.$$

b) From Coulomb's law

$$\frac{(0.0100\ C/s)(19.2\ \text{min}\ \times 60\ s/\text{min})}{96.485\ C/\text{meq}} = 0.1194\ \text{meq}.$$

(The Faraday constant is 96,485 C/eq or 96.485 C/meq, where C is coulombs. Current in amps is equivalent to C/s.) Hence, *after* the passage of the current, we will have

$$[Ag^+] = \frac{(100 \text{ mL})(2.00 \times 10^{-3} \text{ meq/mL}) - 0.1194}{100 \text{ mL}} = 8.06 \times 10^{-4} \underline{M},$$

$$[Fe(CN)_6{}^{3-}] = \frac{(100 \text{ mL})(2.50 \times 10^{-4} \text{ meq/mL}) + 0.1194}{100 \text{ mL}} = 1.444 \times 10^{-3} \underline{M},$$

$$[Fe(CN)_6{}^{4-}] = \frac{(100 \text{ mL})(0.0100 \text{ meq/mL})0.1194}{100 \text{ mL}} = 8.81 \times 10^{-3} \underline{M}.$$

Hence,

$$E_{cell} = [+0.7994 + 0.05916 \log(8.06 \times 10^{-4})] -$$

$$\left(+0.356 - 0.05916 \log \frac{8.81 \times 10^{-3}}{1.444 \times 10^{-3}}\right) - 0.04 = \mathbf{+0.267} \ V.$$

Example 8.12. The emf of the galvanic cell

$$(-)\mathbf{Cu} \mid Cu^{2+}(1.00 \times 10^{-4} \ \underline{M}) \| Ag^+(x\underline{M}) \mid \mathbf{Ag}(+)$$

is +0.362 V. What is the $[Ag^+]$ in the silver half-cell?

Solution. We have

$$E_{cell} = +0.362 = (+0.7994 + 0.05916 \log[Ag^+]) - [+0.337 +$$

$$\frac{0.05916}{2} \log(1.00 \times 10^{-4})] = +0.581 + 0.05916 \log[Ag^+],$$

from which $\log[Ag^+] = -3.70$. Hence, $[Ag^+] = 10^{-3.70} = \mathbf{2.0 \times 10^{-4}} \ \underline{M}$.

Example 8.13. Given that

$$Ox_1 + ne \rightleftharpoons Red_1, \ E^0 = +0.293 \text{ V}$$

$$Ox_2 + e \rightleftharpoons Red_2, \ E^0 = -0.200 \text{ V}$$

and the equilibrium constant of the reaction

$$Ox_1 + nRed_2 \rightleftharpoons Red_2 + nOx_2$$

is 1.00×10^{25}, calculate the value of n.

Solution. We have the relation

$$\log K = \log(1.00 \times 10^{25}) = 25 = \frac{n[+0.293 - (-0.200)]}{0.05916} = 8.33n,$$

from which $n = \mathbf{3}$. Note that since the second reaction is subtracted from the first, the second E^0 is subtracted from the first E^0 to give a positive E^0_{cell} for the spontaneous reaction.

8.4 Applications of the Standard Potentials and the Nernst Equation

Below we study specific examples of various calculations, in which standard potentials and the Nernst equation are used.

Calculation of equilibrium constants. Using the table of standard reduction potentials and the Nernst equation, we can calculate the emf of a galvanic cell (Section 8.3), as well as the equilibrium constant of a redox reaction. Sometimes, more than two half-reactions or reactions should be combined (Example 8.14). We can also calculate the equilibrium constants of displacement reactions, for example, the solubility product (Examples 8.16 and 8.17), the dissociation constant of a weak acid or base (Examples 8.18 and 8.19), and the formation or the instability constant of a complex ion (Examples 8.20 and 8.21).

Example 8.14. Which one of the following two reactions takes place during the dissolution of silver in dilute HNO_3?

$$\mathbf{2Ag} + 2H^+ \rightleftharpoons 2Ag^+ + H_2 \uparrow \qquad (8.17)$$

$$\mathbf{3Ag} + 4H^+ + NO_3^- \rightleftharpoons 3Ag^+ + NO \uparrow + 2H_2O \qquad (8.18)$$

Solution. For the reaction (8.17), we have the following data:

$$Ag^+ + e \rightleftharpoons \mathbf{Ag},\ E^0 = +0.7994\ V \qquad (8.19)$$

$$2H^+ + 2e \rightleftharpoons H_2,\ E^0 = +0.0000\ V \qquad (8.20)$$

By multiplying Equation (8.19) by 2 and subtracting the result from Equation (8.20), we have

$$\mathbf{2Ag} + 2H^+ \rightleftharpoons 2Ag^+ + H_2 \uparrow,\ E^0_{cell} = -0.7994\ V.$$

Hence, on the basis of Equation (8.11*b*), for reaction (8.17) we have

$$K = 10^{16.903 \times 2 \times (-0.7994)} = 9.45 \times 10^{-28}.$$

From the very small value of K, it is concluded that reaction (8.17) proceeds spontaneously from right to left, that is, that the dissolution of silver in dilute HNO_3 can not take place according to Equation (8.17). (We conclude the same from the negative cell emf.)

For the reaction (8.18), we have the following data:

$$NO_3^- + 4H^+ + 3e \rightleftharpoons NO + 2H_2O,\ E^0 = +0.96\ V \qquad (8.21)$$

$$Ag^+ + e \rightleftharpoons \mathbf{Ag},\ E^0 = +0.7994\ V \qquad (8.19)$$

By multiplying Equation (8.19) by 3 and subtracting the result from Equation (8.21), we have

$$3\mathbf{Ag} + 4H^+ + NO_3^- \rightleftharpoons 3Ag^+ + NO\uparrow + 2H_2O,\ E^0_{cell} = +0.1606\ V \tag{8.18}$$

Hence, on the basis of Equation (8.11*b*), for reaction (8.18) we have

$$K = 10^{16.903\times2\times0.1606} = 1.39\times10^8.$$

From the large value of K (and the positive value of E^0_{cell}), it is concluded that reaction (8.18) proceeds spontaneously from left to right. Hence, the dissolution of silver in dilute HNO_3 takes place according to Equation (8.18).

Example 8.15. In the reaction

$$A_{ox} + 2B_{red} \rightleftharpoons A_{red} + 2B_{ox} \tag{8.22}$$

the oxidation number of A is reduced by 2. Calculate the minimal required difference in standard potentials, $E^0_A - E^0_B$, so that the reaction is completed by 99.9%.

Solution. In order for reaction (8.22) to be completed by 99.9%, the equilibrium constant should be equal to at least

$$K = \frac{[A_{red}][B_{ox}]^2}{[A_{ox}][B_{red}]^2} = \frac{(0.999)(0.999)^2}{(0.001)(0.001)^2} = 9.97\times10^8.$$

Substituting the value of K in Equation (8.11*a*), we have

$$\log(9.97\times10^8) = 16.903\times2\times(E^0_A - E^0_B),$$

from which $(E^0_A - E^0_B) = \mathbf{0.2662}$ V.

Determination of the solubility product. By proper selection of half-cells, the measured emf of a galvanic cell can be used for the calculation of the solubility product of slightly soluble (insoluble) electrolytes.

Example 8.16. Given that the emf of the galvanic cell

$$\mathbf{Ag} \mid \mathbf{AgBr},\ Br^-(0.100\ \underline{M}) \| KCl\ (satur.),\ \mathbf{Hg_2Cl_2} \mid \mathbf{Hg}$$

is +0.110 V and that the saturated calomel electrode is the positive electrode (cathode), calculate the solubility product of AgBr.

Solution. *1st method.* The emf of the cell is equal to

$$E_{cell} = +0.110 = E_c - E_a = E_{Hg_2Cl_2,\ Hg} - E_{AgBr,\ Ag} = +0.2412 - E_{AgBr,\ Ag},$$

whereupon

$$E_{AgBr,\ Ag} = +0.2412 - 0.110 = +0.131\ V.$$

Hence,

$$+0.131 = E^0_{Ag^+,\ Ag} + 0.05916 \log[Ag^+] = +0.7994 + 0.05916 \log[Ag^+],$$

whereupon

$$\log[Ag^+] = (+0.131 - 0.7994)/0.05916 = -11.30 \text{ and } [Ag^+] = 5.0 \times 10^{-12}\ \underline{M}.$$

Hence,

$$K_{sp(AgBr)} = [Ag^+][Br^-] = (5.0 \times 10^{-12})(0.100) = \mathbf{5.0 \times 10^{-13}}.$$

2nd method. From appendix B, we have

$$\mathbf{AgBr} + e \rightleftharpoons \mathbf{Ag} + Br^-,\ E^0_1 = 0.095 \text{ V} \tag{8.23}$$

$$Ag^+ + e \rightleftharpoons \mathbf{Ag},\ E^0_2 = +0.7994 \text{ V} \tag{8.24}$$

Subtracting Equation (8.24) from (8.23), we have

$$\mathbf{AgBr} \rightleftharpoons Ag^+ + Br^-,\ E^0_{1-2} = -0.7044 \text{ V}.$$

We have at equilibrium, $\Delta G = 0 = -nFE_{cell}$ (see Equation (8.11)), i.e., $E_{cell} = 0$, or

$$E_{1-2} = 0 = E^0_{1-2} - 0.05916 \log[Ag^+][Br^-] = -0.7044 - 0.05916 \log K_{sp(AgBr)}.$$

Hence,

$$\log K_{sp(AgBr)} = -0.7044/0.05916 = -11.91 \text{ and } K_{sp(AgBr)} = \mathbf{1.2 \times 10^{-12}}.$$

3rd method. Applying the Nernst equation to the half-cell

$$\mathbf{Ag} \mid \mathbf{AgBr},\ Br^-(1.00\ \underline{M})$$

we have

$$E = +0.095 = E^0_{Ag^+,\ Ag} - 0.05916 \log(1/[Ag^+]) = +0.7994 + 0.05916 \log[Ag^+],$$

from which

$$\log[Ag^+] = (+0.095 - 0.7994)/0.05916 = -11.91 \text{ and } [Ag^+] = 1.2 \times 10^{-12}\ \underline{M}.$$

Hence,

$$K_{sp(AgBr)} = [Ag^+][Br^-] = (1.2 \times 10^{-12})(1.00) = \mathbf{1.2 \times 10^{-12}}.$$

Note. The above values are within about a factor of 2, within expected experimental uncertainty. The E^0 values are probably known more accurately.

Example 8.17. A 0.200 M NaX solution is saturated with the slightly soluble compound MX_2, and the saturated solution becomes a part of the galvanic cell

$$\mathbf{M} \mid \mathbf{MX_2},\ NaX(0.200\ \underline{M}) \| H^+(1.00\ \underline{M}) \mid H_2\ (1\text{ atm}),\ \mathbf{Pt}$$

The emf of the cell is +0.080 V, and the M electrode is the anode (negative electrode). Calculate the solubility product of MX_2, given $E^0_{M^{2+}, M} = +0.255$ V.

Solution. The emf of the cell is equal to

$$E_{cell} = E_c - E_a = E_{H^+, H_2} - E_{M^{2+}, M} = 0.000 - E_{M^{2+}, M} = +0.080,$$

whereupon

$$E_{M^{2+}, M} = -0.080 = +0.255 - \frac{0.05916}{2} \log \frac{1}{[M^{2+}]},$$

from which we have

$$\log[M^{2+}] = -11.33.$$

Hence,

$$[M^{2+}] = 4.7 \times 10^{-12} \text{ and } K_{sp(MX_2)} = (4.7 \times 10^{-12})(0.200)^2 = \mathbf{1.88 \times 10^{-13}}.$$

Determination of the dissociation constant of weak acids and bases. By proper selection of half-cells, we can have a galvanic cell, the emf of which depends on the H^+ ion concentration in a solution of a weak acid or a weak base. From the measured emf, $[H^+]$ is calculated, and from this value and the known concentration of the acid or the base, the dissociation constant is calculated (Example 8.18). The dissociation constant can also be calculated by measuring $[H^+]$ in a solution of a salt of the weak acid or the weak base (Example 8.19).

Example 8.18. The emf of the galvanic cell

$$\mathbf{Pt}, H_2(1.00 \text{ atm}) \mid HA(0.100\ \underline{M}) \| KCl(1.00\ \underline{M}),\ \mathbf{Hg_2Cl_2} \mid \mathbf{Hg}$$

is +0.480 V and the calomel electrode is the cathode. Calculate the dissociation constant of the weak acid HA.

Solution. The emf of the cell is equal to

$$E_{cell} = +0.480 = E_c - E_a = E_{Hg_2Cl_2, Hg} - E_{H^+, H_2},$$

whereupon

$$E_{H^+, H_2} = +0.280 - 0.480 = -0.200 \text{ V}.$$

Hence,

$$E_{H^+, H_2} = -0.200 = E^0_{H^+,H_2} - \frac{0.05916}{2} \log \frac{1.00}{[H^+]^2} = 0.000 + 0.05916 \log[H^+],$$

from which

$$\log[H^+] = -0.200/0.05916 = -3.381 \text{ and } [H^+] = 4.16 \times 10^{-4}\ \underline{M}.$$

Hence,

$$[A^-] = [H^+] = 4.16 \times 10^{-4}\ \underline{M},\ [HA] = (0.100 - 4.16 \times 10^{-4})\ \underline{M},$$

and

$$K_{HA} = \frac{[H^+][A^-]}{[HA]} = \frac{(4.16 \times 10^{-4})^2}{0.100 - 0.000416} = \mathbf{1.74 \times 10^{-6}}.$$

Example 8.19. Given that the emf of the galvanic cell

$$(-)\mathbf{Pt}, H_2(1.00\ \text{atm}) \mid NH_4Cl(0.200\ \underline{M}) \| KCl(1.00\ \underline{M}),\ \mathbf{Hg_2Cl_2} \mid \mathbf{Hg}(+)$$

is +0.574 V, calculate the dissociation constant of ammonia.

Solution. The emf of the cell is equal to

$$E_{cell} = +0.574 = E_c - E_a = E_{Hg_2Cl_2,\ Hg} - E_{H^+,\ H_2} = +0.280 - E_{H^+,\ H_2},$$

whereupon $E_{H^+,\ H_2} = +0.280 - 0.574 = -0.294$ V. Hence,

$$E_{H^+,\ H_2} = -0.294 = E^0_{H^+,\ H_2} - \frac{0.05916}{2} \log \frac{1.00}{[H^+]^2} = 0.000 + 0.05916 \log[H^+],$$

whereupon,

$$\log[H^+] = -0.294/0.05916 = -4.970 \text{ and } [H^+] = 1.07 \times 10^{-5}\ \underline{M}.$$

Hence,

$$[NH_3] = 1.07 \times 10^{-5}\ \underline{M}, [NH_4^+] = (0.200 - 1.07 \times 10^{-5})\ \underline{M} \simeq 0.200\ \underline{M},$$

and

$$[OH^-] = 1.0 \times 10^{-14}/1.07 \times 10^{-5} = 9.35 \times 10^{-10}\ \underline{M}.$$

Hence,

$$K_{NH_3} = \frac{[NH_4^+][OH^-]}{[NH_3]} = \frac{(0.200)(9.35 \times 10^{-10})}{1.07 \times 10^{-5}} = \mathbf{1.75 \times 10^{-5}}.$$

Determination of the instability constant of a complex ion. For the determination of the instability constant K_{inst} of the complex ion $[ML_n]^{a-nb}$, a galvanic cell is constructed, the emf of which depends on the concentration of the ion being complexed, M^{n+}. From the measured emf, the concentration of the M^{n+} ion is calculated, and from this value and the known concentrations of the complex ion and the complexing reagent L, the instability constant of the complex ion is calculated.

Example 8.20. The emf of the galvanic cell

$$(-)\mathbf{Ag} \mid [Ag(S_2O_3)_2]^{3-}(0.0100\ \underline{M}),\ S_2O_3^{2-}(1.00\ \underline{M}) \| Ag^+(0.100\ \underline{M}) \mid \mathbf{Ag}(+)$$

is +0.828 V. Calculate the instability constant of the complex $[Ag(S_2O_3)_2]^{3-}$.

Solution. *1st method.* The emf of the cell is equal to

$$E_{cell} = +0.828 = E_c - E_a = E_{Ag^+,\ Ag} - E_{[Ag(S_2O_3)_2]^{3-},\ Ag} = (E^0_{Ag^+,\ Ag} + 0.05916$$

$$\log[Ag^+]) - E_{[Ag(S_2O_3)_2]^{3-},\ Ag} = +0.7994 + 0.05916 \log 0.100 - E_{[Ag(S_2O_3)_2]^{3-},\ Ag}$$

whereupon

$$E_{[Ag(S_2O_3)_2]^{3-},\ Ag} = -0.828 + 0.7994 - 0.05916 = -0.088 \text{ V}.$$

Hence,

$$E_{[Ag(S_2O_3)_2]^{3-},\ Ag} = -0.088 = E^0_{Ag^+,\ Ag} + 0.05916\ \log[Ag^+] = +0.7994 + 0.05916 \log[Ag^+],$$

whereupon,

$$\log[Ag^+] = (-0.088 - 0.7994)/0.05916 = -15.00 \text{ and } [Ag^+] = 1.0 \times 10^{-15}.$$

Hence,

$$K_{inst} = \frac{[Ag^+][S_2O_3{}^{2-}]^2}{[Ag(S_2O_3)_2{}^{3-}]} = \frac{(1.0 \times 10^{-15})(1.00)^2}{(0.0100)} = \mathbf{1.0 \times 10^{-13}}.$$

2nd method. From appendix B, we have

$$[Ag(S_2O_3)_2]^{3-} + e \rightleftharpoons \mathbf{Ag} + 2S_2O_3{}^{2-},\ E^0_1 = +0.01 \text{ V} \tag{8.25}$$

$$Ag^+ + e \rightleftharpoons \mathbf{Ag},\ E^0_2 = +0.7994 \text{ V} \tag{8.26}$$

Subtracting Equation (8.26) from Equation (8.25), we have

$$[Ag(S_2O_3)_2]^{3-} \rightleftharpoons Ag^+ + 2S_2O_3{}^{2-},\ E^0_{1-2} = E^0_1 - E^0_2 = -0.7894 \text{ V}.$$

We have

$$E_{1-2} = E^0_{1-2} - 0.05916 \log \frac{[Ag^+][S_2O_3{}^{2-}]^2}{[Ag(S_2O_3)_2{}^{3-}]} = E^0_{1-2} - 0.05916 \log K_{inst}.$$

At equilibrium, we have $E_{1-2} = 0$. Hence,

$$0 = -0.7894 - 0.05916 \log K_{inst}$$

whereupon

$$\log K_{inst} = -0.7894/0.05916 = -13.34_3.$$

Hence, $K_{inst} = 10^{-13.34_3} = \mathbf{4.5 \times 10^{-14}}$.

3rd method. Applying the Nernst equation to the half-cell

$$\mathbf{Ag} \mid [Ag(S_2O_3)_2]^{3-}(1.00\ \underline{M}),\ S_2O_3{}^{2-}(1.00\ \underline{M})$$

we have

$$E = +0.01 = E^0_{Ag^+,\ Ag} + 0.05916 \log[Ag^+] = +0.7994 + 0.05916 \log$$

$$\frac{K_{inst}[Ag(S_2O_3)_2{}^{3-}]}{[S_2O_3{}^{2-}]^2} = +0.7994 + 0.05916 \log \frac{K_{inst}\ (1.00)}{(1.00)^2},$$

from which

$$\log K_{inst} = (+0.01 - 0.7994)/0.05916 = -13.34_3 \text{ and } K_{inst} = \mathbf{4.5 \times 10^{-14}}.$$

Example 8.21. The emf of the galvanic cell

$$\mathbf{Cu} \mid [CuY]^{2-}(1.0 \times 10^{-4}\ \underline{M}),\ Y^{4-}(2.0 \times 10^{-3}\ \underline{M}) \| H^+(1.00\ \underline{M}) \mid H_2(1\ \text{atm}),\ \mathbf{Pt}$$

is +0.258 V, and the hydrogen electrode is the cathode. Calculate the instability constant of the complex of Cu(II) with ethylenediaminetetraacetic acid, $[CuY]^{2-}$.

Solution. The emf of the cell is equal to

$$E_{cell} = E_c - E_a = E_{H^+,\,H_2} - E_{[CuY]^{2-},\,Cu} = 0.000 - E_{[CuY]^{2-},\,Cu} = +0.258,$$

whereupon,

$$E_{[CuY]^{2-},\,Cu} = -0.258 = E^0_{Cu^{2+},\,Cu} - \frac{0.05916}{2}\log\frac{1}{[Cu^{2+}]} = +0.337 - \frac{0.05916}{2}\log\frac{1}{[Cu^{2+}]},$$

from which

$$\log[Cu^{2+}] = -20.11_5.$$

Hence,

$$[Cu^{2+}] = 7.7 \times 10^{-21} \text{ and } K_{inst[CuY]^{2-}} = \frac{[Cu^{2+}][Y^{4-}]}{[CuY^{2-}]}$$

$$= \frac{(7.7 \times 10^{-21})(2.0 \times 10^{-3})}{1.0 \times 10^{-4}} = \mathbf{1.5 \times 10^{-19}}.$$

Calculation of the standard electrode potential of an unknown half-reaction. The standard electrode (reduction) potential E^0 of an unknown half-reaction can be obtained from the combination of two or more known half-reactions which add to give the desired half-reaction. We multiply the standard reduction potential of each known half-reaction by its number of electrons, and divide the algebraic sum of the products by the number of electrons involved in the unknown half-reaction. This results from the fact that Gibbs free energies of the reactions are additive, i.e., $\Delta G^0_{unk} = \Delta G^0_1 + \Delta G^0_2 + \ldots + \Delta G^0_n$. Since $\Delta G^0 = -nFe^0$, we can write that

$$-n_{unk}FE^0_{unk} = -n_1FE^0_1 - n_2FE^0_2 - \ldots n_nFE^0_n$$

from which

$$E^0_{unk} = (n_1E^0_1 + n_2E^0_2 + \ldots n_nE^0_n)/n_{unk}.$$

Example 8.22. Calculate E^0 for the half-reaction

$$Fe^{3+} + 3e \rightleftharpoons \mathbf{Fe} \qquad (8.27)$$

Solution. Half-reaction (8.27) is equal to the sum of the half-reactions

$$Fe^{3+} + e \rightleftharpoons Fe^{2+}, \; E^0_{Fe^{3+},\,Fe^{2+}} = +0.771 \text{ V} \tag{8.28}$$

$$Fe^{2+} + 2e \rightleftharpoons \mathbf{Fe}, \; E^0_{Fe^{2+},\,Fe} = -0.440 \text{ V} \tag{8.29}$$

Hence, we have

$$E^0_{Fe^{3+},\,Fe} = \frac{+0.771 + 2(-0.440)}{3} = \mathbf{-0.036} \text{ V}$$

Example 8.23. Calculate the standard reduction potential of the half-reaction $2H_2O + 2e \rightleftharpoons H_2 + 2\,OH^-$, given that at 25°C we have

$$E^0_{H^+,\,H_2} = 0.00 \text{ V and } K_w = 1.00 \times 10^{-14}.$$

Solution. For the half-reaction $2H^+ + 2e \rightleftharpoons H_2$, we have the equation

$$E = E^0_{H^+,\,H_2} - \frac{0.05916}{2} \log \frac{P_{H_2}}{[H^+]^2}. \tag{8.30}$$

Combining Equation (8.30) with the relation $K_w = [H^+][OH^-] = 1.00 \times 10^{-14}$, we have

$$E = E^0_{H^+,\,H_2} - \frac{0.05916}{2} \log \frac{P_{H_2}[OH^-]^2}{{K_w}^2} \tag{8.31}$$

or

$$E = (E^0_{H^+,\,H_2} + 0.05916 \log K_w) - \frac{0.05916}{2} \log(P_{H_2}[OH^-]^2) \tag{8.31a}$$

The potential E in Equation (8.31*a*) corresponds to that of the half-reaction $2H_2O + 2e \rightleftharpoons H_2 + 2\,OH^-$, in which the terms $E^0_{H^+,\,H_2} + 0.05916 \log K_w$ correspond to the standard potential of this half-reaction, $E^0_{H_2O,\,H_2}$. Hence, we have

$$E^0_{H_2,\,H_2O} = 0.000 + 0.05916 \log(1.00 \times 10^{-14}) = \mathbf{-0.828} \text{ V}.$$

This value coincides with the one given in Appendix B.

Calculation of concentrations in redox systems at equilibrium. The calculation of the concentration of one or more components of a redox system at equilibrium is carried out using equilibrium constants derived from the standard potentials (Equation (8.11*b*)), in the same manner as in the equilibria studied in chapters 5, 6, and 7, dealing with acid-base reactions, precipitation reactions and complex formation reactions.

Example 8.24. Calculate the concentrations of Ce^{3+}, Ce^{4+}, Fe^{2+}, and Fe^{3+} ions at the equilibrium state, in a mixture that is obtained by adding 25.0 mL of 0.0100 M cerium(IV) sulfate to 50.0 mL of 0.0100 M iron(II) sulfate. Both solutions are 1.00 M in H_2SO_4.

Solution. After mixing, the solution contains

$$(25.0\ \text{mL})(0.0100\ \text{mmol Ce}^{4+}/\text{mL}) = 0.250\ \text{mmol Ce}^{4+}$$

and

$$(50.0\ \text{mL})(0.0100\ \text{mmol Fe}^{2+}/\text{mL}) = 0.500\ \text{mmol Fe}^{2+},$$

which react according to the reaction

$$Ce^{4+} + Fe^{2+} \rightleftharpoons Ce^{3+} + Fe^{3+},\ K = 10^{16.903\times(1.44-0.674)} = 8.9\times10^{12} \quad (8.32)$$

The large value of K denotes that reaction (8.32) is quantitative. Therefore, practically 0.250 mmol of Fe^{3+} is formed, and 0.250 mmol of Fe^{2+} remains unreacted. Hence, after the reaction, at the equilibrium state we have

$$[Fe^{2+}] = 0.250\ \text{mmol}/75.0\ \text{mL} = \mathbf{0.00333}\ \underline{M}$$

$$[Fe^{3+}] = [Ce^{3+}] = 0.250\ \text{mmol}/75.0\ \text{mL} = \mathbf{0.00333}\ \underline{M}.$$

Substituting the above values in the equilibrium constant expression, we have

$$K = 8.9\times10^{12} = \frac{[Ce^{3+}][Fe^{3+}]}{[Ce^{4+}][Fe^{2+}]} = \frac{(0.00333)^2}{[Ce^{4+}](0.00333)}$$

from which

$$[Ce^{4+}] = \mathbf{3.75\times10^{-16}}\ \underline{M}.$$

Example 8.25. The $[I_3]^-$ ions oxidize H_3AsO_3 to H_3AsO_4 according to the reaction

$$[I_3]^- + H_2SO_3 + H_2O \rightleftharpoons H_3AsO_4 + 2H^+ + 3I^-.$$

If at equilibrium we have $[H_3AsO_3] = [H_3AsO_4] = [I^-] = 0.200$ $\underline{M}$, calculate the triodide concentration, when a) pH = 2.00, b) pH = 5.00. Explain the difference in $[I_3^-]$ at the two pH values using Le Châtelier's principle.

Solution. a) We have

$$[I_3]^- + 2e \rightleftharpoons 3I^-, E^0_{[I_3]^-,\ I^-} = +0.536\ V \quad (8.33)$$

$$H_3AsO_4 + 2H^+ + 2e \rightleftharpoons H_3AsO_3 + H_2O,\ E^0_{H_3AsO_4,\ H_3AsO_3} = +0.559\ V \quad (8.34)$$

Subtracting Equation (8.34) from Equation (8.33), we have the reaction

$$[I_3]^- + H_3AsO_3 + H_2O \rightleftharpoons H_3AsO_4 + 2H^+ + 3I^- \quad (8.35)$$

the equilibrium constant of which is equal to

$$K = \frac{[H_3AsO_4][H^+]^2[I^-]^3}{[H_3AsO_3][I_3^-]} = 10^{16.903\times2\times(0.563-0.559)} = 0.167 \quad (8.36)$$

Substituting the data in Equation (8.36), we have

$$0.167 = \frac{(0.200)(1.0\times10^{-2})^2(0.200)^3}{(0.200)[I_3^-]}$$

from which

$$[I_3^-] = \mathbf{4.8 \times 10^{-6}}\ \underline{M}.$$

b) At pH 5.00 we have

$$0.167 = \frac{(0.200)(1.0 \times 10^{-5})^2(0.200)^3}{(0.200)[I_3^-]}$$

from which

$$[I_3^-] = \mathbf{4.8 \times 10^{-12}}\ \underline{M}.$$

The observed decrease in $[I_3^-]$ from 4.8×10^{-6} to 4.8×10^{-12} upon increasing the pH from 2.00 to 5.00 is in accordance with the Le Châtelier's principle, because according to this principle a decrease in $[H^+]$ shifts the equilibrium in reaction (8.35) to the right, whereupon the $[I_3^-]$ is decreased.

Example 8.26. An excess of metallic aluminum is added to a 0.200 M Cu^{2+} solution. Calculate the Cu^{2+} ion concentration at the equilibrium state.

Solution. We have the reaction

$$3Cu^{2+} + \mathbf{2Al} \rightleftharpoons \mathbf{3Cu} + 2Al^{3+}, \tag{8.37}$$

the equilibrium constant of which is equal to

$$K = 10^{16.903 \times 6 \times [0.337-(-1.66)]} = 3.4 \times 10^{202}.$$

From the large value of K, it can be seen that reaction (8.37) is quantitative. Hence, at the equilibrium state we have

$$[Al^{3+}] = \frac{2}{3}(0.200 - [Cu^{2+}]) \simeq 0.1333\ \underline{M},$$

whereupon

$$3.4 \times 10^{202} = \frac{[Al^{3+}]^2}{[Cu^{2+}]^3} = \frac{(0.1333)^2}{[Cu^{2+}]^3},$$

from which

$$[Cu^{2+}] = \mathbf{8.1 \times 10^{-69}}\ \underline{M}.$$

Example 8.27. From data given in the table of standard reduction potentials (Appendix B), calculate a) the solubility of iodine in water at 25°C, and b) the equilibrium constant for the reaction $I_2(aq) + I^- \rightleftharpoons [I_3]^-$.

Solution. a) On the basis of the Nernst equation we have

$$E^0_{I_2,\ I^-} = E^0_{I_2(aq),\ I^-} - \frac{0.05916}{2} \log \frac{[I^-]^2}{[I_2]}$$

or

$$0.5355 = 0.6197 - \frac{0.05916}{2} \log \frac{1^2}{[I_2]}$$

from which $[I_2] = \mathbf{1.42 \times 10^{-3}}$ $\underline{M}$ (solubility of I_2 in H_2O, at 25°C).

b) From the table of standard reduction potentials (Appendix B), we have

$$I_2(aq) + 2e \rightleftharpoons 2I^-, \ E^0 = +0.6197 \text{ V} \tag{8.38}$$

$$[I_3]^- + 2e \rightleftharpoons 3I^-, \ E^0 = +0.536 \text{ V} \tag{8.39}$$

Subtracting Equation (8.39) from Equation (8.38), we obtain the reaction

$$I_2(aq) + I^- \rightleftharpoons [I_3]^-, \ \Delta E^0 = +0.0837 \text{ V},$$

the equilibrium constant of which is equal to

$$K = 10^{16.903 \times 2 \times 0.0837} = \mathbf{6.8 \times 10^2}.$$

Example 8.28. An acidic 0.100 $\underline{M}$ Fe^{2+} solution is left in contact with the atmospheric air. If at equilibrium we have $[H^+] = 0.100$ $\underline{M}$ and $p_{O_2} = 0.20$ atm, calculate the percentage of Fe^{2+} which has not been oxidized.

Solution. We have the half-reactions

$$O_2 + 4H^+ + 4e \rightleftharpoons 2H_2O, \ E^0_{O_2,\ H_2O} = +1.229 \text{ V} \tag{8.40}$$

$$Fe^{3+} + e \rightleftharpoons Fe^{2+}, \ E^0_{Fe^{3+},\ Fe^{2+}} = +0.771 \text{ V} \tag{8.41}$$

Multiplying Equation (8.41) by 4 and subtracting the result from Equation (8.40), we obtain the reaction

$$O_2 + 4Fe^{2+} + 4H^+ \rightleftharpoons 4Fe^{3+} + 2H_2O, \ \Delta E^0 = +0.458 \text{ V}, \tag{8.42}$$

the equilibrium constant of which is equal to

$$K = 10^{16.903 \times 4 \times 0.458} = 9.3 \times 10^{30}.$$

From the large value of K, it is concluded that Equation (8.42) is quantitative. Hence, at the equilibrium state we have

$$K = 9.3 \times 10^{30} = \frac{[Fe^{3+}]^4}{P_{O_2}[Fe^{2+}]^4[H^+]^4} = \frac{(0.100)^4}{(0.20)[Fe^{2+}]^4(0.100)^4},$$

from which

$$[Fe^{2+}] = 2.7 \times 10^{-8} \ \underline{M}.$$

Hence, the percentage of Fe^{2+} that has not been oxidized is equal to

$$(2.7 \times 10^{-8}/0.100)100 = \mathbf{2.7 \times 10^{-5}}\%.$$

Note. This oxidation is kinetically relatively slow. Iron(II) is actually more rapidly oxidized in alkaline medium due to removal of iron(III) as the hydrous oxide.

Example 8.29. Given that the galvanic cell

$$\mathbf{Pt} \mid Fe^{2+}(0.250\ \underline{M}),\ Fe^{3+}(0.00300\ \underline{M}) \|$$

$$MnO_4^-(0.0400\ \underline{M}),\ Mn^{2+}(0.0100\ \underline{M}),\ H^+(1.00\ \underline{M}) \mid \mathbf{Pt},$$

contains equal volumes of solutions in the two half-cells, calculate a) the emf of the cell, b) the potentials of the half-cells and the concentrations of the various ions at the equilibrium state.

Solution. The equations of the half-cell reactions and their standard reduction potentials are

$$MnO_4^- + 8H^+ + 5e \rightleftharpoons Mn^{2+} + 4H_2O,\ E_1^0 = +1.51\ V$$

$$Fe^{3+} + e \rightleftharpoons Fe^{2+},\ E_2^0 = +0.771\ V.$$

For the right-hand and the left-hand half-cells, we have

$$E_1 = +1.51 - \frac{0.05916}{5} \log \frac{0.0100}{(0.0400)(1.00)^8} = +1.517\ V$$

$$E_2 = +0.771 - 0.05916 \log \frac{0.250}{0.00300} = +0.6574\ V,$$

respectively. Hence, the emf of the cell is equal to

$$E_{cell} = E_1 - E_2 = +1.517 - 0.6574 = +\mathbf{0.860}\ V.$$

b) The equilibrium constant for the cell-reaction

$$MnO_4^- + 5Fe^{2+} + 8H^+ \rightleftharpoons Mn^{2+} + 5Fe^{3+} + 4H_2O \tag{8.43}$$

is equal to

$$K = 10^{16.903 \times 5 \times (1.51 - 0.771)} = 2.9 \times 10^{62}.$$

From the large value of K it is concluded that reaction (8.43) is quantitative. Initially, the Fe^{2+} is in excess relative to MnO_4^-. Therefore, the $[MnO_4^-]$ will be decreased during the reaction practically by 0.0400 M, the $[Fe^{2+}]$ by $5 \times 0.0400 = 0.200$ M, and the $[H^+]$ by $8 \times 0.0400 = 0.320$ M, while the $[Fe^{3+}]$ will be increased practically by 0.200 M, and the $[Mn^{2+}]$ by 0.0400 M. Therefore, at equilibrium, the concentrations of the various ions are:

$$[MnO_4^-] = x$$

$$[Fe^{2+}] \simeq 0.250 - 0.200 = 0.050\ \underline{M}$$

$$[H^+] \simeq 1.00 - 0.320 = 0.68\ \underline{M}$$

$$[Mn^{2+}] \simeq 0.0100 + 0.0400 = 0.0500\ \underline{M}$$

$$[Fe^{3+}] \simeq 0.00300 + 0.200 = 0.203\ \underline{M}.$$

Hence, at equilibrium the potential of the left-hand half-cell is equal to

$$E_1' = +0.771 - 0.05916 \log \frac{[Fe^{2+}]}{[Fe^{3+}]} = +0.771 - 0.05916 \log \frac{0.050}{0.203} = \mathbf{+0.807}\ V,$$

and this is also the value of the potential E_2' of the right-hand half-cell (at equilibrium, $E_1' - E_2' = 0$). Substituting the above values of the concentrations in the equilibrium constant expression, we have

$$K = 2.9 \times 10^{62} = \frac{[Mn^{2+}][Fe^{3+}]^5}{[MnO_4^-][Fe^{2+}]^5[H^+]^8} = \frac{(0.0500)(0.203)^5}{x(0.050)^5(0.68)^8},$$

from which

$$x = [MnO_4^-] = \mathbf{4.2 \times 10^{-60}}\ \underline{M}.$$

8.5 Problems

To solve the following problems it is assumed, unless stated otherwise, that the temperature is 25°C, and that the activities of the ions equal their concentrations. The required standard potentials are given in Appendix B.

Oxidizing and reducing strength

8.1. Arrange the following substances in decreasing order of oxidizing strengths:

$$S_2O_8^{2-},\ MnO_4^-,\ Cr_2O_7^{2-},\ H_2O_2, Cd^{2+},\ Cu^{2+},\ K^+,\ HClO.$$

8.2. Arrange the following substances in increasing order of reducing strengths:

$$Ce^{3+},\ K,\ Cl^-,\ Ni,\ H_2S,\ Br^-,\ I^-, H_3AsO_3.$$

The Nernst equation

8.3. Write the Nernst equation for each of the following half-reactions:

(a) $\mathbf{Zn} + 4CN^- \rightleftharpoons [Zn(CN)_4]^{2-} + 2e$
(b) $\mathbf{PbO_2} + 4H^+ + 2e \rightleftharpoons Pb^{2+} + 2H_2O$
(c) $Cu^{2+} + Br^- + e \rightleftharpoons \mathbf{CuBr}$
(d) $NO_2^- + H_2O + e \rightleftharpoons NO + 2\,OH^-$
(e) $\mathbf{CdS} + 2H^+ + 2e \rightleftharpoons \mathbf{Cd} + H_2S\uparrow$

(f) $\mathbf{Hg_2Cl_2} + 2e \rightleftharpoons \mathbf{2Hg} + 2Cl^-$

Standard potential

8.4. From data given in the table of standard reduction potentials (Appendix B), indicate which direction the following reactions will proceed, given that the activities of all substances are 1.00 $\underline{M}$:

(a) $\mathbf{I_2} + 2H_2O \rightleftharpoons H_2O_2 + 2I^- + 2H^+$
(b) $\mathbf{2AgBr} + \mathbf{Zn} \rightleftharpoons \mathbf{2Ag} + Zn^{2+} + 2Br^-$
(c) $V^{3+} + \mathbf{Ag} \rightleftharpoons V^{2+} + Ag^+$
(d) $Ce^{3+} + [Fe(CN)_6]^{3-} + 3SO_4{}^{2-} \rightleftharpoons [Ce(SO_4)_3]^{2-} + [Fe(CN)_6]^{4-}$

8.5. Calculate the standard potentials of the following half-reactions:

(a) $[Ni(NH_3)_4]^{2+} + 2e \rightleftharpoons \mathbf{Ni} + 4NH_3$
(b) $\mathbf{Mn(OH)_2} + 2e \rightleftharpoons \mathbf{Mn} + 2\,OH^-$

8.6. In the reaction

$$A_{ox} + 5B_{red} \rightleftharpoons A_{red} + 5B_{ox}$$

the oxidation number of B is increased by 1. Calculate the minimal required difference in standard potentials, $E^0_A - E^0_B$, so that the reaction is completed by 99.9%.

8.7. Given that $E^0_{Ag^+,\,Ag} = +0.7994$ V and $K_{sp(AgI)} = 8.5 \times 10^{-17}$, calculate the value of $E^0_{AgI,\,Ag}$.

8.8. Given that $E^0_{Pb^{2+},\,Pb} = -0.126$ V and $K_{sp(PbSO_4)} = 1.3 \times 10^{-8}$, calculate the value of $E^0_{PbSO_4,\,Pb}$.

8.9. Given that $E^0_{Ag^+,\,Ag} = +0.7994$ V and$K_{sp(Ag_2S)} = 1.3 \times 10^{-50}$, calculate E^0 for the half-reaction $\mathbf{Ag_2S} + 2e \rightleftharpoons \mathbf{2Ag} + S^{2-}$.

8.10. From the values of $E^0_{Hg_2Cl_2,\,Hg}$ for the normal (standard) and the saturated calomel electrodes, calculate the molar solubility of KCl in water, at 25°C.

8.11. Calculate the standard reduction potential for the half-reaction

$$\mathbf{Cu(OH)_2} + 2e \rightleftharpoons \mathbf{Cu} + 2\,OH^-$$

from

$$K_{sp[Cu(OH)_2]} = 1.6 \times 10^{-19} \text{ and } E^0_{Cu^{2+},\,Cu} = +0.337 \text{ V}.$$

8.12. Calculate the standard reduction potentials of the following half-reactions:

a) $ClO_4{}^- + 8H^+ + 7e \rightleftharpoons 1/2Cl_2 + 4H_2O$

from the standard reduction potentials of the following half-reactions:

$$ClO_4^- + 2H^+ + 2e \rightleftharpoons ClO_3^- + H_2O, E^0_{ClO_4^-,\ ClO_3^-} = +1.19 \text{ V}$$

$$ClO_3^- + 6H^+ + 5e \rightleftharpoons 1/2Cl_2 + 3H_2O,\ E^0_{ClO_3^-,\ Cl_2} = +1.49 \text{ V}$$

b) $[(CH_3COO)_2Hg] + 2e \rightleftharpoons \mathbf{Hg} + 2CH_3COO^-$,
given that the instability constant of the complex $[(CH_3COO)_2Hg]$ is 4.0×10^{-9}.

c) $[HgY]^{2-} + 2e \rightleftharpoons Hg + Y^{4-}$,
where Y^{4-} is the anion of ethylenediaminetetraacetic acid (EDTA), given that the formation constant of the complex $[HgY]^{2-}$ is 6.3×10^{21}.

Half-cell potential

8.13. Calculate the potentials of the following half-cells, using activities:

a) **Ag** | $AgNO_3$(0.100 M)

b) **Ag** | $AgNO_3$(0.100 M), KNO_3(0.175 M)

8.14. How many grams of $CrCl_3$ per liter must be added to a half-cell containing a 0.200 M $K_2Cr_2O_7$-1.00 M HCl solution so that its potential will be +1.14 V?

Single electrode potential

8.15. Calculate the electrode potential of a silver electrode immersed in a) 0.00500 M KI solution that is saturated with AgI, b) a solution that is obtained by mixing 50.0 mL of 0.0500 M KI and 50.0 mL of 0.0800 M $AgNO_3$, c) a solution that is obtained by mixing 50.0 mL of 0.0500 M $AgNO_3$ and 50.0 mL of 0.0800 M KI, d) a solution that is 0.00100 M in $[Ag(S_2O_3)_2]^{3-}$ and 1.00 M in $S_2O_3^{2-}$.

8.16. Calculate the electrode potential of a platinum electrode immersed in a solution a) prepared by mixing 25.0 mL of 0.100 M Ce^{4+} and 50.0 mL of 0.100 M Fe^{2+}, b) prepared by mixing 40.0 mL of 0.250 M KI and 10.0 mL of 0.0100 M H_2O_2, that has a pH of 1.50.

8.17. Convert each of the following electrode potentials to potentials versus the saturated calomel electrode:

(a) $Zn^{2+} + 2e \rightleftharpoons \mathbf{Zn}, E^0 = -0.763$ V
(b) $Ce^{4+} + e \rightleftharpoons Ce^{3+}, E^{0'} = +1.28$ V (in 1 M HCl)
(c) $Sn^{2+} + 2e \rightleftharpoons \mathbf{Sn}, E^0 = -0.136$ V

Cell potential

8.18. Calculate the emf of the galvanic cell

$$\mathbf{Pt},\ H_2\ (0.50\ \text{atm})\ |\ H^+(0.0100\ \underline{M})||KCl(0.100\ \underline{M}),\ \mathbf{AgCl}\ |\ \mathbf{Ag}$$

and the equilibrium constant of its spontaneous reaction, and give the polarity of each electrode.

8.19. Calculate the voltages of the following cells:

(a) $\mathbf{Ag}\ |\ Ag^+(0.050\ \underline{M})||Cu^{2+}(0.200\ \underline{M})\ |\ \mathbf{Cu}$
(b) $\mathbf{Pb}\ |\ \mathbf{PbSO_4},\ SO_4^{2-}(0.050\ \underline{M})||Cl^-(1.00\ \underline{M}),\ \mathbf{AgCl}\ |\ \mathbf{Ag}$
(c) $\mathbf{Ag}\ |\ \mathbf{AgCl},\ KCl(0.100\ \underline{M})||HCl(0.200\ \underline{M})\ |\ H_2(0.50\ \text{atm}),\ \mathbf{Pt}$
(d) $\mathbf{Pt},\ H_2\ (1.00\ \text{atm})\ |\ HCl\ (1.00 \times 10^{-3}\ \underline{M})||HCl(1.00\ \underline{M})\ |\ H_2(1.00\ \text{atm}),\ \mathbf{Pt}$
(e) $\mathbf{Zn}\ |\ ZnSO_4(0.100\ \underline{M})||K_2Cr_2O_7(0.100\ \underline{M}),\ Cr^{3+}(0.100\ \underline{M}),\ H^+(1.00\ \underline{M})\ |\ \mathbf{Pt}$
(f) $\mathbf{Hg}\ |\ \mathbf{Hg_2Cl_2},\ KCl\ (\text{satur.})||Na_2C_2O_4(1.00\ \underline{M}),\ \mathbf{Ag_2C_2O_4}\ |\ \mathbf{Ag}$
(g) $\mathbf{Pt},\ H_2\ (1.00\ \text{atm})\ |\ CH_3COOH\ (0.100\ \underline{M}),\ CH_3COONa\ (0.100\ \underline{M})\ ||\ KCl\ (\text{satur.}),\ \mathbf{Hg_2Cl_2}\ |\ \mathbf{Hg}$

8.20. The emf of the cell

$$\mathbf{Ag}\ |\ \mathbf{AgCl},\ KCl(1.00\ \underline{M})||H^+(x\underline{M})\ |\ H_2(1.00\ \text{atm}),\ \mathbf{Pt}$$

is -0.438 V. Calculate the pH of the unknown solution.

8.21. In each of two beakers 100 mL of a 0.100 $\underline{M}$ Fe^{2+}-0.100 $\underline{M}$ Fe^{3+} solution is added, an electrode is immersed in each solution, and the two solutions are connected by a salt bridge. When 10.0 mL of a reducing solution is added to one of the two beakers, the voltage of the cell is changed from 0.0000 to +0.1183 V. Calculate the normality of the reducing solution.

8.22. In a saturated solution of quinhydrone, which is a molecular compound of quinone and hydroquinone, $Q \cdot H_2Q$, the following half-reaction takes place:

$$Q + 2H^+ + 2e \rightleftharpoons H_2Q,\ E^0_{Q,\ H_2Q} = +0.699V.$$

Calculate the emf of the cell

$$\mathbf{Hg}\ |\ \mathbf{Hg_2Cl_2},\ KCl\ (\text{satur.})||Q\ (\text{satur.}),\ H_2Q\ (\text{satur.}),\ H^+\ (x\underline{M})\ |\ \mathbf{Pt}$$

as a function of pH, for integer values of pH between 0 and 8 (at higher pH values, the quinhydrone electrode does not work).

8.23. At what I^- ion concentration does the emf of the cell

$$\mathbf{Ag}\ |\ \mathbf{AgI},\ KI\ (x\underline{M})||KCl(1.00\ \underline{M}),\ \mathbf{Hg_2Cl_2}\ |\ \mathbf{Hg}$$

become zero?

8.24. Calculate the voltage required to begin deposition of copper from a 0.0200 M $CuSO_4$ solution in H_2SO_4 having $[H^+] = 1.00 \times 10^{-3}$ M.

8.25. The emf of the hypothetical galvanic cell

$$(-)\mathbf{M} \mid M^{2+} \| X^{2+} \mid \mathbf{X}(+)$$

is +0.180 V, when $[M^{2+}]/[X^{2+}] = 10.0$. What will the emf of the cell be if $[M^{2+}] = 0.100$ M, and $[X^{2+}] = 1.00 \times 10^{-5}$ M?

8.26. For the cell

$$\mathbf{Zn} \mid Zn^{2+}(0.200\ \underline{M}) \| Cu^{2+}(0.0200\ \underline{M}) \mid \mathbf{Cu},$$

designate the following: a) the emf of the cell, b) the polarity of the electrodes, c) the direction in which the cell reaction proceeds spontaneously, d) the electrode potentials at equilibrium (after the discharge of the cell).

8.27. The emf of the galvanic cell

$$(-)\mathbf{Hg} \mid \mathbf{Hg_2Cl_2},\ KCl(1.00\ \underline{M}) \| Ag^+(x\underline{M}) \mid \mathbf{Ag}(+)$$

is +0.4010 V, whereas that of the galvanic cell

$$(-)\mathbf{Hg} \mid \mathbf{HgCl_2},\ KCl(1.00\ \underline{M}) \| Ag^+(y\underline{M}) \mid \mathbf{Ag}(+)$$

is +0.4601 V. In what proportion should the x and y solutions be mixed to prepare a solution in which $[Ag^+] = 0.0400$ M?

8.28. a) Calculate the emf of the galvanic cell

$$(-)\mathbf{Pt}, H_2(1.00\ \text{atm}) \mid H^+(1.00\ \underline{M}) \| Ag^+(0.100\ \underline{M}) \mid \mathbf{Ag}(+).$$

b) What will the emf of the cell be, if ammonia is added to the silver half-cell until the concentration of free ammonia becomes 0.100 M?

8.29. The emf of the galvanic cell

$$(-)\mathbf{Ag} \mid Ag^+(0.0100\ \underline{M}) \| AgNO_3(x\underline{M}) \mid \mathbf{Ag}(+)$$

is +0.0784 V. Calculate the molarity of $AgNO_3$.

8.30. Given that the galvanic cell

$$\mathbf{Pt} \mid Fe^{3+}(0.100\ \underline{M}), Fe^{2+}(0.0100\ \underline{M}) \| MnO_4^-(0.0100\ \underline{M}),\ Mn^{2+}(0.00100\ \underline{M}),\ H^+(1.00\ \underline{M}) \mid \mathbf{Pt}$$

contains equal volumes of solutions in the two half-cells, calculate the potential of each half-cell and the voltage of the cell, before the reaction and at equilibrium.

8.31. Calculate the emf of the cell

$$\mathbf{Cu} \mid Cu^{2+}(0.040\ \underline{M}) \| K[Ag(CN)_2](0.020\ \underline{M}),\ HCN(0.20\ \underline{M}),\ \text{buffer pH } 9.4 \mid \mathbf{Ag}.$$

Equilibrium constant

8.32. Calculate the equilibrium constant for the reaction

$$\mathbf{AgBr} + V^{2+} \rightleftharpoons \mathbf{Ag} + V^{3+} + Br^-$$

8.33. Calculate the equilibrium constants for the reactions which are represented by the following equations:

(a) $H_3AsO_3 + [I_3]^- + H_2O \rightleftharpoons H_3AsO_4 + 2H^+ + 3I^-$
(b) $H_2S + 2Fe^{3+} \rightleftharpoons 2Fe^{2+} + 2H^+ + \mathbf{S}$
(c) $MnO_4{}^- + 8H^+ + 5Fe^{2+} \rightleftharpoons Mn^{2+} + 5Fe^{3+} + 4H_2O$
(d) $\mathbf{Cd} + 2H^+ \rightleftharpoons Cd^{2+} + H_2 \uparrow$
(e) $\mathbf{Cd} + 2Ag^+ \rightleftharpoons Cd^{2+} + 2\mathbf{Ag}$
(f) $2\mathbf{AgI} + \mathbf{Zn} \rightleftharpoons 2\mathbf{Ag} + Zn^{2+} + 2I^-$
(g) $2Fe^{3+} + \mathbf{Fe} \rightleftharpoons 3Fe^{2+}$

8.34. From the data given in the table of standard reduction potentials (Appendix B), calculate the equilibrium constant for the reaction

$$2Cu^+ \rightleftharpoons Cu^{2+} + \mathbf{Cu}$$

and the Cu^{2+} ion concentration at equilibrium, when $[Cu^+] = 7.76 \times 10^{-4}$ M.

8.35. Calculate the equilibrium constant for the reaction

$$2H_2 + O_2 \rightleftharpoons 2H_2O(l)$$

Write the half-reactions of the half-cells, and the cell in which this reaction takes place.

8.36. Calculate the equilibrium constant for the reaction

$$2Fe^{3+} + \mathbf{Fe} \rightleftharpoons 3Fe^{2+}$$

from the standard reduction potentials of the following half-reactions:

$$Fe^{3+} + e \rightleftharpoons Fe^{2+}, \quad E^0 = +0.771 \text{ V}$$

$$Fe^{2+} + 2e \rightleftharpoons \mathbf{Fe}, \quad E^0 = -0.440 \text{ V}.$$

8.37. From the data given in the table of standard reduction potentials (Appendix B), calculate the equilibrium constant for the reaction

$$Cr_2O_7{}^{2-} + 6Fe^{2+} + 14H^+ \rightleftharpoons 2Cr^{3+} + 6Fe^{3+} + 7H_2O$$

8.38. From data given in the table of standard reduction potentials (Appendix B), calculate the ion product of water, K_w.

8.39. The emf of the galvanic cell

$$(-)\mathbf{Pt},\ H_2(0.90\ \text{atm})\ |\ OH^-(0.0100\ \underline{M})||H^+(0.100\ \underline{M}),|\ H_2(0.90\ \text{atm}),\ \mathbf{Pt}(+)$$

is +0.6506 V. Calculate the ion product of water, K_w.

Solubility product

8.40. From data given in the table of standard reduction potentials (Appendix B), calculate the solubility products of the following compounds: a) Hg_2Cl_2, b) AgCl, c) Ag_2CrO_4, d) CuI, e) ZnS, f) $Mn(OH)_2$, g) $PbSO_4$.

8.41. Sodium hydroxide is added to the left-hand half-cell of the galvanic cell

$$\mathbf{Cu}\ |\ Cu^{2+}(x\underline{M})||KCl\ (\text{satur.}),\ \mathbf{Hg_2Cl_2}\ |\ \mathbf{Hg}$$

precipitating $Cu(OH)_2$. The final pH is 9.00, and the emf of the cell at this point is +0.164 V with the saturated calomel electrode positive (cathode). Calculate the solubility product of $Cu(OH)_2$.

8.42. The emf of the galvanic cell

$$\mathbf{Ag}\ |\ \mathbf{AgI},\ I^-(0.100\ \underline{M})||KCl\ (\text{satur.})\ \mathbf{Hg_2Cl_2}\ |\ \mathbf{Hg}$$

is +0.331 V with the saturated calomel electrode positive. Calculate the solubility product of AgI.

8.43. Given that the standard potential $E^0_{AgI,\ Ag}$ is -0.150 V, calculate the solubility product of AgI.

8.44. One milliliter of 0.100 $\underline{M}$ KCl is added to 25.00 mL of 0.100 $\underline{M}$ $AgNO_3$ solution. A silver indicator electrode and a reference electrode are immersed in the resulting solution and the emf of the cell is found to be +0.4982 V. After the addition of another 29.00 mL of the KCl solution, the emf becomes +0.1037 V. Calculate the solubility product of AgCl.

Dissociation constant

8.45. The emf of the cell

$$Hg\ |\ \mathbf{Hg_2Cl_2},KCl\ (\text{satur.})||Q\ (\text{satur.}),\ H_2Q\ (\text{satur.}),\ H_3PO_4(0.100\ \underline{M}),\ NaH_2PO_4(0.100\ \underline{M})\ |\ \mathbf{Pt}$$

is +0.329 V. Calculate the dissociation constant K_1 of H_3PO_4.

8.46. The emf of the galvanic cell

$$\mathbf{Pt},\ H_2\ (1.00\ \text{atm})\ |\ HA(1.00\ \underline{M})||KCl(1.00\ \underline{M}),\ \mathbf{Hg_2Cl_2}\ |\ \mathbf{Hg}$$

is +0.420 V with the calomel electrode being the cathode. Calculate the dissociation constant of the weak acid HA.

8.47. The emf of the galvanic cell

$$(-)\mathbf{Pt},\ H_2(1.00\ \text{atm})\ |\ NH_4Cl(0.100\ \underline{M})||KCl(1.00\ \underline{M}),\ \mathbf{Hg_2Cl_2}\ |\ \mathbf{Hg}(+)$$

is +0.582 V. Calculate the dissociation constant of ammonia.

Instability constant

8.48. Using the standard reduction potentials $E^0_{Ag^+, Ag}$ and $E^0_{[Ag(CN)_2]^-, Ag}$, calculate the instability constant of the complex $[Ag(CN)_2]^-$.

8.49. The ions M^{2+} and A^- form the complex $[MA]^+$ according to the reaction

$$M^{2+} + A^- \rightleftharpoons [MA]^+$$

Fifty milliliters of 0.0100 M M^{2+} solution are mixed with 50.0 mL of 0.0100 M A^- solution and the resulting solution is placed in the left-hand half-cell of the galvanic cell

$$(-)\mathbf{M}\ |\ M^{2+}(x\underline{M}),\ [MA]^+\ (0.00500\ \underline{M})||KCl(1.00\ \underline{M}),\mathbf{Hg_2Cl_2}\ |\ \mathbf{Hg}(+)$$

The emf of the cell is +0.267 V. Calculate the instability constant of the complex $[MA]^+$ ($E^0_{M^{2+}, M} = +0.170$ V).

Calculation of concentrations of redox systems at equilibrium

8.50. Sn(II) is often used in analysis for the reduction of Fe(III) to Fe(II), according to the reaction

$$2Fe^{3+} + [SnCl_4]^{2-} + 2Cl^- \rightleftharpoons 2Fe^{2+} + [SnCl_6]^{2-}$$

Calculate the concentration of the Fe^{3+}, $[SnCl_4]^{2-}$, Fe^{2+}, and $[SnCl_6]^{2-}$ ions in a solution which is prepared by mixing 19.0 mL of 0.0050 M Fe(III) and 1.00 mL of 0.050 M Sn(II), after equilibrium, when $[Cl^-] = 1.00$ M.

8.51. Fifty milliliters of 0.1000 M Ce^{4+} solution are mixed with 50.00 mL of 0.1000 M Fe^{2+} solution. Calculate the concentrations of the various ions at equilibrium. Both solutions are 1.00 M in H_2SO_4.

8.52. Twenty-five milliliters of 0.03333 M $K_2Cr_2O_7$ solution are mixed with 25.00 mL of 0.2000 M Fe^{2+} solution. If $[H^+] = 1.00$ M at equilibrium, calculate the concentrations of the other ions after equilibration.

8.53. The emf of the galvanic cell

$$(-)\mathbf{Hg_2Cl_2},KCl\ (\text{satur.})||Fe^{3+}(x\underline{M}),\ Fe^{2+}(0.100 - x)\ \underline{M}\ |\ \mathbf{Pt}(+)$$

is +0.412 V. Calculate the percentage of Fe^{2+} that has been oxidized to Fe^{3+}.

8.54. The emf of the galvanic cell

$$\mathbf{Pb} \mid Pb^{2+}(0.100\ \underline{M}) \| Pb^{2+}(x\underline{M}) \mid \mathbf{Pb}$$

is +0.0089 V. Calculate the value of x.

Miscellaneous redox problems

8.55. A sheet of iron weighing 10.00 g is immersed in a $CuSO_4$ solution and is plated with copper. After equilibration, the sheet weighs 10.10 g. How many grams of copper were plated on the sheet?

8.56. An excess of iron metal is added to one liter of 0.100 $\underline{M}$ Zn^{2+} solution. At equilibrium, what is the Fe^{2+} ion concentration?

8.57. An excess of iron is added to one liter of 0.100 $\underline{M}$ $AgNO_3$ solution. Calculate the concentrations of Ag^+ and Fe^{2+} ions at equilibrium.

8.58. A car battery is a cell designated as

$$\mathbf{Pb} \mid \mathbf{PbSO_4},\ H_2SO_4 \mid \mathbf{PbO_2}$$

Write the half-reactions that take place at the electrodes and the overall cell reaction. Does the density of the electrolyte affect the operation of the battery?

8.59. Equivalent quantities of H_3AsO_4 and $[Fe(CN)_6]^{4-}$ are mixed in water. Calculate the value of the ratio $[H_3AsO_3]/[H_3AsO_4]$ having a) pH = 2.00, b) pH = 6.00.

Chapter 9

An Introduction to Gravimetric Analysis

9.1 Stoichiometry

Gravimetric methods of chemical analysis are either *direct* or *indirect*. In the *direct* methods, the constituent A being determined is separated from the other components of the sample in the form of a pure substance, which can be either A itself or a compound B of known and definite composition, from the weight of which the weight of A can be calculated. The *indirect* methods are based on the fact that two or more compounds, in chemically equivalent quantities, under the same chemical treatment undergo different changes in their weight.

The percentage of constituent A in a sample can be calculated from the experimental data, using the relation

$$\%A = \frac{W \times F \times 100}{S} \tag{9.1}$$

where W is the weight in grams of substance B being weighed in the end (that is, the weight of the precipitate or of its conversion product), S is the weight of the sample in grams, and F is the *gravimetric factor (conversion factor)*, that is, the number by which the weight W must be multiplied to obtain the corresponding weight of component A.

The gravimetric factor F is calculated by the relation

$$F = \frac{a(\text{FW of substance A})}{b(\text{FW of substance B})} \tag{9.2}$$

where a and b are the coefficients of A and B, respectively, that show the stoichiometric relationship between them, and FW is the formula weight. That is, the gravimetric factor is numerically equal to the weight of substance A in grams corresponding to one gram of substance B. For example, in the gravimetric determination of iron in the form of ferric oxide, Fe_2O_3, the gravimetric factor is equal to $F = \frac{2Fe}{Fe_2O_3} = \frac{2 \times 55.847}{159.692} =$ 0.6694.

Example 9.1. Calculate the gravimetric factors for the following conversions: a) Fe_3O_4 to Fe, b) $Mg_2P_2O_7$ to MgO, c) $Mg_2P_2O_7$ to P_2O_5, d) $BaSO_4$ to SO_3, e) $BaSO_4$ to $Al_2(SO_4)_3$, f) AgCl to $KClO_3$, g) aluminum 8-quinolate, $Al(C_9H_6ON)_3$, to Al_2O_3, h) CaO to $Na_2C_2O_4$, i) AgCl to As (As $\rightarrow$ Ag_3AsO_4 $\rightarrow$ 3AgCl).

Solution. a) We have

$$\text{(a)}\ \ F = \frac{3Fe}{Fe_3O_4} = \frac{3 \times 55.847}{231.539} = \mathbf{0.7236}$$

$$\text{(b)}\ \ F = \frac{2MgO}{Mg_2P_2O_7} = \mathbf{0.3623}$$

$$\text{(c)}\ \ F = \frac{P_2O_5}{Mg_2P_2O_7} = \mathbf{0.6377}$$

$$\text{(d)}\ \ F = \frac{SO_3}{BaSO_4} = \mathbf{0.3430}$$

$$\text{(e)}\ \ F = \frac{Al_2(SO_4)_3}{3BaSO_4} = \mathbf{0.4886}$$

$$\text{(f)}\ \ F = \frac{KClO_3}{AgCl} = \mathbf{0.8551}$$

$$\text{(g)}\ \ F = \frac{Al_2O_3}{2Al(C_9H_6ON)_3} = \mathbf{0.11096}$$

$$\text{(h)}\ \ F = \frac{Na_2C_2O_4}{CaO} = \mathbf{2.389}$$

$$\text{(i)}\ \ F = \frac{As}{3AgCl} = \mathbf{0.1742}$$

Example 9.2. A 0.8552-g sample of a copper alloy is treated with 8 M HNO_3 and filtered, and the precipitate is ignited, giving a residue of 0.0632 g of SnO_2. The zinc is determined in one-half of the filtrate by precipitating it as $ZnNH_4PO_4$ and then igniting, whereupon 0.2231 g of $Zn_2P_2O_7$ is formed. In the other half of the filtrate copper is determined as CuSCN, giving a precipitate of 0.5874 g. Calculate the percentage of tin, zinc, and copper in the sample.

Solution. The gravimetric factor for the conversion of SnO_2 to Sn is equal to $118.69/150.69 = 0.7876$. Hence, the precipitate contains $0.0632 \times 0.7876 = 0.0498$ g of Sn. Since this amount of tin is present in 0.8552 g of sample, the per cent tin is equal to

$$\frac{0.0498}{0.8552} \times 100 = \mathbf{5.82\%}\ \text{Sn}$$

or, in brief,

$$\frac{0.0632\ \text{g}\ SnO_2 \times \dfrac{118.69\ \text{g Sn}}{150.69\ \text{g}\ SnO_2}}{0.8552\ \text{g sample}} \times 100 = \mathbf{5.82\%}\ \text{Sn}.$$

Since one-half of the sample was used for the determination of Zn, we have

$$\frac{0.2231 \text{ g } Zn_2P_2O_7 \times 2 \times \dfrac{2 \times 65.37 \text{ g Zn}}{304.68 \text{ g } Zn_2P_2O_7}}{0.8552 \text{ g sample}} \times 100 = \mathbf{22.39\%} \text{ Zn.}$$

Since one-half of the sample was used for the determination of Cu, we have

$$\frac{0.5874 \times 2 \times 0.5224}{0.8552} \times 100 = \mathbf{71.76\%} \text{ Cu.}$$

Example 9.3. A 0.1803-g sample containing only Pb_3O_4 and inert matter gives 0.2378 g of $PbSO_4$. Calculate the per cent of a) Pb_3O_4 and b) Pb, in the sample.

Solution. a) The per cent Pb_3O_4 in the sample is equal to

$$\frac{0.2378 \text{ g } PbSO_4 \times \dfrac{685.57 \text{ g } Pb_3O_4}{(3 \times 303.25) \text{ g } PbSO_4}}{0.1803 \text{ g sample}} \times 100 = \mathbf{99.4\%} \text{ } Pb_3O_4.$$

b) ***1st method.*** The per cent Pb in the sample is equal to

$$\frac{0.2378 \text{ g } PbSO_4 \times \dfrac{207.19 \text{ g Pb}}{303.25 \text{ g } PbSO_4}}{0.1803 \text{ g sample}} \times 100 = \mathbf{90.1\%} \text{ Pb.}$$

2nd method. We have

$$99.4\% \text{ } Pb_3O_4 \times \frac{(3 \times 207.19) \text{ g Pb}}{685.57 \text{ g } Pb_3O_4} = \mathbf{90.1\%} \text{ Pb.}$$

Example 9.4. A series of samples known to contain sulfate in the range between 20.0 and 48.0% is to be analyzed gravimetrically by the method of barium sulfate. a) What is the minimum sample weight that must be taken in order to assure a $BaSO_4$ precipitate that weighs at least 0.3125 g? b) What will the maximum weight of $BaSO_4$ be which can be obtained from the above amount of sample?

Solution. a) The percentage of sulfate in the sample is given by the relation

$$\%SO_4 = \frac{\text{g } BaSO_4 \times F}{\text{g sample}} \times 100 \tag{9.3}$$

where

$$F = \text{gravimetric factor} = \frac{SO_4}{BaSO_4} = \frac{96.06}{233.40} = 0.4116.$$

For a given sample weight, the minimum amount of precipitate will be obtained from the sample that contains the least sulfate (20.0%). Hence, the minimum sample weight that must be taken to assure at least 0.3125 g of $BaSO_4$ is calculated by Equation (9.3), as follows:

$$20.0 = \frac{0.3125 \times 0.4116}{\text{g sample}} \times 100,$$

or g sample = **0.643**.

b) The maximum weight of $BaSO_4$ from 0.643 g of sample will be obtained from the sample that contains the most sulfate (48.0%). Hence, we will obtain

$$\text{g } BaSO_4 = \text{g sample} \times \frac{\%SO_4}{100} \times \frac{BaSO_4}{SO_4} = 0.643 \times 0.480 \times \frac{233.40}{96.06} = \mathbf{0.750} \text{ g } BaSO_4.$$

Example 9.5. Between what limits must the sample weight be for a silver alloy containing 16.5% Ag, so that the weight of AgCl obtained will lie between 0.200 and 0.250 g?

Solution. According to Equation (9.1), we have

$$\%Ag = \frac{W \times \frac{Ag}{AgCl} \times 100}{S}$$

or

$$S = \frac{W \times \frac{107.87}{143.32} \times 100}{16.5} = 4.56W.$$

Hence the sample weight, S, must lie between (4.56 × 0.200) and (4.56 × 0.250) g, that is, between **0.91** and **1.14** g.

Example 9.6. Sodium bicarbonate is converted to sodium carbonate upon ignition, according to the reaction

$$\mathbf{2NaHCO_3} \xrightarrow{\Delta} \mathbf{Na_2CO_3} + CO_2 \uparrow + H_2O \uparrow$$

Ignition of a 0.4827-g sample of impure $NaHCO_3$ yielded a residue weighing 0.3189 g. If the sample impurities are not volatile at the ignition temperature, calculate the per cent of $NaHCO_3$ in the sample.

Solution. The weight loss is due to the CO_2 and H_2O evolved during the ignition. Since 2 mol $NaHCO_3 \equiv 1$ mol CO_2 + 1 mol H_2O, we have

$$\begin{aligned}\%NaHCO_3 &= \frac{\text{g } (CO_2 + H_2O) \times \frac{2NaHCO_3}{CO_2 + H_2O}}{\text{g sample}} \times 100 \\ &= \frac{(0.4827 - 0.3189) \times \frac{2 \times 84.01}{44.01 + 18.015}}{0.4827} \times 100 = \mathbf{91.92}.\end{aligned}$$

Example 9.7. What weight of sample must be taken for analysis so that each 10.0 mg of $BaSO_4$ precipitate obtained represents 1.000% barium in the sample?

Solution. Substituting the data in Equation (9.1), we have

$$1.000\%\ \text{Ba} = \frac{0.0100\ \text{g BaSO}_4 \times \dfrac{137.34\ \text{g Ba}}{233.40\ \text{g BaSO}_4} \times 100}{\text{S}}$$

or S = **0.588** g of sample.

Note. This example is representative of the cases in which a large number of similar samples is to be analyzed, as for example, in industrial laboratories, whereupon it is advantageous to weigh a proper amount of sample so that a simple relation exists between the amount of weighed sample and the percentage of the sought-for constituent. In this way, the calculations are simplified.

Example 9.8. A sample given as an unknown in the gravimetric determination of chloride may consist of NaCl and/or KCl only. a) How many grams of sample must be taken so as to give 0.250–0.500 g of AgCl? b) If a 10% excess of $AgNO_3$ is desired for more complete precipitation of chloride, how many milliliters of 1.00 M $AgNO_3$ solution are required for a sample weighing y g?

Solution. a) The lower limit of sample size is determined from the smallest weight of AgCl (0.250 g) and KCl, which has the smaller percentage of chlorine. A 0.250 g portion of AgCl is obtained from

$$0.250\ \text{g AgCl} \times \frac{74.56\ \text{g KCl}}{143.32\ \text{g AgCl}} = 0.130\ \text{g KCl},$$

whereas, 0.500 g of AgCl is obtained from

$$0.500\ \text{g AgCl} \times \frac{58.44\ \text{g NaCl}}{143.32\ \text{g AgCl}} = 0.204\ \text{g NaCl}.$$

Hence, the sample weight should lie between **0.130** and **0.204** g.

b) Suppose that the sample consists of NaCl (which contains the larger percentage of Cl and therefore requires the largest amount of $AgNO_3$). If V mL is the volume of $AgNO_3$ equivalent to y g of sample, we have the relation

$$\frac{(\text{y g NaCl})(1000\ \text{mg/g})}{58.44\ \text{mg/mmol}} = \text{V mL} \times 1.00\ \text{mmol/mL}$$

or V = 17.1y mL. Since a 10% excess of $AgNO_3$ is desired, a total of (17.1y)(1.1) = **18.8**y mL of 1.00 M $AgNO_3$ is required, that is, 18.8 mL for each gram of sample.

Example 9.9. A sample consisting of only NaCl and NaI contains chlorine and iodine in a weight ratio of Cl:I = 2:1. Calculate the per cent content of sodium in the sample.

Solution. Suppose that the sample contains y% I, and therefore 2y% Cl, and a% NaI, and therefore (100−a)% NaCl. We have

$$\frac{\text{y}}{\text{a}} = \frac{\text{I}}{\text{NaI}} = \frac{126.90}{149.89} \tag{9.4}$$

$$\frac{2y}{100 - a} = \frac{Cl}{NaCl} = \frac{35.45}{58.44} \tag{9.5}$$

Solving the system of Equations (9.4) and (9.5), we find that y = 22.33% I, whereupon %Cl = 2y = 44.66%. Hence, the per cent content of sodium in the sample is equal to

$$22.33 \times \frac{Na}{I} + 44.66 \times \frac{Na}{Cl} = 22.33 \times \frac{22.99}{126.90} + 44.66 \times \frac{22.99}{35.45} = \mathbf{33.01\%}\ Na.$$

Example 9.10. In what weight proportion should KCl and NaCl be mixed so that the sample contains 52.0% chlorine?

Solution. Suppose that a g of KCl should be mixed with b g of NaCl, whereupon the sought-for proportion is a/b. We have

$$\frac{g\ Cl}{g\ KCl + g\ NaCl} \times 100 = 52.0$$

or

$$\frac{a \times \frac{Cl}{KCl} + b \times \frac{Cl}{NaCl}}{g\ KCl + g\ NaCl} \times 100 = 52.0$$

or

$$\frac{a \times \frac{35.45}{74.56} + b \times \frac{35.45}{58.44}}{a + b} \times 100 = 52.0$$

or

$$\frac{\frac{35.45}{74.56} \times \frac{a}{b} + \frac{35.45}{58.44}}{\frac{a}{b} + 1} = 0.520$$

or (a/b) = **1.944**; that is, for each gram of NaCl in the mixture there must be 1.944 g of KCl.

Example 9.11. The chlorine in $NaClO_n$ is converted by proper treatment to chloride ion which is precipitated as AgCl. If 0.2502 g of sample gives 0.3969 g of AgCl, calculate the value of n.

Solution. We have

$$0.3969\ g\ AgCl/(0.14332\ g/mmol) = 2.769\ mmol\ AgCl \equiv 2.769\ mmol\ NaClO_n.$$

The molecular weight of $NaClO_n$ is equal to

$$MW_{NaClO_n} = \frac{250.2\ mg\ NaClO_n}{2.769\ mmol} = 90.36 = 22.99 + 35.45 + 16.00n,$$

whereupon $n = 1.995 \simeq 2$.

Calculations on dry basis and on as-received basis. Sometimes it is important to know the per cent of a constituent in the natural sample, that is, the sample as it is received, which also contains hygroscopic water (moisture), whereupon the results of the analysis are reported on an *as-received basis.* It is usual, however, to dry the samples prior to the analysis, since the exact weighing of such samples is difficult and the moisture of a sample may be altered upon standing because of water exchange with the atmosphere. This is accomplished by heating for 1–2 h at 105–110°C in an electric oven having a thermostat (in a few cases the samples are air-dried at room temperature). In these cases, the results are given on a *dry basis.*

The per cent moisture of a sample, %M, the per cent (by weight) of a constituent A on the as-received basis, $\%A_{a.r.}$, and the per cent of A on a dry basis, $\%A_d$, are interrelated by the equation

$$\%A_d = \frac{\%A_{a.r.} \times 100}{100.00 - \%M}, \tag{9.6}$$

which allows the calculation of any of the three terms, when the other two are known.

Example 9.12. An ore sample contains 1.85% moisture and 28.32% Fe on an as-received basis. What is the per cent Fe on a dry basis?

Solution. Substituting the data in Equation (9.6), we have

$$\%Fe_d = \frac{28.32 \times 100}{100.00 - 1.85} = \mathbf{28.85}.$$

Example 9.13. A soil sample contains 2.60% moisture and 19.88% Al_2O_3, on an as-received basis. What will the % Al_2O_3 be a) after drying the sample, b) after reducing the moisture to 0.55% by air-drying?

Solution. a) Substituting the data in Equation (9.6), we have

$$\%Al_2O_3\ _d = \frac{19.88 \times 100}{100.00 - 2.60} = \mathbf{20.41}.$$

b) From the % Al_2O_3 on a dry basis, the % Al_2O_3 at a moisture of 0.55% can be calculated on the basis of Equation (9.6), as follows:

$$\%Al_2O_3 = \frac{(20.41)(100.00 - 0.55)}{100} = \mathbf{20.30}.$$

Example 9.14. A soil sample contains 20.41% Al_2O_3 on a dry basis and 19.88% Al_2O_3 on an as-received basis. Calculate the per cent moisture of the sample as received.

Solution. Substituting the data in Equation (9.6), we have

$$20.41 = \frac{19.88 \times 100}{100.00 - \%M}$$

or

$$\%M = \mathbf{2.60}.$$

9.2 Indirect Gravimetric Analysis

The simplest type of indirect gravimetric analysis is one in which two pure chemical substances are isolated and weighed together. Subsequently, by further chemical treatment of the chemical substances or by analysis of a new sample, additional data are obtained, from which one of the constituents is calculated, and then the other one is found by difference (Example 9.15). In another general type of indirect analysis, the two substances, after they are isolated and weighed together, are converted to different compounds by a common reagent. Then, either the sum of the weights of both new compounds is determined, or the amount of the reagent required for the conversion is determined (Example 9.18).

Example 9.15. A 0.8904-g sample of limestone gave 0.0426 g of $R_2O_3(Fe_2O_3 + Al_2O_3)$. By volumetric analysis, it was found that the limestone contained 1.75% Fe_2O_3. Calculate the per cent of Al_2O_3 and Al in the sample.

Solution. We have

$$\%R_2O_3 = \frac{0.0426}{0.8904} \times 100 = 4.78.$$

Hence, $\%Al_2O_3 = 4.78 - 1.75 = \mathbf{3.03}$, and

$$\%Al = 3.03 \times \frac{2 \times 26.98}{101.96} = \mathbf{1.60}.$$

Example 9.16. A 0.4828-g sample consisting of only NaCl and KCl gave 1.1280 g of AgCl. Calculate the per cent NaCl and KCl in the sample.

Solution. Suppose that x and y are the weights of NaCl and KCl in 0.4828 g of sample, respectively. From x g of NaCl, (143.32x/58.44) g of AgCl are produced, whereas y g of KCl give (143.32y/74.55) g of AgCl. Hence, we have

$$x + y = 0.4828 \qquad (9.7)$$

$$\frac{143.32x}{58.44} + \frac{143.32y}{74.55} = 1.1280 \qquad (9.8)$$

Solving the system of the two equations, we find that x = 0.3771 g NaCl, and y = 0.1057 g KCl. Hence, the sample contains

$$\frac{0.3771}{0.4828} \times 100 = \mathbf{78.11\%}\ \text{NaCl}$$

and

$$\frac{0.1057}{0.4828} \times 100 = \mathbf{21.89\%}\ \text{KCl}.$$

Note. The indirect analysis gives less accurate results than the direct analysis, because any error made in the determination is multiplied by the coefficients x and y in Equation (9.8). The method can also be applied to any mixture of two salts of the same acid or of the same base, mainly when the two salts coexist in about equal quantities. Also, the method can be applied to any mixture of more than two constituents, provided that we can write as many independent equations as there are unknowns.

Example 9.17. When a 0.5095-g sample consisting of only Fe_2O_3 and Al_2O_3 was heated in a stream of hydrogen, it lost 0.1187 g in weight (Fe_2O_3 was reduced to Fe, whereas Al_2O_3 remained unchanged). Calculate the per cent Fe and Al in the sample.

Solution. Suppose that 0.5095 g of sample contains y g of Fe_2O_3. A loss in weight of 0.1187 g corresponds to y g of Fe_2O_3, whereas a loss of $3 \times 16.00 = 48.00$ g corresponds to 159.69 g of Fe_2O_3 (1 mol of Fe_2O_3). Hence, we have

$$\frac{y}{159.69} = \frac{0.1187}{48.00}$$

or

$$y = 0.3949 \text{ g } Fe_2O_3 \text{ or } \frac{0.3949 \times 100}{0.5095} = 77.51\% \ Fe_2O_3.$$

Hence,

$$\%Al_2O_3 = 100.00 - 77.51 = 22.49.$$

Therefore, we have

$$\%Fe = 77.51 \times \frac{2 \times 55.85}{159.69} = \mathbf{54.22}.$$

and

$$\%Al = 22.49 \times \frac{2 \times 26.98}{101.96} = \mathbf{11.90}.$$

Example 9.18. A 0.2025-g sample consisting of only $BaCl_2$ and KCl required 20.25 mL of 0.1200 $\underline{M}$ $AgNO_3$ solution for the quantitative precipitation of chloride. Calculate the per cent of Ba and K in the sample.

Solution. Suppose that 0.2025 g of sample contains x mmol of $BaCl_2$ and y mmol of KCl, whereupon we have

$$208.25x + 74.56y = 202.5 \tag{9.9}$$

A quantity of x mmol of $BaCl_2$ contains 2x mmol of chloride, whereas y mmol of KCl contains y mmol of chloride. Hence, we have

$$2x + y = (20.25 \text{ mL})(0.1200 \text{ mmol } AgNO_3/\text{mL}) = 2.430 \tag{9.10}$$

Solving the system of Equations (9.9) and (9.10), we find that x = 0.3606 and y = 1.7088. Hence, we have

$$\%Ba = \frac{(0.3606 \text{ mmol } BaCl_2)(137.34 \text{ mg Ba/mmol } BaCl_2)}{202.5 \text{ mg sample}} \times 100 = \mathbf{24.46}$$

and

$$\%K = \frac{(1.7088 \text{ mmol KCl})(39.10 \text{ mg K/mmol KCl})}{202.5 \text{ mg sample}} \times 100 = \mathbf{32.99}.$$

Example 9.19. A 0.5000-g sample consisting of chlorides, bromides, and inert materials, gave 0.9000 g of AgCl + AgBr precipitate. When heated in a stream of chlorine, the precipitate lost 0.1000 g in weight. Calculate the per cent Cl and Br in the sample.

Solution. Suppose that 0.9000 g of precipitate contains x mmol of AgCl and y mmol of AgBr, whereupon we have

$$143.32x + 187.77y = 900.0 \tag{9.11}$$

During the heating of the precipitate in chlorine, AgBr is converted to AgCl ($\mathbf{2AgBr} + Cl_2 \rightleftharpoons \mathbf{2AgCl} + Br_2$). The loss in weight is due to replacement of a Br atom by a Cl atom. Consequently, for each millimole of AgBr we have a loss in weight equal to Br − Cl = 79.90 − 35.45 = 44.45 mg. Hence, we have

$$44.45y = 100.0 \tag{9.12}$$

Solving the system of Equations (9.11) and (9.12), we find that y = 2.250 and x = 3.332. Hence, we have

$$\%\text{Cl} = \frac{(3.332 \text{ mmol AgCl})(35.45 \text{ mg Cl/mmol AgCl})}{500.0 \text{ mg sample}} \times 100 = \mathbf{23.62}$$

and

$$\%\text{Br} = \frac{(2.250 \text{ mmol AgBr})(79.90 \text{ mg Br/mmol AgBr})}{500.0 \text{ mg sample}} \times 100 = \mathbf{35.96}.$$

Example 9.20. When heated, a sample consisting of only $CaCO_3$ and $MgCO_3$ yields a mixture of CaO and MgO. If the weight of the combined oxides is equal to 51.00% of the initial sample weight, calculate the per cent $CaCO_3$ and $MgCO_3$ in the sample.

Solution. The data of the problem are independent of the sample size. Suppose that we have a 1.0000-g sample that contains y g of $CaCO_3$, and therefore (1.0000−y)g of $MgCO_3$. We have

$$\text{g CaO} + \text{g MgO} = 0.5100$$

or

$$y \times \frac{\text{CaO}}{\text{CaCO}_3} + (1.0000 - y) \times \frac{\text{MgO}}{\text{MgCO}_3} = 0.5100$$

or

$$\frac{56.08y}{100.09} + \frac{40.30(1.0000 - y)}{84.31} = 0.5100$$

or y = 0.3865 g of $CaCO_3$. Since we assumed that the sample weighs 1.0000 g, it follows that it contains **38.65%** $CaCO_3$ and 100.00 − 38.65 = **61.35%** $MgCO_3$.

9.3 Calculation of the Empirical Formula

The empirical formula of a compound, which shows the elements present in the compound and their proportion, can be found from the results of quantitative elemental analysis as follows: 1) We divide the percentage of each element by the corresponding atomic weight, and thus we find the number of gram atoms of each element in 100 g of the compound. 2) We divide the numbers of the gram atoms by the smallest number. If the resulting proportion does not contain integers only, we multiply the numbers of the proportion by a proper integer (the smallest possible) so that a proportion having only integers results, which shows the proportion of the various elements in the molecule of the compound.

Example 9.21. In the quantitative analysis of an organic compound which consists of carbon, hydrogen, nitrogen, and oxygen, it was found that the compound contained 19.72% C, 4.89% H, and 22.90% N. If the molecular weight of the compound is 61.0, find the empirical and the molecular formulas of the compound.

Solution. The compound contains $100.00 - (19.72 + 4.89 + 22.90) = 52.49\%$ O. We have

$$C = \frac{19.72}{12.01} = 1.642,\ H = \frac{4.89}{1.008} = 4.85,\ N = \frac{22.90}{14.01} = 1.635,\ O = \frac{52.49}{16.00} = 3.28.$$

Dividing by 1.635, we have N = 1.00, C = 1.00, B = 2.97, O = 2.01. Hence the empirical formula of the compound is $(CH_3NO_2)_v$. We have $(12.01+3\times1.008+14.01+2\times16.00)v = 61.0$ or v = 1. Hence, the molecular formula of the compound is $\mathbf{CH_3NO_2}$.

9.4 Errors in Gravimetric Analysis

In gravimetric analysis, sometimes the weighed precipitate is further analyzed to find out whether it is pure and whether its chemical formula corresponds to the formula used in the stoichiometric calculations. If needed, proper corrections are made in the analytical results.

Example 9.22. In the determination of calcium in a 0.6735-g sample of an ore which also contained manganese, it was observed that the weighed CaO (0.2432 g) was not white. Therefore, it was further analyzed and found to contain 0.0183 g of manganese. Given that the manganese existed as Mn_3O_4 in the ignited precipitate, calculate the per cent of calcium in the ore.

Solution. 0.0183 g of Mn corresponds to

$$0.0183 \text{ g Mn} \times \frac{228.81 \text{ g } Mn_3O_4}{3 \times 54.94 \text{ g Mn}} = 0.0254 \text{ g} Mn_3O_4.$$

Therefore, the weighed precipitate contained 0.2432 − 0.0254 = 0.2178 g of CaO. Hence, the per cent calcium in the ore is equal to

$$\%\text{Ca} = \frac{0.2178 \text{ g CaO} \times \dfrac{40.08 \text{ g Ca}}{56.08 \text{ g CaO}}}{0.6735} \times 100 = \mathbf{23.11}.$$

Example 9.23. What error will result in the calculation of the percentage of MgO in a sample from the presence of 3.00% CaO, which is present in the form of $Ca_3(PO_4)_2$ in the weighed $Mg_2P_2O_7$?

Solution. We have

$$\begin{aligned} 3.00 \text{ g CaO} &\equiv 3.00 \text{ g CaO} \times \frac{310.20 \text{ g Ca}_3(\text{PO}_4)_2}{3 \times 56.08 \text{ g CaO}} = 5.53 \text{ g Ca}_3(\text{PO}_4)_2 \\ &\equiv 5.53 \text{ g Ca}_3(\text{PO}_4)_2 \times \frac{2 \times 40.31 \text{ g MgO}}{222.57 \text{ g Mg}_2\text{P}_2\text{O}_7} = 2.00 \text{ g MgO}. \end{aligned}$$

Hence, a positive error equal to 2.00% will result in the percentage of MgO in the sample.

Example 9.24. A 0.7834-g sample of iron ore containing 18.88% Fe was analyzed gravimetrically by precipitating with ammonia and igniting to Fe_2O_3. If the final precipitate also contains 0.0149 g of Al_2O_3 by mistake, calculate the absolute error in the percentage of iron.

Solution. The weight of the precipitate is equal to

$$\begin{aligned} \text{W} &= \left(\text{g sample} \times 0.1888 \times \frac{\text{Fe}_2\text{O}_3}{2\text{Fe}}\right) + \text{Al}_2\text{O}_3 \\ &= \left(0.7834 \times 0.1888 \times \frac{159.69}{2 \times 55.85}\right) + 0.0149 = 0.2264 \text{ g}. \end{aligned}$$

Hence, it is found experimentally that the ore contains

$$\frac{0.2264 \text{ g Fe}_2\text{O}_3 \times \dfrac{2 \times 55.85 \text{ g Fe}}{159.69 \text{ g Fe}_2\text{O}_3}}{0.7834 \text{ g sample}} \times 100 = 20.21\% \text{ Fe}.$$

Hence, there is a positive error equal to 20.21 − 18.18 = **1.33**% Fe.

Example 9.25. If in the gravimetric determination of chloride, part of the AgCl precipitate is photo-reduced during filtering and drying so that it contains 1.40% elemental Ag, what is the relative error in the percentage of chloride in the sample?

Solution. The data of the problem are independent of the sample size. Suppose that we weigh 1.0000 g of precipitate which, because of the partial decomposition of AgCl, consists of 98.60% AgCl and 1.40% Ag, that is, of 0.9860 g AgCl and 0.0140 g Ag. Then 0.0140 g of Ag results from AgCl equal to (0.0140)(AgCl/Ag) = (0.0140)(143.32/107.87) = 0.0186 g AgCl. Therefore, the weight of the precipitate before the decomposition was

0.9860 + 0.0186 = 1.0046 g. Hence, the relative negative error in the determination of chloride is equal to

$$\frac{1.0000 - 1.0046}{1.0046} \times 100 = \mathbf{-0.46\%}.$$

9.5 Problems

Gravimetric factor

9.1. The gravimetric determination of arsenic is made by oxidizing it to arsenate which is precipitated as Ag_3AsO_4. The latter is dissolved in HNO_3 and the silver is precipitated and weighed as AgCl. Which one of the following relations gives the gravimetric factor for the calculation of the per cent As_2S_3 in the initial sample?

$$\text{a)}\ \frac{As_2S_3}{3AgCl},\ \text{b)}\ \frac{2As_2S_3}{3AgCl},\ \text{c)}\ \frac{As_2S_3}{6AgCl},\ \text{d)}\ \frac{6AgCl}{As_2S_3},\ \text{e)}\ \frac{As_2S_3}{AgCl},\ \text{f)}\ \frac{AgCl}{As_2S_3}$$

9.2. Phosphate is precipitated from its solution with ammonium molybdate, as $(NH_4)_3[PMo_{12}O_{40} \cdot xH_2O]$. Since this precipitate does not have a constant composition, with regard to water content, it is dissolved in ammonia and the molybdate is precipitated with $Pb(NO_3)_2$, as $PbMoO_4$. a) What is the value of the gravimetric factor for the calculation of the percentage of P? b) If the final precipitate weighs 0.1000 g, what is the weight of P in the initial sample?

Stoichiometry

9.3. In the gravimetric determination of iron in an ore containing FeS_2, the final weight of Fe_2O_3 was 0.3117 g. What is the weight of iron expressed as a) Fe, b) FeO, c) FeS_2 in the original sample?

9.4. A 0.2970-g sample containing aluminum gives 0.3227 g of the hydroxyquinolate, $Al(C_9H_6ON)_3$. Calculate the per cent Al in the sample.

9.5. A sample of Mohr's salt, $FeSO_4 \cdot (NH_4)_2SO_4 \cdot 6H_2O$, is analyzed to check its validity as a primary standard for iron. Ignition of a 1.5000-g sample yielded 0.3016 g of Fe_2O_3. Calculate a) the purity of the sample, b) the per cent iron in the sample.

9.6. A 0.5874-g sample of alum, $KAl(SO_4)_2 \cdot 12H_2O$, which contains inert materials, is treated with 8-hydroxyquinoline. The precipitate is ignited, giving 0.0589 g of Al_2O_3. Calculate the per cent Al and K in the sample.

9.7. A 0.1028-g sample of the salt $Na_2S_xO_6$ gives 0.3570 g of $BaSO_4$. Calculate the value of x.

9.8. A 0.9996-g sample of impure Fe_2O_3 is ignited strongly and gives a residue of 0.9784 g. If the loss in weight is due to the conversion of Fe_2O_3 to Fe_3O_4 ($6Fe_2O_3 \rightarrow 4Fe_3O_4 + O_2 \uparrow$), calculate the per cent Fe_2O_3 in the sample.

9.9. How many milliliters of 0.200 $\underline{M}$ $AgNO_3$ are required for the quantitative precipitation of chlorine present in 0.250 g of NaCl, if we want a 10% excess of reagent?

9.10. How many grams of ammonium oxalate, $(NH_4)_2C_2O_4{\cdot}H_2O$, are required for the quantitative precipitation of calcium in 0.500-g sample of impure limestone ($CaCO_3$), if we want a 10% excess of reagent?

9.11. A sample of the inorganic acid H_nA, weighing x g, is neutralized quantitatively with 25.00 mL of 0.1000 $\underline{M}$ $Ca(OH)_2$ solution. The anhydrous neutral salt, Ca_nA_2, obtained after the evaporation of the solution, weighs 0.3000 g. Calculate the per cent hydrogen in the acid.

9.12. A sample of $KClO_4$ is ignited to KCl, evolving 0.1280 g of oxygen. What weight of AgCl will be formed, if the chloride is precipitated quantitatively? For the same amount of evolved oxygen, does the amount of AgCl depend on the nature of the salt $MClO_4$ (M = monovalent metal)?

9.13. A 0.4994-g sample of a hydrate of copper sulfate, $CuSO_4 \cdot xH_2O$, is heated to a constant weight of 0.3184 g (total loss of water). Calculate the value of x.

9.14. The compounds MCl_x and MCl_{x+2} contain 22.55 and 14.87% M, respectively. What element is M?

Indirect analysis

9.15. When ignited, a 0.3200-g sample consisting of only $CaCO_3$ and $MgCO_3$ yielded a residue of CaO and MgO, weighing 0.1664 g. Calculate the per cent CaO and MgO in the residue.

9.16. A 0.2000-g sample consisting of chlorides, iodides, and inert materials, gave 0.3780 g of AgCl + AgI precipitate. When heated in a stream of chlorine, for the conversion of AgI to AgCl, the precipitate lost 0.0488 g in weight. Calculate the per cent Cl and I in the sample.

9.17. A 3.9996-g sample consisting of NaCl and $BaCl_2 \cdot 2H_2O$ weighed 3.7113 g after heating (quantitative removal of water). The anhydrous product was dissolved in water and diluted to 250.0 mL (solution A). How many milliliters of 0.1000 $\underline{M}$ $AgNO_3$ solution are required for the titration of a 10.00-mL aliquot of solution A?

9.18. A sample contains NaCl, NaBr, NaI, and possibly some inert impurities. A 0.3697-g sample is dissolved in water and diluted to 100.0 mL (solution A). A 25.00-mL aliquot of solution A gave a precipitate of PdI_2 weighing 0.0450 g, whereas a 50.00-mL aliquot gave a precipitate of 0.3551 g when treated with excess $AgNO_3$ solution. When the latter precipitate was heated in a stream of chlorine, it gave 0.2866 g of AgCl. Calculate the per cent of each halide in the sample.

9.19. A sample contains only CaC_2O_4 and CaO. A 1.2000-g sample is heated at 900°C, yielding a residue of CaO weighing 0.8400 g. Calculate the per cent CaC_2O_4 and CaO in the sample.

9.20. A 0.3527-g sample containing only CaC_2O_4 and MgC_2O_4 yields a residue of 0.1807 g and 0.1367 g, when heated at 500°C and 900°C, respectively. Given that the following conversions take place upon heating, calculate the per cent of Ca and Mg in the sample:

$$CaC_2O_4 \xrightarrow{500^\circ C} CaCO_3 \xrightarrow{900^\circ C} CaO$$

$$MgC_2O_4 \xrightarrow{500^\circ C} MgO \xrightarrow{900^\circ C} MgO$$

9.21. A 0.2660-g sample consisting of only KCl and NaCl contains 0.1418 g of Cl. Calculate the per cent K and Na in the sample.

9.22. A mixture of KCl and NaBr is converted to the corresponding sulfate salts without any change in weight. Calculate the per cent KCl in the sample.

9.23. A sample consisting of only KCl and $KClO_3$ contains 35.97% Cl. Calculate the per cent KCl and $KClO_3$ in the sample.

9.24. A 0.7500-g sample that contains KCl, KBr, and inert impurities, is titrated with 32.00 mL of 0.1250 $\underline{M}$ $AgNO_3$ solution. A 0.2500-g sample gives a precipitate of AgCl + AgBr that weighs 0.2207 g. Calculate the per cent KCl and KBr in the sample.

9.25. A mixture consisting of only $BaSO_4$ and $PbSO_4$ contains 66.29% as much Ba as Pb. Calculate the per cent $PbSO_4$ in the mixture.

9.26. A 0.4828-g sample that contains 78.11% NaCl and 21.89% MCl gives 1.1280 g of AgCl. Calculate the atomic weight of M.

9.27. Combustion of a 0.2743-g sample that contains only phenol (C_6H_5OH) and glucose ($C_6H_{12}O_6$) gives 0.5281 g of CO_2. Calculate the per cent phenol and glucose in the sample.

Sample size

9.28. A series of samples known to contain 20.0–25.0% $KClO_3$ is to be analyzed gravimetrically, after reducing chlorate to chloride and precipitating the latter as AgCl. a) What is the minimum sample weight that must be taken in order to assure a AgCl precipitate that weighs at least 0.500 g? b) What will the maximum weight of AgCl be which can be obtained from the above amount of sample?

9.29. What weight of steel sample must be taken for analysis so that each milligram of $BaSO_4$ precipitate obtained represents 0.100% sulfur in the sample?

9.30. An iron ore contains 97.4% Fe_2O_3. What weight of sample must be taken for analysis so that each milligram of Fe_2O_3 precipitate obtained represents 0.200% iron in the sample?

9.31. A sample given as an unknown in the gravimetric determination of sulfate may consist of Na_2SO_4 and/or K_2SO_4. a) Between what limits should the sample weight be so as to give 0.250–0.500 g of $BaSO_4$? b) If a 10% excess of $BaCl_2$ is desired for more complete precipitation of sulfate, how many milliliters of 0.200 $\underline{M}$ $BaCl_2$ solution are required for a sample weighing y g?

9.32. What size sample which contains 20.0% Br^- must be taken for analysis so as to obtain a precipitate of AgBr weighing 0.3876 g?

Calculations on dry and on as-received basis

9.33. A zinc ore contains 1.80% moisture as received, and 21.70% Zn after being dried at 110°C. Calculate the per cent of Zn in the initial (as received) sample.

9.34. A copper ore contains 48.10% Cu on a dry basis and 47.08% Cu on an as-received basis. Calculate the per cent moisture of the sample as received.

9.35. A 1.0000-g sample, after being dried at 110°C, weighs 0.9875 g. Another 0.4000-g sample (as received) gives 0.3810 g of AgCl. Calculate the per cent of chlorine in the sample a) on a dry basis, b) on an as-received basis.

Empirical formula

9.36. Combustion of 0.7821 g of an organic compound that contains carbon, hydrogen, bromide, and possibly oxygen, gives 0.3222 g of H_2O and 0.6324 g of CO_2. A 0.1500-g sample of the compound yields 0.2586 g of AgBr. Find the empirical formula of the compound.

Errors in gravimetric analysis

9.37. In the gravimetric determination of iron in a 0.6225-g sample of an ore, it is found that the ore contains 12.69% Fe. Afterwards, the analyst finds that he used a common filter instead of an ashless one. To correct the erroneous result, he ashes a replica common filter which yields a residue weighing 0.0029 g. What is the real percentage of iron in the sample?

9.38. In the gravimetric determination of sulfate in a 0.2841-g sample of chemically pure Na_2SO_4, a $BaSO_4$ precipitate weighing 0.4604 g was obtained. The weight of the precipitate was smaller than the theoretical one, because some $BaSO_4$ was converted to BaS during the ignition process. Calculate a) the per cent of BaS in the precipitate, b) the per cent error of the analysis.

9.39. In the gravimetric determination of iron in a 1.5584-g sample which also contains inert materials, 0.2922 g of Fe_2O_3 was obtained. If the real percentage of iron in the sample is 12.85%, calculate the relative error of the method.

9.40. Calculate the per cent relative error in the gravimetric determination of iron ($Fe^{3+} \rightarrow Fe_2O_3$), if the ignition product contains 70.00% Fe_2O_3 and 30.00% Fe_3O_4, which is due to igniting the precipitate at a much higher temperature than the required one.

9.41. In the gravimetric determination of chloride, 2.40% of the AgCl precipitate is decomposed photochemically (after the filtration) according to the reaction

$$2AgCl \overset{h\nu}{\rightleftharpoons} 2Ag + Cl_2 \uparrow$$

Calculate the per cent error caused by the above decomposition.

9.42. In the gravimetric determination of calcium in the form of $CaCO_3$, the residue contains 2.00% CaC_2O_4 because of incomplete ignition of the CaC_2O_4 precipitate. What will the relative error be if we assume that the residue consists only of $CaCO_3$?

9.43. A student analyzed a chloride sample but forgot to dry it. He found that the sample contained 20.28% Cl, whereas the correct value was 20.35%. Calculate the per cent moisture in the sample.

Chapter 10

An Introduction to Volumetric Methods of Analysis

10.1 Introduction

Titration is the process by which a substance is determined by measuring the quantity of a reagent required for the quantitative reaction with the substance. In volumetric analysis, we use *standard solutions (standards)*, that is, solutions of precisely known concentration, and we measure *precisely* with a buret the volume of the standard solution (titrant) consumed to complete the titration reaction. The amount of the substance being determined is calculated from the volume of the consumed standard solution, its known concentration, and the equivalent weight of the substance. The concentration of the standard solution is usually expressed in chemical units, e.g., moles/liter or equivalents/liter.

The basic goal of a titration is to determine the volume of a standard solution which is chemically equivalent to the substance being titrated (titrand). The titration point at which there is a chemical equivalency between titrant and titrand is called the *equivalence point*, whereas the point at which the addition of titrant is actually terminated is called the *end point*. The end point is detected in various ways, usually with indicators. The difference between the end point and the equivalence point constitutes the *titration error*.

The titration is usually performed by controlled addition of titrant until the end point. Occasionally, e.g., when the titration reaction is slow or there is no proper indicator or an undesirable precipitate is formed, the *technique of the back titration* is applied, that is, a precisely known excess of a standard solution is added to the titrand, and then the unreacted excess of the standard solution is determined by titration with another standard solution (Examples 10.15 and 10.16).

Volumetric methods of analysis are classified in various ways, usually according to the type of the chemical reaction that takes place during the titration. In this way, we have a) *neutralization (acid-base) titrations* (Chapter 11), b) *oxidation-reduction (redox) titrations* (Chapter 12), c) *precipitation titrations* (Chapter 13), and d) *complexometric titrations* (Chapter 14).

10.2 Stoichiometry

Primary standard solutions contain an accurately known amount of a primary standard substance, that is, a highly pure substance, in an accurately measured volume of solution. These solutions are prepared by direct measurement of the weight of the primary standard and the volume of the solution. *Secondary standard solutions* are prepared approximately from chemicals whose exact purity is not known and their concentrations are determined by standardization (titrating a primary standard or a primary standard solution, or gravimetrically).

---

a. Preparation of a primary standard solution from a primary standard

Example 10.1. How much $AgNO_3$ is required for the preparation of 500.0 mL of 0.1000 $\underline{M}$ $AgNO_3$ solution?

Solution. The molecular weight of $AgNO_3$ is 169.87. Hence, we need

$$500.0 \text{ mL} \times 0.1000 \text{ mmol/mL} \times 0.16987 \text{ g } AgNO_3\text{/mmol} = \mathbf{8.494} \text{ g } AgNO_3.$$

Example 10.2. A 4.9236-g portion of $K_2Cr_2O_7$ is dissolved in water and diluted to exactly one liter (solution A). Calculate the normality of standard solution A [for Cr(VI) $\rightarrow$ Cr(III)].

Solution. The $Cr_2O_7^{2-}$ ions are reduced to Cr^{3+} ions according to the half-reaction

$$Cr_2O_7^{2-} + 14H^+ + 6e \rightleftharpoons 2Cr^{3+} + 7H_2O$$

The change in oxidation number of each Cr atom is from +6 to +3, that is, a total change of 6 per $Cr_2O_7^{2-}$ ion. Hence, the equivalent weight of $K_2Cr_2O_7$ is equal to 294.19/6, and the normality of the standard solution is equal to

$$N = \frac{4.9236 \text{ g } K_2Cr_2O_7\text{/L}}{(294.19/6) \text{ g } K_2Cr_2O_7\text{/eq}} = \mathbf{0.1004} \text{ eq/L}.$$

---

b. Preparation of a standard solution by dilution of another standard solution

Example 10.3. What volume of standard solution A of Example 10.2 is required to prepare 500.0 mL of 0.0500 $\underline{N}$ $K_2Cr_2O_7$ solution?

Solution. Suppose that y mL of solution A is required. By diluting solution A, the number of milliequivalents remains constant. Hence, we have

$$(\text{y mL})(0.1004 \text{ meq/mL}) = (500.0 \text{ mL})(0.0500 \text{ meq/mL})$$

or y = **249.0** mL.

---

c. Preparation of a solution of approximately known concentration

Example 10.4. What volume of HCl solution, that has a density of 1.19 g/mL and contains 37% HCl by weight, should be diluted with water to prepare one liter of an approximately 0.100 N HCl solution?

Solution. The normality of the initial HCl solution is equal to

$$N_{HCl} = \frac{1000 \text{ mL/L} \times 1.19 \text{ g solution/mL} \times 0.37 \text{ g HCl/g solution}}{36.46 \text{ g HCl/eq}} = 12.1 \text{ eq/L}.$$

Suppose that y mL of HCl solution is required, whereupon we have

$$\text{y mL} \times 12.1 \text{ meq/mL} = 1000 \text{ mL} \times 0.100 \text{ meq/mL}$$

or y = **8.3** mL.

---

d. Standardization of a secondary standard solution against a primary standard substance

Example 10.5. In the standardization of an iodine solution, 39.16 mL were consumed to titrate a solution containing 0.1941 g of As_2O_3 (primary standard substance). Calculate the normality of the iodine solution.

Solution. The titration reaction is

$$H_3AsO_3 + H_2O + [I_3]^- \rightleftharpoons HAsO_4^{2-} + 4H^+ + 3I^-$$

The total change in oxidation number is 4 per As_2O_3 molecule. At the equivalence point we have the relation

$$\text{meq } I_2 = \text{meq } As_2O_3.$$

Hence, we have

$$39.16 \text{ mL} \times \text{N meq } I_2/\text{mL} = \frac{194.1 \text{ mg } As_2O_3}{(197.84/4) \text{ mg } As_2O_3/\text{meq}},$$

from which

$$N_{I_2} = \mathbf{0.1002} \text{ meq/mL}.$$

Example 10.6. A 0.4355-g portion of $CaCO_3$ (primary standard substance) is dissolved in 1+1 HCl and the solution is diluted with water to 250.0 mL (solution A). In the standardization of an EDTA solution (see Example 7.6 for EDTA formula), 43.60 mL were consumed to titrate 25.00 mL of solution A. Calculate the molarity of the EDTA solution.

Solution. The titration reaction is

$$Ca^{2+} + HY^{3-} \rightleftharpoons [CaY]^{2-} + H^+ \quad (HY^{3-} = \text{anion of EDTA})$$

At the equivalence point we have the relation

$$\text{mmol EDTA} = \text{mmol } CaCO_3$$

Hence, we have

$$43.60 \text{ mL} \times M \text{ mmol EDTA/mL} = \frac{\left(435.5 \times \frac{25.00}{250.0}\right) \text{ mg } CaCO_3}{100.09 \text{ mg } CaCO_3\text{/mmol}},$$

from which

$$M_{EDTA} = \mathbf{0.00998} \text{ mmol/mL}.$$

---

e. Standardization of a secondary standard solution with another standard solution

Example 10.7. What is the normality of a NaOH solution, a 50.00-mL portion of which required 42.19 mL of 0.1184 N HCl for titration?

Solution. At the equivalence point we have the relation

$$\text{meq NaOH} = \text{meq HCl}$$

Hence, we have

$$50.00 \text{ mL} \times N \text{ meq NaOH/mL} = 42.19 \text{ mL} \times 0.1184 \text{ meq HCL/mL},$$

from which

$$N_{NaOH} = \mathbf{0.0999} \text{ meq/mL}.$$

---

f. Gravimetric standardization of a standard solution

Example 10.8. What is the molarity of an HCl solution, 50.00 mL of which yields 0.7238 g of AgCl?

Solution. We have the relation

$$\frac{0.7238 \text{ g AgCl}}{0.14332 \text{ g/mmol}} = 50.00 \text{ mL} \times \text{M mmol HCl/mL},$$

from which we have

$$M_{HCl} = \mathbf{0.1010} \text{ mmol/mL}.$$

---

g. Calculation of sample size to consume desired volume of standard solution

Example 10.9. The volume of titrant consumed in a standardization or in a titration should be from 35 to 45 mL, in order to minimize the error in reading the volume (with a 50-mL buret). How many grams of substance should be taken a) when standardizing a 0.1000 N NaOH solution with potassium acid phthalate (MW 204.23), b) when titrating a sample containing 40% potassium acid phthalate?

Solution. a) The titration reaction is

$$KHC_8H_4O_4 + NaOH \rightleftharpoons KNaC_8H_4O_4 + H_2O$$

The potassium acid phthalate acts as a monoprotic acid, therefore 1 meq $KHC_8H_4O_4$ = 0.20423 g $KHC_8H_4O_4$. Hence, from (35 × 0.1000 × 0.20423) to (45 × 0.1000 × 0.20423) = **0.71** to **0.92** g of potassium acid phthalate should be taken.

b) From (0.71/0.40) to (0.92/0.40) = **1.8** to **2.3** g of sample should be taken.

---

h. Calculation of the per cent content by direct titration

Representative examples are given below, one for each type of titration, that is, neutralization titrations (Example 10.10), redox titrations (Example 10.11), precipitation titrations (Example 10.12), and complexometric titrations (Example 10.13).

Example 10.10. A 0.3344-g sample containing $NaHCO_3$ is dissolved in water and titrated with 38.14 mL of HCl solution. In the standardization of the HCl solution, 37.83 mL of the acid were consumed to titrate 0.2001 g of Na_2CO_3 to the methyl orange end point. Calculate the per cent $NaHCO_3$ in the sample.

Solution. In the standardization of HCl the following reaction takes place:

$$CO_3^{2-} + 2H^+ \rightleftharpoons H_2CO_3 \rightleftharpoons H_2O + CO_2 \uparrow$$

Hence, the normality of the HCl solution is equal to

$$N_{HCl} = \frac{200.1 \text{ mg } Na_2CO_3}{(105.99/2) \text{ mg } Na_2CO_3/\text{meq} \times 37.83 \text{ mL}} = 0.0998 \text{ meq/mL}.$$

In the analysis of $NaHCO_3$, the following reaction takes place:

$$HCO_3^- + H^+ \rightleftharpoons H_2CO_3 \rightleftharpoons H_2O + CO_2 \uparrow$$

Hence, the per cent of $NaHCO_3$ in the sample is equal to

$$\%NaHCO_3 = \frac{38.14 \text{ mL} \times 0.0998 \text{ meq/mL} \times 84.01 \text{ mg } NaHCO_3/\text{meq}}{334.4 \text{ mg sample}} \times 100 = \mathbf{95.6}.$$

Example 10.11. A 1.5380-g sample of iron ore is dissolved in acid, the iron is reduced to the +2 oxidation state quantitatively and titrated with 43.50 mL of $KMnO_4$ solution ($Fe^{2+} \rightarrow Fe^{3+}$), 1.000 mL of which is equivalent to 11.17 mg of iron. Express the results of the analysis as a) % Fe, b) % Fe_2O_3, c) % Fe_3O_4.

Solution. a) The normality of the $KMnO_4$ solution is equal to

$$N_{KMnO_4} = \frac{11.17 \text{ mg Fe/mL}}{55.85 \text{ mg Fe/meq}} = 0.2000 \text{ meq/mL}.$$

Hence, the per cent content of iron in the sample is equal to

$$\%Fe = \frac{43.50 \text{ mL} \times 0.2000 \text{ meq/mL} \times 55.85 \text{ mg Fe/meq}}{1538.0 \text{ mg sample}} \times 100 = \mathbf{31.6}.$$

$$\text{b) } \%Fe_2O_3 = \frac{43.50 \text{ mL} \times 0.2000 \text{ meq/mL} \times (159.69/2) \text{ mg } Fe_2O_3/\text{meq}}{1538.0 \text{ mg sample}} \times 100 = \mathbf{45.2}.$$

$$\text{c) } \%Fe_3O_4 = \frac{43.50 \text{ mL} \times 0.2000 \text{ meq/mL} \times (231.54/3) \text{ mg } Fe_3O_4/\text{meq}}{1538.0 \text{ mg sample}} \times 100 = \mathbf{43.7}.$$

Example 10.12. A 0.9832-g sample of silver ore is dissolved in HNO_3 and titrated with 35.13 mL of potassium thiocyanate solution that has a titer equal to 0.01000 g Ag/mL, in the presence of Fe^{3+} as an indicator. Calculate the per cent silver in the ore.

Solution. The sample contains

$$35.13 \text{ mL} \times 0.01000 \text{ g Ag/mL} = 0.3513 \text{ g Ag}.$$

Hence, the per cent silver in the ore is equal to

$$\frac{0.3513}{0.9832} \times 100 = \mathbf{35.73\%} \text{ Ag}.$$

Example 10.13. A 50.00-mL portion of drinkable water is titrated with 14.75 mL of EDTA solution at pH 10.0. In the standardization of EDTA, 24.88 mL of the EDTA solution was consumed for the titration of 25.00 mL of a standard 0.01002 M $CaCl_2$ solution. Calculate the hardness of the water in French degrees, that is, in mg $CaCO_3$/100 mL solution.

Solution. The reaction between Ca^{2+} ion and EDTA is

$$Ca^{2+} + HY^{3-} \rightleftharpoons [CaY]^{2-} + H^+$$

The molarity of the EDTA solution is equal to

$$M_{EDTA} = \frac{25.00 \text{ mL} \times 0.01002 \text{ mmol/mL}}{24.88 \text{ mL}} = 0.01007 \text{ mmol/mL}.$$

Hence, the hardness of the drinkable water is equal to 14.75 mL × 0.01007 mmol/mL × 100.09 mg $CaCO_3$/mmol × 100/50 = **29.7** mg $CaCO_3$/100 mL water ≡ **29.7** French degrees (≡ 297 mg $CaCO_3$/L ≡ 297 ppm $CaCO_3$ or American degrees of hardness).

Example 10.14. What weight of sample must be taken for analysis so that each milliliter of 0.1000 N NaOH solution consumed in the titration represents 1.000% of potassium acid phthalate in the sample?

Solution. We have

$$\%KHC_8H_4O_4 = \frac{\text{mL NaOH} \times 0.1000 \text{ meq/mL} \times 204.23 \text{ mg } KHC_8H_4O_4\text{/meq}}{\text{mg sample}} \times 100 \quad (10.1)$$

Since $\%KHC_8H_4O_4$ = mL NaOH, Equation (10.1) is simplified to

$$\text{mg sample} = 0.1000 \times 204.23 \times 100 = 2042.$$

Hence, 2.042 g of sample must be taken for analysis.

Note. The example is representative of the cases in which a large number of similar samples is to be analyzed, e.g., in industrial laboratories (as in Example 9.7).

---

i. Calculation of the per cent content by back titration

Example 10.15. A 0.2759-g sample of chromite is fused with sodium peroxide. After the decomposition of excess Na_2O_2, the fused material is extracted with water, the resulting solution containing Cr(VI) is acidified and 50.00 mL of 0.0999 N Fe^{2+} solution is added to it. After the complete reduction of Cr(VI), 18.77 mL of 0.1003 N $KMnO_4$ solution is consumed to titrate the excess of Fe^{2+} ions. Calculate the per cent content of chromium in the sample.

Solution. During the analysis, the following reactions take place:

$$2Fe(CrO_2)_2 + 7Na_2O_2 \rightleftharpoons Fe_2O_3 + 4Na_2CrO_4 + 3Na_2O$$

$$2CrO_4^{2-} + 2H^+ \rightleftharpoons Cr_2O_7^{2-} + H_2O$$

$$Cr_2O_7^{2-} + 6Fe^{2+} + 14H^+ \rightleftharpoons 2Cr^{3+} + 6Fe^{3+} + 7H_2O$$

$$MnO_4^- + 5Fe^{2+} + 8H^+ \rightleftharpoons Mn^{2+} + 5Fe^{3+} + 4H_2O$$

We have the relation

$$\text{meq Cr} = \text{meq } Fe^{2+} = \text{meq added } Fe^{2+} - \text{meq } KMnO_4.$$

Therefore,

$$\text{meq Cr} = (50.00 \text{ mL} \times 0.0999 \text{ meq/mL}) - (18.77 \text{ mL} \times 0.1003 \text{ meq/mL}) = 3.112$$

Hence, the per cent chromium in the sample is equal to

$$\frac{3.112 \text{ meq} \times (52.00/3) \text{ mg Cr/meq}}{275.9 \text{ mg sample}} \times 100 = \mathbf{19.55\%} \text{ Cr}.$$

Example 10.16. A 0.1821-g sample of an aluminum ore is dissolved, 50.00 mL of 0.0508 M EDTA solution is added and the excess of unreacted EDTA is titrated with 3.81 mL of 0.0502 M magnesium solution. Calculate the per cent aluminum in the sample.

Solution. The per cent of aluminum in the sample is equal to

$$\%\text{Al} = \frac{[(50.00 \text{ mL} \times 0.0508 \text{ mmol/mL}) - (3.81 \text{ mL} \times 0.0502 \text{ mmol/mL})] \times 26.98 \text{ mg Al/mmol}}{182.1 \text{ mg}} \times 100 = \mathbf{34.8}.$$

---

j. Indirect volumetric analysis

The simplest case of indirect volumetric analysis is one in which a mixture of two pure chemical substances having a common ion is titrated (Section 9.2).

Example 10.17. A 0.2414-g sample consisting of only NaCl and KCl is titrated with 39.35 mL of 0.1000 M $AgNO_3$. Calculate the per cent NaCl and KCl in the sample.

Solution. Suppose that y = g NaCl in the sample, whereupon 0.2414 - y = g KCl. We have

$$\frac{y \text{ g NaCl}}{0.05844 \text{ g/mmol}} + \frac{(0.2414 - y) \text{ g KCl}}{0.07456 \text{ g/mmol}} = 39.35 \text{ mL} \times 0.1000 \text{ mmol/mL}$$

or y = 0.1885 g NaCl and 0.2414 − 0.1885 = 0.0529 g KCl. Hence, the sample contains

$$\frac{0.1885}{0.2414} \times 100 = \mathbf{78.09\%}\ \text{NaCl}$$

and

$$\frac{0.0529}{0.2414} \times 100 = \mathbf{21.91\%}\ \text{KCl}.$$

---

k. Calculation of concentrations in redox systems at equilibrium

Example 10.18. A solution containing 4.00 meq of As(III) is titrated with a standard 0.1000 N triiodide solution (0.0500 M $[I_3]^-$ − 0.20 M KI) in the presence of starch as an indicator. If the pH of the solution is maintained at 8.00, its final volume is 140 mL and the end-point blue color appears at $[I_3^-] = 5 \times 10^{-7}$ M, calculate the As(III) concentration at the end point.

Solution. The oxidation of As(III) by $[I_3]^-$ is done according to the reaction

$$H_3AsO_3 + [I_3]^- + H_2O \rightleftharpoons H_3AsO_4 + 3I^- + 2H^+, \tag{10.2}$$

for which we have

$$K = \frac{[H_3AsO_4][I^-]^3[H^+]^2}{[H_3AsO_3][I_3^-]} = 10^{\frac{2(0.536-0.559)}{0.05916}} = 0.167 \tag{10.3}$$

At the end point, the solution contains a total of 14 mmol I^-, i.e., 6 mmol produced in Reaction (10.2), and 8 mmol from KI (40.0 mL × 0.20 mmol/mL). Hence, at the end point we have the following concentrations: $[I^-] = 14/140 = 0.100$ M, $[I_3^-] = 5 \times 10^{-7}$ M, and $C_{As} = [As(V)] + [As(III)] = 2.00/140 = 0.0143$ M, whereupon $[As(V)] = 0.0143 - [As(III)]$. We also have the relations (Section 5.4)

$$\begin{aligned} \alpha_0 &= \frac{[H_3AsO_4]}{[As(V)]} \\ &= (1.0 \times 10^{-8})^3/[(1.0 \times 10^{-8})^3 + (6 \times 10^{-3})(1.0 \times 10^{-8})^2 + (6 \times 10^{-3}) \\ &\quad (1 \times 10^{-7})(1.0 \times 10^{-8}) + (6 \times 10^{-3})(1 \times 10^{-7})(3 \times 10^{-12})] \\ &= 1.5 \times 10^{-7} \end{aligned} \tag{10.4}$$

or

$$\begin{aligned} [H_3AsO_4] &= 1.5 \times 10^{-7}[As(V)] = 1.5 \times 10^{-7}(0.0143 - [As(III)]) \\ &= 2.15 \times 10^{-9} - 1.5 \times 10^{-7}[As(III)] \end{aligned} \tag{10.4a}$$

and

$$\alpha_0{}' = \frac{[H_3AsO_3]}{[As(III)]} = \frac{1.0 \times 10^{-8}}{1.0 \times 10^{-8} + 6 \times 10^{-10}} = 0.94 \tag{10.5}$$

or

$$[H_3AsO_3] = 0.94[As(III)] \qquad (10.5a)$$

Substituting the above values in Equation (10.3), we have

$$0.167 = \frac{(2.15 \times 10^{-9} - 1.5 \times 10^{-7}[As(III)])(0.100)^3(1.0 \times 10^{-8})^2}{0.94[As(III)](5.0 \times 10^{-7})},$$

from which $[As(III)] = 2.7 \times 10^{-21}$ $\underline{M}$. Hence, the oxidation of As(III) is quantitative.

10.3 Errors in Volumetric Analysis

Example 10.19. A 50.00-mL portion of a) 0.1000 $\underline{M}$ HCl, b) 0.1000 $\underline{M}$ CH_3COOH, is titrated with 0.1000 $\underline{M}$ NaOH solution, in the presence of methyl orange as an indicator, at a fixed end-point pH of 4.00. What is the per cent determinate titration error in each case?

Solution. a) *1st method.* The titration error is negative, because the end point (pH = 4.00) precedes the equivalence point (pH 7.00). Let V mL be the volume of NaOH added until the end point. We have

$$\begin{aligned}[H^+] &= 1.0 \times 10^{-4} = \frac{\text{mmol HCl} - \text{mmol NaOH}}{(50.00 + V)\text{ mL}} \\ &= \frac{(50.00\text{ mL} \times 0.1000\text{ mmol/mL}) - (V\text{ mL} \times 0.1000\text{ mmol/mL})}{(50.00 + V)\text{ mL}}\end{aligned}$$

from which V = 49.90 mL. Hence, the titration error is equal to

$$\frac{49.90 - 50.00}{50.00} \times 100 = \mathbf{-0.2\%}.$$

2nd method. Let V mL be the volume of NaOH added until the end point, whereupon according to the principle of electroneutrality, we have

$$[Na^+] + [H^+] = [Cl^-] + [OH^-]$$

or

$$\frac{V \times 0.1000}{50.00 + V} + 1.0 \times 10^{-4} = \frac{50.00 \times 0.1000}{50.00 + V} + 1.0 \times 10^{-10},$$

from which V = 49.90 mL. Hence, the titration error is equal to

$$\frac{49.90 - 50.00}{50.00} \times 100 = \mathbf{-0.2\%}.$$

b) 50.00 mL of 0.1000 M NaOH are required until the equivalence point. Let V mL be the volume of NaOH added until the end point. We have from K_a

$$1.8 \times 10^{-5} = \frac{(1.0 \times 10^{-4}) \cdot \dfrac{(\text{V ml} \times 0.1000 \text{ mmol/mL})}{(50.00 + \text{V}) \text{ mL}}}{\dfrac{(50.00 \text{ mL} \times 0.1000 \text{ mmol/mL}) - (\text{V mL} \times 0.1000 \text{ mmol/mL})}{(50.00 + \text{V}) \text{ mL}}}$$

from which V = 7.63 mL. Hence, the titration error is equal to

$$\frac{7.63 - 50.00}{50.00} \times 100 = \mathbf{-84.7\%}.$$

Note. From the above large negative value it can be seen that methyl orange cannot be used as indicator in the titration of weak acids.

Example 10.20. A 50.00-mL portion of 0.1000 M Cl^- solution is titrated with 0.1000 M $AgNO_3$ solution, in the presence of CrO_4^{2-} ions as indicator (Mohr method). If we assume that the final concentration of CrO_4^{2-} ions is 0.0020 M and that the Ag_2CrO_4 precipitate becomes visible at 1.0×10^{-5} mol of solid Ag_2CrO_4 per liter of solution, calculate the per cent titration error.

Solution. The equivalence-point volume of the solution is 100.0 mL, and the end-point volume is about 100 mL. The concentration of Ag^+ ions required to start the precipitation of Ag_2CrO_4, that is, the concentration of Ag^+ ions at the end point, $[Ag^+]_{en}$, is, from K_{sp} of Ag_2CrO_4

$$[Ag^+]_{en} = \sqrt{1.9 \times 10^{-12}/0.0020} = 3.08 \times 10^{-5} \text{ M},$$

whereas the concentration of Ag^+ ions at the equivalence point, $[Ag^+]_{eq}$, is, from K_{sp} of AgCl

$$[Ag^+]_{eq} = [Cl^-]_{eq} = \sqrt{1.8 \times 10^{-10}} = 1.34 \times 10^{-5} \text{ M}.$$

The concentration of Cl^- ions at the end point is equal to

$$[Cl^-]_{en} = 1.8 \times 10^{-10}/3.08 \times 10^{-5} = 0.58 \times 10^{-5} \text{ M}.$$

In order to increase the Ag^+ ion concentration from 1.34×10^{-5} M to 3.08×10^{-5} M, and at the same time reduce the Cl^- ion concentration from 1.34×10^{-5} M to 0.58×10^{-5} M, we need

$$(3.08 \times 10^{-5} - 0.58 \times 10^{-5})100 = 2.50 \times 10^{-3} \text{ mmol } Ag^+.$$

Given that the Ag_2CrO_4 precipitate becomes visible at a quantity of 1.00×10^{-5} mol/L, for the formation of this quantity of precipitate, we need

$$2 \times 1.0 \times 10^{-5} \times 100 = 2.0 \times 10^{-3} \text{ mmol } Ag^+,$$

that is, we need a total excess of $(2.50 + 2.0)10^{-3} = 4.5 \times 10^{-3}$ mmol Ag^+, which corresponds to

$$(4.5 \times 10^{-3} \text{ mmol})/(0.1000 \text{ mmol/mL}) = 0.045 \text{ mL } 0.1000 \text{ M } AgNO_3.$$

Hence, the titration error is positive and equal to

$$\frac{\text{mL AgNO}_3 \text{ in excess}}{\text{mL AgNO}_3 \text{ until the equivalence point}} \times 100 = \frac{0.045 \text{ mL}}{50.00 \text{ mL}} \times 100 = \mathbf{+0.09\%}.$$

Example 10.21. A 1.4906-g sample of about 99% pure NaCl is dissolved in water and diluted to exactly 250.0 mL. In titrating 10.00-mL, 25.00-mL, and 50.00-mL aliquots of the NaCl solution with 0.1020 $\underline{N}$ $AgNO_3$ solution by the Mohr method, an average of 9.95 mL, 24.80 mL, and 49.58 mL of the standard solution was consumed, respectively (N = 4). Calculate the titration error. (**Note.** In titrating the 10.00-mL and 25.00-mL aliquots, 80 mL and 50 mL of water were added, respectively, so that in all titrations the volume at the end point was about the same, that is, 100 mL.)

Solution. Let x mL be the titration error. We have the relations

$$\frac{24.80 - x}{9.95 - x} = \frac{25.00}{10.00}, \text{ from which } x = 0.05 \text{ mL},$$

$$\frac{49.58 - x}{9.95 - x} = \frac{50.00}{10.00}, \text{ from which } x = 0.04 \text{ mL, and}$$

$$\frac{49.58 - x}{24.80 - x} = \frac{50.00}{25.00}, \text{ from which } x = 0.02 \text{ mL}.$$

The average of the calculated titration errors is taken as the experimental titration error, and is equal to

$$\frac{0.05 + 0.04 + 0.02}{3} = \mathbf{0.04} \text{ mL}.$$

Example 10.22. Are the following statements true or false?

a) If the standard substance used for the standardization of an HCl solution contains impurities that are inert with respect to HCl, the normality of HCl will be found smaller than the actual value.

b) If the Na_2CO_3 used for the standardization of an HCl solution also contains $NaHCO_3$, the normality of HCl will be found larger than the actual value.

c) In the standardization of a NaOH solution with a standard HCl solution, if a volume of NaOH solution twice the one given in the procedure is taken, the normality of NaOH will be found double the actual value.

d) Let V and v be the volumes of standard solution A consumed in the standardization of solution B and the titration of the blank, respectively. If by mistake the volume V instead of the correct one, V−v, is used for calculating the normality of solution B, the normality N_B will be found larger than the actual value.

Solution. a) The statement is false. In the standardization, the volume of HCl used will be smaller than the volume which would have been used in the absence of inert impurities (for the same sample size); hence, $N_{calc} > N_{real}$.

b) The statement is true. We have $V_{exp} < V_{theor}$; hence, $N_{exp(HCl)} > N_{theor}$.

c) The statement is false. The normality of NaOH will be found equal to the real one (provided we know twice the volume *was* used and this is employed in the calculation), because the normality is independent of the volume of the solution used in the standardization.

d) The statement is true. We have the relations

$$N_{B_c} V_B = N_A (V - v) \text{ (correct calculation of N)} \tag{10.6}$$

$$N_{B_{inc}} V_B = N_A V \text{ (incorrect calculation of N)} \tag{10.7}$$

Dividing Equations (10.6) and (10.7), we have

$$N_{B_c} V = N_{B_{inc}} (V - v) \tag{10.8}$$

Since $V > V{-}v$, it follows that $N_{B_{inc}} > N_{B_c}$.

10.4 Problems

Stoichiometry

10.1. How much NaCl is required for the preparation of 250.0 mL of 0.100 M NaCl solution?

10.2. A 4.8893-g portion of $K_2Cr_2O_7$ is dissolved in water and diluted to exactly one liter (solution A). Calculate the molarity and the normality of solution A [for Cr(VI) → Cr(III)].

10.3. In the standardization of an HCl solution, 45.18 mL were consumed to titrate a solution containing 0.5384 g of tris(hydroxymethyl)aminomethane, $(CH_2OH)_3CNH_2$, known as TRIS or THAM (primary standard substance). Calculate the normality of the HCl solution.

10.4. If 25.00 mL of solution A of oxalic acid is titrated with 30.00 mL of 0.1800 N NaOH solution, and if 20.50 mL of solution A react with 44.28 mL of a $KMnO_4$ solution in acidic medium ($MnO_4^- \rightarrow Mn^{2+}$), calculate the normality and the molarity of the $KMnO_4$ solution.

10.5. What is the molarity of a H_2SO_4 solution, 25.00 mL of which yields 0.5870 g of $BaSO_4$?

10.6. For the titration of a solution containing 0.1875 g of a mixture, consisting of $NaHC_2O_4$, $Na_2C_2O_4$, and inert materials in acidic medium, 20.10 mL of 0.1200 N $KMnO_4$ solution is required ($MnO_4^- \rightarrow Mn^{2+}$). For the titration of a solution containing twice the above amount of sample, 12.50 mL of 0.0999 N NaOH solution is required. Calculate the per cent $NaHC_2O_4$ and $Na_2C_2O_4$ in the sample.

10.7. In the titration of a 1.0580-g sample with a standard $AgNO_3$ solution, it was found that the percentage of NaCl in the sample was 200 times the normality of the titrant. How many milliliters of $AgNO_3$ solution were consumed?

10.8. What must the normality of a standard $AgNO_3$ solution be so that the milliliters of $AgNO_3$ divided by 2 equals the per cent KCl in 0.3728 g of sample?

10.9. If a 25.00-mL portion of a KHC_2O_4 solution requires 34.02 mL of 0.1003 N $KMnO_4$ solution ($MnO_4^- \rightarrow Mn^{2+}$), how many milliliters of 0.1000 N NaOH are required for the titration of the above solution?

10.10. What volume of H_2SO_4 solution, that has a density of 1.835 g/mL and contains 93.1% H_2SO_4 by weight, must be diluted with water to prepare one liter of an approximately 1.00 N H_2SO_4 solution?

Back titration

10.11. A 0.2108-g sample of pyrolusite is treated with 50.00 mL of 0.1000 M Fe^{2+} solution, whereupon MnO_2 is reduced to Mn^{2+}. After the complete reduction of Mn(IV), the solution is acidified and the excess of unreacted Fe^{2+} ions is titrated with 13.31 mL of 0.1002 N $KMnO_4$. Calculate the per cent MnO_2 in the sample.

10.12. A 0.2400-g sample of impure NaCl is dissolved in water, 50.00 mL of 0.1006 M $AgNO_3$ solution is added to precipitate Cl^- as AgCl, and the excess of unreacted Ag^+ ions is titrated with 9.58 mL of 0.0998 M KSCN solution. Calculate the per cent NaCl in the sample.

10.13. In the Kjeldahl method for the determination of nitrogen, the sample is treated with concentrated H_2SO_4 in the presence of a catalyst to convert nitrogen to NH_4^+ (digestion). The solution is made strongly alkaline, whereupon the NH_4^+ ion is converted to NH_3, which is then distilled into a known excess of a standard acid solution. The excess of unreacted acid is titrated with a standard base solution. From the following data, calculate the per cent nitrogen in a fertilizer: 0.5874-g sample, 50.00 mL of 0.1000 M HCl, 22.36 mL of 0.1064 M NaOH.

10.14. A 0.2352-g sample of an iron ore is dissolved in acid, 50.00 mL of 0.0512 M EDTA solution is added and the excess of unreacted EDTA is back-titrated with 17.16 mL of 0.0504 M Zn^{2+}. solution. Calculate the per cent Fe_2O_3 in the sample.

Sample size

10.15. If the volume of titrant consumed in a standardization or in a titration is required to be from 35 to 45 mL, how many grams of substance should be taken a) when standardizing a 0.1000 N HCl solution with Na_2CO_3, b) when titrating a sample containing 50% Na_2CO_3, in the presence of methyl orange as an indicator?

10.16. What weight of sample must be taken for analysis so that each milliliter of 0.1000 N $AgNO_3$ solution consumed in the titration represents 2.000% of KCl in the sample?

Indirect volumetric analysis

10.17. A 0.2284-g sample of a pharmaceutical preparation containing $NaHCO_3$, MgO, and inert materials, is dissolved in 20.00 mL of 0.5000 M HCl, the solution is boiled to remove CO_2, and the excess of unreacted HCl is titrated with 40.00 mL of 0.1500 M NaOH solution. Another 0.1142-g sample is dissolved in HCl, the pH is adjusted at 10.00 with an NH_3-NH_4Cl buffer, and the solution is titrated with 25.00 mL of 0.02000 M EDTA solution, in the presence of Eriochrome Black T as an indicator. Calculate the per cent of inert materials in the preparation.

10.18. The barium in a 0.2508-g sample is precipitated as $BaCrO_4$. The precipitate is filtered, washed, dissolved in acid, and the Cr(VI) is titrated with 32.17 mL of 0.1001 M Fe^{2+}. solution. Calculate the per cent barium in the sample.

10.19. A 0.2500-g sample of fuming sulfuric acid (a solution of SO_3 in anhydrous H_2SO_4) is diluted with water and titrated with 26.74 mL of 0.2000 M NaOH solution. Calculate the per cent SO_3 in the sample.

10.20. A 0.2958-g sample consisting of only $NaHC_2O_4$ and $Na_2C_2O_4$ is titrated in acidic medium with 40.00 mL of 0.1206 N $KMnO_4$ solution ($MnO_4^- \rightarrow Mn^{2+}$). Calculate the per cent $NaHC_2O_4$ and $Na_2C_2O_4$ in the sample.

Titration errors

10.21. A 50.00-mL portion of 0.1000 M NaOH solution is titrated with 0.1000 M HCl solution in the presence of an indicator, at a fixed end-point pH of 10.00. What is the per cent determinate titration error?

10.22. A 25.00-mL portion of 0.1000 M HCl solution is standardized with 0.1000 M NaOH solution in the presence of methyl orange as an indicator. If the addition of titrant is terminated at an end-point pH of 3.80, calculate the experimental value of the concentration of the HCl solution.

10.23. A student uses $H_2C_2O_4 \cdot 2H_2O$ to standardize a $KMnO_4$ solution, but by mistake believes that he uses $Na_2C_2O_4$. Using this $KMnO_4$ solution, he finds that an iron ore contains 25.80% Fe_2O_3. Calculate the real per cent Fe_2O_3 in the ore.

Chapter 11

Acid-Base Titrations

11.1 Introduction

Bases and acids are determined volumetrically. Regardless of the identity of the base or the acid, the neutralization reaction that takes place in aqueous solutions is always the same ($H^+ + OH^- \rightleftharpoons H_2O$). Organic compounds usually are weak electrolytes, and often too weak to titrate in water. Hence, their titration is usually carried out in nonaqueous solvents that are less acidic or basic than water.

The most important characteristics of a neutralization titration can be summarized in the *titration curve* (usually pH as a function of the volume V of titrant). The titration curve may be calculated theoretically, whereupon conclusions can be drawn from it for the feasibility and the expected accuracy of a titration, and the proper protolytic indicator can be selected (Section 5.3). Alternatively, the curve is determined experimentally with a proper instrument, usually a pH meter. The end point is determined from the curve graphically, or an appropriate indicator can be selected based on the curve, for future titrations.

The pH range within which the indicator changes color must include the equivalence point pH of the solution, because otherwise large errors can result, which make the indicator unsuitable for the particular case (Examples 10.9 and 11.9).

11.2 Titration Curves

The main types of acid-base titrations are studied below.

a. Titration of a strong acid with a strong base

As an example, the titration of an HCl solution with a standard NaOH solution is examined. In this case, there are three regions of the titration curve. If V_a and M_a are the initial volume and the initial concentration of the acid, and V_b and M_b the volume and the concentration of the base added, respectively, different calculations are made in each of the three regions, as follows:

1. Before reaching the equivalence point, the pH is determined by the excess of H^+ ions in the solution. For a 1:1 reaction, the millimoles H^+ remaining will equal the millimoles acid taken (V_aM_a), less the millimoles base added (V_bM_b). Division by the total volume ($V_a + V_b$) gives the concentration. Thus, we have

$$[H^+] = \frac{V_aM_a - V_bM_b}{V_a + V_b} \tag{11.1}$$

2. At the equivalence point, the acid is just neutralized and we have

$$V_{b_{eq}} = V_aM_a/M_b \tag{11.2}$$

and

$$[H^+] = [OH^-],$$

whereupon

$$[H^+] = \sqrt{K_w}. \tag{11.3}$$

3. After the equivalence point, the pH is determined by the excess of OH^- ions in the solution. The millimoles of excess OH^- will equal the total millimoles added (V_bM_b), less the millimoles acid (V_aM_a) reacted. Thus, we have

$$[OH^-] = \frac{V_bM_b - V_aM_a}{V_a + V_b} \tag{11.4}$$

Example 11.1. Calculate the titration curve for the titration of 50.00 mL of 0.1000 M HCl with a standard 0.1000 M NaOH solution. Assume that activities equal concentrations (this assumption is also made for all examples and problems of this book which concern titration curves, unless otherwise stated).

Solution. Before any base is added, we have $[H^+] = 0.1000$ M and pH = − log(0.1000) = **1.00** (pH is usually given to only a hundredth of a unit, since it cannot be *accurately* measured any closer—although it may be measured *reproducibly* to within a thousandth of a unit).

After 10.00 mL of base have been added, we have

$$[H^+] = \frac{(50.00 \times 0.1000) - (10.00 \times 0.1000)}{60.00} = 0.0667 \text{ M}$$

and

$$pH = -\log(0.0667) = \mathbf{1.18}.$$

At the equivalence point ($V_{NaOH} = 50.00$ mL), we have

$$[H^+] = \sqrt{1.00 \times 10^{-14}} = 1.00 \times 10^{-7} \text{ M (at 25°C) and pH} = \mathbf{7.00}.$$

Table 11.1. *Titration of a strong acid and a weak acid with NaOH (50.00 mL of 0.1000 M acid are titrated with 0.1000 M NaOH).*

NaOH, mL	pH, HCl	pH, CH_3COOH
0.00	1.00	2.87
1.00	1.02	3.18
5.00	1.09	3.80
10.00	1.18	4.14
20.00	1.37	4.57
30.00	1.60	4.92
40.00	1.95	5.35
45.00	2.28	5.70
49.00	3.00	6.43
49.90	4.00	7.44
50.00	7.00	8.72
50.10	10.00	10.00
51.00	11.00	11.00
55.00	11.68	11.68
60.00	11.96	11.96
70.00	12.22	12.22

After 55.00 mL of the base have been added,

$$[OH^-] = \frac{(55.00 \times 0.1000) - (50.00 \times 0.1000)}{105.00} = 4.76 \times 10^{-3}\ \underline{M}.$$

Hence,

$$[H^+] = 1.00 \times 10^{-14}/4.76 \times 10^{-3} = 2.10 \times 10^{-12} \text{ and } pH = -\log(2.10 \times 10^{-12}) = \mathbf{11.68}.$$

The results of the calculations are summarized in Table (11.1), and the corresponding titration curve is given in Figure (11.1). The titration curve for a strong base titrated with a strong acid, e.g., NaOH with HCl, would be identical to the titration curve in Figure (11.1), if pOH is plotted instead of pH. If pH is plotted, as in Figure (11.1), the titration curve is inverted, starting at a high pH value and dropping to a low pH after the equivalence point.

---

b. Titration of a weak acid with a strong base

Example 11.2. Calculate the titration curve for the titration of 50.00 mL of 0.1000 M CH_3COOH with a standard 0.1000 M NaOH solution.

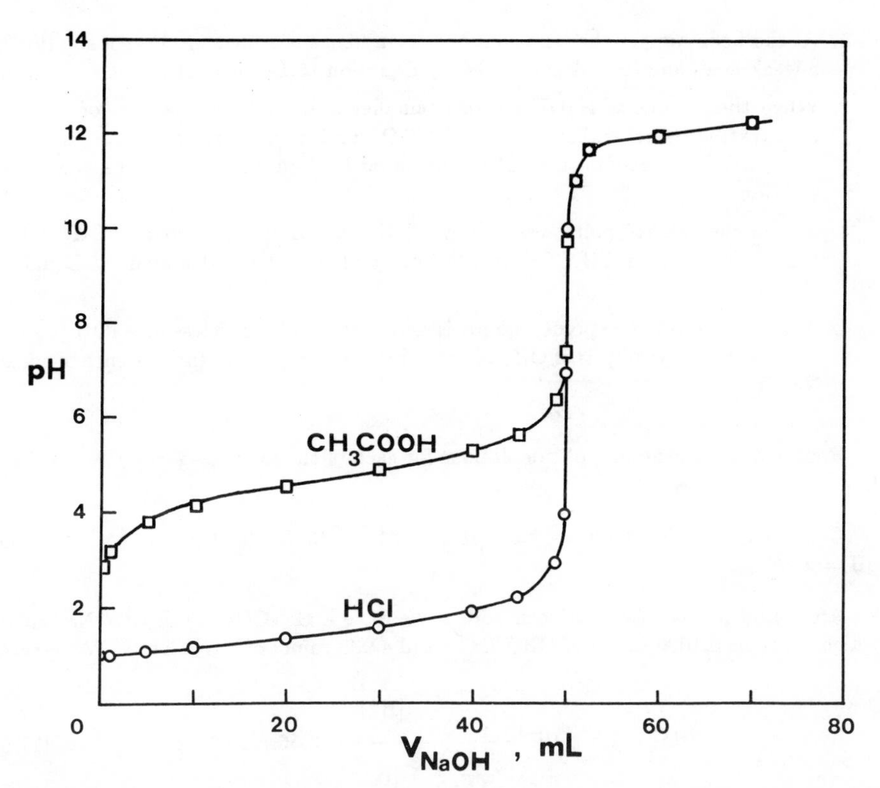

Figure 11.1. *Titration curves for a strong acid and a weak acid. 50.00 mL of 0.1000 M acid are titrated with 0.1000 M NaOH.*

Solution. The equilibrium constant for the titration reaction

$$CH_3COOH + OH^- \rightleftharpoons CH_3COO^- + H_2O \tag{11.5}$$

is equal to

$$K = \frac{[CH_3COO^-]}{[CH_3COOH][OH^-]} = \frac{K_a}{K_w} = \frac{1.8 \times 10^{-5}}{1.00 \times 10^{-14}} = 1.8 \times 10^9 \tag{11.6}$$

From the large value of K, it is concluded that reaction (11.5) takes place quantitatively after each addition of OH^-.

As in the case of the titration of a strong acid with a strong base (Equation (11.2)), we have

$$V_{b_{eq}} = (50.00 \text{ mL} \times 0.1000 \text{ mmol/mL})/(0.1000 \text{ mmol/mL}) = 50.00 \text{ mL} \tag{11.7}$$

The pH calculations at various points of the titration curve fall into the following four categories:

1. At the beginning of the titration, before any base is added, we have a solution of a weak acid, and $[H^+]$ is calculated by Equation (5.19) or (5.20).
2. When the volume of NaOH added is smaller than 50.00 mL, a part of the acid is converted by neutralization to CH_3COO^- and a part remains as undissociated CH_3COOH, whereupon the pH is calculated by Equation (5.82); this is a buffer solution.
3. At the equivalence point, essentially all CH_3COOH is converted to CH_3COO^-, that is, we have a CH_3COONa solution, and $[OH^-]$ is calculated by Equation (5.100).
4. After the equivalence point, the pH is determined by the excess of NaOH (we can disregard the hydrolysis of CH_3COO^-, which is negligible in the presence of excess NaOH).

Representative examples of calculations for each of the four categories for this titration are given below.

Before any base is added, we have $[H^+] \simeq \sqrt{1.8 \times 10^{-5} \times 0.1000} = 1.34 \times 10^{-3}$, and pH = **2.87**.

After 1.00 mL of base has been added, we have a CH_3COOH-CH_3COONa buffer, which contains 0.1000 mmol of CH_3COO^- and 4.900 mmol of CH_3COOH. We have

$$[H^+] = 1.8 \times 10^{-5} \cdot \frac{\dfrac{4.900}{51.00} - [H^+]}{\dfrac{0.1000}{51.00} + [H^+]} = 6.56 \times 10^{-4}\ \underline{M} \tag{11.8}$$

and pH = **3.18**. We solve the quadratic equation, because estimation indicates $[H^+]$ is not negligible in comparison to the other term in the denominator (ca. 2×10^{-3}) of Equation (11.8).

After 25.00 mL of base have been added, we have

$$pH = pK_a + \log \frac{[CH_3COO^-]}{[CH_3COOH]} = 4.74 + \log \frac{(2.500/75.00) + [H^+]}{(2.500/75.00) - [H^+]} \simeq \mathbf{4.74}.$$

We neglected $[H^+]$ in the log term because we know it will be smaller than above and the other term in the denominator is larger now. It should be noted that the pK_a of a monoprotic acid is essentially equal to the pH value, which corresponds to a base volume equal to $V_{b_{eq}}/2$.

After 50.00 mL of base have been added, we have from Equation (5.100)

$$[OH^-] \simeq \sqrt{(0.0500)(1.00 \times 10^{-14})/1.8 \times 10^{-5}} = 5.27 \times 10^{-6}\ \underline{M},\ pOH = 5.28,$$

and pH = 14.00 − 5.28 = **8.72**.

After 55.00 mL of base have been added, we have $[OH^-] = 4.76 \times 10^{-3}\ \underline{M}$ (Example 11.1), and pH = **11.68**.

The results of the calculations are summarized in Table (11.1), and the corresponding titration curve is given in Figure (11.1). From the comparison of the two curves, it can be seen that the pH values up to the equivalence point are higher for the weak acid, whereas after the equivalence point they are essentially identical for the two acids.

The titration curve for a weak base titrated with a strong acid, say NH_3 with HCl, is calculated likewise, as in the case of the weak acid being titrated with a strong base.

Example 11.3. A 50.00-mL portion of a 0.1000 M solution of the weak acid HA is titrated with 0.1000 M NaOH solution. Calculate the minimum value of the equilibrium constant K for the neutralization reaction so that, when 49.90 mL of base have been added, the neutralization reaction is practically complete, and the pH changes by 3.00 units on the addition of 0.20 mL more of base.

Solution. A volume of 50.00 mL of NaOH solution is required to reach the equivalence point. After 49.90 + 0.20 = 50.10 mL of base have been added, i.e., after an excess of 0.10 mL base, we have $[OH^-] = (0.10 \times 0.100)/100.10 = 1.0 \times 10^{-4}$ M, pOH = 4.00, and pH = 10.00. Consequently, after the addition of 49.90 mL of base, the pH will be equal to 10.00 − 3.00 = 7.00. At this point, only 0.10 × 0.100 = 0.010 mmol of HA has not reacted. Hence, we have

$$pH = pK_a + \log\frac{[A^-]}{[HA]} = pK_a + \log\frac{4.99}{0.010} = 7.00$$

or $pK_a = 4.30$, whereupon $K_a = 5.0 \times 10^{-5}$. Hence, according to Equation (11.6) we have

$$K = \frac{[A^-]}{[HA][OH^-]} = \frac{K_a}{K_w} = \frac{5.0 \times 10^{-5}}{1.00 \times 10^{-14}} = \mathbf{5.0 \times 10^9}.$$

Example 11.4. A solution of CH_3COOH is titrated with standard NaOH solution. Calculate the percentage of acid neutralized at the following pH values: a) 4.14, b) 7.44.

Solution. a) We have

$$pH = 4.14 = pK_a + \log\frac{[CH_3COO^-]}{[CH_3COOH]} = 4.74 + \log\frac{[CH_3COO^-]}{[CH_3COOH]}$$

or

$$\frac{[CH_3COO^-]}{[CH_3COOH]} = 0.25,$$

whereupon

$$\frac{[CH_3COO^-]}{[CH_3COOH] + [CH_3COO^-]} = \frac{0.25}{1.00 + 0.25} = 0.20.$$

Hence, at pH 4.14, **20%** of CH_3COOH has been neutralized.

b) We have

$$\frac{[CH_3COO^-]}{[CH_3COOH]} = 10^{7.44-4.74} = 501,$$

$$\frac{[CH_3COO^-]}{[CH_3COOH] + [CH_3COO^-]} = \frac{501}{1 + 501} = 0.998.$$

Hence, at pH 7.44, **99.8%** of CH_3COOH has been neutralized.

Example 11.5. A 50.00-mL portion of 0.1000 M solution of the weak acid HA is titrated with 0.1000 M NaOH solution. After 30.00 mL of base have been added, the pH of the solution was 4.92. Calculate the dissociation constant K_a of HA.

Solution. After 30.00 mL of 0.1000 M NaOH solution have been added, we have an HA-NaA buffer containing $30.00 \times 0.1000 = 3.000$ mmol of A^- and $(50.00 \times 0.1000) - 3.000 = 2.000$ mmol of HA. We have

$$pH = 4.92 = pK_a + \log \frac{[A^-]}{[HA]} = pK_a + \log \frac{3.000}{2.000}$$

or

$$pK_a = 4.74, \text{ and } K_a = \mathbf{1.8 \times 10^{-5}}.$$

Example 11.6. A 0.3050-g sample containing the weak acid HA (MW 60.05) is dissolved in water and diluted to 50.00 mL (solution D). Solution D is titrated with 0.1000 M NaOH solution. When half the acid has been neutralized, the pH is 4.74, whereas at the equivalence point the pH is 8.72. Calculate the per cent HA in the sample.

Solution. We have

$$pK_a \simeq pH \text{ (at } V_{eq}/2) = 4.74, \text{ and } K = 10^{-4.74}.$$

At the equivalence point we have

$$[OH^-] = \sqrt{CK_w/K_a} = \sqrt{10^{-14.00}C/10^{-4.74}},$$

whereupon C = 0.0501 M. Let V mL be the volume of NaOH consumed at the equivalence point, whereupon we have

$$V \text{ mL} \times 0.1000 \text{ mmol/mL} = 0.1000 \text{ V mmol } OH^- = 0.1000 \text{ V mmol } A^-$$

and

$$0.1000 \text{ V}/(50.00 + \text{V}) = 0.0501$$

or

$$V = 50.2 \text{ mL}.$$

Hence, the percentage of HA in the sample is equal to

$$\%HA = \frac{(0.1000 \text{ mmol/mL})(50.2 \text{ mL})(60.05 \text{ mg HA/mmol})}{305.0 \text{ mg sample}} \times 100 = \mathbf{98.8}.$$

Example 11.7. A 0.4884-g sample of a 100.0% pure aromatic acid, which contains only C, H, and O, is titrated potentiometrically with 0.1000 M NaOH solution using a pH meter (Chapter 15). From the titration curve, it is concluded that the acid is monoprotic and that the end point occurs at 40.00 mL of base. After 20.00 mL of base have been added, the pH was 4.18. Calculate the dissociation constant K_a and the molecular weight of the acid, and write a possible formula for the acid.

Solution. After 20.00 mL of base have been added, one-half of the acid has been neutralized; hence, $pK_a = pH$ (at $V_{eq}/2) = 4.18$, and $K_a = 10^{-4.18} = 6.6 \times 10^{-5}$.

Let x be the molecular weight of the acid, whereupon we have

$$(40.00 \text{ mL})(0.1000 \text{ mmol/mL})(x \text{ mg HA/mmol}) = 488.4 \text{ mg}$$

or $x = \mathbf{122.1}$. The acid has the formula ArCOOH; hence, the weight of Ar $= 122.1 - 45.0 = 77.1$, which corresponds to the group C_6H_5-. Hence, a possible formula for the acid is C_6H_5COOH (benzoic acid).

---

c. Titration of polyprotic acids

If the ratio of the successive dissociation constants of a polyprotic acid H_nA is about 10^4 or larger and the product of K_a and the concentration of the acid being titrated (K_aC) is about 10^{-8} or larger, the acid can be titrated stepwise, with distinct equivalence points for the acids H_nA, $H_{n-1}A^-$, $H_{n-2}A^{2-}$... An example is the titration of 50.00 mL of 0.1000 M H_3PO_4 with 0.1000 M NaOH solution, in which two distinct equivalence points are observed, while the titration of the acid HPO_4^{2-} is impossible, because of the small value of its dissociation constant $K_3(1 \times 10^{-12})$. The calculation of pH at the various points of the titration curve is made as in the case of the titration of a weak acid HA with a strong base (Example 11.2). Characteristic points of the titration curve are those corresponding to volumes of 25.00 mL NaOH, where $pH = pK_1$, 50.00 mL, where $pH \simeq (pK_1 + pK_2)/2$, 75.00 mL, where $pH = pK_2$, and 100.00 mL, where $pH \simeq (pK_2 + pK_3)/2$. In the case of oxalic acid, we have $K_1/K_2 = 3.8 \times 10^{-2}/5.0 \times 10^{-5} = 760 << 10^4$, and therefore $H_2C_2O_4$ cannot be titrated in two clearly separate steps with distinct equivalence points.

General formula for the calculation of titration curves. It is possible to calculate the pH at any point of a titration, if we know the relative volumes and the concentrations, using the methods already described. However, it is also possible to follow the reverse procedure for the calculation of the titration curve, that is, to calculate the required volume of titrant for a given $[H^+]$ (pH). In this method, we first derive a general equation which relates the volume V of titrant to $[H^+]$ in the solution. Then we substitute in this equation a series of values for $[H^+]$ and calculate the corresponding values of V. This method leads to equations that contain $[H^+]$ raised in the third power (for the acid HA), fourth power (H_2A), fifth power (H_3A), and so forth. With this method, in contrast to the previously described methods in which, many times, approximate solutions are calculated, the exact value of $[H^+]$ can be found.

Such an equation is derived below for the titration of the diprotic acid H_2A. Let V_a be the solution volume of the acid H_2A, having dissociation constants K_1 and K_2, C_a the acid concentration, V_b the volume of the strong base (titrant) and C_b its concentration, and K_w the ion product of water. The titration takes place in two steps, according to the reactions

$$H_2A + OH^-(+Na^+) \rightleftharpoons H_2O + HA^-(+Na^+) \qquad (11.9)$$

$$HA^-(+Na^+) + OH^-(+Na^+) \rightleftharpoons H_2O + A^{2-}(+2Na^+) \tag{11.10}$$

(It is assumed that the salts dissociate completely.) At any point of the titration the following relations are valid:

$$\text{Electroneutrality:} \quad [Na^+] + [H^+] = [OH^-] + [HA^-] + 2[A^{2-}] \tag{11.11}$$

$$[Na^+] = \frac{V_b C_b}{V_a + V_b} \tag{11.12}$$

$$[OH^-] = K_w/[H^+] \tag{11.13}$$

$$[HA^-] = \alpha_1 C_a \cdot \frac{V_a}{V_a + V_b} \tag{11.14}$$

$$\alpha_1 = \frac{K_1[H^+]}{[H^+]^2 + K_1[H^+] + K_1K_2} \tag{11.15}$$

$$[A^{2-}] = \alpha_2 C_a \cdot \frac{V_a}{V_a + V_b} \tag{11.16}$$

$$\alpha_2 = \frac{K_1K_2}{[H^+]^2 + K_1[H^+] + K_1K_2} \tag{11.17}$$

Combining Equations (11.11), (11.12), (11.13), (11.14), and (11.16), we have

$$\frac{V_b C_b}{V_a + V_b} + [H^+] = \frac{K_w}{[H^+]} + \frac{\alpha_1 C_a V_a}{V_a + V_b} + \frac{2\alpha_2 C_a V_a}{V_a + V_b} \tag{11.18}$$

Solving Equation (11.18) for V_b, we finally have

$$V_b = \frac{\left(\alpha_1 C_a + 2\alpha_2 C_a + \frac{K_w}{[H^+]} - [H^+]\right) V_a}{C_b - \frac{K_w}{[H^+]} + [H^+]} \tag{11.19}$$

(It is assumed that the ionic strength remains constant.) Equation (11.19), in combination with Equations (11.15) and (11.17), allows the calculation of volume V_b of base required to achieve a specific value of H^+ ion concentration, $[H^+]$, or conversely, to calculate $[H^+]$ at a given volume of base.

For the general case of the polyprotic acid H_nA, the following general equation is derived in a similar manner.

$$V_b = \frac{\left(\alpha_1 C_a + 2\alpha_2 C_a + \ldots + n\alpha_n C_a + \frac{K_w}{[H^+]} - [H^+]\right) V_a}{C_b - \frac{K_w}{[H^+]} + [H^+]} \tag{11.20}$$

In the case of the weak monoprotic acid HA, Equation (11.20) is simplified to

$$V_b = \frac{\left(\alpha_1 C_a + \frac{K_w}{[H^+]} - [H^+]\right) V_a}{C_b - \frac{K_w}{[H^+]} + [H^+]} = \frac{\left(\frac{KC_a}{[H^+] + K} + \frac{K_w}{[H^+]} - [H^+]\right) V_a}{C_b - \frac{K_w}{[H^+]} + [H^+]} \tag{11.21}$$

In the case of a strong monoprotic acid, we have $K/([H^+]+K) \simeq 1$ and Equation (11.21) is simplified to

$$V_b = \frac{\left(C_a + \frac{K_w}{[H^+]} - [H^+]\right) V_a}{C_b - \frac{K_w}{[H^+]} + [H^+]}. \tag{11.22}$$

To check Equation (11.22), let V_a = 50.00 mL, $C_a = C_b$ = 0.1000 M, and $[H^+] = 1.0 \times 10^{-4}$ M, whereupon we have

$$V_b = \frac{\left(0.1000 + \frac{1.00 \times 10^{-14}}{1.0 \times 10^{-4}} - 1.0 \times 10^{-4}\right) 50.00}{0.1000 - \frac{1.00 \times 10^{-14}}{1.00 \times 10^{-4}} + 1.0 \times 10^{-4}} = 49.90.$$

The value 49.90 is identical to the one given in Table 11.1, where V = 49.90 mL corresponds to pH 4.00. Using Equations (11.21) and (11.22), we can also check the other data given in Table 11.1.

It should be noted that the general Equation (11.20) and the simplified Equations (11.19), (11.21), and (11.22) are valid for the entire titration curve, and so it is *not* necessary to use different equations for the various regions of the titration curve. Similar general equations can be derived for the titrations of bases with acids. In all these equations, it is assumed that the ionic strength remains constant during the titration, whereas it is actually changing continuously. The accurate calculation of a titration curve in which the real ionic strength would be taken into account is very tedious, but it can easily be accomplished using an appropriate computer program.

The titration curve for the titration of 50.00 mL of 0.1000 M H_2SO_3 solution with 0.1000 M NaOH was calculated by Equation (11.19), and it is given in Figure 11.2.

d. Titration of the anion of a weak acid with a strong acid

The anion of a very weak acid is a strong base, and therefore it can be titrated with a strong acid. Polyvalent anions of very weak polyprotic acids can be titrated stepwise with a strong acid.

Example 11.8. Calculate the titration curve for the titration of 50.00 mL of 0.1000 M Na_2CO_3 solution with 0.1000 M HCl solution.

Solution. The titration takes place in two steps, according to the reactions

$$CO_3^{2-} + H^+ \rightleftharpoons HCO_3^-, \; K = \frac{1}{K_{a_2}} = \frac{1}{4.8 \times 10^{-11}} = 2.1 \times 10^{10} \tag{11.23}$$

$$HCO_3^- + H^+ \rightleftharpoons H_2CO_3, \; K' = \frac{1}{K_{a_1}} = \frac{1}{4.2 \times 10^{-7}} = 2.4 \times 10^{6} \tag{11.24}$$

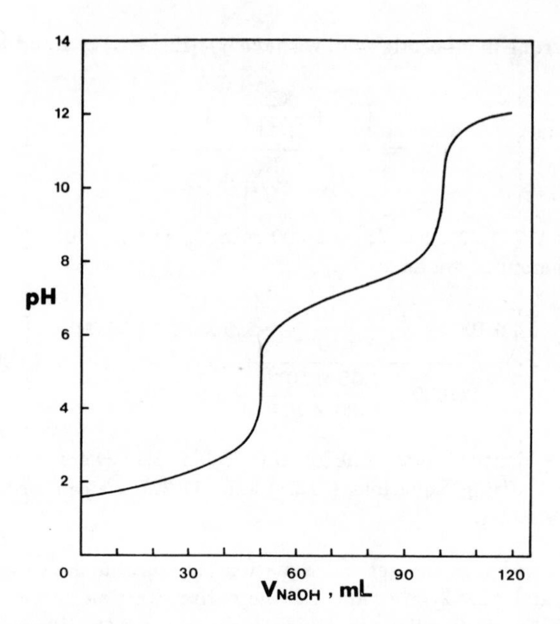

Figure 11.2. *Titration curve for the titration of 50.00 mL of 0.1000 M H_2SO_3 with 0.1000 M NaOH (Equation 11.19).*

From the large values of K and K′, it is concluded that reactions (11.23) and (11.24) take place quantitatively after each addition of H^+. The two equivalence points of the titration correspond to volumes of 50.00 mL and 100.00 mL HCl. Representative examples of calculations for the various regions of the titration curve are given below.

Before any acid is added, for the 0.1000 M Na_2CO_3 solution, which is hydrolyzed according to the reaction

$$CO_3^{2-} + H_2O \rightleftharpoons HCO_3^- + OH^- \tag{11.25}$$

we have the relation

$$\frac{[OH^-]^2}{C - [OH^-]} = \frac{K_w}{K_{a_2}} = \frac{1.00 \times 10^{-14}}{4.8 \times 10^{-11}} \simeq \frac{[OH^-]^2}{0.1000} \tag{11.26}$$

from which we have $[OH^-] = 4.56 \times 10^{-3}$ M, pOH = 2.34, and pH = **11.66**.

After 10.00 mL of acid have been added, we have pH = 10.32 + log(4.000/1.000) = **10.92**.

After 25.00 mL of acid have been added, we have pH = pK_2 = **10.32**.

After 50.00 mL of acid have been added, we have a 0.0500 M $NaHCO_3$ solution. According to Equation (5.118), we have $[H^+] = \sqrt{(4.2 \times 10^{-7})(4.8 \times 10^{-11})} = 4.5 \times 10^{-9}$ M, and pH = **8.35**.

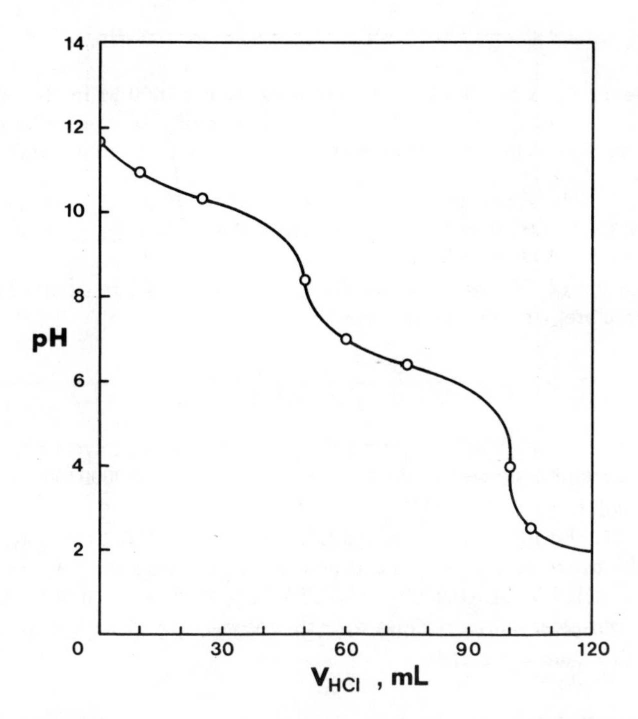

Figure 11.3. *Titration curve for the titration of 50.00 mL of 0.1000 $\underline{M}$ Na_2CO_3 with 0.1000 $\underline{M}$ HCl.*

After 60.00 mL of acid have been added, we have pH = 6.38 + log(4.000/1.000) = **6.98**.

After 75.00 mL of acid have been added, we have pH = pK_1 = **6.38**.

After 100.00 mL of acid have been added, we have a 0.0333 $\underline{M}$ H_2CO_3 solution. Hence, $[H^+] = \sqrt{(4.2 \times 10^{-7})(3.33 \times 10^{-2})} = 1.18 \times 10^{-4}$ $\underline{M}$ and pH = **3.93**.

After the addition of 105.00 mL of acid, $[H^+] \simeq (5.00 \times 0.1000)/155.00 = 3.2 \times 10^{-3}$ $\underline{M}$, and pH = **2.49**.

The titration curve is given in Figure 11.3.

11.3 Errors in Acid-Base Titrations

Large errors can arise in an acid-base titration, if the proper indicator is not selected for the determination of the end point (Example 10.19).

Example 11.9. A 50.00-mL portion of a solution 0.1000 M in HCl and 0.0600 M in CH_3COOH is titrated with 0.2000 M NaOH. Are methyl orange (color change at pH 4.0) and phenolphthalein (color change at pH 8.0) suitable as indicators?

Solution. HCl is titrated first, and then CH_3COOH. We consume $(50.00 \times 0.1000)/0.2000 = 25.00$ mL of 0.2000 M NaOH to the first equivalence point and $(50.00 \times 0.0600)/0.2000 = 15.00$ mL more of base to the second equivalence point. After the neutralization of HCl, we have a $(50.00 \times 0.0600/75.00) = 0.0400$ M CH_3COOH solution. Therefore, we have the relation

$$\frac{[H^+]^2}{0.0400 - [H^+]} = 1.8 \times 10^{-5},$$

from which $[H^+] = 8.40 \times 10^{-4}$ M, and pH = 3.08. Hence, we have a positive error for HCl. At the second equivalence point we have a $(50.00 \times 0.0600/90.00) = 0.0333$ M CH_3COONa solution. Hence, $[OH^-] = \sqrt{0.333(1.00 \times 10^{-14}/1.8 \times 10^{-5})} = 4.3 \times 10^{-6}$ M, pOH = 5.37, and pH = 8.63. Let y mL be the volume of 0.2000 M NaOH required to increase the pH from 8.0 (second end point, $[OH^-] = 1.0 \times 10^{-6}$ M) to 8.63 (second equivalence point). We have $0.2000y = 90.00 \times 3.3 \times 10^{-6}$ or y = 0.001 mL. Hence, the second end point essentially coincides with the second equivalence point.

At pH 4.0 we have

$$\alpha_1 = \frac{[CH_3COO^-]}{C_{CH_3COOH}} = \frac{[CH_3COO^-]}{0.0400} = \frac{1.8 \times 10^{-5}}{1.0 \times 10^{-4} + 1.8 \times 10^{-5}} = 0.1525.$$

Hence, at the first end point (pH 4.0), $0.1525 \times 0.0400 \times 75.00 = 0.4575$ mmol of CH_3COOH have been cotitrated with HCl. So, the per cent titration error is equal to

$$\frac{0.4575}{50.00 \times 0.1000} \times 100 = 9.15\% \text{ for HCl}$$

and

$$\frac{0.4575}{50.00 \times 0.0600} \times 100 = -15.25\% \text{ for } CH_3COOH.$$

Hence, methyl orange is unsuitable as indicator for the titration of HCl, whereas phenolphthalein is suitable as indicator for the titration of the total acidity of the solution ($HCl + CH_3COOH$).

11.4 Applications of Acid-Base Titrations in Aqueous Solutions

The main application of acid-base titrations is in the determination of substances having acidic or basic properties. Typical examples of applications in stoichiometric analysis and in the determination of inorganic and organic substances, as well as of functional groups, are given below.

Quantitative elemental analysis. Many elements, e.g., C, H, N, Cl, Br, and P, are determined by converting them to acids or bases and titrating with a standard base or acid, respectively.

Example 11.10. In the determination of nitrogen in a fertilizer by the Kjeldahl method (Chapter 10, Problem 10.13), 0.5696 g of sample, 50.00 mL of 0.1000 $\underline{M}$ HCl solution, and 25.38 mL of 0.1003 $\underline{M}$ NaOH solution were used. Calculate the per cent nitrogen in the sample.

Solution. We have

$$\%N = \frac{[(50.00 \times 0.1000) - (25.38 \times 0.1003)] \text{ mmol} \times 14.007 \text{ mg N/mmol}}{569.6 \text{ mg}} \times 100 = \mathbf{6.04}.$$

Example 11.11. A 2.8750-g sample containing $(NH_4)_2SO_4$, NH_4NO_3, and inert materials was dissolved in water and diluted to 250.0 mL (solution A). A 25.00-mL aliquot of solution A was made strongly basic, whereupon NH_4^+ was converted to NH_3, which was distilled into an unknown excess of boric acid ($HBO_2 + NH_3 \rightleftharpoons NH_4^+ + BO_2^-$). The BO_2^- ions produced were titrated with 30.00 mL of 0.1000 $\underline{M}$ HCl (Winkler modification of the Kjeldahl method). Another 25.00-mL aliquot of solution A was treated with Devarda's alloy, whereupon NO_3^- was reduced to NH_3:

$$3NO_3^- + 8\mathbf{Al} + 5\,OH^- + 18H_2O \overset{\Delta}{\rightleftharpoons} 3NH_3\uparrow + 8[Al(OH)_4]^-.$$

The total NH_3 (from NH_4^+ and NO_3^-) was distilled into boric acid and 40.00 mL of the standard HCl were consumed to titrate the BO_2^- ions. Calculate the per cent $(NH_4)_2SO_4$ and NH_4NO_3 in the sample.

Solution. Suppose that the sample contains y mmol $(NH_4)_2SO_4$ and z mmol NH_4NO_3. We have the relations

$$0.2000y + 0.1000z = 30.00 \text{ mL} \times 0.1000 \text{ mmol/mL} = 3.000 \qquad (11.27)$$

$$0.2000y + 0.2000z = 40.00 \text{ mL} \times 0.1000 \text{ mmol/mL} = 4.000 \qquad (11.28)$$

Solving the system of Equations (11.27) and (11.28), we find that y = z = 10.00. Hence, the sample contains

$$\frac{10.00 \text{ mmol} \times 132.14 \text{ mg } (NH_4)_2SO_4/\text{mmol}}{2875.0 \text{ mg}} \times 100 = \mathbf{45.96\%}\ (NH_4)_2SO_4$$

and

$$\frac{10.00 \text{ mmol} \times 80.04 \text{ mg } NH_4NO_3/\text{mmol}}{2875.0 \text{ mg}} \times 100 = \mathbf{27.84\%}\ NH_4NO_3.$$

Table 11.2. *Volume relation in the analysis of $NaOH$-Na_2CO_3-$NaHCO_3$ mixtures.*

Sample components	Relation between V_m and V_p[a]
$NaOH$	$V_m = V_p$
Na_2CO_3	$V_m = 2V_p$
$NaHCO_3$	$V_m > V_p = 0$
$NaOH + Na_2CO_3$	$V_m < 2V_p$
$Na_2CO_3 + NaHCO_3$	$V_m > 2V_p$

a) V_m and V_p are the volumes of standard acid required in the presence of methyl orange and phenolphthalein, respectively.

Analysis of $NaOH$-Na_2CO_3-$NaHCO_3$ mixtures. A solution prepared by dissolving $NaOH$ and/or Na_2CO_3 and/or $NaHCO_3$ cannot contain significant amounts of more than two of the three components, because a neutralization reaction takes place between $NaOH$ and $NaHCO_3$. Thus, when a mixture of $NaOH$ and $NaHCO_3$ is dissolved in water, we obtain a solution containing either a) only Na_2CO_3 (equivalent amounts of $NaOH$ and $NaHCO_3$ in the initial sample), b) a mixture of $NaOH$ and Na_2CO_3 (the amount of $NaOH$ is larger than the stoichiometrically equivalent amount of $NaHCO_3$), or c) a mixture of $NaHCO_3$ and Na_2CO_3 (the amount of $NaHCO_3$ is larger than the stoichiometrically equivalent amount of $NaOH$). Therefore, if a solution is prepared by dissolving various amounts of one or more of the above compounds ($NaOH$, Na_2CO_3, $NaHCO_3$), the possible solutions that could result are 1) $NaOH$, 2) Na_2CO_3, 3) $NaHCO_3$, 4) $NaOH + Na_2CO_3$, or 5) $NaHCO_3 + Na_2CO_3$.

The composition of the sample can be determined (Table 11.2) from the relative volumes of standard acid solution required for the titration of equal sample volumes in the presence of a) an indicator with a transition range of pH 8–10, e.g., phenolphthalein, and b) an indicator with a transition range of pH 3.5–4.5, e.g., methyl orange or methyl red. After the qualitative composition of the mixture has been deduced, the percentage of each component is calculated from the volumes of standard acid consumed in the two titrations. The composition of the sample can also be determined from the titration curve of a single solution obtained with a pH meter.

Example 11.12. Each of the following solutions contains one or more of the following substances: $NaOH$, $NaHCO_3$, Na_2CO_3. The volumes of 0.1000 M HCl required for the titration of 25.00 mL of each of the five solutions to reach the methyl orange and phenolphthalein end points are given below. Which compound or compounds and in what amounts, in milligrams, are contained in each of the five solutions?

Sample	V_{HCl}, mL	
	Methyl orange (V_m)	Phenolphthalein (V_p)
a	30.03	22.50
b	42.16	21.08
c	18.17	0.00
d	38.76	14.71
e	32.13	32.13

Solution. Since the same HCl solution is used in all titrations, the relations between the HCl volumes required are also the relations between equivalents. Hence, we have:

a) Since $V_m < 2V_p$ ($30.03 < 2 \times 22.50$), it is deduced that the solution contains Na_2CO_3 and NaOH (Table 11.2), in the following amounts:

$$\text{mg } Na_2CO_3 = (30.03 - 22.50) \times 0.1000 \times 105.99 = \mathbf{79.8}$$
$$\text{mg NaOH} = [22.50 - (30.03 - 22.50)] \times 0.1000 \times 40.00 = \mathbf{59.9}.$$

b) Since $V_m = 2V_p$, it is deduced that the solution contains only Na_2CO_3, in the following amount:

$$\text{mg } Na_2CO_3 = 21.08 \times 0.1000 \times 105.99 = \mathbf{223.4}$$

or

$$\text{mg } Na_2CO_3 = 42.16 \times 0.1000 \times (105.99/2) = \mathbf{223.4}$$

c) Since $V_m > V_p = 0$, it is deduced that the solution contains only $NaHCO_3$, in the following amount:

$$\text{mg}NaHCO_3 = 18.17 \times 0.1000 \times 84.01 = \mathbf{152.6}.$$

d) Since $V_m > 2V_p$, it is deduced that the solution contains Na_2CO_3 and $NaHCO_3$, in the following amounts:

$$\text{mg } Na_2CO_3 = 14.71 \times 0.1000 \times 105.99 = \mathbf{155.9}$$
$$\text{mg } NaHCO_3 = [38.76 - (2 \times 14.71)] \times 0.1000 \times 84.01 = \mathbf{78.5}$$

e) Since $V_m = V_p$, it is deduced that the solution contains only NaOH, in the following amount:

$$\text{mg NaOH} = 32.13 \times 0.1000 \times 40.00 = \mathbf{128.5}.$$

Example 11.13. A sample containing Na_2CO_3, $NaHCO_3$, and inert material, was analyzed as follows: A 0.2982-g sample was dissolved in water and 39.88 mL of 0.0998 $\underline{N}$ HCl solution was consumed to reach the methyl orange end point. To another 0.2412 g of sample, 25.00 mL of 0.1004 $\underline{N}$ NaOH solution and 10 mL of 10% $BaCl_2$ were added, and 8.95 mL of the same HCl solution were consumed to titrate the excess unreacted NaOH to the phenolphthalein end point. Calculate the per cent Na_2CO_3 and $NaHCO_3$ in the sample.

Solution. The following reactions take place:

$$CO_3^{2-} + 2H^+ \rightleftharpoons H_2CO_3 \rightleftharpoons H_2O + CO_2\uparrow \tag{11.29}$$

$$HCO_3^- + H^+ \rightleftharpoons H_2CO_3 \rightleftharpoons H_2O + CO_2\uparrow \tag{11.30}$$

$$HCO_3^- + OH^- \rightleftharpoons CO_3^{2-} + H_2O \tag{11.31}$$

$$CO_3^{2-} + Ba^{2+} \rightleftharpoons \mathbf{BaCO_3} \tag{11.32}$$

$$OH^-\ (\text{exc.}) + H^+ \rightleftharpoons H_2O \tag{11.33}$$

The amount of $NaHCO_3$ is equivalent to the amount of NaOH that was consumed to convert HCO_3 to CO_3^{2-} (which was removed as $BaCO_3$, Equation 11.32), that is, to the difference between the amount of NaOH initially added and the excess NaOH that was titrated with the HCl (Equation 11.33). Hence, the sample contains

$$\frac{[(25.00\text{ mL} \times 0.1004\text{ meq/mL}) - (8.95\text{ mL} \times 0.0998\text{ meq/mL})]84.01\text{ mg/meq}}{241.2\text{ mg}} \times 100 = \mathbf{56.3\%}\ NaHCO_3,$$

and

$$\frac{((39.88 \times 0.0998) - [(25.00 \times 0.1004) - (8.95 \times 0.0998)]0.2982/0.2412)(105.99/2)\text{ mg}}{298.2\text{ mg}} \times 100 = \mathbf{35.2\%}\ Na_2CO_3.$$

Example 11.14. The sample given in Example (11.13) was also analyzed in the following way: A 0.2982-g sample was dissolved in water and titrated with 9.92 mL of 0.0998 N HCl solution to the phenolphthalein end point. Methyl orange was then added, and 29.96 mL more of the same acid were consumed to change the color of this indicator. Calculate the per cent Na_2CO_3 and $NaHCO_3$ in the sample.

Solution. Only the CO_3^{2-} ions are titrated in the presence of phenolphthalein as indicator. Hence, we have

$$\%\ Na_2CO_3 = \frac{9.92\text{ mL} \times 0.0998\text{ meq/mL} \times 105.99\text{ mg } Na_2CO_3\text{/meq}}{298.2\text{ mg}} \times 100 = \mathbf{35.2}.$$

From the 29.96 mL of HCl consumed between the first and second end points, 9.92 mL were consumed to titrate the HCO_3^- ions produced from CO_3^{2-} ions, and the remaining $29.96 - 9.92 = 20.04$ mL for the titration of HCO_3^- present in the sample from the beginning. Hence, we have

$$\%NaHCO_3 = \frac{20.04\text{ mL} \times 0.0998\text{ meq/mL} \times 84.01\text{ mg } NaHCO_3\text{/meq}}{298.2\text{ mg}} \times 100 = \mathbf{56.3}.$$

Example 11.15. A 0.2424-g sample containing NaOH, Na_2CO_3, and inert materials was titrated with 30.68 mL of 0.0999 M solution in the presence of phenolphthalein as indicator. After the solution turned colorless, methyl orange was added to the solution, and 5.17 mL more of the acid were consumed before the indicator changed color. Calculate the per cent NaOH and Na_2CO_3 in the sample.

Solution. The OH^- and CO_3^{2-} are cotitrated until the first end point, whereas between the first and the second end point the HCO_3^- ions produced in the titration of CO_3^{2-} are titrated. Hence, 5.17 mL of acid are required to titrate CO_3^{2-} to HCO_3^-, whereas $30.68 - 5.17 = 25.51$ mL of acid are required to titrate the OH^- ions. Hence, we have

$$\%\text{NaOH} = \frac{25.51 \text{ mL} \times 0.0999 \text{ mmol/mL} \times 40.00 \text{ mg NaOH/mmol}}{242.4 \text{ mg}} \times 100 = \mathbf{42.05}$$

and

$$\%\ \text{Na}_2\text{CO}_3 = \frac{5.17 \text{ mL} \times 0.0999 \text{ mmol/mL} \times 105.99 \text{ mg Na}_2\text{CO}_3\text{/mmol}}{242.4 \text{ mg}} \times 100 = \mathbf{22.59}.$$

Example 11.16. A standard NaOH solution is contaminated with Na_2CO_3. In the titration of 50.00 mL of 0.1002 M HCl solution, 32.00 and 32.50 mL of base are required to reach the methyl orange and phenolphthalein end points, respectively. Calculate the grams of NaOH and Na_2CO_3, in one liter of solution.

Solution. Suppose that the solution contains y mol NaOH and z mol Na_2CO_3 in one liter. The CO_3^{2-} ions act as a diacidic base in the presence of methyl orange and as a monoacidic base in the presence of phenolphthalein. Hence, we have

$$32.00y + 64.00z = 50.00 \times 0.1002 = 5.01$$
$$32.50y + 32.50z = 50.00 \times 0.1002 = 5.01.$$

Solving the above system of two equations, we find that $y = 0.1517$, and $z = 0.002413$. Hence, one liter of solution contains

$$0.1517 \text{ mol/L} \times 40.00 \text{ g NaOH/mol} = \mathbf{6.07} \text{ g NaOH/L}$$

and

$$0.002413 \text{ mol/L} \times 105.99 \text{ g Na}_2\text{CO}_3\text{/mol} = \mathbf{0.256} \text{ g Na}_2\text{CO}_3\text{/L}.$$

Analysis of phosphate mixtures. The possibility of titrating phosphoric acid as either a monoprotic or a diprotic acid allows the qualitative and quantitative analysis of phosphate mixtures, either by performing two titrations with appropriate indicators, e.g., methyl orange and phenolphthalein, or from the titration curve obtained with a pH meter, as in the analysis of carbonates. Suppose, for example, that a solution is prepared by dissolving any or all of the following substances in water: H_3PO_4, NaH_2PO_4, Na_2HPO_4, Na_3PO_4, HCl, NaOH. A reaction may take place or not, depending on the substances present, and the final composition of the solution is expressed in terms of

Table 11.3. *Relation of equivalents in the analysis of phosphate mixtures.*

Sample components	Relation among B_p, B_m, A_p, and A_m [a]	
	Titration with base	Titration with acid
H_3PO_4	$B_p = 2B_m$	
NaH_2PO_4	$B_p > B_m = 0$	$A_m = 0$
Na_2HPO_4	$B_p = 0$	$A_m > A_p = 0$
Na_3PO_4		$A_m = 2A_p$
HCl	$B_p = B_m$	
NaOH		$A_m = A_p$
HCl, H_3PO_4	$B_p < 2B_m$	
H_3PO_4, NaH_2PO_4	$B_p > 2B_m$	
NaH_2PO_4, Na_2HPO_4	$B_p > B_m = 0$	$A_m > A_p = 0$
Na_2HPO_4, Na_3PO_4		$A_m > 2A_p$
Na_3PO_4, NaOH		$A_m < 2A_p$

a) B_m and B_p are the equivalents of base required in the presence of methyl orange and phenolphthalein, respectively. A_m and A_p are the equivalents of acid required in the presence of methyl orange and phenolphthalein, respectively.

at most two of the above substances being present in significant amounts (if NaCl is formed, it is not taken into consideration, because it is an inert material not titratable with the acids and bases used in the titration). The composition of the sample may be qualitatively deduced from the relative equivalents of standard acid and/or base (Table 11.3) required for the titration of equal sample volumes in the presence of a) methyl orange, or b) phenolphthalein as indicator. After the qualitative composition of the mixture has been deduced, the percentage of each component is calculated from the volumes of the standard solutions consumed in the two titrations. The composition of the sample can also be determined from the titration curve obtained with a pH meter.

Example 11.17. Each of the following solutions contains one or more of the following substances: NaOH, Na_3PO_4, Na_2HPO_4. The volumes of 0.1000 $\underline{M}$ HCl required for the titration of 25.00 mL of each of the four solutions to reach the methyl orange and the phenolphthalein end points are given below. Which compound or compounds and in what amounts, in milligrams, are contained in each of the four solutions?

Sample	V_{HCl}, mL	
	Methyl orange (V_m)	Phenolphthalein (V_p)
a	41.30	17.09
b	40.92	24.33
c	37.44	18.72
d	29.08	0.00

Solution. Since the same HCl solution is used in all titrations, the relations between the HCl volumes required are also the relations between equivalents. Hence, we have:

a) Since $V_m > 2V_p$ ($41.30 > 2 \times 17.09$), it is deduced that the solution contains Na_3PO_4 and Na_2HPO_4 (Table 11.3), in the following amounts:

$$\text{mg } Na_3PO_4 = 17.09 \times 0.1000 \times 163.94 = \mathbf{280.2}$$
$$\text{mg } Na_2HPO_4 = [41.30 - (2 \times 17.09)] \times 0.1000 \times 141.96 = \mathbf{101.1}.$$

b) Since $V_m < 2V_p$, it is deduced that the solution contains Na_3PO_4 and NaOH, in the following amounts:

$$\text{mg } Na_3PO_4 = (40.92 - 24.33) \times 0.1000 \times 163.94 = \mathbf{272.0}$$
$$\text{mg NaOH} = [24.33 - (40.92 - 24.33)] \times 0.1000 \times 40.00 = \mathbf{31.0}.$$

c) Since $V_m = 2V_p$, it is deduced that the solution contains only Na_3PO_4, in the following amount:

$$\text{mg } Na_3PO_4 = 18.72 \times 0.1000 \times 163.94 = \mathbf{306.9}$$

or

$$\text{mg } Na_3PO_4 = 37.44 \times 0.1000 \times (163.94/2) = \mathbf{306.9}.$$

d) Since $V_m > V_p = 0$, it is deduced that the solution contains only Na_2HPO_4, in the following amount:

$$\text{mg } Na_2HPO_4 = 29.08 \times 0.1000 \times 141.96 = \mathbf{412.8}.$$

Example 11.18. Each of the following solutions contains one or more of the following substances: HCl, H_3PO_4, NaH_2PO_4. The volumes of 0.1000 $\underline{M}$ NaOH solution required for the titration of 25.00 mL of each of the five solutions to reach the methyl orange and the phenolphthalein end points are given below. Which compound or compounds and in what concentrations ($\underline{M}$) does each of the five solutions contain?

Sample	V_{NaOH}, mL	
	Methyl orange (V_m)	Phenolphthalein (V_p)
a	15.98	24.82
b	24.17	24.17
c	0.00	32.89
d	14.63	46.05
e	19.24	38.48

Solution. Since the same NaOH solution is used in all titrations, the relations between the NaOH volumes required are also the relations between equivalents. Hence, we have:

a) Since $V_p < 2V_m$ ($24.82 < 2 \times 15.98$), it is deduced that the solution contains H_3PO_4 and HCl (Table 11.3), at concentrations equal to

$$\underline{M}_{H_3PO_4} = \frac{(24.82 - 15.98) \times 0.1000}{25.00} = \mathbf{0.0354}$$

$$\underline{M}_{HCl} = \frac{[(15.98 - (24.82 - 15.98)] \times 0.1000}{25.00} = \mathbf{0.0286.}$$

b) Since $V_p = V_m$, it is deduced that the solution contains only HCl, at a concentration of

$$\underline{M}_{HCl} = \frac{24.17 \times 0.1000}{25.00} = \mathbf{0.0967.}$$

c) Since $V_p > V_m = 0$, it is deduced that the solution contains only NaH_2PO_4, at a concentration of

$$\underline{M}_{Na_2HPO_4} = \frac{32.89 \times 0.1000}{25.00} = \mathbf{0.1316.}$$

d) Since $V_p > 2V_m$, it is deduced that the solution contains H_3PO_4 and NaH_2PO_4, at concentrations equal to

$$\underline{M}_{H_3PO_4} = \frac{14.63 \times 0.1000}{25.00} = \mathbf{0.0585}$$

$$\underline{M}_{NaH_2PO_4} = \frac{[46.05 - (2 \times 14.63)] \times 0.1000}{25.00} = \mathbf{0.0672.}$$

e) Since $V_p = 2V_m$, it is deduced that the solution contains only H_3PO_4, at a concentration of

$$\underline{M}_{H_3PO_4} = \frac{19.24 \times 0.1000}{25.00} = \mathbf{0.0770.}$$

Example 11.19. A 0.5027-g sample containing Na_3PO_4, Na_2HPO_4 and inert materials is titrated with 16.16 mL of 0.1044 $\underline{M}$ HCl solution in the presence of phenolphthalein as indicator. After the solution turned colorless, methyl orange was added to the solution, and 28.31 mL more of the acid were consumed before the indicator changed color. Calculate the per cent content of the sample in Na_3PO_4, Na_2HPO_4, and inert materials.

Solution. The PO_4^{3-} ions are titrated to HPO_4^{2-} at the first end point, whereas from the first to the second end point the sum of HPO_4^{2-} is titrated, i.e., the HPO_4^{2-} ions that were present from the beginning and those derived from the titration of the PO_4^{3-} ions. Hence, 16.16 mL of acid are required to titrate PO_4^{3-} to HPO_4^{2-}, whereas

28.31 − 16.16 = 12.15 mL of acid are required to titrate the HPO_4^{2-} ions initially present. Hence, the sample contains

$$\frac{16.16 \text{ mL} \times 0.1044 \text{ mmol/mL} \times 163.94 \text{ mg } Na_3PO_4\text{/mmol}}{502.7 \text{ mg}} \times 100 = \mathbf{55.02\%} \; Na_3PO_4,$$

$$\frac{12.15 \text{ mL} \times 0.1044 \text{ mmol/mL} \times 141.96 \text{ mg } Na_2HPO_4\text{/mmol}}{502.7 \text{ mg}} \times 100 = \mathbf{35.82\%} \; Na_2HPO_4,$$

and

$$100 - (55.02 + 35.82) = \mathbf{9.16\%} \text{ inert materials.}$$

Example 11.20. A 0.3035-g sample containing P_2O_5 and H_3PO_4 impurity is dissolved in water ($P_2O_5 + 3H_2O \rightarrow 2H_3PO_4$) and the resulting solution is titrated with 42.00 mL of 0.2000 N NaOH solution in the presence of phenolphthalein as indicator. Calculate the percentage composition of the sample.

Solution. Suppose that the sample contains y mg of P_2O_5, and consequently (303.5−y) mg H_3PO_4. Two moles of NaOH correspond to one mole of H_3PO_4, whereas four moles of NaOH correspond to one mole of P_2O_5 ($P_2O_5 \equiv 2H_3PO_4 \equiv 4NaOH$). Hence, we have

$$\frac{\text{y mg } P_2O_5}{(141.94/4) \text{ mg/meq}} + \frac{(303.5 - \text{y}) \text{ mg } H_3PO_4}{(98.00/2) \text{ mg/meq}} = 42.00 \text{ mL} \times 0.2000 \text{ meq/mL}$$

or

$$\text{y} = 283.83 \text{ mg } P_2O_5.$$

Hence, the sample contains

$$\frac{283.83}{303.5} \times 100 = \mathbf{93.50\%} \; P_2O_5$$

and

$$100.00 - 93.50 = \mathbf{6.50\%} \; H_3PO_4.$$

Example 11.21. Solution A is obtained by mixing 50.00 mL of 0.2000 M H_3PO_4 solution with 50.00 mL of 0.3000 M Na_3PO_4. Calculate a) the pH of solution A, b) the volume of standard 0.1000 N acid or base solution required to titrate 20.00 mL of solution A, in the presence of 1) methyl orange, and 2) phenolphthalein as indicator.

Solution. a) Initially, we have (50.00 mL × 0.2000 mmol/mL) = 10.00 mmol of H_3PO_4 and (50.00 mL × 0.3000 mmol/mL) = 15.00 mmol of Na_3PO_4. H_3PO_4 reacts with Na_3PO_4 according to the reaction

$$H_3PO_4 + PO_4^{3-} \rightleftharpoons H_2PO_4^- + HPO_4^{2-} \tag{11.34}$$

for which we have

$$K = \frac{[H_2PO_4^-][HPO_4^{2-}]}{[H_3PO_4][PO_4^{3-}]} \cdot \frac{[H^+]}{[H^+]} = \frac{K_1}{K_3} = \frac{7.5 \times 10^{-3}}{1 \times 10^{-12}} = 7.5 \times 10^9 \tag{11.35}$$

From the large value of K it is concluded that Reaction (11.34) is quantitative. Hence, after the end of the reaction, the solution contains 10.00 mmol of $H_2PO_4^-$, 10.00 mmol of HPO_4^{2-}, and 5.00 mmol of unreacted PO_4^{3-}. The acid reacts with the base according to the reaction

$$H_2PO_4^- + PO_4^{3-} \rightleftharpoons 2HPO_4^{2-} \qquad (11.36)$$

for which we have

$$K' = \frac{[HPO_4^{2-}]^2}{[H_2PO_4^-][PO_4^{3-}]} \cdot \frac{[H^+]}{[H^+]} = \frac{K_2}{K_3} = \frac{6.2 \times 10^{-8}}{1 \times 10^{-12}} = 6.2 \times 10^4 \qquad (11.37)$$

From the large value of K′ it is concluded that Reaction (11.36) is quantitative. Hence, after the end of the reaction, solution A contains 20.00 mmol of HPO_4^{2-} and 5.00 mmol of $H_2PO_4^-$. That is, we have a $H_2PO_4^-$-HPO_4^{2-} buffer having a

$$pH = pK_2 + \log\frac{[HPO_4^{2-}]}{[H_2PO_4^-]} = 7.21 + \log\frac{20.00/100.00}{5.00/100.00} = \mathbf{7.81}.$$

b) There are 4.00 mmol of HPO_4^{2-} and 1.00 mmol of $H_2PO_4^-$. Hence, 4.00/0.1000 = 40.00 mL of 0.1000 N acid are required to titrate 20.00 mL of solution A in the presence of methyl orange, whereas 10.00 mL of 0.1000 N base are required in the presence of phenolphthalein.

Determination of functional groups. Many organic functional groups are determined on the basis of quantitative chemical reactions, in which either an acid or a base is produced or consumed. An example is the determination of hydroxyl groups, e.g., in alcohols, by the following method: An accurately known excess of acetic anhydride is added to a sample of alcohol ROH. After the acetylation reaction (11.38) is complete, water is added to hydrolyze the unreacted anhydride (11.39), and then the resulting CH_3COOH is titrated (11.40):

$$(CH_3CO)_2O + ROH \rightleftharpoons CH_3COOR + CH_3COOH \qquad (11.38)$$

$$(CH_3CO)_2O + H_2O \rightleftharpoons 2CH_3COOH \qquad (11.39)$$

$$CH_3COOH + OH^- \rightleftharpoons CH_3COO^- + H_2O \qquad (11.40)$$

A blank is also carried out, whereupon only reactions (11.39) and (11.40) take place. The difference in volumes of base consumed in the blank and in the titration of the sample corresponds to the amount of alcohol in the sample.

Example 11.22. A 0.2208-g sample of ethyl alcohol was treated with an excess of acetic anhydride, and the solution resulting after the addition of water was heated and titrated with 36.76 mL of ethanolic 0.5000 M solution of NaOH. A blank determination required 45.54 mL of base. Calculate the per cent (w/w) content of the sample in ethanol.

Solution. We have

$$\frac{(45.54 - 36.76) \text{ mL} \times 0.5000 \text{ mmol/mL} \times 46.07 \text{ mg } C_2H_5OH/\text{mmol}}{220.8 \text{ mg}} \times 100$$

$= \mathbf{91.6\%}$ C_2H_5OH.

The neutralization titrations are also used for the determination of the equivalent weights of organic acids after their purification, and thus facilitate their identification.

Example 11.23. A 0.3657-g sample of organic acid that contains only C, H, and O (MW 170), is titrated with 43.21 mL 0.1000 M NaOH solution in the presence of phenolphthalein as indicator. Calculate a) the equivalent weight of the acid, b) the number of titratable acid groups in each molecule.

Solution. a) We have

$$\text{equivalent weight} = \frac{0.3657 \text{ g} \times 1000 \text{ meq/eq}}{43.21 \text{ mL} \times 0.1000 \text{ meq/mL}} = \mathbf{84.6} \text{ g/eq}.$$

b) Let n be the number of acid groups in each molecule of the acid. We have

$$n = \frac{170 \text{ g/mol}}{84.6 \text{ g/eq}} = 2.01 \text{ eq/mol} \simeq \mathbf{2} \text{ eq/mol}.$$

11.5 Neutralization Titrations in Nonaqueous Solvents

Most organic acids or bases, which are soluble in water, are very weak and therefore can not be titrated in aqueous solutions. The strength of such weak electrolytes is greatly increased in certain organic solvents, and they can be accurately titrated. In such titrations, the degree of completion of a neutralization reaction increases with increasing dissociation constant (K_a or K_b) of the substance being titrated and with decreasing autoprotolysis constant K_s of the solvent (if we denote an amphiprotic solvent as SH, the autoprotolysis constant for $2SH \rightleftharpoons SH_2^+ + S^-$ is equal to $K_s = [SH_2{}^+][S^-]$). Examples are the volumetric determination of amines, amino acids, alkaloids, phenols, and salts. Calculations involving these titrations are similar to those for aqueous titrations.

Example 11.24. The analysis of a sample containing aniline, N-ethylaniline, N,N-diethylaniline, and substances unreactive to $HClO_4$, was carried out as follows: A 1.000-g sample was dissolved in glacial acetic acid and the solution was divided into four equal portions, A, B, C, and D. Solution A was titrated with 18.05 mL of 0.1000 M $HClO_4$ in glacial acetic acid. Solution B was treated with acetic anhydride and then titrated with 11.62 mL of the acid. Solution C was treated with salicylaldehyde and then titrated with 15.50 mL of the acid. Calculate the percentage composition of the sample.

Solution. In the various steps of the analysis, the following reactions took place:

Solution A

$$C_6H_5NH_2 + HClO_4 \rightleftharpoons C_6H_5NH_3ClO_4$$
$$C_6H_5NHC_2H_5 + HClO_4 \rightleftharpoons C_6H_5NH_2C_2H_5ClO_4$$
$$C_6H_5N(C_2H_5)_2 + HClO_4 \rightleftharpoons C_6H_5NH(C_2H_5)_2ClO_4$$

Solution B

$$C_6H_5NH_2 + (CH_3CO)_2O \rightleftharpoons C_6H_5NHOCCH_3 + CH_3COOH$$
$$C_6H_5NHC_2H_5 + (CH_3CO)_2O \rightleftharpoons C_6H_5NC_2H_5OCCH_3 + CH_3COOH$$
$$C_6H_5N(C_2H_5)_2 + HClO_4 \rightleftharpoons C_6H_5NH(C_2H_5)_2ClO_4$$

Solution C

$$C_6H_5NH_2 + OHCC_6H_4OH \rightleftharpoons C_6H_5N{=}CHC_6H_4OH + H_2O$$
$$C_6H_5NHC_2H_5 + HClO_4 \rightleftharpoons C_6H_5NH_2C_2H_5ClO_4$$
$$C_6H_5N(C_2H_5)_2 + HClO_4 \rightleftharpoons C_6H_5NH(C_2H_5)_2ClO_4$$

Hence, 11.62 mL of $HClO_4$ were consumed for the titration of $C_6H_5N(C_2H_5)_2$, (18.05 − 15.50) = 2.55 mL of $HClO_4$ for the titration of $C_6H_5NH_2$, and (18.05 − 11.62 − 2.55) = 3.88 mL of $HClO_4$ for the titration of $C_6H_5NHC_4H_5$. The above quantities of acid were consumed to titrate 1/4 of the sample. Hence, the sample contains

$$\frac{(11.62\ \text{mL} \times 0.1000\ \text{mmol/mL} \times 149.24\ \text{mg/mmol}) \times 4}{1000\ \text{mg}} \times 100 = \mathbf{69.37\%}\ C_6H_5N(C_2H_5)_2$$

$$\frac{(2.55\ \text{mL} \times 0.1000\ \text{mmol/mL} \times 93.13\ \text{mg/mmol}) \times 4}{1000\ \text{mg}} \times 100 = \mathbf{9.50\%}\ C_6H_5NH_2$$

$$\frac{(3.88\ \text{mL} \times 0.1000\ \text{mmol/mL} \times 121.18\ \text{mg/mmol}) \times 4}{1000\ \text{mg}} \times 100 = \mathbf{18.81\%}\ C_6H_5NHC_2H_5$$

and

$$100.00 - (69.37 + 9.50 + 18.81) = \mathbf{2.32\%}\ \text{unreactive substances.}$$

11.6 Problems

Sample size

11.1. What weight of sample should be taken for analysis, so that each milliliter of 0.1000 N HCl solution consumed in the titration in the presence of methyl orange as indicator represents 1.000% Na_2CO_3 in the sample?

11.2. A series of Na_2CO_3-NaCl samples containing 25–40% Na_2CO_3 is given to students as unknowns for the volumetric determination of carbonates. What range should the sample size be for analysis so that 24–48 mL of 0.1000 M HCl solution will be consumed in the presence of methyl orange as indicator?

11.3. What weight (mg) of vinegar should be taken for analysis, so that the volume of 0.1000 M NaOH titrant in milliliters consumed in the titration is five times larger than the percentage of acetic acid in the sample?

Standardization of acids and bases

11.4. For the standardization of an HCl solution, 0.2120 g of Na_2CO_3 is dissolved in water, and 50.00 mL of the acid are added to the solution. Then, the solution is boiled, and the excess of the unreacted acid is titrated with 5.00 mL of a NaOH solution, 25.00 mL of which are equivalent to 50.00 mL of the HCl solution. Calculate the normalities of the HCl and NaOH solutions.

11.5. In the standardization of an HCl solution, 0.2970 g of chemically pure $CaCO_3$ was dissolved in 30.00 mL of acid, and the excess of unreacted acid was titrated with 3.36 mL of a NaOH solution, 20.25 mL of which are equivalent to 25.00 mL of the acid. Calculate the normality of the HCl.

11.6. A 0.5700-g sample of the weak acid H_nA required 23.04 mL of a NaOH solution for the complete titration of the acid. A 21.20-mL portion of the base was equivalent to 20.00 mL of 0.1260 N HCl solution. Calculate a) the normality of NaOH, b) the equivalent weight of the acid, and c) the per cent of titratable hydrogen in the acid.

11.7. What is the normality of a solution prepared by mixing 21.12 mL of 0.0510 M NaOH solution with 28.28 mL of 0.1000 M Na_2CO_3 solution, with respect to a) phenolphthalein, b) methyl orange indicator?

Selection of an acid-base indicator

11.8. Which one of the following indicators is the most suitable for the titration of a 0.2000 M solution of the weak acid HA ($K_a = 1.00 \times 10^{-5}$) with 0.2000 M standard NaOH solution? a) phenolphthalein (pH color transition range = 8.0–9.8), b) thymolphthalein (9.3–10.5), c) phenol red (6.8–8.2), d) bromthymol blue (6.0–7.6).

Titrations curves

11.9. Give schematically the titration curves for the following titrations, drawn to scale for the relative volumes:

a) Titration of 25.00 mL of 0.1000 M Na_2HPO_4-0.1500 M NaH_2PO_4 solution with 0.1000 M HCl and 0.1000 M NaOH. b) Titration of 40.00 mL of 0.2000 M NaOH-0.1000 M Na_2CO_3 solution with 0.2500 M HCl. c) Titration of 25.00 mL of 0.1200 M Na_2CO_3-0.1000 M $NaHCO_3$ solution with 0.2000 M HCl. d) Titration of 25.00 mL of 0.1000 M H_3PO_4-0.1200 M NaH_2PO_4 solution with 0.2500 M NaOH.

11.10. A 0.2000-g sample of the very strong acid HA (MW 100.0) is dissolved in water and diluted to 50.0 mL. The final solution is titrated with 0.1000 M NaOH standard solution. Calculate the pH of the solution, when the volume of the base added corresponds to the following percentages of the stoichiometric quantity: a) 10.0%, b) 99%, c) 101%.

11.11. Calculate the titration curve for the titration of 50.00 mL of 0.1000 M NaA solution with 0.1000 M HCl solution (dissociation constant of HA = 2.00×10^{-11}).

11.12. A 50.00-mL portion of 0.1000 M solution of the weak diprotic acid H_2A ($K_1 = 1.00 \times 10^{-4}$, $K_2 = 1.00 \times 10^{-8}$) is titrated with 0.1000 M NaOH solution. Calculate the pH of the solution after the addition of the following volumes of titrant: a) 25.00, b) 50.00, c) 75.00, and d) 100.00 mL.

Errors in acid-base titrations

11.13. A 50.00-mL portion of 0.1000 M NH_3 solution is titrated with 0.1000 M HCl. What will the titration error be, if we use as indicator a) phenolphthalein (discoloration at pH 8.0), b) methyl orange (color change at pH 4.0)? What conclusion can be drawn for the suitability of these indicators for the titration of NH_3 (weak base) with HCl (strong acid)?

Quantitative elemental analysis

11.14. A 0.2250-g sample of impure $(NH_4)_2SO_4$ is dissolved in water, the solution is made strongly alkaline and heated, and the ammonia produced is distilled into 25.00 mL of 0.2020 M HCl solution. The excess of unreacted acid is titrated with 16.75 mL of 0.1040 M NaOH solution. Calculate the per cent $(NH_4)_2SO_4$ in the sample.

11.15. The percentage of nitrogen in a large number of proteins is about the same (16%), and therefore the conversion factor of nitrogen to protein is equal to 100/16 = 6.25. For the determination of proteins in serum, a 2.00-mL sample is treated according to the Winkler modified Kjeldahl method (Example 11.11), the ammonia is distilled into excess boric acid, and 21.18 mL of 0.0996 M HCl solution are consumed to titrate the

the ammonium borate ($BO_2^- + H^+ \rightleftharpoons HBO_2$). Calculate the per cent (w/v) of protein in serum.

Titration of mixture of two acids

11.16. A 50.00-mL sample of solution D containing H_2SO_4 and H_3PO_4 required 32.00 mL of 0.2000 M NaOH solution to reach the methyl orange end point, and 11.20 mL more of base to reach the second equivalence point (phenolphthalein). Calculate the molarities of H_2SO_4 and H_3PO_4 in solution D.

Analysis of carbonate mixtures

11.17. The normality of a sodium hydroxide solution that has been contaminated with Na_2CO_3 because of CO_2 absorption, was found equal to 0.1000 N, when titrated with a standard HCl solution in the presence of methyl orange as indicator, and 0.0990 N the presence of phenolphthalein as indicator. How many grams of NaOH and Na_2CO_3 are there in one liter of solution?

11.18. A 2.0230-g sample containing Na_2CO_3, $NaHCO_3$, and inert materials, was dissolved in water and diluted to 100.0 mL (solution A). A 25.00-mL aliquot of solution A consumed 41.24 mL of 0.1004 N HCl solution to reach the methyl orange end point. Twenty five milliliters of a NaOH solution were added to another 25.00-mL aliquot of sample A, then 10 mL of a 10% $BaCl_2$ solution were added and the excess of unreacted NaOH was titrated with 9.24 mL of the HCl solution in the presence of phenolphthalein as indicator. A blank determination required 24.85 mL of the HCl solution. Calculate the per cent $NaHCO_3$ and Na_2CO_3 in the sample.

11.19. A 0.4706-g sample containing only $NaHCO_3$ and Na_2CO_3 is dissolved in water and diluted to 100.0 mL (solution A) and is titrated with standard HCl solution in the presence of phenolphthalein as indicator. The normality of the base was found to be 0.0410 N. What normality would be found for solution A if methyl orange is used as the indicator?

11.20. Solution A, that contains 0.2650 g of Na_2CO_3 and 0.1260 g of $NaHCO_3$, is titrated with 0.2000 M HCl solution. How many milliliters of acid are required in the presence of a) phenolphthalein, b) methyl orange, as indicator?

11.21. A 0.3800-g sample that contains only $NaHCO_3$ and Na_2CO_3 in a molar ratio of 3:1 is dissolved in water and titrated with 0.1000 M HCl solution. Calculate the volumes V_p and V_m of acid required to reach the phenolphthalein and methyl orange end point, respectively.

11.22. A 0.2400-g sample that contains only NaOH and Na_2CO_3 in a molar ratio 2:1 is dissolved in water and titrated with 0.1000 M HCl solution. Calculate the volumes V_p and V_m of acid required to reach the phenolphthalein and methyl orange end point, respectively.

11.23. A 0.2000-g sample that consists of only Na_2CO_3 and $NaHCO_3$ requires 24.25 mL of 0.1000 N HCl solution to reach the methyl orange end point. Calculate the per cent Na_2CO_3 and $NaHCO_3$ in the sample.

11.24. A 100.0-mL sample of natural water is titrated with 2.80 mL of 0.0250 N HCl solution in the presence of phenolphthalein, and 29.60 mL of the acid in the presence of methyl orange, as indicators. Calculate the CO_3^{2-} and HCO_3^- concentration of the sample, in ppm.

11.25. A 0.2528-g sample was titrated with 14.34 mL of 0.0998 N HCl solution in the presence of phenolphthalein, and 35.68 mL in the presence of methyl orange, as indicators. Which of the three substances—NaOH, Na_2CO_3, $NaHCO_3$—are present in the sample and in what percentage?

11.26. The solutions listed below contain one or more of the following substances: NaOH, $NaHCO_3$, Na_2CO_3. Below are also listed the volumes of 0.1000 M HCl solution consumed in the titration of 25.00 mL of each of the five solutions, in the presence of a) methyl orange, b) phenolphthalein, as indicators. Which substance or substances and in what quantities, in g, are present in each of the five solutions?

Solution	V_{HCl}, mL	
	Methyl orange (V_m)	Phenolphthalein (V_p)
a	19.76	19.76
b	42.37	18.01
c	32.23	0.00
d	43.80	21.90
e	38.24	23.72

11.27. Each of the samples listed below contains the indicated number of millimoles of NaOH, $NaHCO_3$, and Na_2CO_3. Calculate the volumes of 0.1000 M HCl solution which would be consumed for each sample, in the presence of a) phenolphthalein (V_p), b) methyl orange (V_m), as indicators.

Solution	mmol		
	NaOH	$NaHCO_3$	Na_2CO_3
a	0	1.42	0
b	2.31	0	0
c	0	0	2.15
d	1.48	2.48	0
e	2.48	1.48	0
f	1.48	1.48	0
g	0	0.83	1.19
h	2.56	0	1.00

Analysis of phosphate mixtures

11.28. Solution A contains two sodium phosphate salts and has a pH of about 7. A 25.00-mL portion of solution A is titrated with 49.20 mL of 0.1000 N base in the presence of phenolphthalein as indicator, and with 30.90 mL of 0.1000 N acid in the presence of methyl orange. Calculate the per cent by weight of each phosphate salt in solution A and its pH (density of solution, d = 1.00 g/mL).

11.29. The solutions listed below contain one or more of the following substances: HCl, H_3PO_4, NaH_2PO_4. Below are also listed the volumes of 0.1000 M NaOH solution required for the titration of 25.00 mL of each of the five solutions, in the presence of a) methyl orange, b) phenolphthalein, as indicators. Which substance or substances, and in what concentrations, are present in each of the five solutions?

Solution	V_{NaOH}, mL	
	Methyl orange (V_m)	Phenolphthalein (V_p)
a	19.12	49.45
b	37.35	37.35
c	0.00	39.16
d	16.92	28.77
e	22.46	44.92

11.30. The solutions listed below contain one or more of the following substances: NaOH, Na_3PO_4, Na_2HPO_4. Below are also listed the volumes of 0.1000 M HCl solution required for the titration of 25.00 mL of each of the four solutions in the presence of a) methyl orange, b) phenolphthalein, as indicators. Which substance or substances and in what quantities, in mg, are present in each of the four solutions?

Solution	V_{HCl}, mL	
	Methyl orange (V_m)	Phenolphthalein (V_p)
a	19.77	0.00
b	40.10	20.05
c	41.28	25.74
d	48.26	22.49

11.31. Solution A of phosphates consumed 11.22 mL of standard acid solution when titrated in the presence of phenolphthalein and an additional 33.66 mL for the subsequent titration to the methyl orange end point. Calculate the pH of solution A.

Determination of functional groups

11.32. A method for the determination of hydroxyl groups in phenols and in primary and secondary alcohols is the following: The sample is acetylated with an excess of acetic anhydride, which produces an equivalent amount (1 mole) of CH_3COOH (compared to 2 moles from hydrolysis of acetic anhydride). Water is added to hydrolyze the excess acetic anhydride. The CH_3COOH produced in the acetylation and in the hydrolysis of excess anhydride is titrated with NaOH. A blank is also carried out. Calculate the grams of phenol in a sample, on the basis of the following experimental data: The sample was dissolved in an inert solvent and acetylated with acetic anhydride, then water was added and the solution was titrated with 8.12 mL of ethanolic 0.5040 $\underline{M}$ NaOH solution. A blank determination with the same volume of acetic anhydride required 18.37 mL of the NaOH solution.

11.33. Primary and secondary amines interfere with the determination of hydroxyl groups by esterification (Problem 11.32), because they too react with the acetic anhydride. The determination of hydroxyl groups in the presence of amines is carried out as follows: The sum of the hydroxyl groups and the primary and secondary amine groups is determined by acetylation. In another portion of the sample, the amino groups are titrated with acid, and the hydroxyl groups are calculated by difference. Calculate the amount of methyl alcohol present in sample A, which also contains methylamine, on the basis of the following experimental data: One fourth of the sample was dissolved in an inert solvent and acetylated with acetic anhydride. Then, water was added and the solution was titrated with 16.10 mL of ethanolic 0.5000 $\underline{M}$ NaOH solution. A blank determination required 26.10 mL of NaOH. Twenty five milliliters of 0.2000 $\underline{M}$ $HClO_4$ solution was added to another fourth of the sample, and the excess of unreacted $HClO_4$ was titrated with 20.00 mL of 0.1250 $\underline{M}$ CH_3COONa solution in glacial acetic acid.

Neutralization titrations in nonaqueous solvents

11.34. Calculate the per cent of unreacted NH_4^+ at the equivalence point, in the titration of a) an aqueous 0.1000 $\underline{M}$ NH_4^+ solution with an aqueous 0.1000 $\underline{M}$ NaOH solution, and b) an ethanolic 0.1000 $\underline{M}$ NH_4^+ solution with an ethanolic 0.1000 $\underline{M}$ C_2H_5ONa solution (the dissociation constant of the NH_4^+ ions in C_2H_5OH is 1×10^{-10}, and the autoprotolysis constant of ethanol is equal to $K_S = [C_2H_5OH_2^+][C_2H_5O^-] = 3 \times 10^{-20}$). Compare the suitability of water and ethanol as solvents in the titration of NH_4^+ ions.

11.35. A 0.0941-g sample which consists of methanol and ethanol is acetylated with acetic anhydride, and the resulting solution is titrated with 20.96 mL of ethanolic 0.1000 $\underline{M}$ NaOH solution. A blank determination required 45.96 mL of base. Calculate the per cent of CH_3OH and C_2H_5OH in the sample.

11.36. A 2.000-g sample that consists of aniline, N-methylaniline, N,N-dimethylaniline, and material unreactive to $HClO_4$ was dissolved in glacial acetic acid, and the resulting solution was divided in four equal portions, A, B, C, and D. Solution A was titrated with 33.25 mL of 0.1000 $\underline{M}$ $HClO_4$ solution in glacial acetic acid. Solution

B was treated with acetic anhydride and then titrated with 17.40 mL of the $HClO_4$. Solution C was treated with salicylaldehyde and then titrated with 23.25 mL of the acid. Calculate the percentage composition of the sample.

11.37. Amino acids can be determined by titrating the amino group with a $HClO_4$ solution in glacial acetic acid, and also by measuring the volume of nitrogen evolved during their treatment with nitrous acid (Van Slyke method). Twenty five milliliters of 0.2008 M $HClO_4$ were added to a 0.2208-g sample of alanine, and the excess of unreacted $HClO_4$ was titrated with 26.89 mL of 0.1004 M CH_3COONa solution. a) Calculate the per cent alanine in the sample. b) If the same amount of sample had been used in the Van Slyke method, what volume of nitrogen would have been evolved, when measured at a pressure of 758 mm of Hg and a temperature of 25°C?

11.38. A 0.6486-g sample of an aromatic alcohol, which contains one –OH group in its molecule, was dissolved in an inert solvent and acetylated with acetic anhydride. The resulting solution was diluted with water and titrated with 36.20 mL of ethanolic 0.2000 M NaOH solution. A blank determination with the same amount of anhydride required 6.20 mL of the base. Calculate the molecular weight of the alcohol.

Miscellaneous acid-base titrations

11.39. Guanine, 2-amino-6-hydroxy-purine, $C_5H_5N_5O$, is insoluble in water but soluble in acids. The equivalent weight of guanine equals its molecular weight (151.13). Calculate the per cent guanine in a sample on the basis of the following experimental data: A 0.1650-g sample was dissolved in 25.00 mL of 0.1000 M HCl solution, and the excess of unreacted acid was titrated with 15.32 mL of 0.1000 M NaOH solution.

11.40. The determination of SO_2 in the air of a factory was performed as follows: The air was bubbled at a rate of 10 L/min through a trap containing H_2O_2. The H_2SO_4 produced in 30 min ($SO_2 + H_2O_2 \rightleftharpoons 2H^+ + SO_4^{2-}$) was titrated with 5.62 mL of 0.01000 M NaOH solution. Calculate the concentration of SO_2 in the air, in ppm (mL $SO_2/10^6$ mL air), given that the density of SO_2 is 2.86 mg/mL.

11.41. A 1.800-g sample of a solution of the monoprotic acid HA (MW 60.0) requires 15.00 mL of 0.1000 M NaOH solution for the titration of the acid. Which of the following values represents the percentage of HA in the sample (w/w)? a) 10.0, b) 25.0, c) 2.50, d) 5.00 (density of the HA solution, d = 1.00 g/mL).

11.42. A mixture of $CaCO_3$ and $MgCO_3$ weighs x grams. Forty milliliters of 0.1500 N HCl solution were required to dissolve half of the mixture ($\mathbf{MCO_3} + 2H^+ \rightarrow M^{2+} + H_2O + CO_2$). The oxides formed during the ignition of the remaining mixture (x/2) weighed 0.1525 g. Calculate the value of x.

11.43. A 0.9300-g sample of partly hydrated sodium carbonate, $Na_2CO_3 \cdot yH_2O$, was dissolved in water and diluted to 500.0 mL (solution A). A 50.00-mL portion of solution A was titrated with 18.75 mL of 0.0800 M HCl solution using methyl orange indicator. a) Calculate the value of y. b) How many milliliters of acid would have been required for the entire sample if it consisted of $Na_2CO_3 \cdot 10H_2O$?

11.44. Twenty five grams of a soil sample were kept in a closed space (under a glass bell). Also under the glass bell was a dish containing 100.0 mL of NaOH solution to absorb any CO_2. After 48 hours, 25.00 mL of the NaOH solution were titrated with 12.17 mL of 0.1250 M HCl solution in the presence of phenolphthalein as indicator. A blank (without soil) required 23.31 mL of the acid. Calculate the rate of CO_2 evolution from the soil, which is the result of bacterial action, in mg CO_2 per gram of soil per hour.

11.45. A 0.3228-g sample of a chemically pure dibasic organic acid requires 42.67 mL of 0.1008 M NaOH solution for complete neutralization. Calculate the molecular weight of the acid.

11.46. The determination of boric acid in pharmaceutical preparations is carried out by ashing the sample, dissolving the ash in H_2SO_4, neutralizing the solution, adding CH_3OH and H_2SO_4, and distilling the borate methyl ester formed into a NaOH solution. Then, the codistilled methanol is removed, the solution is neutralized, mannitol is added to convert boric acid to a stronger acid (mannitol-boric acid complex), which is titrated potentiometrically with standard NaOH solution. A 0.8755-g sample required 14.75 mL of 0.1075 M NaOH solution for the titration. Calculate the per cent content of H_3BO_3 in the sample.

11.47. A 0.8242-g benzaldehyde sample was mixed with 25.00 mL of 0.6 M hydroxylamine hydrochloride solution, and the HCl released was titrated with 15.12 mL of 0.5020 M NaOH. A blank determination required 0.12 mL of NaOH. Calculate the per cent content of C_6H_5CHO in the sample.

Chapter 12

Oxidation-Reduction Titrations

12.1 Electrode Potentials—Strength of Oxidizing and Reducing Agents

Oxidation-reduction (redox) titrations are possibly the widest used of the volumetric techniques, and they are used for the determination of most of the elements, as well as of functional groups in organic analysis. Standard solutions of relatively few substances are needed, because a strong oxidant (or reductant) can be used for the determination of a large number of reductants (or oxidants). Care should be exercised in stoichiometric calculations for the correct definition of the equivalent weight of an oxidant or a reductant, because it can vary depending on the conditions for the reaction (Chapter 3).

The degree of completion, i.e., the equilibrium constant, of a redox reaction depends on the potentials of the half-reactions (Example 12.1). As a rule, the reaction for which the product, nE_{cell} of the galvanic cell (Section 8.3), exceeds 0.5 V is quantitative. If there is a difference of at least 0.2 V between the standard potentials of two substances, they can be titrated stepwise.

In any redox titration, the substance being titrated should exist quantitatively (100.00%) in a single oxidation state. If it is not, then it must be reduced or oxidized prior to the titration (prereduction or preoxidation) by a suitable reducing or oxidizing agent, A, which is added in excess, since the amount of the substance being determined is not known in advance (the unreacted A is removed or destroyed prior to the titration so it does not interfere (Problem 12.11).

The most important characteristics of a redox titration can be summarized in the *titration curve* (usually, potential E as a function of the volume V of titrant). The titration curve is either calculated theoretically, whereupon conclusions can be drawn from it for the feasibility and the probable accuracy of a titration and the proper redox indicator is selected, or is determined experimentally with a suitable potential measuring instrument, such as a pH meter or a pIon meter (Section 15.1), and the end point is determined from the curve graphically (Section 11.2). In certain cases, e.g., with $KMnO_4$, no indicator is required, because the titrant colors the solution after the equivalence point. The transition potential of an indicator must be close to the potential of the solution at the equivalence point (ideally, it is identical to it).

12.2 Titration Curves

The calculation of a titration curve is made using the Nernst equation (Equation (8.2) or (8.2*a*)). For calculating E, we can use the conjugate redox couple of either the titrand (analyte) or the titrant. To simplify the calculations, we use the redox couple of the titrand before the equivalence point and the redox couple of the titrant, after the equivalence point. The potential of the system at the equivalence point is a function of the standard (or formal) potentials of the two redox couples (Equation (12.6)), but in certain cases it also depends on pH (Equation (12.11)) or on the concentration of various substances (Equation (12.14)). Below are given representative examples of calculating titration curves, and subsequently various categories of redox titration are studied (with $KMnO_4$, $K_2Cr_2O_7$, etc.)

Example 12.1. a) Construct a titration curve for the titration of 50.00 mL of 0.0500 $\underline{N}$ Fe(II) solution with 0.1000 $\underline{N}$ Ce(IV) solution. Both solutions are 1.0 $\underline{M}$ in H_2SO_4. b) Calculate the concentrations of Fe^{2+}, Fe^{3+}, Ce^{3+}, and Ce^{4+} ions at the equivalence point.

Solution. a) For simplicity, we assume that iron and cerium exist in the solution as simple cations. Since the titration is carried out in 1.0 $\underline{M}$ H_2SO_4 solution, formal potentials will be used. The equilibrium constant for the reaction

$$Fe^{2+} + Ce^{4+} \rightleftharpoons Fe^{3+} + Ce^{3+} \tag{12.1}$$

at the experimental conditions, is equal to

$$K = 10^{(1.44-0.674)/0.05916} = 8.9 \times 10^{12}.$$

Hence, Reaction (12.1) is quantitative. Below are given representative examples for the calculation of E in each of the regions of the titration curve.

1. *Initial potential.* The solution does not contain cerium ions, whereas it contains a small but unknown concentration of Fe(III) due to air oxidation of Fe(II). Therefore, the calculation of the initial potential is impossible.

2. *Before the equivalence point.* For example, after 10.00 mL of Ce(IV) were added, we have

$$\begin{aligned} E &= E^{0'}_{Fe^{3+},\,Fe^{2+}} - 0.05916 \log \frac{[Fe^{2+}]}{[Fe^{3+}]} \\ &= 0.674 - 0.05916 \log \frac{\dfrac{(50.00 \times 0.0500 - 10.00 \times 0.1000)}{60.00} + [Ce^{4+}]}{\dfrac{10.00 \times 0.1000}{60.00} - [Ce^{4+}]} \\ &= 0.674 - 0.05916 \log \frac{1.50}{1.00} = \mathbf{0.66}\ \text{V} \ ([Ce^{4+}] \text{ is negligible}). \end{aligned}$$

3. *At the equivalence point.* Potential E_{eq} at the equivalence point is given by the relation

$$E_{eq} = E^{0'}_{Fe^{3+},\ Fe^{2+}} - 0.05916 \log\frac{[Fe^{2+}]}{[Fe^{3+}]}, \qquad (12.2)$$

and also by the relation

$$E_{eq} = E^{0'}_{Ce^{4+},\ Ce^{3+}} - 0.05916 \log\frac{[Ce^{3+}]}{[Ce^{4+}]}. \qquad (12.3)$$

Adding Equations (12.2) and (12.3), we have

$$2E_{eq} = E^{0'}_{Fe^{3+},\ Fe^{2+}} + E^{0'}_{Ce^{4+},\ Ce^{3+}} - 0.05916 \log\frac{[Fe^{2+}][Ce^{3+}]}{[Fe^{3+}][Ce^{4+}]}. \qquad (12.4)$$

At the equivalence point we have

$$[Fe^{2+}] = [Ce^{4+}],\ [Fe^{3+}] = [Ce^{3+}],$$

and consequently, we also have

$$\frac{[Fe^{2+}]}{[Fe^{3+}]} = \frac{[Ce^{4+}]}{[Ce^{3+}]} \qquad (12.5)$$

Combining Equations (12.4) and (12.5), we have

$$E_{eq} = \frac{E^{0'}_{Fe^{3+},\ Fe^{2+}} + E^{0'}_{Ce^{4+},\ Ce^{3+}}}{2} = \frac{0.674 + 1.44}{2} = \mathbf{1.06}\ \text{V}. \qquad (12.6)$$

Potential E_{eq} is independent of the concentrations of iron and cerium.

4. *After the equivalence point.* For example, after 25.05 mL of Ce(IV) were added, we have

$$\begin{aligned} E &= E^{0'}_{Ce^{4+},\ Ce^{3+}} - 0.05916 \log \frac{[Ce^{3+}]}{[Ce^{4+}]} \\ &= 1.44 - 0.05916 \log \frac{\dfrac{25.00 \times 0.1000}{75.05} - [Fe^{2+}]}{\dfrac{(25.05 \times 0.1000 - 50.00 \times 0.0500)}{75.05} + [Fe^{2+}]} \\ &= 1.44 - 0.05916 \log \frac{2.50}{0.005} = \mathbf{1.28}\ \text{V} \ ([Fe^{2+}] \text{ is negligible}). \end{aligned}$$

The results of the calculations are summarized in Table (12.1), and the corresponding titration curve is given in Figure (12.1).

b) *1st method.* At the equivalence point we have $C_{Fe} = 50.00 \times 0.0500/75.00 = 0.0333$ M. Since Reaction (12.1) is quantitative, at the equivalence point we have $[Fe^{3+}] = [Ce^{3+}] = 0.0333 - Fe^{2+} \simeq$ **0.0333** M. We also have the relation

$$E = 1.06 = 0.674 - 0.05916 \log \frac{[Fe^{2+}]}{[Fe^{3+}]}$$

Table 12.1. *Electrode Potentials (vs. SHE) During Titration of 50.00 mL of 0.0500 N Fe(II) with a) 0.1000 N Ce(IV), b) 0.1000 N $KMnO_4$ ($[H_2SO_4]$ = 1.0 M).*

V_{ox}, mL	E, V	
	Ce(IV)	$KMnO_4$
5.00	0.64	0.64
10.00	0.66	0.66
12.50	0.674	0.674
20.00	0.71	0.71
24.00	0.76	0.76
24.95	0.83	0.83
25.00	1.06	1.37
25.05	1.28	1.48
25.50	1.34	1.49
30.00	1.40	1.50
50.00	1.44	1.51

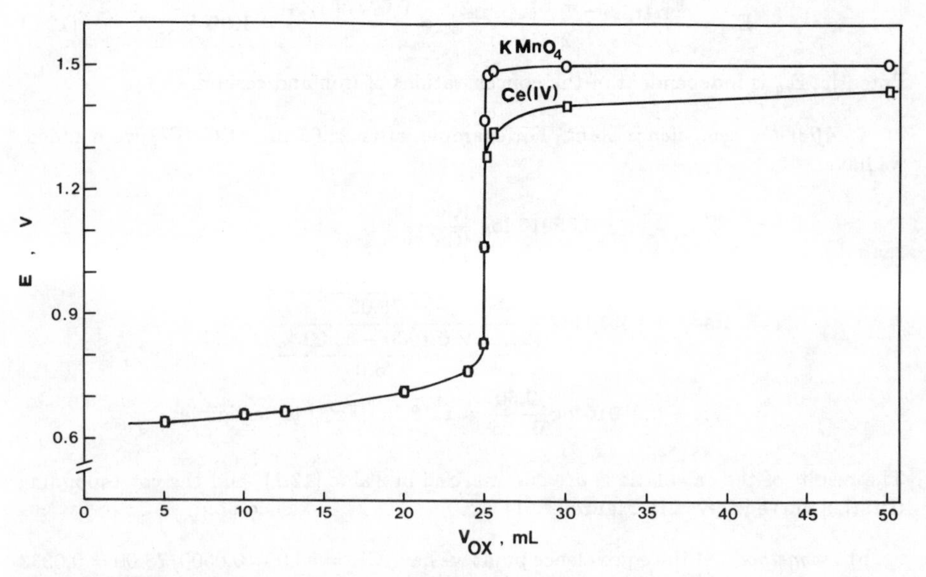

Figure 12.1. *Titration curve for the titration of 50.00 mL of 0.0500 N Fe(II) with a) 0.1000 N Ce(IV), b) 0.1000 N (0.02000 M) $KMnO_4$.*

or

$$[Fe^{2+}]/[Fe^{3+}] = 3.4 \times 10^{-7},$$

whereupon,

$$[Fe^{2+}] = 3.4 \times 10^{-7} \times 0.0333 = \mathbf{1.1 \times 10^{-8}}\ \underline{M} = [Ce^{4+}].$$

Note. From the very small value of $[Fe^{2+}]$ at the equivalence point, it can be seen that Reaction (12.1) is indeed quantitative. The same results are found by applying the Nernst equation to the half-reaction $Ce^{4+} + e \rightleftharpoons Ce^{3+}$.

2nd method. Let x be the molarity of Fe^{2+} ions at the equivalence point, whereupon $[Fe^{2+}] = [Ce^{4+}] = x$, and $[Fe^{3+}] = [Ce^{3+}] = 0.0333 - x$. Substituting these values in the expression for the equilibrium constant for Reaction (12.1), we have

$$\frac{[Fe^{3+}][Ce^{3+}]}{[Fe^{2+}][Ce^{4+}]} = \frac{(0.0333 - x)^2}{x^2} = 8.9 \times 10^{12}$$

or $x = 1.1 \times 10^{-8}$. Hence,

$$[Fe^{2+}] = [Ce^{4+}] = \mathbf{1.1 \times 10^{-8}}\ \underline{M}$$

and

$$[Fe^{3+}] = [Ce^{3+}] = 0.0333 - 1.1 \times 10^{-8} \simeq \mathbf{0.0333}\ \underline{M}.$$

Example 12.2. Construct a titration curve for the titration of 50.00 mL of 0.0500 N Fe(II) solution with 0.1000 N (0.02000 M) $KMnO_4$ solution. Both solutions are 1.0 M in H_2SO_4 (neglect the change in $[H^+]$ during the titration).

Solution. The titration reaction is

$$5Fe^{2+} + MnO_4^- + 8H^+ \rightleftharpoons 5Fe^{3+} + Mn^{2+} + 4H_2O \text{ (in 1.0 } \underline{M}\ H_2SO_4) \tag{12.7}$$

The equilibrium constant for reaction (12.7) is equal to

$$K = 10^{5[(1.51-0.674)/0.05916]} = 5 \times 10^{70}.$$

Hence, Reaction (12.7) is quantitative. Below are given representative examples for the calculation of potential E.

1. *Before the equivalence point.* As in Example (12.1), the potentials of the system before the equivalence point are calculated on the basis of the redox couple $Fe^{2+} - Fe^{3+}$, and therefore, they are identical to the corresponding values in Example (12.1).

2. *At the equivalence point.* Potential E_{eq} at the equivalence point is given by the relation

$$E_{eq} = E^{0'}_{Fe^{3+},\ Fe^{2+}} - 0.05916 \log \frac{[Fe^{2+}]}{[Fe^{3+}]} \tag{12.8}$$

and by the relation

$$E_{eq} = E^{0}_{MnO_4^-,\ Mn^{2+}} - \frac{0.05916}{5} \log \frac{[Mn^{2+}]}{[MnO_4^-][H^+]^8} \tag{12.9}$$

Multiplying Equation (12.9) by 5 and adding the product to Equation (12.8), we finally have

$$6E_{eq} = E^{0'}_{Fe^{3+}, Fe^{2+}} + 5E^{0}_{MnO_4^-, Mn^{2+}} - 0.05916 \log \frac{[Fe^{2+}][Mn^{2+}]}{[Fe^{3+}][MnO_4^-][H^+]^8} \qquad (12.10)$$

Because of the stoichiometry of Reaction (12.7), at the equivalence point we have $[Fe^{2+}] = 5[MnO_4^-]$, and $[Fe^{3+}] = 5[Mn^{2+}]$. Substituting in Equation (12.10), we finally have

$$E_{eq} = \frac{E^{0'}_{Fe^{3+}, Fe^{2+}} + 5E^{0}_{MnO_4^-, Mn^{2+}}}{6} - \frac{0.05916}{6} \log \frac{1}{[H^+]^8} \qquad (12.11)$$

In 1.0 $\underline{M}$ H_2SO_4 we have $[H^+] \simeq 1.0$ $\underline{M}$. Substituting the data of the problem in Equation (12.11), we have

$$E_{eq} = \frac{0.674 + (5 \times 1.51)}{6} - \frac{0.05916}{6} \log \frac{1}{1^8} = \mathbf{1.37} \text{ V}.$$

Potential E_{eq} depends on the concentration of H^+ ions.

3. *After the equivalence point.* For example, after 25.05 mL of $KMnO_4$ were added, we have

$$\begin{aligned} E &= E^{0}_{MnO_4^-, Mn^{2+}} - \frac{0.05916}{5} \log \frac{[Mn^{2+}]}{[MnO_4^-][H^+]^8} \\ &= 1.51 - \frac{0.05916}{5} \log \frac{\frac{1}{5}\left(\frac{50.00 \times 0.0500}{75.05} - [Fe^{2+}]\right)}{\frac{1}{5}\left(\frac{25.05 \times 0.1000 - 50.00 \times 0.0500}{75.05} + [Fe^{2+}]\right)(1)^8} \\ &= \mathbf{1.48} \; V. \end{aligned}$$

The results of the calculations are summarized in Table (12.1), and the corresponding titration curve is given in Figure (12.1). From the comparison of the two curves, it can be seen that the titration curve is symmetric for Ce(IV) (because both half-reactions involve the same number of electrons) and unsymmetric for $KMnO_4$ (because the two reduction half-reactions involve a different number of electrons: 1e for the Fe^{3+}-Fe^{2+} couple and 5e for the MnO_4^--Mn^{2+} couple). The change of potential in the region of the equivalence point is larger for $KMnO_4$, because of the larger equilibrium constant [5×10^{70} for Reaction (12.7) (Example 12.2)), compared to 8.9×10^{12} for reaction (12.1) (Example (12.1))].

Example 12.3. Calculate the potential of the system at the equivalence point, for the reaction

$$6Fe^{2+} + Cr_2O_7^{2-} + 14H^+ \rightleftharpoons 6Fe^{3+} + 2Cr^{3+} + 7H_2O \text{ (in 1 } \underline{M} \text{ } H_2SO_4) \qquad (12.12)$$

(It is assumed that $[H^+]_{eq} = 1.0$ $\underline{M}$.)

Solution. Working as in Example (12.2), for reaction (12.12) we obtain the relation

$$7E_{eq} = E^{0'}_{Fe^{3+}, Fe^{2+}} + 6E^{0}_{Cr_2O_7^{2-}, Cr^{3+}} - 0.05916 \log \frac{[Fe^{2+}][Cr^{3+}]^2}{[Fe^{3+}][CrO_7^{2-}][H^+]^{14}} \qquad (12.13)$$

At the equivalence point we have $[Fe^{2+}] = 6[Cr_2O_7{}^{2-}]$ and $[Fe^{3+}] = 3[Cr^{3+}]$. Substituting in Equation (12.13), we finally have

$$E_{eq} = \frac{E^{0'}_{Fe^{3+}, Fe^{2+}} + 6E^0_{Cr_2O_7{}^{2-}, Cr^{3+}}}{7} - \frac{0.05916}{7}\log\frac{2[Cr^{3+}]_{eq}}{[H^+]^{14}} \tag{12.14}$$

Potential E_{eq} cannot be calculated by Equation (12.14), because the value of $[Cr^{3+}]_{eq}$ is unknown (unless the initial $[Fe^{2+}]$ is known, from which we can also calculate $[Cr^{3+}]_{eq}$).

Notes. 1. Generally, whenever the ratio between the number of atoms in the oxidized and reduced compounds of the two oxidation states of an element in one of the redox couples of the reaction is not 1:1 (e.g., Cr^{3+} and $Cr_2O_7{}^{2-}$), potential E_{eq} depends on the concentration of this element. Therefore, it cannot be calculated, unless this concentration is known. Such is the case of the Cr^{3+} ion concentration in Equation (12.14) (a similar case is the use of $[I_3]^-$ as oxidizing agent, whereupon $3I^-$ are formed, Problem 12.28). On the other hand, when this ratio is 1:1, as for example when using Ce^{4+} and $MnO_4{}^-$, which are reduced to Ce^{3+} and Mn^{2+}, respectively, potential E_{eq} is independent of the concentrations of these reagents and can be calculated [Equations (12.6) and (12.11)].

2. Generally, for the reaction

$$aOx_1 + bRed_2 + cH^+ \rightleftharpoons aRed_1 + bOx_2 + c/2(H_2O), \tag{12.15}$$

where the ratio between the two oxidation states of the same element in a redox couple is 1:1, in a similar way we obtain the relation

$$E_{eq} = \frac{bE^0_{Ox_1, Red_1} + aE^0_{Ox_2, Red_2}}{a+b} - \frac{0.05916}{a+b}\log\frac{1}{[H^+]^c} \tag{12.16}$$

3. In any redox titration, we can have a general idea of the shape of the titration curve from the values of the potential E of the system at the following three characteristic points of the curve: 50%, 100% (equivalence point), and 200%, that is, the points at which the amount of titrant added is equal to 50%, 100%, and 200%, respectively, of the stoichiometric amount (the amount of titrant that is chemically equivalent to the amount of the titrated substance). Potential $E_{50\%}$ is equal to the standard (or formal) potential of the titrand, $E_{100\%}$ is equal to E_{eq}, which is calculated as described above, and $E_{200\%}$ is equal to the standard (or formal) potential of the titrant.

12.3 Errors in Oxidation-Reduction Titrations

Example 12.4. A 50.00-mL portion of 0.0500 $\underline{N}$ Fe(II) solution is titrated with 0.1000 $\underline{N}$ Ce(IV) solution. Both solutions are 1.0 $\underline{M}$ in H_2SO_4. Calculate the titration error, if the sodium salt of diphenylamine sulfonic acid is used as indicator (color change midpoint at potential of 0.85 V).

Solution. At the end point we have $E_{en} = 0.85 < 1.06 = E_{eq}$ (Example 12.1); therefore, we have a negative error. We have

$$E_{en} = E^{0'}_{Fe^{3+}, Fe^{2+}} - 0.05916 \log \frac{[Fe^{2+}]}{[Fe^{3+}]} = 0.85 = 0.674 - 0.05916 \log \frac{[Fe^{2+}]}{[Fe^{3+}]}$$

or

$$\frac{[Fe^{2+}]}{[Fe^{3+}]} = 1.06 \times 10^{-3},$$

whereupon

$$\frac{[Fe^{2+}]}{[Fe^{3+}] + [Fe^{2+}]} = \frac{1.06 \times 10^{-3}[Fe^{3+}]}{[Fe^{3+}] + 1.06 \times 10^{-3}[Fe^{3+}]} = 1.06 \times 10^{-3}.$$

Hence, the titration error is equal to $-1.06 \times 10^{-3} \times 100 = \mathbf{-0.11\%}$.

12.4 Titrations Involving Permanganate

Potassium permanganate, $KMnO_4$, is a widely used oxidizing agent for the titration of reducing agents, because it is a strong oxidant and at the same time acts as indicator. Permanganate ions can be reduced to various products, depending on the conditions and, especially, the acidity of the solution.

Example 12.5. What is the per cent purity of sodium oxalate, if 0.5208 g of sample requires 38.82 mL of $KMnO_4$ solution, given that 1.000 mL of $KMnO_4$ is equivalent to 11.17 mg of iron [Fe(II) → Fe(III)]?

Solution. The normality of the $KMnO_4$ solution is equal to

$$N_{KMnO_4} = \frac{11.17 \text{ mg Fe/mL}}{55.85 \text{ mg Fe/meq}} = 0.2000 \text{ meq/mL}.$$

Therefore, the percentage of sodium oxalate in the sample is equal to

$$\% \, Na_2C_2O_4 =$$

$$\frac{38.82 \text{ mL} \times 0.2000 \text{ meq/mL} \times (134.00/2) \text{ mg } Na_2C_2O_4\text{/meq}}{520.8 \text{ mg}} \times 100 = \mathbf{99.9}.$$

Example 12.6. A 2.9500-g sample, containing anhydrous sodium oxalate, a hydrate of oxalic acid ($H_2C_2O_4 \cdot 2H_2O$) and inert materials, is dissolved in water and diluted to 100.0 mL (solution A). A 25.00-mL aliquot of solution A is titrated with 25.00 mL of 0.1080 M NaOH solution ($H_2C_2O_4 + 2\,OH^- \rightarrow C_2O_4^{2-} + 2H_2O$). Another 25.00-mL aliquot of solution A is acidified and titrated with 49.42 mL of 0.1040 N $KMnO_4$ solution. Calculate the per cent $Na_2C_2O_4$ and $H_2C_2O_4 \cdot 2H_2O$ in the sample.

Solution. Suppose that the sample contains y mmol of $Na_2C_2O_4$ and z mmol of $H_2C_2O_4 \cdot 2H_2O$. We have

$$z \text{ mmol} \times 2 \text{ meq/mmol} \times (25.00/100.00) = 25.00 \text{ mL} \times 0.1080 \text{ meq/mL}$$

or

$$z = 5.400$$

and

$$(5.400 \text{ mmol} \times 2 \text{ meq/mmol} \times 0.2500) + (y \text{ mmol} \times 2 \text{ meq/mmol} \times 0.2500)$$

$$= 49.42 \text{ mL} \times 0.1040 \text{ meq/mL}$$

or y = **4.879**. Hence, the sample contains

$$\frac{5.400 \text{ mmol} \times 126.07 \text{ mg } H_2C_2O_4 \cdot 2H_2O/\text{mmol}}{2950.0 \text{ mg}} \times 100 = \mathbf{23.08\%}\ H_2C_2O_4 \cdot 2H_2O,$$

and

$$\frac{4.879 \text{ mmol} \times 134.00 \text{ mg } Na_2C_2O_4/\text{mmol}}{2950.0 \text{ mg}} \times 100 = \mathbf{22.16\%}\ Na_2C_2O_4.$$

Example 12.7. A 0.1922-g sample of calcium carbonate is dissolved in hydrochloric acid and the calcium is precipitated as CaC_2O_4. The precipitate is filtered and dissolved in dilute H_2SO_4, and the resulting solution is titrated with 36.42 mL of $KMnO_4$ solution, 39.12 mL of which are equivalent to 0.2621 g of $Na_2C_2O_4$. A blank determination requires 0.10 mL $KMnO_4$. Calculate the per cent calcium in the sample.

Solution. The following reactions take place during the analysis:

$$\mathbf{CaCO_3} + 2H^+ \rightleftharpoons Ca^{2+} + H_2O + CO_2 \uparrow$$

$$Ca^{2+} + C_2O_4^{2-} \rightleftharpoons \mathbf{CaC_2O_4}$$

$$\mathbf{CaC_2O_4} + 2H^+ \rightleftharpoons Ca^{2+} + H_2C_2O_4$$

$$2MnO_4^- + 5H_2C_2O_4 + 6H^+ \rightleftharpoons 2Mn^{2+} + 8H_2O + 10CO_2 \uparrow$$

Let N be the normality of the $KMnO_4$ solution. We have

$$262.1 \text{ mg}/(134.00/2) \text{ mg } Na_2C_2O_4/\text{meq} = 39.12 \text{ mL} \times N_{\text{meq/mL}}$$

or N = 0.1000. Hence, the percentage of calcium in the sample is equal to

$$\% \text{ Ca} = \frac{(36.42 - 0.10) \text{ mL} \times 0.1000 \text{ meq/mL} \times 20.04 \text{ mg Ca/meq}}{192.2 \text{ mg}} \times 100 = \mathbf{37.87}.$$

Example 12.8. A student's limestone unknown sample contains 33.00% CaO. The student precipitates the calcium as CaC_2O_4, dissolves the precipitate in H_2SO_4, titrates the resulting solution with standard $KMnO_4$ solution and finds that the sample contains 35.75% CaO. If the error is caused by MgC_2O_4 contamination of the precipitate, calculate the number of milligrams of magnesium coprecipitated with each gram of calcium.

Solution. Let y be the number of milligrams of magnesium coprecipitated with each gram of calcium. We have

$$\text{mmol } C_2O_4 \text{ precipitated/g Ca} = \frac{1000.0 \text{ mg Ca}}{40.08 \text{ mg Ca/mmol } C_2O_4} + \frac{\text{y mg Mg}}{24.30 \text{ mg Mg/mmol } C_2O_4} = 24.95 + 0.04115\, y$$

Therefore, we have

$$\frac{24.95 + 0.04115 \text{ y}}{24.95} = \frac{35.75}{33.00}$$

from which

$$\text{y} = \mathbf{50.5} \text{ mg Mg/g Ca.}$$

Example 12.9. A 50.00-mL portion of solution A, which contains $NaNO_2$ and $NaNO_3$, is acidified and titrated with 40.00 mL of 0.2500 N $KMnO_4$ solution ($2MnO_4^- + 5HNO_2 + 6H^+ \rightleftharpoons 2Mn^{2+} + 5NO_3^- + 3H_2O$). In another 50.00-mL portion of solution A, NO_2^- and NO_3^- are reduced to ammonia, which is collected and titrated with 40.00 mL of 0.2500 M HCl solution. How many grams of $NaNO_2$ and $NaNO_3$ are there in one liter of solution A?

Solution. Suppose that solution A contains y g of $NaNO_2$ and z g of $NaNO_3$ in one liter. We have

$$[(\text{y g NaNO}_2/\text{L})/0.06900 \text{ g/mmol}] \times 2 \text{ meq/mmol} \times 0.5000 \text{ L}$$

$$= 40.00 \text{ mL} \times 0.2500 \text{ meq/mL or y} = \mathbf{6.900} \text{ g NaNO}_2/\text{L}$$

and

$$\left(\frac{6.900 \text{ g NaNO}_2/\text{L}}{0.06900 \text{ g/mmol}} + \frac{\text{z g NaNO}_3/\text{L}}{0.08499 \text{ g/mmol}}\right) \times 0.05000 \text{ L} = 40.00 \text{ mL} \times 0.2500 \text{ mmol/mL}$$

or

$$\text{z} = \mathbf{8.499} \text{ g NaNO}_3/\text{L.}$$

Example 12.10. A 0.1074-g sample of pure element X is dissolved in acid to give X^{3+} ions, which are titrated to XO_4^{2-} ions with 30.00 mL of 0.1062 N $KMnO_4$ solution. Calculate the atomic weight of element X.

Solution. Let y be the atomic weight of element X. Since in the oxidation of X^{3+} ion to XO_4^{2-} the oxidation number of X increases by 3, the equivalent weight of X is equal to y/3. Hence, we have

$$\frac{107.4 \text{ mg X}}{\text{y/3 mg/meq}} = 30.00 \text{ mL} \times 0.1062 \text{ meq/mL}$$

or

$$y = \mathbf{101.1}\ (\mathrm{Ru}).$$

Example 12.11. $Na_2C_2O_4$ and $KHC_2O_4 \cdot H_2C_2O_4$ (three replaceable H) are mixed in such a proportion that each gram of the mixture will react with equal volumes of 0.2000 $\underline{M}$ $KMnO_4$ solution ($MnO_4^- \rightarrow Mn^{2+}$) and 0.2000 $\underline{M}$ NaOH. Calculate the grams of $Na_2C_2O_4$ per gram of $KHC_2O_4 \cdot H_2C_2O_4$ in the mixture.

Solution. Suppose that the mixture contains y mmol $Na_2C_2O_4$/1 mmol $KHC_2O_4 \cdot H_2C_2O_4$. The following reactions take place during the titrations:

$$5Na_2C_2O_4 + 2MnO_4^- + 16H^+ \rightleftharpoons 10Na^+ + 2Mn^{2+} + 8H_2O + 10CO_2$$

$$5KHC_2O_4 \cdot H_2C_2O_4 + 4MnO_4^- + 17H^+ \rightleftharpoons 5K^+ + 4Mn^{2+} + 16H_2O + 20CO_2$$

$$KHC_2O_4 \cdot H_2C_2O_4 + 3\,OH^- \rightleftharpoons K^+ + 2C_2O_4^{2-} + 3H_2O$$

On the basis of the stoichiometry of the above reactions, we have

$$2y/5 + 4/5 = 3$$

or

$$\begin{aligned} y &= \frac{5.500 \text{ mmol } Na_2C_2O_4}{\text{mmol } KHC_2O_4 \cdot H_2C_2O_4} \equiv \frac{5.500 \text{ mmol} \times 0.1340 \text{ g } Na_2C_2O_4/\text{mmol}}{0.2181 \text{ g } KHC_2O_4 \cdot H_2C_2O_4} \\ &= \mathbf{3.379} \text{ g } Na_2C_2O_4/\text{g } KHC_2O_4 \cdot H_2C_2O_4. \end{aligned}$$

12.5 Titrations Involving Dichromate

Potassium dichromate, $K_2Cr_2O_7$, is of limited use in volumetric analysis compared to $KMnO_4$, because it is not as strong an oxidizing agent and many of its reactions are slow. $K_2Cr_2O_7$ has also the disadvantage that its color is not sufficiently strong so that it can serve as its own indicator. Despite these disadvantages, however, $K_2Cr_2O_7$ is preferred to $KMnO_4$ in certain applications, especially in organic analysis, since dichromate can be boiled without decomposition and, therefore, can be used for the oxidation of organic compounds, which usually require prolonged heating at high temperature.

Example 12.12. A 5.00-mL sample of cognac is diluted with water to 500.0 mL (solution A). A 10.00-mL aliquot of solution A is distilled and the ethanol is collected into 50.00 mL of acidic (H_2SO_4) 0.1000 $\underline{N}$ $K_2Cr_2O_7$ solution, where it is oxidized to acetic acid. The excess of unreacted $K_2Cr_2O_7$ is titrated with 16.24 mL of 0.1006 $\underline{N}$ Fe(II) solution. Calculate the percentage (w/v) of ethanol in the sample.

Solution. The following reactions take place during the analysis:

$$2Cr_2O_7^{2-} + 3C_2H_5OH + 16H^+ \rightleftharpoons 4Cr^{3+} + 3CH_3COOH + 11H_2O$$

$$Cr_2O_7^{2-} + 6Fe^{2+} + 14H^+ \rightleftharpoons 2Cr^{3+} + 6Fe^{3+} + 7H_2O$$

Hence, the percentage of ethanol in the cognac sample is equal to

$$\% \ C_2H_5OH \ (w/v) =$$

$$\frac{(50.00 \times 0.1000 - 16.24 \times 0.1006) \text{ meq } (0.04607/4) \text{ g } C_2H_5OH/\text{meq}}{5.00 \text{ mL } (10.00/500.0)} \times 100 = \mathbf{38.77}.$$

Example 12.13. How many milliliters of solution A, containing 6.129 g of $K_2Cr_2O_7$/L, should be added to 1000 mL of solution B, containing 86.284 g of $FeSO_4 \cdot (NH_4)_2SO_4 \cdot 6H_2O$/L, so that the resulting solution C will have the same normality of remaining Fe(II) as the initial $K_2Cr_2O_7$ solution?

Solution. The normalities of solutions A and B are equal to

$$\underline{N}_A = \underline{N}_{K_2Cr_2O_7} = \frac{6.129 \text{ g } K_2Cr_2O_7/L}{(294.19/6) \text{ g/eq}} = 0.1250 \text{ eq/L}$$

$$\underline{N}_B = \underline{N}_{Fe(II)} = \frac{86.284 \text{ g } FeSO_4 \cdot (NH_4)_2SO_4 \cdot 6H_2O/L}{392.14 \text{ g/eq}} = 0.2200 \text{ eq/L}$$

Suppose that y mL of solution A should be added to 1000 mL of solution B, whereupon part of Fe(II) is oxidized to Fe(III) and (1000 + y) mL of solution C are produced [containing 0.1250 $\underline{N}$ Fe(II)]. Hence, the milliequivalents of remaining Fe(II) in solution C is equal to the milliequivalents of Fe(II) in solution B, less the milliequivalents of Cr(VI) reacted from solution A, i.e.,

$$(1000 + y) \text{ mL } (0.1250 \text{ meq/mL}) = (1000 \times 0.2200 - 0.1250y) \text{ meq}$$

or

$$y = \mathbf{380.0} \text{ mL of solution A.}$$

Example 12.14. Fe(III) in 25.00 mL of solution A, containing 38.577 g of $(NH_4)_2SO_4 \cdot Fe_2(SO_4)_3 \cdot yH_2O$ (ferric alum)/L, is reduced to Fe(II), and then is titrated with 25.03 mL of $K_2Cr_2O_7$ solution, containing 3.9176 g of $K_2Cr_2O_7$/L. Calculate the value of y.

Solution. *1st method.* The normality of the $K_2Cr_2O_7$ solution is equal to

$$\underline{N}_{K_2Cr_2O_7} = \frac{3.9176 \text{ g } K_2Cr_2O_7/L}{(294.19/6) \text{ g/eq}} = 0.0799 \text{ eq/L}.$$

The per cent of $(NH_4)_2SO_4 \cdot Fe_2(SO_4)_3$ in the alum is equal to

$$\frac{(25.03 \times 0.0799) \text{ meq } (0.53202/2 \text{ g } (NH_4)_2SO_4 \cdot Fe_2(SO_4)_3/\text{meq})(1000/25.00)}{38.577 \text{ g}}$$

$$\times 100 = 55.16\%.$$

Hence, the alum contains 100.00 − 55.16 = 44.84% H_2O, whereupon we have

$$\frac{18.015y}{18.015y + 532.02} \times 100 = 44.84$$

or

$$y = 24.008 \simeq \mathbf{24}.$$

2nd method. We have

$$(0.0799 \times 0.02503)\ \text{eq} = \frac{38.577\ \text{g}\ \times (25.00/1000)}{[(532.02 + 18.015\ y)/2]\ \text{g/eq}}$$

or

$$y = \mathbf{24}.$$

12.6 Titrations Involving Cerium(IV)

Standard solutions of Ce(IV) combine many of the advantages of standard $KMnO_4$ and $K_2Cr_2O_7$ solutions, and therefore, they are preferred in many applications, especially in micro analysis.

Example 12.15. What is the per cent purity of potassium ferrocyanide, $K_4[Fe(CN)_6]\cdot 3H_2O$, if a 1.6801-g sample requires 39.63 mL of Ce(IV) solution in H_2SO_4, given that 50.00 mL of an As(III) solution, containing 4.848 g of As_2O_3/L, are equivalent to 49.06 mL of the Ce(IV) solution?

Solution. The titration reaction is

$$[Ce(SO_4)_3]^{2-} + [Fe(CN)_6]^{4-} \rightleftharpoons [Ce(SO_4)_3]^{3-} + [Fe(CN)_6]^{3-}$$

The normality of the As(III) solution is equal to

$$\underline{N}_{As(III)} = \frac{4.848\ \text{g}\ As_2O_3/\text{L}}{(197.84/4)\ \text{g}\ As_2O_3/\text{eq}} = 0.0980,$$

whereas the normality of the Ce(IV) solution is equal to

$$\underline{N}_{Ce(IV)} = \frac{50.00\ \text{mL} \times 0.0980\ \text{meq/mL}}{49.06\ \text{mL}} = 0.0999.$$

Hence, the per cent of $K_4[Fe(CN)_6]_3 \cdot 3H_2O$ in the sample is equal to

$$\%\ K_4[Fe(CN)_6] \cdot 3H_2O =$$

$$\frac{39.63\ \text{mL} \times 0.0999\ \text{meq/mL} \times 422.41\ \text{mg}\ K_4[Fe(CN)_6] \cdot 3H_2O/\text{meq}}{1680.1\ \text{mg}}$$

$$\times 100 = \mathbf{99.5}.$$

12.7 Iodimetry-Iodometry

The standard reduction potential E^0 of the $[I_3]^-$-I^- couple is +0.5355 V. This value of E^0, about in the middle of the table of standard reduction potentials (Appendix B), allows the ***direct*** determination of reducing agents with a standard iodine solution (*iodimetry*), as well as the ***indirect*** determination of oxidizing agents by reacting them with an excess of I^- ions and titrating the liberated iodine, which is chemically equivalent to the oxidizing agent, with a standard solution of a reducing agent, usually $Na_2S_2O_3$ (*iodometry*), i.e.,

$$\underset{\text{(standard)}}{Red_1 + [I_3]^-} \rightleftharpoons Ox_1 + 3I^- \qquad \text{(iodimetry)} \tag{12.17}$$

$$\left.\begin{aligned} &\underset{\text{(excess)}}{Ox_2 + I^-} \rightleftharpoons Red_2 + [I_3]^- && (12.18) \\ &\underset{\text{(standard)}}{[I_3]^- + 2S_2O_3^{2-}} \rightleftharpoons 3I^- + S_4O_6^{2-} && (12.19) \end{aligned}\right\} \text{(iodometry)}$$

By adjusting the conditions, particularly pH, it is often possible to shift at will the direction of reactions involving the $[I_3]^-$-I^- couple so that they can be used in volumetric analysis (Example 10.18).

Example 12.16. A 0.5258-g sample, containing As_2O_3, As_2O_5 and inert materials, is dissolved in a NaOH solution, forming sodium arsenite and sodium arsenate. The pH of the solution is adjusted at about 8, and the As(III) is titrated to As(V) with 26.33 mL of 0.1002 N triiodide solution (Example 10.18). The solution is then strongly acidified, excess KI is added, and the liberated iodine is titrated with 39.68 mL of 0.1005 N $Na_2S_2O_3$ solution to determine the total As (as As(V)). Calculate the per cent As_2O_3, As_2O_5, and As in the sample.

Solution. The following reactions take place during the analysis:

$$H_3AsO_3 + [I_3]^- + H_2O \rightleftharpoons HAsO_4^{2-} + 4H^+ + 3I^-$$

$$H_3AsO_4 + 3I^- + 2H^+ \rightleftharpoons H_3AsO_3 + [I_3]^- + H_2O$$

$$[I_3]^- + 2S_2O_3^{2-} \rightleftharpoons 3I^- + S_4O_6^{2-}$$

The per cent As_2O_3, As_2O_5, and As in the sample is equal to

$$As_2O_3 = \frac{26.33 \text{ mL} \times 0.1002 \text{ meq/mL } (197.84/4) \text{ mg } As_2O_3/\text{meq}}{525.8 \text{ mg}}$$

$$\times 100 = \mathbf{24.82},$$

$$As_2O_5 = \frac{(39.68 \times 0.1005 - 26.33 \times 0.1002) \text{ meq } (229.84/4) \text{ mg } As_2O_5/\text{meq}}{525.8 \text{ mg}}$$

$$\times 100 = \mathbf{14.75},$$

$$\%As = \frac{39.68 \text{ mL} \times 0.1005 \text{ meq/mL } (74.92/2) \text{ mg As/meq}}{525.8 \text{ mg}}$$

$$\times 100 = \mathbf{28.41}.$$

Example 12.17. A 0.5674-g sample of bleaching powder, containing Ca(OCl)Cl, is dissolved in water, the solution is acidified, excess KI is added, and the liberated iodine is titrated with 36.26 mL of 0.1052 N $Na_2S_2O_3$ solution. Calculate the percentage of active chlorine in the sample.

Solution. The following reactions take place during the analysis:

$$HOCl + 3I^- + H^+ \rightleftharpoons [I_3]^- + Cl^- + H_2O$$

$$[I_3]^- + 2S_2O_3^{2-} \rightleftharpoons 3I^- + S_4O_6^{2-}$$

Hence, the percentage of active chlorine in the sample is equal to

$$\%Cl = \frac{36.26 \text{ mL} \times 0.1052 \text{ meq/mL } (35.45/2) \text{ mg Cl/meq}}{567.4 \text{ mg}} \times 100 = \mathbf{11.92}.$$

Example 12.18. A 0.2004-g sample, containing ethylmercaptan and propylmercaptan, is treated with 50.00 mL of 0.1002 N iodine solution. The excess of unreacted iodine is titrated with 20.00 mL of 0.1005 N $Na_2S_2O_3$ solution. Calculate the per cent of each mercaptan in the sample.

Solution. The following reactions take place during the analysis:

$$2C_2H_5SH + [I_3]^- \rightleftharpoons C_2H_5SSC_2H_5 + 2H^+ + 3I^-$$

$$2C_3H_7SH + [I_3]^- \rightleftharpoons C_3H_7SSC_3H_7 + 2H^+ + 3I^-$$

$$[I_3]^- + 2S_2O_3^{2-} \rightleftharpoons 3I^- + S_4O_6^{2-}$$

Suppose that the sample contains y meq of C_2H_5SH and z meq of C_3H_7SH. We have

$$y + z = (50.00 \text{ mL} \times 0.1002 \text{ meq/mL}) - (20.00 \text{ mL} \times 0.1005 \text{ meq/mL}) = 3.000 \quad (12.20)$$

$$62.13y + 76.16z = 200.4 \quad (12.21)$$

Solving the system of Equations (12.20) and (12.21), we find that y = 2.000 and z = 1.000. Hence, the sample contains

$$\frac{2.000 \text{ meq} \times 62.13 \text{ mg } C_2H_5SH/\text{meq}}{200.4 \text{ mg}} \times 100 = \mathbf{62.0\%}\ C_2H_5SH$$

and

$$\frac{1.000 \text{ meq} \times 76.16 \text{ mg } C_3H_7SH/\text{meq}}{200.4 \text{ mg}} \times 100 = \mathbf{38.0\%}\ C_3H_7SH.$$

12.8 Titrations Involving Periodate

A characteristic property of periodate is the *selective* oxidative cleavage of C–C bonds in compounds having hydroxyl or carbonyl or amine groups in vicinal carbon atoms. The simple stoichiometry of these reactions, which are known as the Malarpade reactions, allows their use for analytical purposes. Since most Malarpade reactions are slow, the back-titration technique is used in analysis.

Example 12.19. The determination of ethylene glycol, CH_2OHCH_2OH, with periodic acid is carried out as follows: Ethylene glycol is oxidized with a known excess of periodic acid to give formaldehyde, while periodate ions are reduced to iodate ions. The excess of unreacted periodic acid is reduced to iodic acid with a known excess of As(III), in a weakly alkaline solution (pH 7–9). The excess of unreacted As(III) is titrated with a standard iodine solution. Calculate the number of milligrams of ethylene glycol present in one milliliter of solution A, on the basis of the following experimental data: A 10.00-mL aliquot of solution A is diluted with water to 100.0 mL (solution B). To 10.00 mL of solution B are added 10.00 mL of H_5IO_6 and 25.00 mL of As(III) solution. Consumed volumes of 0.1005 $\underline{N}$ iodine solution: 40.32 mL for the titration of ethylene glycol (volume S), 20.32 mL for the blank (volume B).

Solution. The following reactions take place during the analysis:

$$\underset{}{CH_2OHCH_2OH} + \underset{\text{(excess)}}{IO_4^-} \xrightarrow{\text{pH } 1} 2CH_2O + IO_3^- + H_2O$$

$$\underset{\text{(unreacted)}}{IO_4^-} + \underset{\text{(excess)}}{H_3AsO_3} \xrightleftharpoons{\text{pH } 7\text{–}9} IO_3^- + HAsO_4^{2-} + 2H^+$$

$$\underset{\text{(unreacted)}}{H_3AsO_3} + [I_3]^- + H_2O \xrightleftharpoons{\text{pH } 7\text{–}8} HAsO_4^{2-} + 4H^+ + 3I^-$$

The content of the sample in mg ethylene glycol/mL solution A is equal to

$$\text{mg glycol/mL A} = (\text{S} - \text{B})\ \text{mL}\ (\underline{\text{N}}_{\text{I}_2}\ \text{meq/mL})(62.07/2)\ \text{mg}\ CH_2OHCH_2OH/\text{meq}$$

$$= (40.32 - 20.32)\ \text{mL}\ (0.1005\ \text{meq/mL})(62.07/2)\ \text{mg/meq} = \mathbf{62.4}.$$

Note. It is not necessary to know the concentrations of the H_5IO_6 and As(III) solutions.

12.9 Titrations Involving Bromate

Bromate ion is a strong oxidizing agent in acid solution, in which it is reduced to Br^- or Br_2. Standard $KBrO_3$ solutions, also containing KBr, are used for the determination of organic substances, by bromination, e.g., of phenol, which is converted to tribromophenol. Of special importance is the volumetric determination of cations forming insoluble salts with 8-hydroxyquinoline, e.g., Al^{3+}, Mg^{2+}.

Example 12.20. Ten tablets of aspirin (acetylsalicylic acid, $HOOCC_6H_4OCOCH_3$), weighing 6.255 g, are pulverized. A 0.1251-g sample of the powder is brominated by treatment with 20.00 mL of 0.0400 M $KBrO_3$ solution, which also contains 75 g KBr/L. After the bromination is complete, the solution is treated with excess potassium iodide and the liberated iodine is titrated with 14.12 mL of 0.1039 N $Na_2S_2O_3$ solution. How many grams of aspirin are there in each tablet, on the average?

Solution. The following reactions take place during the analysis:

$$BrO_3^- + 5Br^- + 6H^+ \rightleftharpoons 3Br_2 + 3H_2O$$

$$HOOCC_6H_4OCOCH_3 + 3Br_2 \rightleftharpoons HOOCC_6HBr_3OCOCH_3 + 3H^+ + 3Br^-$$

$$Br_2 + 3I^- \rightleftharpoons 2Br^- + [I_3]^-$$

$$[I_3]^- + 2S_2O_3^{2-} \rightleftharpoons 3I^- + S_4O_6^{2-}$$

In 0.1251 g of pulverized tablets, there is

$$[(20.00\ \text{mL})(0.0400\ \text{mmol/mL})(6\ \text{meq/mmol}) - (14.12\ \text{mL})(0.1039\ \text{meq/mL})]$$

$$\times(0.1802/6\ \text{g aspirin/meq}) = 0.1001\ \text{g aspirin.}$$

The average weight of each tablet is equal to (6.255/10) = 0.6255 g. Hence, on the average, each tablet contains

$$(0.6255/0.1251)(0.1001) = \mathbf{0.5005}\ \text{g of aspirin.}$$

Example 12.21. A 0.4858-g sample, containing $MgSO_4$, is dissolved and magnesium is precipitated as the 8-hydroxyquinolate. The precipitate is filtered, washed, dissolved in acid, and 50.00 mL of 0.0400 M $KBrO_3$ solution, also containing 75 g KBr/L, are added. After the bromination is complete, excess KI is added to the solution and the liberated iodine is titrated with 32.00 mL of 0.1000 N $Na_2S_2O_3$ solution. Calculate the per cent $MgSO_4$ in the sample.

Solution. The following reactions take place during the analysis:

$$Mg^{2+} + 2C_9H_6(OH)N \rightleftharpoons \mathbf{Mg(C_9H_6ON)_2} + 2H^+$$

$$\mathbf{Mg(C_9H_6ON)_2} + 2H^+ \rightleftharpoons Mg^{2+} + 2C_9H_6(OH)N$$

$$BrO_3^- + 5Br^- + 6H^+ \rightleftharpoons 3Br_2 + 3H_2O$$

$$C_9H_6(OH)N + 2Br_2 \rightleftharpoons C_9H_4Br_2(OH)N + 2H^+ + 2Br^-$$

$$Br_2 + 3I^- \rightleftharpoons 2Br^- + [I_3]^-$$

$$[I_3]^- + 2S_2O_3^{2-} \rightleftharpoons 3I^- + S_4O_6^{2-}$$

We have 1 $Mg^{2+} \equiv 2C_9H_6(OH)N \equiv 4Br_2 \equiv 8$ electrons. Hence, the equivalent weight of $MgSO_4$ is equal to $MgSO_4/8 = 120.37/8 = 15.046$ g/eq. Therefore, in 0.4858 g of sample there are

$$[(50.00 \text{ mL})(0.0400 \text{ mmol/mL})(6 \text{ meq/mmol}) - (32.00 \text{ mL})$$

$$(0.1000 \text{ meq/mL})] \times (0.015046 \text{ g } MgSO_4/\text{meq}) = 0.1324 \text{ g } MgSO_4.$$

Hence, the per cent of $MgSO_4$ in the sample is equal to

$$(0.1324/0.4858)100 = \mathbf{27.25}.$$

12.10 Problems

Titration curves

12.1. Show that for the titration reaction

$$Br_2 + 2Fe^{2+} \rightleftharpoons 2Br^- + 2Fe^{3+}$$

the potential of the system at the equivalence point, E_{eq}, is given by the relation

$$E_{eq} = \frac{E^0_{Fe^{3+},\ Fe^{2+}} + 2E^0_{Br_2,\ Br^-}}{3} - \frac{RT}{3F} \ln 2[Br^-]_{eq}$$

Titrations involving $KMnO_4$

12.2. In the potentiometric titration of $KMnO_4$ with a $FeSO_4$ solution in acid medium using a Pt-saturated calomel pair of electrodes, the pH of the solution at the point 50% to the equivalence point was 1.00. a) What is the emf of the cell, E_{cell}, at this point? b) What is the value of E_{cell}, when a 40% excess of $FeSO_4$ solution has been added?

12.3. A 50.00-mL portion of 0.140 $\underline{M}$ Sn^{2+} solution is titrated potentiometrically with 0.140 $\underline{M}$ $KMnO_4$ solution ($MnO_4^- \rightarrow Mn^{2+}$). Calculate the concentration of Mn^{2+}, MnO_4^-, Sn^{2+}, and Sn^{4+} ions and the potential of the system at the equivalence point. Assume $[H^+]_{eq} = 1.00$ $\underline{M}$.

12.4. In standardizing $KMnO_4$ solution A with $Na_2C_2O_4$, its normality was found to be 0.1005 $\underline{N}$ ($MnO_4^- \rightarrow Mn^{2+}$). A 0.8857-g sample of a manganese ore was dissolved and the solution was titrated with 35.87 mL of solution A (Volhard method: $2MnO_4^- + 3Mn^{2+} + 4\,OH^- \rightleftharpoons \mathbf{5MnO_2} + 2H_2O$). Calculate the per cent of manganese in the sample.

12.5. A 0.1914-g sample of calcium carbonate is dissolved in hydrochloric acid and the calcium is precipitated as CaC_2O_4. The precipitate is dissolved in dilute H_2SO_4, and the resulting solution is titrated with 36.50 mL of $KMnO_4$ solution, 35.57 mL of which are equivalent to 0.2383 g of $Na_2C_2O_4$. A blank determination required 0.08 mL $KMnO_4$. Calculate the per cent CaO in the sample.

12.6. A tungsten solution, containing 11.74 g of W/L, is reduced by an amalgam to solution A. A 25.00-mL aliquot of solution A requires 48.00 mL of 0.1000 $\underline{N}$ $KMnO_4$ solution to oxidize tungsten to an ion in which the oxidation number of the metal is +6. Calculate the oxidation number of tungsten in solution A.

12.7. A 0.2000-g sample of the oxide M_xO_y, containing 51.04% $\underline{M}$ (atomic weight of M = 50.94), is dissolved and reduced, and the solution (A) is oxidized with 40.00 mL of 0.1005 $\underline{N}$ $KMnO_4$ solution, whereupon element M is oxidized to an ion in which the oxidation number of M is +5. What is the oxidation number of M in solution A?

12.8. Solution A contains 5.084 g of a hydrate of an oxalate potassium salt/L. A 50.00-mL portion of solution A is titrated either with 40.00 mL of 0.1000 $\underline{N}$ $KMnO_4$ solution (as reducing agent), or with 25.00 mL of 0.1200 $\underline{N}$ NaOH solution (as acid). What is the formula of the oxalate salt?

12.9. By mistake, a standard $FeSO_4$ solution (A) was left exposed to atmospheric air. A 50.00-mL portion of solution A is titrated with 36.27 mL of 0.1006 $\underline{N}$ $KMnO_4$ solution. Then Fe^{3+} is reduced to Fe^{2+} and the resulting solution is titrated with 40.89 mL of the same $KMnO_4$ solution. What percentage of $FeSO_4$ was oxidized during its exposure to the air?

12.10. A 0.8700-g sample of manganese ore is dissolved and the manganese is titrated with 46.65 mL of a $KMnO_4$ solution ($2MnO_4^- + 3Mn^{2+} + 4\,OH^- \rightleftharpoons \mathbf{5MnO_2} + 2H_2O$), 42.68 mL of which are equivalent to 0.2860 g of $Na_2C_2O_4$. Calculate the per cent manganese in the sample.

12.11. An iron ore contains the iron in the forms FeO and Fe_2O_3. A 2.5000-g sample is dissolved in acid in a nitrogen atmosphere, and the solution is titrated with 29.08 mL of 0.1000 N $KMnO_4$ solution. Another 1.0000-g sample is dissolved in acid, the solution is treated with $SnCl_2$ to reduce Fe^{3+} to Fe^{2+} (the excess of Sn^{2+} is destroyed with $HgCl_2$) and titrated with 31.98 mL of the same $KMnO_4$ solution (Zimmermann-Reinhardt method). Calculate the per cent Fe, FeO, and Fe_2O_3 in the sample.

12.12. A 0.1604-g sample of arsenic ore is dissolved, and arsenic is precipitated as As_2S_3, which is treated with 50.00 mL of 0.2000 N $KMnO_4$ solution ($5As_2S_3 + 28MnO_4^- + 54H^+ \rightleftharpoons 10H_3AsO_4 + 15SO_4^{2-} + 28Mn^{2+} + 12H_2O$). The excess of unreacted $KMnO_4$ is titrated with 18.24 mL 0.1000 M $FeSO_4$ solution. Calculate the per cent arsenic in the sample.

12.13. A 1.0000-g sample of a meteorite is dissolved in acid in a nitrogen atmosphere in order to prevent atmospheric oxidation of Fe(II) during the dissolution process ($Fe_3O_4 + 8H^+ \rightleftharpoons 2Fe^{3+} + Fe^{2+} + 4H_2O$), and the resulting solution is titrated with 30.00 mL of 0.1250 N $KMnO_4$ solution. a) Calculate the per cent Fe_3O_4 in the meteorite. b) What volume of the $KMnO_4$ solution will be used if the total iron is determined by the Zimmermann-Reinhardt method, in which all iron is brought to the Fe(II) oxidation state and then is oxidized by $KMnO_4$ (Problems 12.11 and 12.48)?

12.14. The determination of hydrogen peroxide is carried out by titrating with a standard $KMnO_4$ solution ($5H_2O_2 + 2MnO_4^- + 6H^+ \rightleftharpoons 5O_2 + 2Mn^{2+} + 8H_2O$). a) Calculate the percentage of H_2O_2 (w/v) in a sample, on the basis of the following experimental data: A 10.00-mL portion of the sample is diluted to 100.0 mL (solution A). Five milliliters of 50% H_2SO_4 are added to 20.00 mL of solution A and the solution is titrated with 32.85 mL of $KMnO_4$ solution, 23.41 mL of which correspond to 0.1683 g of $Na_2C_2O_4$. A blank determination required 0.20 mL of $KMnO_4$. b) What is the volume of oxygen produced during the titration (STP)?

12.15. Solution A of $KHC_2O_4 \cdot H_2C_2O_4$ is 0.1035 N as an acid. Calculate the normality of a $KMnO_4$ solution, given that 34.12 mL of this solution were consumed for the titration of 25.00 mL of solution A.

12.16. A 0.1790-g sample, containing $Na_2C_2O_4$, $NaHC_2O_4$ and inert materials, was dissolved in water, the solution was acidified and titrated with 19.20 mL of 0.1250 N $KMnO_4$ solution. A sample twice the above amount required 12.00 mL of 0.0950 N NaOH solution to titrate $NaHC_2O_4$, in the presence of phenolphthalein as indicator. Calculate the per cent of $Na_2C_2O_4$ and $NaHC_2O_4$ in the sample.

12.17. A 0.1790-g sample, consisting of only $Na_2C_2O_4$ and $NaHC_2O_4$, is dissolved in water, the solution is acidified and titrated with 24.00 mL of 0.1250 N $KMnO_4$ solution. Calculate the per cent $Na_2C_2O_4$ in the sample.

12.18. A 0.4800-g sample, consisting of only BaC_2O_4 and CaC_2O_4, requires 49.50 mL of 0.1040 N $KMnO_4$ solution for their titration. Calculate the per cent of each salt in the sample.

12.19. A 0.2350-g sample of iron wire (primary standard substance) was dissolved in acid under reducing conditions and titrated with 42.82 mL of a $KMnO_4$ solution.

A blank determination required 0.15 mL of $KMnO_4$. a) What is the normality of the $KMnO_4$ solution? b) If the wire contained 2.00% Fe_2O_3, what is the real normality of the $KMnO_4$ solution, given that Fe_2O_3 dissolves to give Fe^{3+}, which reacts with metallic iron ($2Fe^{3+} + \mathbf{Fe} \rightleftharpoons 3Fe^{2+}$)?

Titrations involving $K_2Cr_2O_7$

12.20. A 0.5697-g sample of iron ore is dissolved, reduced (Fe^{2+}) and titrated with $K_2Cr_2O_7$ solution, each milliliter of which corresponds to 1.000% Fe in the ore. What is the normality and the molarity of the $K_2Cr_2O_7$ solution?

12.21. A 0.2157-g sample of 100.00% $K_2Cr_2O_7$ is dissolved in water, H_2SO_4 and excess KI are added, and the liberated iodine is titrated with 35.20 mL of 0.1250 $\underline{N}$ $Na_2S_2O_3$ solution. What is the equivalent weight of $K_2Cr_2O_7$?

12.22. A 0.2683-g sample of chromite is cofused with sodium peroxide. After the excess Na_2O_2 is destroyed, the fused residue is extracted with water, the resulting solution is acidified, and 50.00 mL of 0.1022 $\underline{N}$ Fe^{2+} are added. After the reduction of Cr(VI) is complete, the excess of unreacted Fe^{2+} is titrated with 18.04 mL of 0.0999 $\underline{N}$ $K_2Cr_2O_7$ solution. Calculate the per cent chromium in the sample.

Titrations involving Ce(IV)

12.23. A solution containing 2.50 mmol of Fe^{2+} is titrated potentiometrically with 0.1000 $\underline{M}$ Ce(IV) solution. Both solutions are 1.0 $\underline{M}$ in H_2SO_4 (assume that activities equal concentrations). Calculate the volume of added Ce(IV) solution, when the potential of the indicator electrode versus a standard hydrogen electrode is a) 0.65, b) 1.06, and c) 1.39 V.

12.24. A 50.00-mL portion of 0.1600 $\underline{M}$ Fe^{2+} solution is titrated with a standard Ce(IV) solution. Both solutions are 1.0 $\underline{M}$ in H_2SO_4 (assume that activities equal concentrations). Calculate the potential of the system versus a standard hydrogen electrode, after the addition of a) 3.20, b) 8.00, and c) 16.00 mmol of Ce(IV).

12.25. A 1.9168-g sample of a Ce(IV) compound of 99.0% purity is titrated with 30.00 mL of 0.1000 $\underline{N}$ Fe^{2+} solution. What is the equivalent weight of the Ce(IV) compound?

12.26. A 0.2150-g sample, containing sodium azide (NaN_3) and inert materials, requires 24.25 mL of 0.1064 $\underline{N}$ Ce(IV) solution for the oxidation of the N_3^- ion to molecular nitrogen. Calculate the per cent NaN_3 in the sample.

12.27. A 0.0708-g sample of a potassium compound is dissolved in water, the potassium is precipitated as $K_2Na[Co(NO_2)_6]$, the precipitate is dissolved in acid, and the nitrite ions are titrated with 40.00 mL of 0.1050 $\underline{N}$ Ce(IV) solution. Calculate the per cent potassium in the sample.

Iodimetry-Iodometry

12.28. A 50.00-mL portion of 0.1000 N Sn^{2+} solution is titrated potentiometrically with 0.1000 N $[I_3]^-$ solution. Calculate the potential of an inert indicator electrode, e.g., Pt, versus a standard hydrogen electrode, after the addition of a) 49.50, b) 50.00, and c) 50.50 mL of $[I_3]^-$ solution.

12.29. Samples of an arsenic compound weighing 0.5000 g are titrated with a standard iodine solution. What should be the normality of the iodine solution so that each mL of titrant represents 1.000% As_2O_3 in the sample?

12.30. A 0.9703-g sample of an insecticide, containing calcium arsenate, is dissolved in acid, excess KI is added, and the liberated iodine is titrated with 31.14 mL of $Na_2S_2O_3$ solution, each milliliter of which is equivalent to 1.678 mg of $K_2Cr_2O_7$. Calculate the per cent $Ca_3(AsO_4)_2$ in the sample.

12.31. A 0.5700-g sample, consisting of only KI and KBr, is treated with excess $K_2Cr_2O_7$ and acid. The liberated bromine and iodine are distilled into excess potassium iodide solution, and the total liberated iodine is titrated with 40.00 mL of 0.1000 N $Na_2S_2O_3$ solution. Calculate the per cent iodine in the sample.

12.32. In the iodometric determination of copper, 35.66 mL of 0.1004 N $Na_2S_2O_3$ solution were consumed for the titration of a sample containing 0.3407 g of the sample. Calculate the per cent copper in the sample.

12.33. In the iodometric titration of 0.2418 g of As_2O_3 (primary standard), 43.76 mL of an iodine solution were consumed. In the titration of a 0.2084-g sample of impure As_2O_3, 36.09 mL of the same iodine solution were required. What is the per cent purity of the As_2O_3 sample?

12.34. A 0.9400-g sample, containing $NaAsO_2$ and inert materials, is titrated with 29.38 mL of standard iodine solution, 35.45 mL of which are equivalent to 0.1767 g of As_2O_3. Calculate the per cent $NaAsO_2$ in the sample.

12.35. V mL of an aqueous hydrogen sulfide solution are titrated with 3V mL of 0.0200 M iodine solution ($[I_3]^- + H_2S \rightarrow 3I^- + S + 2H^+$). Calculate the H_2S content of the sample, in g/L.

12.36. A 25.00-mL portion of solution A, containing 5.3800 g of $K_2Cr_2O_7$/L, is acidified, excess KI is added, and the liberated iodine is titrated with 28.00 mL of standard sodium thiosulfate solution (solution B). If 12.214 g of solid $Na_2S_2O_3 \cdot 5H_2O$ were used to prepare 500.0 mL of solution B, what is the purity of this substance?

12.37. An excess of KI is added to 25.00 mL of acid XeO_3 solution, and the liberated iodine during the reduction of XeO_3 to Xe is titrated with 44.00 mL of 0.0341 M $Na_2S_2O_3$ solution. Calculate the molarity of the XeO_3 solution.

12.38. The molecular weight of an oxidizing agent is 122.55. A 0.2450-g sample of the oxidizing agent is treated with excess KI and the liberated iodine is titrated with 48.00 mL of 0.2500 N $Na_2S_2O_3$ solution. How many electrons are gained by each molecule of the oxidizing agent during this reaction?

Titrations involving periodate

12.39. A 25.00-mL portion of solution A, containing IO_4^- and IO_3^- ions, is buffered at pH 8, excess KI is added, whereupon IO_4^- is reduced to IO_3^-, and the liberated iodine is titrated with 40.00 mL of 0.1000 N $Na_2S_2O_3$ solution. Another 5.00-mL portion of solution A is made strongly acidic (pH $\simeq$ 1), excess KI is added, whereupon IO_4^- and IO_3^- ions are reduced to I_2, which is titrated with 47.00 mL of the $Na_2S_2O_3$ solution. Calculate the molarity of the IO_3^- ions in solution A.

12.40. A 0.4832-g sample, containing ethylene glycol, ethanolamine and water, was diluted with water to 100.0 mL (solution A). A 50.00-mL aliquot of solution A was titrated with 20.85 mL of 0.1007 N HCl solution. A 10.00-mL aliquot of solution A was mixed with 20.00 mL of about 0.075 M $NaIO_4$ solution. After the reaction was complete, the pH was adjusted to about 7 with $NaHCO_3$, excess KI was added, and the liberated iodine was titrated with 15.47 mL of 0.0983 N $Na_2S_2O_3$ solution. A blank determination required 30.74 mL of the $Na_2S_2O_3$ solution. Calculate the percentage composition of the sample.

Titrations involving $KBrO_3$

12.41. A 0.0550-g phenol sample was dissolved in water and the solution was treated with 20.00 mL of 0.0500 M $KBrO_3$ solution, which also contained 75 g KBr/L. After the bromination was complete, the solution was treated with excess KI and the liberated iodine was titrated with 27.15 mL of 0.1012 N $Na_2S_2O_3$ solution. Calculate the per cent phenol in the sample.

12.42. In titrations with potassium bromate, in which Br_2 is an intermediate product, it is possible to increase the reduction potential of bromine, $E_{Br_2,\,Br^-}$, by adding Hg^{2+} ions, which form very stable complexes with Br^- ions. If by adding Hg^{2+} ions, $E_{Br_2,\,Br^-}$ increased by 0.4732 V, calculate the decrease in $[Br^-]$.

12.43. Hydrazine is oxidized by bromine to gaseous nitrogen ($N_2H_4 + 2Br_2 \rightarrow N_2 + 4H^+ + 4Br^-$). The bromine is produced by reacting potassium bromate with excess of Br^- ions ($BrO_3^- + 5Br^- + 6H^+ \rightleftharpoons 3Br_2 + 3H_2O$). If 10.55 mL of 0.1008 M $KBrO_3$ are required to oxidize the hydrazine in a 25.00-mL sample, calculate the content of hydrazine in the sample, in g/L.

12.44. How many milliliters of a 0.0500 M $KBrO_3$ solution are needed to furnish sufficient bromine to react with the magnesium 8-hydroxyquinolate obtained from 0.0300 g of magnesium?

Miscellaneous redox titrations

12.45. Calculate the potential of the system at the equivalence point, E_{eq}, in the titration of Hg_2^{2+} ions with a standard Cl^- solution, as a function of the constants $E^0_{Hg_2Cl_2,\ Hg}$ and $K_{sp(Hg_2Cl_2)}$.

12.46. Each half-cell of a galvanic cell consists of a platinum electrode immersed in 100.0 mL of 0.100 $\underline{M}$ Ce^{3+}-0.100 $\underline{M}$ Ce^{4+} solution. Ten milliliters of reducing solution A are added to one half-cell, whereupon the emf of the cell changes from 0.0000 to +0.1083 V. Calculate the normality of solution A.

12.47. How many grams of solid $K_2C_2O_4 \cdot 2H_2O$ should be added to 200 mL of a 0.100 $\underline{M}$ solution of the salt $KHC_2O_4 \cdot H_2C_2O_4 \cdot 2H_2O$ so that the normality of the solution as a reducing agent is three times its normality as an acid (assume that the volume of the solution remains constant during the addition of $K_2C_2O_4 \cdot 2H_2O$)?

12.48. In the volumetric determination of iron by the Zimmermann-Reinhardt method, at first Fe(III) is reduced to Fe(II) with a small excess of Sn(II). Then the excess of unreacted Sn(II) is oxidized to Sn(IV) with a large excess of Hg(II), and finally Fe(II) is titrated with a standard solution of an oxidizing agent, e.g., $KMnO_4$. If we assume that only 2% excess Sn(II) is added, that added Hg(II) corresponds to a 0.01 $\underline{M}$ concentration, that after equilibration we have $[Cl^-] = 6$ $\underline{M}$, that the formal potentials of the half-reactions in 1 $\underline{M}$ HCl can be used for the real experimental conditions, i.e., for $[Cl^-] =$ 6 $\underline{M}$, and that the total iron concentration is 0.100 $\underline{M}$, calculate a) the concentrations of Fe^{3+} and Fe^{2+} ions after the addition of Sn(II), and b) the concentrations of $[SnCl_6]^{2-}$ and $[SnCl_4]^{2-}$ after the addition of $HgCl_2$, given the following formal potentials:

$$Fe^{3+} + e \rightleftharpoons Fe^{2+}\ (1\ \underline{M}\ HCl),\ E^{0\prime} = 0.700\ V$$

$$[SnCl_6]^{2-} + 2e \rightleftharpoons [SnCl_4]^{2-} + 2Cl^-,\ E^{0\prime} = 0.14\ V$$

$$2[HgCl_4]^{2-} + 2e \rightleftharpoons \mathbf{Hg_2Cl_2} + 6Cl^-,\ E^{0\prime} = 0.69\ V$$

12.49. A 0.1600-g sample of 100.00% KIO_3 is dissolved in water, acid and excess KI is added to the solution, and the liberated iodine is titrated with 40.72 mL of $Na_2S_2O_3$ solution. Calculate the normality of the $Na_2S_2O_3$ solution.

12.50. A 0.8976-g sample of chemically pure KCl, containing KI impurity, is treated with bromine in a weakly acidic solution to oxidize I^- to IO_3^- ions. After removal of excess bromine, HCl and excess KI are added, and the liberated iodine is titrated with 4.182 mL of sodium arsenite solution, containing 0.2563 g As_2O_3/L. Calculate the per cent KI in the sample.

Chapter 13

Precipitation Titrations

13.1 Introduction

Precipitation titrations represent the smallest group of the volumetric methods, and they are mainly used for the titration of halides with standard $AgNO_3$ solution. As in the case of the other categories of titrations, the most important characteristics of a precipitation titration can be summarized in the *titration curve* (usually pX (X = ion of titrand or titrant) or potential E as a function of the volume V of titrant). From this, conclusions can be drawn for the accuracy of a titration and the selection of the method of end point detection.

13.2 Titration Curves

The titration curves for precipitation titrations are similar to the titration curves for strong acid-strong base neutralization titrations, in which pH is substituted by pX and the ion product of water K_w by the solubility product K_{sp} of the substance being precipitated.

Example 13.1. Calculate the titration curve for the titration of 50.00 mL of 0.1000 M NaCl solution with standard 0.1000 M $AgNO_3$ solution.

Solution. 1. *Initially.* Before any $AgNO_3$ has been added, we have $[Cl^-] = 0.1000$ M, and $pCl = -\log(0.1000) =$ **1.00** (two significant figures suffice for the construction of the titration curve).

2. *Before the equivalence point.* When $AgNO_3$ is added, the following precipitation reaction takes place:

$$Cl^- + Ag^+ \rightleftharpoons AgCl \qquad (13.1)$$

Up to the equivalence point, $[Cl^-]$ is determined from the excess of Cl^- ions in the solution, which is the sum of the untitrated Cl^- ions and the Cl^- ions from the dissolution of AgCl ($[Cl^-]_{\text{from AgCl}} = [Ag^+] = K_{sp(AgCl)}/[Cl^-]$). For example, after 10.00 mL of $AgNO_3$ were added, we have

$$\begin{aligned}[Cl^-] &= \frac{\text{mmol } Cl^- \text{ taken} - \text{mmol } Ag^+ \text{ added}}{\text{total solution volume in mL}} + \frac{K_{sp(AgCl)}}{[Cl^-]} \\ &= \frac{(50.00 \text{ mL} \times 0.1000 \text{ mmol/mL}) - (10.00 \text{ mL} \times 0.1000 \text{ mmol/mL})}{60.00 \text{ mL}} \\ &+ \frac{1.8 \times 10^{-10}}{[Cl^-]} = 0.0667 + \frac{1.8 \times 10^{-10}}{[Cl^-]} \simeq 0.0667 \; \underline{M}.\end{aligned} \tag{13.2}$$

There is no need to solve the quadratic Equation (13.2), except for points very close to the equivalence point, because the second term on its right-hand side is neglected as too small compared to 0.0667. Hence, $pCl = -\log(0.0667) = \mathbf{1.18}$.

3. *At the equivalence point.* At the equivalence point, i.e., after the addition of 50.00 mL of $AgNO_3$ solution, we have

$$[Cl^-] = [Ag^+] = \sqrt{1.8 \times 10^{-10}} = 1.34 \times 10^{-5} \; \underline{M}, \text{ and } pCl = \mathbf{4.87}.$$

4. *After the equivalence point.* The excess of Ag^+ ions in the solution is calculated, which is the sum of the excess $AgNO_3$ added and the Ag^+ ions from the dissolution of AgCl ($[Ag^+]_{\text{from AgCl}} = [Cl^-] = K_{sp(AgCl)}/[Ag^+]$). For example, after 51.00 mL of $AgNO_3$ were added, we have

$$\begin{aligned}[Ag^+] &= \frac{\text{mmol } AgNO_3 \text{ beyond eq. point}}{\text{total solution volume in mL}} + \frac{K_{sp(AgCl)}}{[Ag^+]} \\ &= \frac{1.00 \text{ mL} \times 0.1000 \text{ mmol/mL}}{101.00 \text{ mL}} + \frac{1.8 \times 10^{-10}}{[Ag^+]} \simeq 9.90 \times 10^{-4} \; \underline{M}.\end{aligned} \tag{13.3}$$

As above, there is no need to solve the quadratic Equation (13.3). We have

$$[Cl^-] = 1.8 \times 10^{-10}/9.90 \times 10^{-4} = 1.8 \times 10^{-7} \; \underline{M}, \text{ and } pCl = \mathbf{6.74}.$$

The results of the calculations are summarized in Table (13.1), and the corresponding titration curve is given in Figure (13.1). The calculation of the titration curve can also be made using the Nernst equation (Equation 8.2 or 8.2a) to compute emf of cell [with a silver electrode (cathode) and a saturated calomel electrode (anode)] versus the volume V of titrant (Problem 13.3).

Titration curve for halide mixtures. In a solution containing a mixture of halides, e.g., I^-, Br^-, and Cl^-, in not too greatly differing concentrations, each ion can, in principle, be titrated successively by fractional precipitation with standard $AgNO_3$ solution.

Table 13.1. *Calculation of the titration curve for the titration of 50.00 mL 0.1000 M NaCl with 0.1000 M $AgNO_3$.*

$AgNO_3$, mL	$[Cl^-]$	pCl
0.00	0.1000	1.00
10.00	0.0667	1.18
20.00	0.0429	1.37
30.00	0.0250	1.60
40.00	0.0111	1.95
49.00	0.0010	3.00
49.90	1.00×10^{-4}	4.00
50.00	1.34×10^{-5}	4.87
50.10	1.8×10^{-6}	5.74
51.00	1.8×10^{-7}	6.74
60.00	2.0×10^{-8}	7.70
70.00	1.1×10^{-8}	7.96

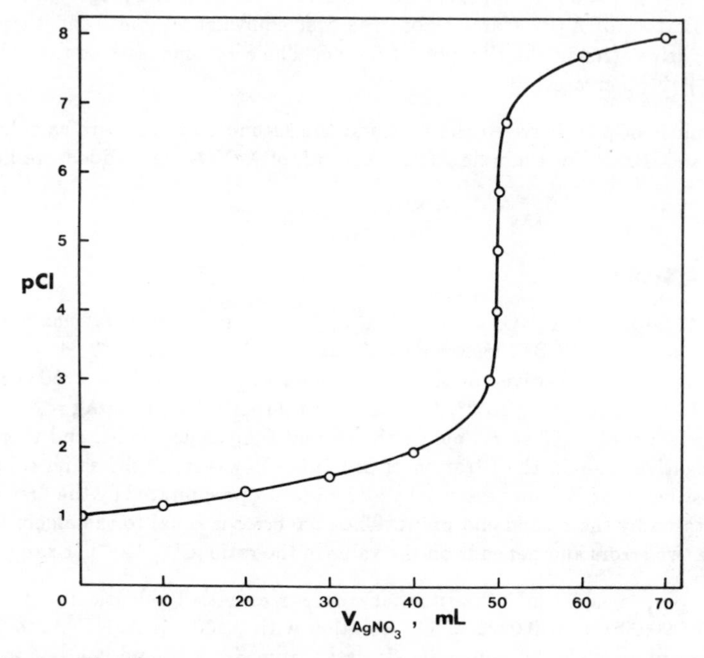

Figure 13.1. *Titration curve for the titration of 50.00 mL of 0.1000 M NaCl with 0.1000 M $AgNO_3$.*

Example 13.2. Calculate the titration curve for the titration of 50.00 mL of a solution which is 0.0400 $\underline{M}$ each in NaI, NaBr, and NaCl with a 0.1000 $\underline{M}$ $AgNO_3$ solution.

Solution. The solution contains 2.00 mmol each of NaI, NaBr, and NaCl. Up to the first end point, the values of pAg are calculated from $K_{sp(AgI)}$ and $[I^-]$. For example, after 0.01 mL $AgNO_3$ was added, AgI started to precipitate, because $[Ag^+][I^-] = [(0.01 \times 0.1000)/50.01](0.0400) > 8.5 \times 10^{-17} = K_{sp(AgI)}$. We have $[Ag^+] = K_{sp(AgI)}/[I^-] = 8.5 \times 10^{-17}/(1.999/50.01) = 2.13 \times \times 10^{-15}$ $\underline{M}$ and and pAg $= -\log(2.13 \times 10^{-15}) =$ **14.67**. This value can be essentially considered as the value of pAg at the start of the titration. After 5.00 mL $AgNO_3$ were added, we have

$$[Ag^+] = \frac{8.5 \times 10^{-17}}{(50.00 \times 0.0400 - 5.00 \times 0.1000)/55.00} = 3.12 \times 10^{-15} \ \underline{M},$$

and pAg = **14.51**. The other values of pAg up to the equivalence point are calculated similarly.

After 20.00 mL of $AgNO_3$ were added, i.e., at the first equivalence point, if the solution contained only I^- ions, we would have $[Ag^+]_{eq_1} = \sqrt{8.5 \times 10^{-17}} = 9.2 \times 10^{-9}$ $\underline{M}$, and pAg = 8.04. However, at this point we have $[Br^-] = (0.0400 \times 50.00)/70.00 = 2.86 \times 10^{-2}$, $[Ag^+] = 5 \times 10^{-13}/2.86 \times 10^{-2} = 1.75 \times 10^{-11}$ $\underline{M}$, and pAg = **10.76**. Hence, the precipitation of AgBr starts before the first equivalence point, and therefore, we have a negative error in the titration of iodide. The error increases as the value of the ratio $[Br^-]/[I^-]$ increases.

The values of pAg between the first and the second end point are calculated from $K_{sp(AgBr)}$ and $[Br^-]$. For example, after 25.00 mL of $AgNO_3$ were added, we have

$$[Ag^+] = \frac{5 \times 10^{-13}}{1.50/75.00} = 2.5 \times 10^{-11} \ \underline{M},$$

and pAg = **10.60**.

After 40.00 mL of $AgNO_3$ were added, i.e., at the second equivalence point, if the solution contained only Br^- ions, we would have $[Ag^+]_{eq_2} = \sqrt{5 \times 10^{-13}} = 7.07 \times 10^{-7}$ $\underline{M}$, and pAg = 6.15. However, at this point we have $[Cl^-] = (0.0400 \times 50.00)/90.00 = 2.22 \times 10^{-2}$ $\underline{M}$, $[Ag^+] = 1.8 \times 10^{-10}/2.22 \times 10^{-2} = 8.11 \times 10^{-9}$ $\underline{M}$, and pAg = **8.09**. Hence, the precipitation of AgCl starts before the second equivalence point, and therefore, we have a negative error in the titration of bromide. However, at the same time we also have a positive error, because essentially all iodide not precipitated by the first end point is precipitated by the second end point. The final error is equal to the algebraic sum of the above two errors and depends on the value of the ratio $[Cl^-]/[Br^-]$ (Example 13.12).

For $V_{AgNO_3} > 40.00$ mL, the titration curve is essentially identical to that for the titration of 90.00 mL of 0.0222 $\underline{M}$ Cl^- solution with 0.1000 $\underline{M}$ $AgNO_3$, and its points until the third equivalence point are calculated from $K_{sp(AgCl)}$ and $[Cl^-]$. For example, when 45.00 mL of $AgNO_3$ were added, we have

$$[Ag^+] = \frac{1.8 \times 10^{-10}}{1.50/95.00} = 1.14 \times 10^{-8} \ \underline{M},$$

Table 13.2. *Calculation of the titration curve for the titration of 50.00 mL of 0.4000 M NaI-0.0400 M NaBr-0.0400 M NaCl with 0.1000 M $AgNO_3$.*

$AgNO_3$, mL	$[Ag^+]$	pAg
0.01	2.13×10^{-15}	14.67
5.00	3.12×10^{-15}	14.51
10.00	5.10×10^{-15}	14.29
15.00	1.10×10^{-14}	13.96
20.00	1.75×10^{-11}	10.76
25.00	2.5×10^{-11}	10.60
30.00	4.0×10^{-11}	10.40
35.00	8.5×10^{-11}	10.07
40.00	8.11×10^{-9}	8.09
45.00	1.14×10^{-8}	7.94
50.00	1.8×10^{-8}	7.74
55.00	3.8×10^{-8}	7.42
60.00	1.34×10^{-5}	4.87
61.00	9.0×10^{-4}	3.05
65.00	4.35×10^{-3}	2.36
70.00	8.3×10^{-3}	2.08

and pAg = **8.09**.

After 60.00 mL of $AgNO_3$ were added, i.e., at the third equivalence point, we have $[Ag^+] = \sqrt{1.8 \times 10^{-10}} = 1.34 \times 10^{-5}$ M, and pAg = **4.87**. After the third equivalence point, pAg is calculated from the known excess of Ag^+ ions. For example, after 61.00 mL of $AgNO_3$ were added, we have $[Ag^+] = (1.00 \times 0.1000)/111.00 = 9.0 \times 10^{-4}$ M, and pAg = **3.05**. The results of the calculations are summarized in Table (13.2), and the corresponding titration curve is given in Figure (13.2).

Note. Because of formation of mixed salts between AgI and AgBr or between AgBr and AgCl, normally only mixtures of I^- and Cl^- are accurately titrated.

Example 13.3. A 50.00-mL portion of 0.1000 M NaX solution is titrated with 0.1000 M $AgNO_3$ solution. Calculate the required values of the equilibrium constant for the precipitation reaction and of $K_{sp(AgX)}$ so that when 49.90 mL of titrant have been added the reaction is essentially complete, and the pX increases by 3.00 units on the addition of 0.20 mL of additional titrant. The salt NaX is dissociated completely.

Solution. The titration reaction is

$$Ag^+ + X^- \rightleftharpoons \mathbf{AgX}, \ K = 1/K_{sp(AgX)} \tag{13.4}$$

Fifty milliliters (5.000 mmol) of $AgNO_3$ solution are required to reach the equivalence point. After 49.90 mL of titrant were added, $49.90 \times 0.1000 = 4.990$ mmol X^- have been precipitated. Therefore, 0.0100 mmol X^- remains in solution, and we have $[X^-]$

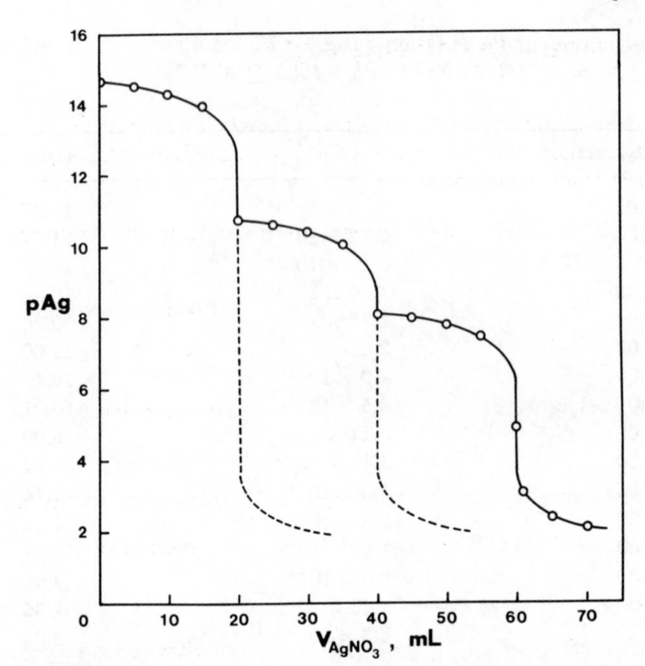

Figure 13.2. *Titration curve for the titration of 50.00 mL of a solution which is 0.0400 M each in NaI, NaBr, and NaCl with a 0.1000 M $AgNO_3$ solution. The solid line is the calculated curve for mixtures, whereas the dotted lines are the calculated curves for a solution containing only iodide or only bromide.*

= 0.0100/99.90 = 1.00×10^{-4} M and pX = 4.00. Hence, when 50.10 mL of titrant were added, we have $[Ag^+] = (0.10 \times 0.1000)/100.10 = 1.0 \times 10^{-4}$ M, pX = 4.00 + 3.00 = 7.00 and $[X^-] = 1.0 \times 10^{-7}$ M. Hence, according to Equation (13.4), we have

$$K = \frac{1}{[Ag^+][X^-]} = \frac{1}{(1.0 \times 10^{-4})(1.0 \times 10^{-7})} = \mathbf{1.0 \times 10^{11}}$$

and

$$K_{sp(AgX)} = 1/K = \mathbf{1.0 \times 10^{-11}}.$$

Example 13.4. How many milliliters of 0.100 M $AgNO_3$ solution should be added to 50.00 mL of 0.100 M NaBr solution so that the Br^- ion concentration is decreased to one tenth of the original concentration?

Solution. Suppose that y mL of $AgNO_3$ should be added, whereupon we have

$$[Br^-] = \frac{(50.00 \times 0.100) - 0.100y}{50.0 + y} + \frac{5 \times 10^{-13}}{0.0100} = 0.0100$$

or

$$y = \mathbf{40.9}\ \text{mL}.$$

13.3 Titrations Involving Ag(I)

The volumetric determination of Cl^- ions is performed either by direct titration with standard $AgNO_3$ solution (Mohr method and Fajans method), or indirectly by the back-titration technique (Volhard method).

The Mohr method is based on the fractional precipitation of Cl^- ions in the presence of chromate ions, which act as indicator:

$$Ag^+ + Cl^- \rightleftharpoons \underset{\text{(white)}}{\mathbf{AgCl}} \qquad \text{titration reaction} \tag{13.5}$$

$$2Ag^+ + CrO_4{}^{2-} \rightleftharpoons \underset{\text{(brick red)}}{\mathbf{Ag_2CrO_4}} \qquad \text{indicator reaction} \tag{13.6}$$

In the titration of chloride by the Volhard method, the following reactions take place:

$$\underset{\text{(excess)}}{Ag^+} + Cl- \rightleftharpoons \mathbf{AgCl} + \text{unreacted } Ag^+ \qquad \text{titration reaction} \tag{13.7}$$

$$SCN^- + Ag^+ \text{ (unreacted)} \rightleftharpoons \mathbf{AgSCN} \qquad \text{back-titration reaction} \tag{13.8}$$

$$SCN^- + Fe^{3+} \rightleftharpoons \underset{\text{(blood red)}}{[Fe(SCN)]^{2+}} \qquad \text{indicator reaction} \tag{13.9}$$

Example 13.5. In the "standardization" of a primary standard 0.1000 M $AgNO_3$ solution (solution A), 48.10 mL of this solution were consumed for the titration of a solution containing 0.2805 g of NaCl (primary standard substance). If 38.13 mL of solution A were consumed for the titration of a 0.4526-g sample, containing NaCl and $NaNO_3$, calculate the titration error and the per cent NaCl in the sample.

Solution. The theoretical volume of 0.1000 M $AgNO_3$ solution required to titrate 0.2805 g of NaCl is equal to

$$V = \frac{0.2805 \text{ g NaCl}}{(0.05844 \text{ g/mmol})(0.1000 \text{ mmol/mL})} = 48.00 \text{ mL}.$$

Hence, the titration error is equal to $48.10 - 48.00 = \mathbf{0.10}$ mL. Since the titration error is constant at 0.10 mL, then 0.10 mL of $AgNO_3$ should be subtracted in all titrations. Hence, the per cent of NaCl in the sample is equal to

$$\%\text{NaCl} = \frac{(38.13 - 0.10) \text{ mL} \times 0.1000 \text{ mmol/mL} \times 58.44 \text{ mg NaCl/mmol}}{452.6 \text{ mg}} \times 100 = \mathbf{49.10}.$$

Example 13.6. A NaCl solution is titrated with standard 0.1000 M $AgNO_3$ solution, in the presence of Na_3A as precipitation indicator, which forms the precipitate Ag_3A of different color from that of AgCl ($K_{sp(Ag_3A)} = 1 \times 10^{-22}$). What should the concentration of indicator A^{3-} ion be so that the precipitation of Ag_3A will start at the equivalence point?

Solution. At the equivalence point we have

$$[Ag^+] = \sqrt{K_{sp(AgCl)}} = \sqrt{1.8 \times 10^{-10}} = 1.34 \times 10^{-5}\ \underline{M}.$$

Hence, the concentration of indicator A^{3-} ion should be equal to

$$[A^{3-}] = K_{sp(Ag_3A)}/[Ag^+]^3 = 1 \times 10^{-22}/(1.34 \times 10^{-5})^3 = \mathbf{4 \times 10^{-8}}\ \underline{M}.$$

Example 13.7. A 0.7439-g sample containing chloride is dissolved in water, 50.00 mL of 0.1007 $\underline{M}$ $AgNO_3$ are added, and the excess of unreacted $AgNO_3$ is titrated with 7.86 mL of 0.0996 $\underline{M}$ KSCN solution. Calculate the per cent chloride in the sample.

Solution. We have

$$\%Cl^- = \frac{(\text{mmol } AgNO_3 - \text{mmol KSCN}) \times 35.45 \text{ mg } Cl^-/\text{mmol}}{743.9 \text{ mg}} \times 100$$

$$= \frac{(50.00 \times 0.1007 - 7.86 \times 0.0996) \text{ mmol} \times 35.45 \text{ mg } Cl^-/\text{mmol}}{743.9 \text{ mg}} \times 100$$

$$= \mathbf{20.26}.$$

Example 13.8. A 0.2502-g sample containing arsenic is treated with an oxidizing agent, and the arsenate formed is precipitated as Ag_3AsO_4. The precipitate is dissolved in acid and the Ag^+ ions are titrated with 26.25 mL of 0.1040 $\underline{M}$ KSCN solution. Calculate the per cent of arsenic in the sample.

Solution. The percentage of arsenic in the sample is equal to

$$\frac{(26.25 \times 0.1040) \text{ mmol KSCN} \times \dfrac{1 \text{ mmol Ag}}{\text{mmol KSCN}} \times \dfrac{1 \text{ mmol As}}{3 \text{ mmol Ag}} \times \dfrac{74.92 \text{ mg As}}{\text{mmol As}}}{250.2 \text{ mg}} \times 100$$

$$= \mathbf{27.25\%} \text{ As}.$$

Example 13.9. Calculate the volume of standard 0.1000 $\underline{M}$ $AgNO_3$ solution required to titrate y g of $C_6H_5NH_3^+Cl^-$ by the Mohr method.

Solution. The titration reaction is

$$C_6H_5NH_3^+Cl^- + Ag^+ \rightleftharpoons C_6H_5NH_3^+ + \mathbf{AgCl}$$

Hence, we have the relation

$$\frac{1000y \text{ mg } C_6H_5NH_3^+Cl^-}{129.59 \text{ mg/mmol}} = V \text{ ml} \times 0.1000 \text{ mmol/mL},$$

from which

$$V = \mathbf{77.17y} \text{ mL}.$$

13.4 Errors in Precipitation Titrations

Example 13.10. A 50.00-mL portion 0.1000 M KI solution is titrated with standard 0.1000 M $AgNO_3$ solution in the presence of 0.0200 M Na_2X as indicator. Within what range of values should the solubility product $K_{sp(AgX)}$ be so that the indicator Ag_2X precipitate will form within ±0.2% of the equivalence point?

Solution. At the equivalence point, the solution volume is 100.0 mL, $[Na_2X] = 0.0100$ M, and there are (50.00 mL × 0.1000 mmol/mL) = 5.00 mmol of I^- and, therefore, 5.00 mmol of silver. For a +0.2% error, we have

$$[Ag^+]_{exc} = \frac{0.2}{100} \times \frac{5.00 \text{ mmol}}{100.0 \text{ mL}} = 1.000 \times 10^{-4} \text{ M} = [Ag^+]_{en} - [I^-]_{en}$$
$$= [Ag^+]_{en} - \frac{8.5 \times 10^{-17}}{[Ag^+]_{en}}$$

or

$$[Ag^+]^2_{en} - 1.000 \times 10^{-4}\,[Ag^+]_{en} - 8.5 \times 10^{-17} = 0 \qquad (13.10)$$

Solving Equation (13.10), we find that $[Ag^+]_{en} = 1.00 \times 10^{-4}$ M. Hence,

$$K_{sp(Ag_2X)} = [Ag^+]^2_{en}[X^{2-}] = (1.000 \times 10^{-4})^2(0.0100) = 1.00 \times 10^{-10}.$$

For a −0.2% error, we have

$$[I^-]_{exc} = \frac{0.2}{100} \times \frac{5.00 \text{ mmol}}{100.0 \text{ mL}} = 1.000 \times 10^{-4} \text{ M} = [I^-]_{en} - [Ag^+]_{en} = \frac{8.5 \times 10^{-17}}{[Ag^+]_{en}} - [Ag^+]_{en}$$

or

$$[Ag^+]^2_{en} + 1.000 \times 10^{-4}\,[Ag^+]_{en} - 8.5 \times 10^{-17} = 0 \qquad (13.11)$$

Solving Equation (13.11), we find that $[Ag^+]_{en} = 8.5 \times 10^{-13}$ M. Hence,

$$K_{sp(Ag_2X)} = (8.5 \times 10^{-13})^2(0.0100) = 7.2 \times 10^{-27}.$$

Therefore, we should have the relation $\mathbf{7.2 \times 10^{-27}} \leq K_{sp(Ag_2X)} \leq \mathbf{1.00 \times 10^{-10}}$.

Example 13.11. What must $[Fe^{3+}]$ be so that in the determination of chloride by the Volhard method the titration error will be equal to zero (the formation constant of the complex $[Fe(SCN)]^{2+}$, K_f, is 138 (for ionic strength $\mu = 0.5$), and its detection limit is 6.4×10^{-6} M)?

Solution. For a zero titration error, the total moles of silver must be equal to the sum of the moles of chloride plus thiocyanate. If x represents the number of moles, we should have

$$x_{Ag^+} + x_{AgCl} + x_{AgSCN} = x_{Cl^-} + x_{SCN^-} + x_{AgCl} + x_{AgSCN} + x_{[Fe(SCN)]^{2+}}$$

or $x_{Ag^+} = x_{Cl^-} + x_{SCN^-} + x_{[Fe(SCN)]^{2+}}$ or, in terms of concentrations,

$$[Ag^+] = [Cl^-] + [SCN^-] + [Fe(SCN)^{2+}] \qquad (13.12)$$

We also have the relations

$$[Ag^+][Cl^-] = 1.8 \times 10^{-10} \quad (13.13)$$

$$[Ag^+][SCN^-] = 1 \times 10^{-12} \quad (13.14)$$

$$\frac{[Fe(SCN)^{2+}]}{[Fe^{3+}][SCN^-]} = 138 \quad (13.15)$$

Combining Equations (13.13) and (13.14), we have

$$[SCN^-] = [Cl^-]/180 \quad (13.16)$$

Substituting the values of $[Ag^+]$ and $[SCN^-]$ from Equations (13.13) and (13.16) in Equation (13.12), as well as the value of $[Fe(SCN^-]^{2+}$ at the detection limit, we have

$$\frac{1.8 \times 10^{-10}}{[Cl^-]} = [Cl^-] + \frac{[Cl^-]}{180} + 6.4 \times 10^{-6} \quad (13.17)$$

or

$$1.0056[Cl^-]^2 + 6.4 \times 10^{-6}[Cl^-] - 1.8 \times 10^{-10} = 0 \quad (13.17a)$$

Solving Equation (13.17*a*), we find that $[Cl^-] = 1.06 \times 10^{-5}$ $\underline{M}$. Therefore, from Equation (13.16) we have $[SCN^-] = 1.06 \times 10^{-5}/180 = 5.89 \times 10^{-8}$ $\underline{M}$, whereupon from Equation (13.15) we have

$$[Fe^{3+}] = 6.4 \times 10^{-6}/(5.89 \times 10^{-8} \times 138) = \mathbf{0.79}\ \underline{M}.$$

Example 13.12. What must be the value of the ratio C_{Cl^-}/C_{Br^-} so that in the titration with Ag^+ of a solution containing I^-, Br^-, and Cl^- ions, the titration error for bromide will be equal to zero?

Solution. In the second (Br^-) end point of the titration, in which Cl^- starts to coprecipitate with Br^- prior to the equivalence point (Example 13.2), we have two errors. One is a positive error, e_1, due to the cotitration with bromide of the I^- not precipitated at the first end point. And one is a negative error, e_2, due to the start of AgCl precipitation before the second equivalence point. We have

$$e_1\% = \frac{[I^-]_{en_1} - [I^-]_{en_2}}{C_{Br^-}} \times 100 \simeq \frac{[I^-]_{en_1}}{C_{Br^-}} \times 100$$

$$= \frac{\dfrac{K_{sp(AgI)}}{K_{sp(AgBr)}/C_{Br^-}}}{C_{Br^-}} \times 100 = \frac{K_{sp(AgI)}}{K_{sp(AgBr)}} \times 100$$

and

$$e_2\% = -\frac{[Br^-]_{en_2} - [Br^-]_{eq_2}}{C_{Br^-}} \times 100 \simeq -\frac{[Br^-]_{en_2}}{C_{Br^-}} \times 100$$

$$= -\frac{K_{sp(AgBr)}}{K_{sp(AgCl)}} \times \frac{C_{Cl^-}}{C_{Br^-}} \times 100.$$

The titration error for bromide is equal to the algebraic sum of e_1 and e_2. Hence, we have

$$e_{Br^-} = \left[\frac{K_{sp(AgI)}}{K_{sp(AgBr)}} - \frac{K_{sp(AgBr)}}{K_{sp(AgCl)}} \times \frac{C_{Cl^-}}{C_{Br^-}}\right] \times 100 = 0$$

or

$$C_{Cl^-}/C_{Br^-} = \mathbf{0.061}.$$

Note. For $C_{Cl^-}/C_{Br^-} < 0.061$, we have a positive error for the titration of bromide, whereas for $C_{Cl^-}/C_{Br^-} > 0.061$, we have a negative error.

13.5 Problems

Sample size

13.1. A series of NaCl-$NaNO_3$ samples given to students as unknowns for the determination of chloride contain 25–30% NaCl. What size (range of) sample must be taken for analysis, so that 30–45 mL of 0.1000 M $AgNO_3$ solution are consumed in the titration?

13.2. A sample given as unknown for the determination of chloride may contain KCl and/or $BaCl_2$. Within what range should the weight of sample taken for analysis fall so that 25.0–45.0 mL of 0.2000 M $AgNO_3$ are required for titration?

Titration curves

13.3. Construct the titration curve, emf of cell—with a silver electrode (cathode) and a saturated calomel electrode—as a function of the volume of titrant, in the titration of 50.00 mL of 0.0500 M NaCl solution with standard 0.1000 M $AgNO_3$ solution.

13.4. Fifty milliliter portions of a) 0.0400 M BaI_2, b) 0.1000 M BaI_2, c) 0.0500 M $CaCl_2$, d) 0.0750 M Na_3AsO_4, are titrated with standard 0.1000 M $AgNO_3$ solution. Calculate the concentration of the ion being titrated at the equivalence point, for each of the four cases (assume that As is present as AsO_4^{3-}).

Titration of mixtures

13.5. A 0.5000-g sample, containing $KClO_4$ and KCl, is dissolved in water, the ClO_4^- is reduced to Cl^-, and the solution is titrated with 32.00 mL of 0.1377 M $AgNO_3$ solution. Calculate the per cent KCl and $KClO_4$ in the sample.

13.6. A 0.3953-g sample, containing only NaCl and NaBr, is dissolved in water and titrated with 49.00 mL of 0.1000 M $AgNO_3$ solution. Calculate the per cent NaCl and NaBr in the sample.

13.7. How many grams of a sample, containing 60.0% NaCl and 40.0% KCl, should be taken for the determination of chloride by the Volhard method, so that when 50.00

mL of 0.1000 M $AgNO_3$ solution are added to the solution, the excess of unreacted Ag^+ will require 8.00 mL of 0.1250 M KSCN for titration?

Titrations involving Ag(I)

13.8. A 50.00-mL portion of 0.1000 M KIO_3 solution is titrated with 0.1000 M $AgNO_3$ solution in the presence of CrO_4^{2-} indicator. Within what range should the CrO_4^{2-} concentration fall so that the chromate forms a precipitate within ±0.1% of the equivalence point?

13.9. A 0.7900-g sample of disinfectant is dissolved in ethanol and the iodoform is decomposed with concentrated HNO_3 and 8.40 mL of 0.1690 M $AgNO_3$ solution, according to the reaction

$$CHI_3 + 3Ag^+ + H_2O \rightleftharpoons 3\mathbf{AgI} + 3H^+ + CO$$

After the reaction is complete, the excess of unreacted Ag^+ is titrated with 1.79 mL of 0.0950 M KSCN solution.Calculate the percent iodoform in the sample.

13.10. A NaCl solution is titrated with standard 0.1000 M $AgNO_3$ solution, in the presence of Na_3X as precipitation indicator, which forms the precipitate Ag_3X of different color from that of AgCl ($K_{sp(Ag_3X)} = 1.8 \times 10^{-18}$). What should $[X^{3-}]$ be so that the precipitation of Ag_3X will start at the equivalence point?

13.11. A 0.2142-g sample of the salt MCl_2 is dissolved in water and titrated with 36.00 mL of 0.1250 M $AgNO_3$ solution. Calculate the atomic weight of metal M.

13.12. Calculate the per cent of silver in an alloy from the following data: A 2.0000-g sample of the alloy is dissolved in HNO_3 and the solution is diluted with water to 100.0 mL (solution A). A 25.00-mL aliquot of solution A is titrated with 31.52 mL of KSCN solution, 40.00 mL of which are equivalent to 50.00 mL of 0.1000 M $AgNO_3$ solution.

13.13. How many milliliters of 0.2500 M $AgNO_3$ solution are required to titrate a) 6.000 meq of $FeCl_3$, b) 40.00 mL of 0.1000 M $K_2C_2O_4$ solution?

13.14. In the Volhard method for the determination of bromide, if W grams is the sample weight, V_A and M_A are the volume in milliliters and the molarity of the $AgNO_3$ solution used, and V_T and M_T are the volume in milliliters and the molarity of the KSCN solution used, derive a general equation giving the percentage of bromide as a function of the above parameters.

13.15. What must be the molarity of a $AgNO_3$ solution so that the volume in milliliters consumed in the titration of chloride will be equal to the percentage of chloride in the sample when using a 1.000-g sample?

13.16. What is the $[Cl^-]$ at the end point in determining chloride by the Mohr method, if the $[CrO_4^{2-}]$ at the end point is 0.0020 M?

13.17. A 0.4165-g bromide sample is dissolved in water, the solution is acidified with HNO_3 and 25.00 mL of 0.1013 M $AgNO_3$ solution are added. After the precipitation of

AgBr, a few drops of Fe^{3+} indicator solution are added and the solution is titrated with 20.12 mL of 0.0502 $\underline{M}$ KSCN solution. Calculate the per cent bromide in the sample.

13.18. In order to find the volume of a reservoir of irregular shape, a chemist added 108.2 g of NaCl in the reservoir and filled it with water. After the NaCl was dissolved, a 100.0-mL portion of the solution was titrated with 9.39 mL of 0.1008 $\underline{M}$ $AgNO_3$ solution. What is the volume of the reservoir?

13.19. Arsenic in a 7.150-g sample of a pesticide is converted to arsenate ions, which are precipitated with 25.00 mL of 0.05000 $\underline{M}$ $AgNO_3$ solution. The excess of unreacted Ag^+ is titrated with 3.85 mL of 0.05000 $\underline{M}$ KSCN solution. Calculate the per cent As in the sample.

Errors in precipitation titrations

13.20. The soluble salt NaX (MW 120) of an acid forms the insoluble salt AgX. In the titration of chloride with standard $AgNO_3$ solution, what is the maximum amount of NaX that can coexist with chloride, so that the titration error will be zero, given that the final volume is 100 mL ($K_{sp(AgX)} = 1.34 \times 10^{-8}$)?

13.21. A solution containing 4.00 mmol of NaA is titrated with standard $AgNO_3$ solution in the presence of CrO_4^{2-} as indicator. If we assume that at the end point $[CrO_4^{2-}] = 2.50 \times 10^{-3}$ $\underline{M}$ and the solution volume is 100 mL, within which limits must the solubility product $K_{sp(AgA)}$ fall so that the titration error will not exceed $\pm 0.05\%$?

13.22. What must be the value of the ratio C_{A^-}/C_{B^-}, so that in the titration of a solution, containing A^-, B^-, and C^- ions, the titration error for B^- ions will be zero? ($K_{sp(AgA)} = 1.0 \times 10^{-10}$, $K_{sp(AgB)} = 1.0 \times 10^{-13}$, $K_{sp(AgC)} = 5.0 \times 10^{-18}$).

13.23. A 50.00-mL portion of 0.1000 $\underline{M}$ $AgNO_3$ solution is titrated with 0.1000 $\underline{M}$ KSCN solution in the presence of Fe^{3+} as indicator (Volhard method). If $[Fe^{3+}]$ at the end point is equal to a) 1.0 $\underline{M}$, b) 0.1 $\underline{M}$, calculate the titration error in each case.

13.24. If 50.00 mL of 0.1000 $\underline{M}$ NaCl is titrated with 0.1000 $\underline{M}$ $AgNO_3$, what is the percent error if an indicator having $pCl = 4.20$ is used?

Miscellaneous precipitation titrations

13.25. Calculate pAg and $pCrO_4$ when 10.00 mL of 0.4375 $\underline{M}$ $AgNO_3$ solution is mixed with 24.00 mL of 0.0531 $\underline{M}$ K_2CrO_4 solution.

13.26. The determination of SO_2 in the air of a factory was carried out as follows: The air was bubbled at a rate of 20 L/min through a trap containing H_2O_2. The sulfates formed in 30 minutes ($SO_2 + H_2O_2 \rightleftharpoons 2H^+ + SO_4^{2+}$) were titrated with 5.62 mL of 0.01000 $\underline{M}$ $BaCl_2$ solution. Calculate the concentration of SO_2 in the air, in ppm (mL $SO_2/10^6$ mL air), given that the density of SO_2 is 2.86 mg/mL.

Chapter 14

Complexometric Titrations

14.1 Introduction

Complexometric titrations are based on the formation of complex ions and are widely used in quantitative analysis. It is possible to determine volumetrically most of the elements with a standard solution of only one complexing agent, EDTA (Example 7.6). Of limited use are the complexometric titrations of cyanide with Ag^+ ions and chloride with Hg^{2+} ions.

14.2 Titration of Cyanide with Ag(I)

Example 14.1. In the Liebig method, cyanide ions are titrated with standard $AgNO_3$ solution, whereupon the following reactions take place:

$$2CN^- + Ag^+ \rightleftharpoons [Ag(CN)_2]^- \quad \text{titration reaction} \tag{14.1}$$

$$[Ag(CN)_2]^- + Ag^+ \rightleftharpoons \mathbf{Ag[Ag(CN)_2]} \text{ (or } \mathbf{2AgCN}) \quad \text{indicator reaction} \tag{14.2}$$

A 0.4029-g sample of NaCN is dissolved in water and titrated with 20.25 mL of 0.1012 $\underline{M}$ $AgNO_3$ solution. Calculate the per cent NaCN in the sample.

Solution. The per cent of NaCN in the sample is equal to

$$\%\text{NaCN} = \frac{40.25 \text{ mL} \times 0.1012 \text{ mmol Ag/mL} \times 2 \text{ mmol NaCN/mmol Ag} \times 49.01 \text{ mg NaCN/mmol}}{402.9 \text{ mg}} \times 100 = \mathbf{99.1}.$$

14.3 Titration of Halides with Hg(II)

Anions such as Cl^-, Br^-, I^-, and SCN^-, which form slightly dissociated compounds with mercury, can be titrated with standard $Hg(NO_3)_2$ or $Hg(ClO_4)_2$ solution in the presence of a suitable indicator.

Example 14.2. The volumetric determination of chloride in biological fluids is carried out with standard Hg(II) solution, in the presence of diphenylcarbazide as indicator, whereupon the following reactions take place:

$$2Cl^- + Hg^{2+} \rightleftharpoons HgCl_2 \tag{14.3}$$

$$2[O{=}C(NHNHC_6H_5)_2] + Hg^{2+} \rightleftharpoons \left(\begin{array}{l} \quad\quad NHNHC_6H_5 \\ \quad / \\ O{=}C \\ \quad \backslash \\ \quad\quad NHNC_6H_5 \end{array} \right)_2 Hg + 2H^+ \tag{14.4}$$

purple complex

A 10.00-mL portion of a urine sample is titrated with 15.04 mL of $Hg(NO_3)_2$ solution. In the standardization of the $Hg(NO_3)_2$ solution, 35.67 mL were consumed to titrate a solution containing 0.2045 g of NaCl. Calculate the content of chloride in the sample, in mg Cl^-/mL.

Solution. The molarity of the $Hg(NO_3)_2$ solution is equal to

$$M_{Hg(NO_3)_2} = \frac{204.5 \text{ mg NaCl} \times \dfrac{1 \text{ mmol } Hg(NO_3)_2}{2 \text{ mmol NaCl}}}{58.44 \text{ mg/mmol NaCl} \times 35.67 \text{ mL}} = 0.04905 \text{ mmol/mL}.$$

Hence, we have

$$\frac{15.04 \text{ mL} \times 0.04905 \text{ mmol } Hg^{2+}/\text{mL} \times 2 \text{ mmol } Cl^-/\text{mmol } Hg^{2+} \times 35.45 \text{ mg } Cl^-/\text{mmol}}{10.00 \text{ mL}}$$

$= \mathbf{5.23}$ mg Cl^-/mL urine.

14.4 Complexometric Titrations with EDTA

The most widely used of the polyaminocarboxylic acids is ethylenediaminetetraacetic acid, $(HOOCCH_2)_2NCH_2CH_2N(CH_2COOH)_2$ (abbreviated as EDTA or H_4Y), and its disodium salt, $Na_2H_2Y \cdot 2H_2O$, having the following dissociation constants: $K_1 = 1.02 \times 10^{-2}$, $K_2 = 2.14 \times 10^{-3}$, $K_3 = 6.92 \times 10^{-7}$, $K_4 = 5.50 \times 10^{-11}$.

Table 14.1. *α_4 and $-\log\alpha_4$ for EDTA as a function of pH at 20° C.*

pH	α_4	$-\log\alpha_4$	pH	α_4	$-\log\alpha_4$
0.00	8.26×10^{-22}	21.08	7.00	4.81×10^{-4}	3.32
0.50	8.06×10^{-20}	19.09	7.50	1.66×10^{-3}	2.78
1.00	7.57×10^{-18}	17.12	8.00	5.38×10^{-3}	2.27
1.50	6.17×10^{-16}	15.21	8.50	1.71×10^{-2}	1.77
2.00	3.72×10^{-14}	13.43	9.00	5.21×10^{-2}	1.28
2.50	1.30×10^{-12}	11.89	9.50	0.148	0.83
3.00	2.51×10^{-11}	10.60	10.00	0.355	0.45
3.50	3.30×10^{-10}	9.48	10.50	0.633	0.20
4.00	3.61×10^{-9}	8.44	11.00	0.847	0.07
4.50	3.66×10^{-8}	7.44	11.50	0.943	0.03
5.00	3.55×10^{-7}	6.45	12.00	0.980	0.01
5.50	3.12×10^{-6}	5.51	12.50	0.990	0.00
6.00	2.25×10^{-5}	4.65	13.00	0.998	0.00
6.50	1.19×10^{-4}	3.92	14.00	1.000	0.00

The fraction, α_4, of EDTA present in the form of the tetravalent ion Y^{4-} is of special importance in calculations involving chemical equilibria and titration curves, since it is this form that is directly involved in complexation.

The magnitude of α_4 at various pH values is given in Table 14.1. These values were calculated on the basis of the relation

$$\alpha_4 = \frac{K_1K_2K_3K_4}{[H^+]^4 + K_1[H^+]^3 + K_1K_2[H^+]^2 + K_1K_2K_3[H^+] + K_1K_2K_3K_4} \tag{14.5}$$

$[Y^{4-}]$ is calculated from the relation

$$\alpha_4 = [Y^{4-}]/[Y'] \tag{14.6}$$

where Y′ is the total (analytical) concentration of EDTA, i.e.,

$$[Y'] = [H_4Y] + [H_3Y^-] + [H_2Y^{2-}] + [HY^{3-}] + [Y^{4-}] \tag{14.7}$$

In metal-EDTA complexation calculations, $[Y']$ represents the uncomplexed portion of EDTA in equilibrium with the complex.

Formation of M-EDTA complexes. The general reaction for the formation of a complex ion between the metal ion M^{n+} and EDTA is

$$M^{n+} + Y^{4-} \rightleftharpoons [MY]^{(n-4)+} \tag{14.8}$$

Table 14.2. *Formation constants of M-EDTA complexes.*

Metal ion	log K_{MY}	Metal ion	log K_{MY}
Co^{3+}	36	Cd^{2+}	16.46
Fe^{3+}	25.1	Co^{2+}	16.31
Th^{4+}	23.2	Al^{3+}	16.13
Cr^{3+}	23	Fe^{2+}	14.33
Bi^{3+}	22.8	Mn^{2+}	13.79
Sn^{2+}	22.1	Ca^{2+}	10.70
Hg^{2+}	21.80	Mg^{2+}	8.69
Cu^{2+}	18.80	Sr^{2+}	8.63
Ni^{2+}	18.62	Ba^{2+}	7.76
Pb^{2+}	18.04	Ag^{+}	7.3
Zn^{2+}	16.50	Li^{+}	2.8

The formation constant of the complex ion $[MY]^{(n-4)+}$ is given by the relation

$$K_{MY} = \frac{[MY^{(n-4)+}]}{[M^{n+}][Y^{4-}]} \tag{14.9}$$

Table 14.2 lists the formation constants of various M-EDTA complexes.

Most of the complexometric titrations are carried out in neutral or weakly acidic or weakly alkaline solutions, in which $\alpha_4 < 1$. We have

$$K_{MY} = \frac{[MY^{(n-4)+}]}{[M^{n+}][Y^{4-}]} = \frac{[MY^{(n-4)+}]}{[M^{n+}]\alpha_4[Y']} \tag{14.10}$$

or

$$K_{MY}\alpha_4 = \frac{[MY^{(n-4)+}]}{[M^{n+}][Y']} = K_{MY'} \tag{14.11}$$

where $K_{MY'}$ is a new formation constant, the *conditional formation constant*, the value of which depends on pH.

In complexometric titrations with EDTA, usually buffers are used to maintain the pH constant. In some cases, the buffer prevents the precipitation of metal ions as hydroxides, by masking them as soluble complexes, less stable than the corresponding complexes with EDTA. For example, an NH_3-NH_4Cl buffer precludes the precipitation of metal ions that form ammine complexes, such as Cd^{2+}, Cu^{2+}, Ni^{2+}, Zn^{2+}, Co^{2+}. However, such a complexation of these ions results in a decrease of the conditional formation constants of their EDTA complexes, i.e., makes formation of the EDTA complex more difficult. In these cases we define a new conditional formation constant, $K_{M'Y'}$, which also includes the term β_0 (Section 7.2). The value of $K_{M'Y'}$ now depends on the pH and the concentration of the complexing agent (ligand) of the buffer. Since $\beta_0 = [Mn^{n+}]/[M']$, where $[M']$ is the total (analytical) concentration of the metal ion in equilibrium with the metal-EDTA complex and not complexed with EDTA, Equation (14.10) becomes

$$K_{MY} = \frac{[MY^{(n-4)+}]}{[M^{n+}][Y^{4-}]} = \frac{[MY^{(n-4)+}]}{\beta_0[M']\alpha_4[Y']} \tag{14.12}$$

or

$$K_{MY}\alpha_4\beta_0 = \frac{[MY^{(n-4)+}]}{[M'][Y']} = K_{M'Y'} \tag{14.13}$$

Example 14.3. Calculate $[Ni^{2+}]$ in a solution of the complex $[NiY]^{2-}$ having an analytical concentration of 1.00×10^{-3} $\underline{M}$, at pH a) 4.00, b) 8.00.

Solution. a) At pH 4.00, we have $\alpha_4 = 3.61 \times 10^{-9}$ (Table 14.1). We have

$$K_{NiY} = 10^{18.62} = 4.2 \times 10^{-18}.$$

Hence,

$$K_{NiY'} = \alpha_4 K_{NiY} = (3.61 \times 10^{-9})(4.2 \times 10^{-18}) = 1.5 \times 10^{-10}.$$

Since the conditional formation constant $K_{NiY'}$, is very large, we assume that

$$[NiY^{2-}] = C_{NiY} - [Ni^{2+}] \simeq C_{NiY} = 1.00 \times 10^{-3}\ \underline{M}.$$

Since the complex $[NiY]^{2-}$ is the only source of Ni^{2+} ions and EDTA species, we have

$$[Ni^{2+}] = [Y^{4-}] + [HY^{3-}] + [H_2Y^{2-}] + [H_3Y^{-}] + [H_4Y] = [Y'].$$

Substituting the above values in the expression for the conditional formation constant, we have

$$\frac{1.00 \times 10^{-3}}{[Ni^{2+}]^2} = 1.5 \times 10^{10}$$

or

$$[Ni^{2+}] = \sqrt{1.00 \times 10^{-3}/1.5 \times 10^{10}} = \mathbf{2.6 \times 10^{-7}}\ \underline{M}.$$

Since $2.6 \times 10^{-7} << 1.00 \times 10^{-3}$, the assumption made is correct.

b) At pH 8.00, $\alpha_4 = 5.38 \times 10^{-3}$, whereupon $K_{NiY'} = (5.38 \times 10^{-3})(4.2 \times 10^{18}) = 2.3 \times 10^{16}$, and $[Ni^{2+}] = \sqrt{1.00 \times 10^{-3}/2.3 \times 10^{16}} = \mathbf{2.1 \times 10^{-10}}\ \underline{M}$.

Example 14.4. At pH 4.00, calculate a) $[Ni^{2+}]$ in a solution having an analytical nickel concentration (C_{Ni}) of 1.00×10^{-3} $\underline{M}$ and being 1.00×10^{-6} $\underline{M}$ in uncomplexed EDTA, b) $[Ca^{2+}]$ in a solution 1.00×10^{-3} $\underline{M}$ in total calcium (C_{Ca}) and 1.00×10^{-6} $\underline{M}$ in uncomplexed EDTA.

Solution. a) At pH 4.00, $K_{NiY'} = 1.5_2 \times 10^{10}$ (Example 14.3). Since the conditional formation constant is very large, we assume that

$$[NiY^{2-}] = C_{Ni} - [Ni^{2+}] \simeq C_{Ni} = 1.00 \times 10^{-3}\ \underline{M}.$$

Hence,

$$[Ni^{2+}] = \frac{[NiY^{2-}]}{[Y']K_{NiY'}} = \frac{1.00 \times 10^{-3}}{(1.00 \times 10^{-6})(1.5_2 \times 10^{10})} = \mathbf{6.6 \times 10^{-8}}\ \underline{M}.$$

Since $6.6\times10^{-8} << 1.00\times10^{-3}$, the assumption made is is correct. Hence, the percentage of Ni^{2+} that has *not* been complexed with EDTA is equal to $(6.6\times10^{-8}/1.00\times10^{-3})100 = 6.6\times10^{-3}\%$, i.e., is negligible.

b) $K_{CaY'} = \alpha_4 K_{CaY} = (3.61\times10^{-9})(5.0\times10^{10}) = 181$. We have $[CaY^{2-}] = 1.00\times 10^{-3} - [Ca^{2+}]$. Hence,

$$[Ca^{2+}] = \frac{[CaY^{2-}]}{[Y']K_{CaY'}} = \frac{1.00\times10^{-3} - [Ca^{2+}]}{(1.00\times10^{-6})181} \qquad (14.14)$$

Solving Equation (14.14), we find that $[Ca^{2+}] = \mathbf{9.998\times10^{-4}}$ $\underline{M}$. Hence, the percentage of Ca^{2+} that has been complexed is equal to

$$\frac{(1.00\times10^{-3}) - (9.998\times10^{-4})}{1.00\times10^{-3}}\times100 = 0.02\%$$

i.e., negligible.

Note. From the results, it can be seen that the selective titration of Ni^{2+} in the presence of Ca^{2+} is possible, under the conditions given in the example.

Example 14.5. Calculate the conditional formation constant of the complex $[NiY]^{2-}$, $K_{Ni'Y'}$, in a buffer containing 0.050 $\underline{M}$ NH_3 and 0.090 $\underline{M}$ NH_4Cl. The successive formation constants of the nickel ammines are: $K_1 = 10^{2.75}$, $K_2 = 10^{2.20}$, $K_3 = 10^{1.69}$, $K_4 = 10^{1.15}$, $K_5 = 10^{0.71}$, $K_6 = 10^{-0.01}$.

Solution. We have $[OH^-] = 1.8\times10^{-5}\times0.050/0.090 = 1.0\times10^{-5}$ $\underline{M}$. Hence, pOH $= 5.00$, pH $= 9.00$, and $\alpha_4 = 5.21\times10^{-2}$ (See Table 14.1). We have

$$\beta_0 = \frac{1}{1+10^{2.75}(0.05)+10^{4.95}(0.05)^2+10^{6.64}(0.05)^3+10^{7.79}(0.05)^4+10^{8.50}(0.05)^5+10^{8.49}(0.05)^6} = 7.8\times10^{-4}.$$

Hence,

$$K_{Ni'Y'} = \alpha_4\beta_0 K_{NiY} = (5.11\times10^{-2})(7.8\times10^{-4})(4.2\times10^{18}) = \mathbf{1.7\times10^{14}}.$$

Example 14.6. Calculate the minimum permissible pH for the complexometric titration of a) 1.00×10^{-3} $\underline{M}$ Ni^{2+}, b) 1.00×10^{-3} $\underline{M}$ Ca^{2+}, with 0.100 $\underline{M}$ EDTA solution, if we accept as a criterion 99.9% completion of the reaction with 0.1% excess of EDTA (0.1% error). (Assume the buffer contains no auxiliary complexing agents, i.e., that $\beta_0 = 1$.)

Solution. a) Since the concentration of EDTA is much larger than the concentration of Ni^{2+}, the dilution of the solution during the titration is not taken into account. Since $C_{Ni} = 1.00\times10^{-3}$ $\underline{M}$, we have $[Y'] = (1.00\times10^{-3}\times0.1)/100 = 1.00\times10^{-6}$ $\underline{M}$, and $[NiY^{2-}]/[Ni^{2+}] = 999$. Hence,

$$K_{NiY'} = \frac{[NiY^{2-}]}{[Ni^{2+}][Y']} = \frac{999}{1.0\times10^{-6}} = 1.0\times10^{9}$$

and

$$\alpha_4 = K_{NiY'}/K_{NiY} = 1.00 \times 10^9/4.2 \times 10^{18} = 2.4 \times 10^{-10},$$

whereupon

$$-\log \alpha_4 = -\log(2.4 \times 10^{-10}) = 9.62.$$

From Table 14.1, we find that

$$-\log \alpha_4 = 9.62 \text{ at pH} \simeq \mathbf{3.4}.$$

b) We have

$$\alpha_4 = K_{CaY'}/K_{CaY} = 1.00 \times 10^9/5.0 \times 10^{10} = 0.020,$$

and

$$-\log \alpha_4 = -\log(0.020) = 1.70.$$

From Table 14.1, we find that

$$-\log \alpha_4 = 1.70 \text{ at pH} \simeq \mathbf{8.6}.$$

Note. From the above it is concluded that 1) the larger the formation constant K_{MY}, the lower the minimum permissible pH in which the quantitative titration of a metal ion is still feasible (which allows the selective titration of certain ions in the presence of other ions by adjusting the pH), 2) the larger the concentration of the titrated ion, the lower the minimum permissible pH, and 3) the smaller the permissible error (the larger the desired accuracy), the higher the minimum permissible pH.

Titration curves with EDTA. Titration curves for complexometric titrations with EDTA, for given experimental conditions, are calculated on the basis of the conditional stability constant $K_{M'Y'}$ and the initial concentration of the ion being titrated, C_M^0. They are graphically presented by a plot of pM as a function of the EDTA volume or of the per cent titrated. Under usual experimental conditions, in which the concentrations of the metal ions are small compared to the concentrations of the buffer and the auxiliary complexing agents, the fractions α_4 and β_0 remain essentially constant throughout the titration. Therefore, $K_{M'Y'}$ is also considered constant.

General equations are derived below for calculating pM on the basis of the known K_{MY} and C_M^0, along with the calculated α_4 and β_0 (for a given set of conditions). Typical examples for calculating titration curves are given for the titration of Ca^{2+} and Ni^{2+} with EDTA (the dilution of the solution is considered negligible).

1. *Initially.* Before the addition of EDTA, we have $[M^{n+}] = \beta_0 C_M^0$. Hence,

$$pM = -\log[M^{n+}] = -\log \beta_0 - \log C_M^0 \tag{14.15}$$

2. *Before the equivalence point.* We have $[M'] = C_M^0(1 - X)$, where X is the fraction of the stoichiometric quantity of EDTA added. Hence,

$$[M^{n+}] = \beta_0 C_M^0(1 - X)$$

and

$$pM = -\log \beta_0 - \log C_M^0(1 - X) \tag{14.16}$$

i.e., before the equivalence point, pM is independent of α_4, but it depends on β_0.

3. *At the equivalence point.* From the equilibrium

$$[MY]^{(n-4)+} \rightleftharpoons M' + Y',$$

we have

$$[M'] = [Y'], \text{ and } [MY^{(n-4)+}] = C_M^0 - [M^{n+}] \simeq C_M^0.$$

Hence,

$$K_{M'Y'} = [MY^{(n-4)+}]/[M'][Y'] = C_M^0/[M']^2 \text{ or } [M'] = \sqrt{C_M^0/K_{M'Y'}}.$$

Hence,

$$[M^{n+}] = \beta_0\sqrt{C_M^0/K_{M'Y'}} = \sqrt{C_M^0\beta_0/K_{MY}\alpha_4}$$

and

$$pM = (\log K_{MY} + \log \alpha_4 - \log C_M^0 - \log \beta_0)/2, \tag{14.17}$$

i.e., at the equivalence point, pM depends on α_4 and β_0, but the dependence on β_0 is smaller than before the equivalence point.

4. *After the equivalence point.* We have

$$[Y'] = (X - 1)C_M^0$$

and

$$[MY^{n-4)+}] = C_M^0 - [M^{n+}] \simeq C_M^0.$$

Hence,

$$[M'] = C_M^0/[Y']K_{M'Y'} = C_M^0/(X - 1)C_M^0K_{M'Y'} = 1/(X - 1)K_{M'Y'}.$$

Therefore,

$$[M^{n+}] = \beta_0[M'] = \beta_0/(X - 1)K_{MY}\alpha_4\beta_0 = 1/(X - 1)K_{MY}\alpha_4$$

and

$$pM = \log(X - 1) + \log K_{MY} + \log \alpha_4, \tag{14.18}$$

i.e., after the equivalence point, pM is independent of β_0, whereas it depends on α_4.

Example 14.7. Calculate the titration curve for the titration of 50.00 mL of 1.00×10^{-3} M Ca^{2+} solution with 0.0500 M EDTA solution, the pH of which is maintained constant at 10.0 with an NH_3-NH_4Cl buffer.

Solution. We have $\log K_{CaY} = 10.70$ (Table 14.2), and $\log \alpha_4 = -0.45$ (Table 14.1). Since the concentration of EDTA is much larger than the Ca^{2+} concentration, the solution volume is considered constant during the titration. The Ca^{2+} ion is not complexed by NH_3, therefore the various points of the titration curve are calculated by using Equations (14.15), (14.16), (14.17), and (14.18), for $\beta_0 = 1$, i.e.,

$$\text{Initially}: \; pCa = -\log(0.00100) = \mathbf{3.00}.$$

Before the equivalence point: At points corresponding to various percentages of EDTA added, we have

- for 50% EDTA, $pCa = -\log C^0_{Ca}(1 - X) = -\log 0.00100(1 - 0.50) = \mathbf{3.30}$
- for 90% EDTA, $pCa = -\log 0.00100(1 - 0.90) = \mathbf{4.00}$
- for 99% EDTA, $pCa = -\log 0.00100(1 - 0.99) = \mathbf{5.00}$
- for 99.9% EDTA, $pCa = -\log 0.00100(1 - 0.999) = \mathbf{6.00}$.

At the equivalence point: $pM = (\log K_{CaY} + \log \alpha_4 - \log C^0_{Ca})/2 = (10.70 - 0.45 + 3.00)/2 = \mathbf{6.62}$.

After the equivalence point: We have

- for 100.1% EDTA, $pCa = \log(1 - X) + \log K_{CaY} + \log \alpha_4 = \log(1.001 - 1) + 10.70 - 0.45 = \mathbf{7.25}$
- for 101% EDTA, $pCa = \log(1.01 - 1) + 10.70 - 0.45 = \mathbf{8.25}$
- for 110% EDTA, $pCa = \log(1.10 - 1) + 10.70 - 0.45 = \mathbf{9.25}$
- for 200% EDTA, $pCa = \log(2.00 - 1) + 10.70 - 0.45 = \mathbf{10.25} = -pK_{CaY'}$.

The results of the calculations are summarized in Table (14.3), and the corresponding titration curve is given in Figure (14.1).

The various points of the titration curve can also be calculated as follows:

Before the equivalence point, $[Ca^{2+}]$ is calculated on the basis of % EDTA added. For example, for 90% EDTA we have $[Ca^{2+}] = (1 - 0.90)(1.00 \times 10^{-3}) = 1.00 \times 10^{-4}$ $\underline{M}$ and $pCa = \mathbf{4.00}$.

At the equivalence point, we have $[Ca^{2+}] = [Y']$. We have $K_{CaY'} = \alpha_4 K_{CaY} = 0.355(5.0 \times 10^{10}) = 1.8 \times 10^{10}$. Hence,

$$\frac{[CaY^{2-}]}{[Ca^{2+}][Y']} = 1.8 \times 10^{10} \simeq \frac{1.00 \times 10^{-3}}{[Ca^{2+}]^2}$$

or

$$[Ca^{2+}] = \sqrt{1.00 \times 10^{-3}/1.8 \times 10^{10}} = 2.3_6 \times 10^{-7} \text{ and } pCa = \mathbf{6.62}.$$

After the equivalence point, $[Ca^{2+}]$ is calculated by Equation (14.11). For example, for 110% EDTA added, we have

$$[Ca^{2+}] = \frac{[CaY^{2-}]}{[Y']K_{CaY'}} \simeq \frac{1.00 \times 10^{-3}}{(1.00 \times 10^{-4})(1.8 \times 10^{10})} = 5.6 \times 10^{-10} \ \underline{M}$$

and

$$pCa = \mathbf{9.25}.$$

Example 14.8. Calculate the titration curve for the titration of 50.00 mL of 1.00×10^{-3} $\underline{M}$ Ni^{2+} solution with 0.0500 $\underline{M}$ EDTA solution, the pH of which is maintained constant at 9.0 with a buffer containing 0.050 (free) NH_3 and 0.090 $\underline{M}$ NH_4Cl.

Table 14.3. *Calculation of the titration curve for the titration of 50.00 mL* 1.00×10^{-3} *M* Ca^{2+} *or* Ni^{2+} *with 0.0500 M EDTA.*

% EDTA	pCa	pNi
0	3.00	6.11
50	3.30	6.41
90	4.00	7.11
99	5.00	8.11
99.9	6.00	9.11
100	6.62	11.72
100.1	7.25	14.34
101	8.25	15.34
110	9.25	16.34
200	10.25	17.34

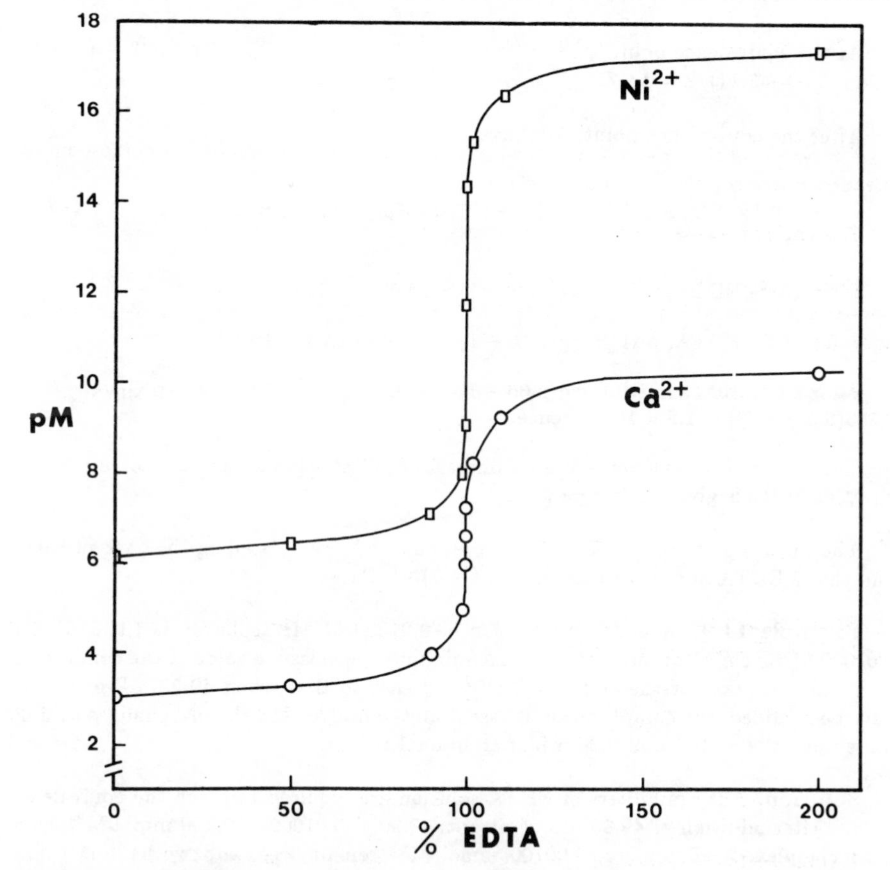

Figure 14.1. *Titration curve for the titration of 50.00 mL* 1.00×10^{-3} *M* Ca^{2+} *or* Ni^{2+} *with 0.0500 M EDTA.* $[NH_3] + [NH_4^+] = 0.14$ *M.*

Solution. We have $\log K_{NiY} = 18.62$ (Table 14.2), $\log \alpha_4 = -1.28$ (Table 14.1), $\beta_0 = 7.8 \times 10^{-4}$ (Example 14.5), and $\log \beta_0 = -3.11$. Working as in Example (14.7), we have

$$\text{Initially}: \ \text{pNi} = -\log \beta_0 - \log C^0_{Ni} = 3.11 - \log(1.00 \times 10^{-3}) = \mathbf{6.11}.$$

Before the equivalence point: We have

- for 50% EDTA, $\text{pNi} = -\log \beta_0 - \log C^0_{Ni}(1-X) = 3.11 - \log 0.00100(1-0.50) = \mathbf{6.41}$
- for 90% EDTA, $\text{pNi} = 3.11 - \log 0.00100(1-0.90) = \mathbf{7.11}$
- for 99% EDTA, $\text{pNi} = 3.11 - \log 0.00100(1-0.99) = \mathbf{8.11}$
- for 99.9% EDTA, $\text{pNi} = 3.11 - \log 0.00100(1-0.999) = \mathbf{9.11}$

At the equivalence point: $\text{pNi} = (\log K_{NiY} + \log \alpha_4 - \log C^0_{Ni} - \log \beta_0)/2 = (18.62 - 1.28 + 3.00 + 3.11)/2 = \mathbf{11.72}$.

After the equivalence point: We have

- for 100.1% EDTA, $\text{pNi} = \log(X-1) + \log K_{NiY} + \log \alpha_4 = \log(1.001-1) + 18.62 - 1.28 = \mathbf{14.34}$
- for 101% EDTA, $\text{pNi} = \log(1.01-1) + 18.62 - 1.28 = \mathbf{15.34}$
- for 110% EDTA, $\text{pNi} = \log(1.10-1) + 18.62 - 1.28 = \mathbf{16.34}$
- for 200% EDTA, $\text{pNi} = \log(2.00-1) + 18.62 - 1.28 = \mathbf{17.34} = -\text{p}K_{NiY'}$.

The results of the calculations are summarized in Table (14.3), and the corresponding titration curve is given in Figure (14.1).

The various points of the titration curve can also be calculated on the basis of $K_{Ni'Y'}$ and the % EDTA added, as in Example (14.7) for Ca^{2+}.

Example 14.9. A 50.00-mL portion of 0.01000 M M^{2+} solution is titrated with 0.01000 M EDTA solution. Calculate the minimum permissible value of the conditional formation constant of the complex $[MY]^{2-}$, $K_{MY'}$, so that when 49.90 mL of titrant have been added, the complexation is essentially complete, and the pM changes by 3.00 units upon the addition of 0.20 mL of additional titrant.

Solution. Fifty milliliters of EDTA solution are required to reach the equivalence point. After addition of 49.90 mL of titrant, $49.90 \times 0.01000 = 0.499$ mmol M^{2+} have been complexed. Therefore, 0.00100 mmol M^{2+} remain free, and we have $[M^{2+}] = 0.00100/99.90 = 1.00 \times 10^{-5}$ M and pM = 5.00. Hence, when 50.10 mL of titrant are added, we have pM = 5.00 + 3.00 = 8.00, and $[M^{2+}] = 1.0 \times 10^{-8}$ M, $[Y'] =$

$(0.10 \times 0.01000)/100.10 = 1.0 \times 10^{-5}$ $\underline{M}$, and $[MY^{2-}] = 0.500/100.10 = 5.00 \times 10^{-3}$ $\underline{M}$. Hence,

$$K_{MY'} = \frac{[MY^{2-}]}{[M^{2+}][Y']} = \frac{5.00 \times 10^{-3}}{(1.0 \times 10^{-8})(1.00 \times 10^{-5})} = \mathbf{5.0 \times 10^{10}}.$$

Applications of EDTA titrations. EDTA is widely used for the determination of water hardness. The end point is usually detected with metal indicators or potentiometrically (mostly with ion-selective electrodes) or spectrophotometrically. The determination of total water hardness is performed by titrating the sum of Ca^{2+} and Mg^{2+} ions with EDTA at pH 10, in the presence of Eriochrome Black T (EBT) as indicator. The determination of the separate Ca^{2+} and Mg^{2+} contents of water is carried out as follows: Calcium is determined in the presence of magnesium as insoluble $Mg(OH)_2$, at pH 12–13 with murexide or calcein or calcon as indicator, the sum of calcium and magnesium is determined at pH 10 with EBT indicator, and magnesium is calculated by difference. By proper adjustment of pH, in certain cases it is possible to titrate one ion selectively in the presence of other ions.

Example 14.10. A 0.2054-g sample of $CaCO_3$ (primary standard substance) is dissolved in hydrochloric acid and the solution is diluted with water to 250.0 mL (solution A). A 50.00-mL aliquot of solution A is titrated with 41.12 mL of EDTA solution. Calculate the molarity of the EDTA solution.

Solution. Since Ca^{2+} ions react with EDTA in a proportion of 1:1, we have

$$\underline{M}_{EDTA} = \frac{205.4 \text{ mg } CaCO_3 \times (50.00/250.00)}{100.09 \text{ mg/mmol} \times 41.12 \text{ mL}} = \mathbf{0.00998} \text{ mmol/mL}.$$

Example 14.11. A 100.0-mL sample of water containing Ca^{2+} and Mg^{2+} is titrated with 22.74 mL of the EDTA solution in Example (14.10), at pH 10.0. Another 100.0-mL sample is treated with NaOH to precipitate $Mg(OH)_2$, and then titrated at pH 13 with 15.86 mL of the same EDTA solution. Calculate a) the total water hardness in French degrees, F^0 (mg $CaCO_3$/100 mL), in German degrees, D^0 (mg CaO/100 mL), and in American degrees (mg $CaCO_3$/L, i.e., ppm $CaCO_3$), b) the ppm of $CaCO_3$ and $MgCO_3$ in the sample.

Solution. a) We have

$$22.74 \text{ mL} \times 0.00998 \text{ mmol/mL} \times 100.09 \text{ mg } CaCO_3\text{/mmol} = \mathbf{22.71}\ F^0$$

$$22.74 \text{ mL} \times 0.00998 \text{ mmol/mL} \times 56.08 \text{ mg CaCO/mmol} = \mathbf{12.73}\ D^0$$

$$(22.74 \times 0.00998) \text{ mmol/mL} \times 100.09 \text{ mg } CaCO_3\text{/mmol} \times (1000/100) =$$

$$\mathbf{227.1} \text{ ppm } CaCO_3 \text{ (Ca + Mg)}.$$

b) 15.86 mL of EDTA solution are consumed for the titration of Ca^{2+} and 22.74 − 15.86 = 6.88 mL for the titration of Mg^{2+}. Hence, we have

$$(15.86 \times 0.00998) \text{ mmol} \times 100.09 \text{ mg } CaCO_3/\text{mmol} \times (1000/100) = \mathbf{158.4} \text{ ppm}CaCO_3$$

$$(6.88 \times 0.00998) \text{ mmol} \times 84.31 \text{ mg } MgCO_3/\text{mmol} \times (1000/100) = \mathbf{57.9} \text{ ppm}MgCO_3.$$

14.5 Metal Ion Buffers

Metal ion buffers contain a complex ion and excess of ligand. They maintain their pM essentially unchanged, when small amounts of metal ion or ligand are added (analogous to maintaining pH constant in a buffer). For example, suppose that we have a $[CaY]^{2-}$-EDTA metal ion buffer, which is prepared by mixing Ca^{2+} and excess EDTA. The following equations are valid:

$$\frac{[CaY^{2-}]}{[Ca^{2+}][Y']} = K_{CaY'} \qquad (14.19)$$

$$pCa = \log K_{CaY'} - \log \frac{[CaY^{2-}]}{[Y']} \qquad (14.20)$$

Equation (14.20) can be compared to the Henderson-Hasselbalch Equation (5.82), which refers to the common acid-base buffers.

Example 14.12. A metal ion buffer is prepared by mixing 100 mL of 0.200 $\underline{M}$ Ca^{2+} solution and 800 mL of 0.100 $\underline{M}$ EDTA solution, adjusting the pH at 10.0 and diluting to exactly one liter. a) Calculate the pCa of the buffer. b) Calculate the change in pCa, when 1.00 mmol of Ca^{2+} is added to one liter of the buffer.

Solution. a) When the two solutions are mixed, Ca^{2+} ions react with EDTA, whereupon $100 \times 0.200 = 20.0$ mmol $[CaY]^{2-}$ are formed and $(800 \times 0.100) - 20.0 = 60.0$ mmol EDTA remain uncomplexed. At pH 10.0, we have $\log K_{CaY'} = \log K_{CaY} + \log \alpha_4 = 10.70 - 0.45 = 10.25$. Substituting these values in Equation (14.20), we have

$$pCa = 10.25 - \log \frac{20.0/1000}{60.0/1000} = \mathbf{10.73}.$$

b) When calcium is added, 1.00 mmol $[CaY]^{2-}$ is formed, and 60.0 − 1.00 = 59.0 mmol EDTA remain uncomplexed. Therefore, we have

$$pCa = 10.25 - \log \frac{21.0/1000}{59.0/1000} = 10.70.$$

Hence, when 1.00 mmol Ca^{2+} was added, the pM decreased by 10.73 − 10.70 = **0.03** units.

Note. In preparing a metal ion buffer, a metal complex should be selected whose $\log K_{M'Y'}$ is close to the desired pM.

14.6 Errors in Complexometric Titrations

In order to study the topic of titration errors in complexometric titrations, we will examine the volumetric determination of Mg^{2+} ions with standard EDTA solution in the presence of the metal indicator Eriochrome Black T (EBT), as a representative example. EBT is a triprotic acid (H_3In). The first ionization constant of the indicator refers to the ionization of the sulfonic group and is very large, and it is not taken into account in the study of the equilibria of EBT. In a solution of EBT, we have the following equilibria:

$$pK_2 = 6.3 \quad pK_3 = 11.5$$

$$\underset{\text{red}}{H_2In^-} \underset{}{\overset{-H^+}{\rightleftharpoons}} \underset{\text{blue}}{HIn^{2-}} \overset{-H^+}{\rightleftharpoons} \underset{\text{orange}}{In^{3-}} \tag{14.21}$$

The reaction of Mg^{2+} ions with EBT at pH 10 is represented by

$$\underset{\text{blue}}{HIn^{2-}} + Mg^{2+} \rightleftharpoons \underset{\text{red}}{[MgIn]^-} + H^+ \tag{14.22}$$

From Equations (14.21) and (14.22) it can be seen that the indicator color change is affected by pH. The conditional formation constant of the complex $[MgIn]^-$ is given by

$$K_{MgIn'} = \frac{[MgIn^-]}{[Mg^{2+}][In']} = K_{MgIn}\alpha_3 \tag{14.23}$$

where $[In']$ is the total concentration of the indicator not complexed with Mg^{2+}, i.e.,

$$[In'] = [H_3In] + [H_2In^-] + [HIn^{2-}] + [In^{3-}] \tag{14.24}$$

We have

$$[In^{3-}] = \alpha_3[In'] \tag{14.25}$$

and

$$\alpha_3 = \frac{K_1K_2K_3}{[H^+]^3 + K_1[H^+]^2 + K_1K_2[H^+] + K_1K_2K_3} \tag{14.26}$$

or, since $K_1 >> K_2$ or K_3,

$$\alpha_3 \simeq \frac{K_2K_3}{[H^+]^2 + K_2[H^+] + K_2K_3} \tag{14.27}$$

From Equation (14.23), we have

$$\log K_{MgIn'} = pMg + \log\frac{[MgIn^-]}{[In']} = \log K_{MgIn} + \log\alpha_3 \tag{14.28}$$

In order to avoid a titration error, the pMg at the equivalence point, pMg_{eq}, should fall within the pMg range in which the complex $[MgIn]^-$ is destroyed, because of the indicator reaction

$$\underset{\text{red}}{[MgIn]^-} + HY^{3-} \rightleftharpoons \underset{\text{blue}}{[MgY]^{2-}} + HIn^{2-} \tag{14.29}$$

(HY^{3-} is the main EDTA species at pH 10). Usually, we observe color changes for values of the ratio $[MgIn^-]/[In']$ between 10/1 and 1/10, i.e., between the pMg values given by

$$pMg = \log K_{MgIn'} - \log(10/1) = \log K_{MgIn'} - 1 \qquad (14.30)$$

$$pMg = \log K_{MgIn'} - \log(1/10) = \log K_{MgIn'} + 1 \qquad (14.31)$$

The range of pMg values in which the color change occurs (Equation 14.29), which is defined by Equations (14.30) and (14.31), is calculated from the formation constant K_{MgIn} (8.9×10^6) and the dissociation constants K_2 and K_3 of the indicator. For example, for pH 10, we calculate from Equation (14.27) that $\alpha_3 = 0.0307$, whereupon from Equation (14.28) we have that $\log K_{MgIn'} = 6.95 - 1.51 = 5.44$. Hence, the indicator color change occurs in the pMg range from about 4.4 to 6.4. (The values of $\log K_{MgIn'}$ for other pH values are calculated in the same manner.)

For known titration conditions, the titration error can be calculated from the values of the conditional formation constants $K_{MgY'}$ and $K_{MgIn'}$ (Example 14.13).

Example 14.13. Calculate the theoretical titration error for the titration of 1.00×10^{-3} M Mg^{2+} solution with 0.0500 M EDTA solution at pH 10.0, in the presence of Eriochrome Black T as indicator. Assume that at the end point the following percentage of the indicator has been converted from $[MgIn]^-$ to In′. a) 9.0%, b) 50%, c) 91% (neglect the dilution of the solution).

Solution. a) At pH 10.0, we have $\log K_{MgY'} = 8.69 - 0.45 = 8.24$ (Tables 14.1 and 14.2) and $\log K_{MgIn'} = 5.44$. From Equation (14.17) we find that at the equivalence point we have $pMg_{eq} = (8.69 - 0.45 + 3.00)/2 = 5.62$ (since $\beta_0 = 1$). From Equation (14.28) we find that for a 9% conversion of the indicator from $[MgIn]^-$ to In′, we have $pMg_{en} = 5.44 - \log(91/9.0) = 4.44 < 5.62 = pMg_{eq}$. Hence, we have a negative error, which is calculated as follows: The concentration $[Mg^{2+}]_{en}$ is equal to the sum of the concentration of the untitrated Mg^{2+}, $C^0_{Mg}(1 - X)$, and the concentration of the magnesium coming from the dissociation of the complex $[MgY]^{2-}$, which is equal to $[Y']$. We have

$$[Y'] = (C^0_{Mg} - [Mg^{2+}]_{en})/K_{MgY'}[Mg^{2+}]_{en} \simeq C^0_{Mg}/K_{MgY'}[Mg^{2+}].$$

Hence,

$$[Mg^{2+}]_{en} = C^0_{Mg}(1 - X) + \frac{C^0_{Mg}}{K_{MgY'}[Mg^{2+}]_{en}}$$

or

$$(1 - X) = \frac{[Mg^{2+}]_{en}}{C^0_{Mg}} - \frac{1}{K_{MgY'}[Mg^{2+}]_{en}} \qquad (14.32)$$

Equation (14.32) is a general equation for the calculation of negative titration errors. In the present case, we have

$$(1 - X) = \frac{10^{-4.44}}{1.00 \times 10^{-3}} - \frac{1}{10^{8.24} \cdot 10^{-4.44}} = 0.0361.$$

Hence, we have a negative error of **−3.61%**.

b) For a 50% conversion of the indicator, we have $pMg_{en} = 5.44 < 5.62 = pMg_{eq}$. Therefore, we have

$$(1 - X) = \frac{10^{-5.44}}{1.00 \times 10^{-3}} - \frac{1}{10^{8.24} \cdot 10^{-5.44}} = 0.0020.$$

Hence, we have a negative error of **−0.20%**.

c) For a 91% conversion of the indicator, we have $pMg_{en} = 5.44 - \log(9/91) = 6.44 > 5.62 = pMg_{eq}$. Hence, we have a positive error, which is calculated as follows: The concentration $[Y']$ is equal to the sum of the concentration of excess EDTA added, i.e., $C^0_{Mg}(X - 1)$, and the concentration of EDTA coming from the dissociation of the complex $[MgY]^{2-}$, which is equal to $[Mg^{2+}]_{en}$. Hence, we have

$$[Y'] = C^0_{Mg}(X - 1) + [Mg^{2+}]_{en}$$

or

$$\frac{C^0_{Mg}}{K_{MgY'}[Mg^{2+}]_{en}} = C^0_{Mg}(X - 1) + [Mg^{2+}]_{en}$$

or

$$(X - 1) = \frac{1}{K_{MgY'} \cdot [Mg^{2+}]_{en}} - \frac{[Mg^{2+}]_{en}}{C^0_{Mg}} \tag{14.33}$$

Equation (14.33) is a general equation for the calculation of positive titration errors. In the present case, we have

$$(X - 1) = \frac{1}{10^{8.24} \cdot 10^{-6.44}} - \frac{10^{-6.44}}{1.00 \times 10^{-3}} = 0.0155.$$

Hence, we have a positive error of **+1.55%**.

14.7 Problems

Titration of cyanide with Ag(I)

14.1. A 0.4565-g sample containing only NaCN and KCN is dissolved in water and titrated with 40.00 mL of 0.1000 M $AgNO_3$ solution (Liebig method). Calculate the per cent of NaCN and KCN in the sample.

14.2. A 0.4000-g sample, containing NaCN, NaCl, and inert materials, is dissolved in water and titrated with 20.00 mL of 0.1000 M $AgNO_3$ solution to a faint turbidity. Then 50.00 mL more of the $AgNO_3$ solution are added, the precipitates of $Ag[Ag(CN)_2]$ and AgCl are filtered and the excess of unreacted Ag^+ ions is titrated in the filtrate with 4.00 mL of 0.1250 M KSCN solution, in the presence of Fe^{3+} as indicator. Calculate the per cent of NaCN, NaCl, and inert materials in the sample.

Titer

14.3. Calculate the titer of 0.01000 $\underline{M}$ EDTA solution in mg $CaCO_3$/mL.

Formation of M-EDTA complexes

14.4. At pH 7.00, calculate the $[Zn^{2+}]$ in a solution having an analytical zinc concentration of 1.00×10^{-3} $\underline{M}$ and 1.00×10^{-6} $\underline{M}$ in uncomplexed EDTA.

14.5. Calculate the $[Zn^{2+}]$ in a solution of the complex $[ZnY]^{2-}$ having an analytical concentration of 1.00×10^{-3} $\underline{M}$, at pH a) 3.00, b) 7.00.

14.6. Calculate the minimum permissible pH for the complexometric titration of 1.00×10^{-3} $\underline{M}$ Zn^{2+} solution with 0.100 $\underline{M}$ EDTA solution, if we accept as a criterion 99.9% completion of the reaction with 0.1% excess of EDTA (0.1% error). (Assume the buffer contains no auxiliary complexing agents, i.e., that $\beta_0 = 1$.)

Formation constant

14.7. Calculate the minimum value for the formation constant K_{ML} (K_f) of the complex ML, so that the reaction of ML formation from solutions having $[M^{n+}] = [L^{n-}] = 0.010$ $\underline{M}$ can be used for the complexometric titration of M^{n+} with L^{n-}, that is, so that the titration reaction is completed by 99.9% at the equivalence point.

14.8. Calculate the conditional formation constant of the complex $[ZnY]^{2-}$, $K_{Zn'Y'}$, in a buffer containing 0.100 $\underline{M}$ NH_3, 0.180 $\underline{M}$ NH_4Cl. The successive formation constants of the zinc amines are: $K_1 = 10^{2.27}$, $K_2 = 10^{2.34}$, $K_3 = 10^{2.40}$, $K_4 = 10^{2.05}$.

14.9. A 25.00-mL portion of 0.01000 $\underline{M}$ M^{2+} solution is titrated with 0.01000 $\underline{M}$ EDTA solution. Calculate the minimum permissible value of the conditional formation constant of the complex $[MY]^{2-}$, $K_{MY'}$, so that when 24.95 mL of titrant have been added, the complexation reaction is complete, and the pM changed by 4.00 units on the addition of 0.10 mL of additional titrant.

Mixtures of solutions

14.10. The pH of a solution containing 0.0200 $\underline{M}$ Ba^{2+} and 0.0100 $\underline{M}$ Pb^{2+} is maintained at 7.00. Show that, when the solution is made 0.050 $\underline{M}$ in EDTA, it is possible to separate the two cations by precipitating the Ba^{2+} ions in the form of $BaSO_4$.

Titration curves

14.11. The formation constant of the complex [ML] is 2.0×10^8. Calculate the degree of completion of the titration reaction at the equivalence point, when a 0.0100 $\underline{M}$ M^{n+} solution is titrated with standard 0.0100 $\underline{M}$ L^{n-} solution.

14.12. Calculate the titration curve for the titration of 50.00 mL of 2.00×10^{-3} M Ca^{2+} solution with 0.100 M EDTA solution, the pH of which is maintained constant at 10.0 with an NH_3-NH_4Cl buffer.

14.13. Calculate the titration curve for the titration of 50.00 mL of 2.00×10^{-3} M Zn^{2+} solution with 0.100 M EDTA solution, the pH of which is maintained constant at 9.0 with a buffer containing 0.100 M (free) NH_3, 0.180 M NH_4Cl.

Selective titration

14.14. If a solution 0.0100 M in Ba^{2+} and 0.0100 M in Ni^{2+} is made 0.0300 M in EDTA, calculate the permissible pH range for the quantitative complexation of the Ni^{2+} ions (at least by 99.99%) with simultaneous complexation of the Ba^{2+} ions by less than 0.01%, so that the Ba^{2+} ions can react with another reagent without any interference from the Ni^{2+} ions.

Masking

14.15. A 50.00-mL portion of solution A, which contains Ca^{2+} and Zn^{2+} ions, requires 46.78 mL of 0.01046 M EDTA solution to titrate both metal ions. Another 50.00-mL portion of solution A is treated with KCN to mask the Zn^{2+} ions, and then the calcium is titrated with 26.39 mL ot the same EDTA solution. Calculate the concentration (M) of the Ca^{2+} and Zn^{2+} ions in solution A.

Applications of EDTA titrations

14.16. For the standardization of an EDTA solution (solution A), a 0.2500-g sample of $CaCO_3$ (primary standard substance) was dissolved in hydrochloric acid and diluted with water to 500.0 mL (solution B). A 25.00-mL portion of solution B was titrated with 31.25 mL of solution A. In the determination of the total hardness of water at pH 10.0, 30.00 mL of solution A were used for 25.00 mL of water. The calcium hardness was found equal to 180 ppm $CaCO_3$ by flame spectrophotometry. Calculate the total hardness of the water in French, German, and American degrees of hardness (Example 14.11), as well as the magnesium hardness expressed in ppm $CaCO_3$.

14.17. A 0.2184-g sample of a pharmaceutical preparation that contains MgO, $NaHCO_3$ and inert materials is dissolved in 20.00 mL of 0.5000 M HCl solution. The solution is boiled to expel CO_2, and the excess of the unreacted acid is titrated with 30.00 mL of 0.2000 M NaOH solution, in the presence of methyl orange as indicator. Another 0.1092-g sample is dissolved in excess HCl, the pH is adjusted at 10.0 with an NH_3-NH_4Cl buffer, and the solution is titrated with 25.00 mL of 0.02000 M EDTA solution, in the presence of Eriochrome Black T as indicator. Calculate the per cent of MgO, $NaHCO_3$, and inert materials in the sample.

14.18. For the determination of calcium in human serum by microtitration, 2 drops of 2 M NaOH solution are added to a 0.100-mL sample, and the calcium is titrated with 0.252 mL of 0.00130 M EDTA solution, in the presence of calcon as indicator, using a microburet. Calculate the calcium content in the serum sample, in mg Ca/100 mL and in meq Ca/L.

14.19. A 0.2574-g sample of limestone is dissolved in HCl and diluted with water to 100.0 mL (solution A). A 25.00-mL aliquot of solution A is titrated at pH 13 with 30.04 mL of an EDTA solution, having a titer of 1.600 mg $CaCO_3$/mL, in the presence of calcon as indicator. Another 25.00-mL aliquot of solution A is titrated at pH 10.0 with 32.75 mL of the same EDTA solution, in the presence of Eriochrome Black T as indicator. Calculate the per cent Ca and Mg in the sample.

14.20. A 0.7362-g sample, containing only $Ca(NO_3)_2 \cdot 4H_2O$, NaCl, and KCl, is dissolved in water and diluted to 100.0 mL (solution A) . A 25.00-mL aliquot of solution A is titrated with 25.00 mL of 0.01000 M EDTA solution. Another 25.00-mL aliquot of solution A is treated with 25.00 mL of 0.1000 M $AgNO_3$ solution, and the excess of unreacted Ag^+ ions is back-titrated with 4.00 mL of 0.1250 M KSCN solution. Calculate the percentage of each of the three components in the sample.

14.21. A 0.2767-g sample of an alloy containing principally iron and zinc is dissolved in acid and diluted with water to 500.0 mL (solution A). The pH of a 50.00-mL aliquot of solution A is adjusted to about 1, and the iron is titrated with 32.00 mL of 0.01250 M EDTA solution. The pH of the solution is then increased to 6 and the zinc is titrated with 6.40 mL of the same EDTA solution. Calculate the per cent Fe and Zn in the sample.

14.22. Fifty milliliters of 0.01000 M EDTA solution are added to a solution containing 0.0777 g of lead, and the excess of unreacted EDTA is back-titrated with V mL of 0.01210 M $MgSO_4$ solution. Calculate the value of V.

14.23. A 0.1455-g sample of sulfates was dissolved in water, the sulfates were precipitated as $BaSO_4$, the precipitate was dissolved in 25.00 mL of 0.02004 M EDTA solution and the excess of unreacted EDTA was titrated with 3.25 mL of $MgSO_4$ solution. A blank determination with the same volume of EDTA required 45.34 mL of $MgSO_4$ solution. Calculate a) the molarity of the $MgSO_4$ solution, b) the per cent of Na_2SO_4 in the sample.

Metal ion buffer

14.24. A metal ion buffer is prepared by mixing 100 mL of 0.100 M Ca^{2+} solution and 600 mL of 0.100 M EDTA solution, adjusting the pH at 10.0 and diluting to exactly one liter. a) Calculate the pCa of the buffer. b) Calculate the change in pCa, when 1.20 mmol of Ca^{2+} is added to one liter of the buffer.

Titration error

14.25. Calculate the theoretical error for the titration of 2.00×10^{-3} M Mg^{2+} solution with 0.1000 M EDTA solution of pH 10.0, in the presence of Eriochrome Black T as indicator. Assume that at the end point the following percentage of the indicator has been converted from $[MgIn]^-$ to In′: a) 9.0%, b) 50%, c) 91% (neglect the dilution of the solution).

Chapter 15

Potentiometry

15.1 Introduction

Potentiometry has to do with the measurement of the electromotive force (emf) of an electrochemical cell, E_{cell}. The cell consists of an indicator electrode, with a potential of E_{ind}, which is a function of the activity (concentration) of the ion being determined and the experimental conditions, plus a reference electrode, with a potential of E_{ref}, which is known, constant and independent of the chemical composition of the solution into which the two electrodes are immersed (Section 8.3).

The measurement of the emf of the cell is carried out with special potential-measuring instruments, known as *electrometers*, which draw a current of negligible (practically zero) intensity, so that the composition of the measured solution remains essentially unchanged. Special types of electrometers are *pH meters*, which are used for the measurement of pH, and *pIon meters*, which are used more generally for the measurement of activity of various ions.

The emf of the cell is given by

$$E_{cell} = E_{ind} - E_{ref} + E_j \tag{15.1}$$

where E_j is the liquid-junction potential due to the different rates of mobility of ions across a liquid junction between a salt bridge and the test solution, which we try to keep constant as much as possible during the measurements, so that it can be included in the constant term E_{ref}.

Potentiometric methods of analysis are classified as *direct potentiometric methods* (direct or absolute potentiometry), and *potentiometric titrations*.

15.2 Direct (Absolute) Potentiometry

Direct potentiometry is based on the linear relationship between the potential of the indicator electrode, E_{ind} or E, and consequently of E_{cell}, and the logarithm of the activity of the measured ion or the ratio of the activities of the electroactive ions, which are present at the surface of the electrode. This relationship is given by the Nernst equation (Equation 8.2 or 8.2*a*). As a rule, electrodes used as indicator electrodes are membrane electrodes, e.g., ion selective electrodes (ISE) and gas electrodes.

Ion-selective electrodes. The potential measured with an ISE is called *potential of the selective electrode*, but as a matter of fact, what is measured is the emf of the galvanic cell that consists of the ISE and the reference electrode and is given by the general equation

$$E = E_{const} + S \log a_A = E' + S \log[A] \tag{15.2}$$

where

E_{const} is a sum of constants that includes the standard potential of the indicator electrode, the potential of the reference electrode and the liquid-junction potential (in the case of a series of solutions with constant ionic strength, the logarithm of the activity coefficient of the measured ion can also be included in E_{const}; then instead of the activity a_A, the concentration [A] is used, and $E' = E_{const} + S \log f_A$),

S is a prelogarithmic term, usually referred to as the *slope coefficient* or Nernst coefficient,

a_A is the activity of ion A to which the selective electrode responds.

The slope coefficient S is theoretically equal to the term $2.303RT/z_aF$ (Equation 8.2*a*), where z_a is the charge of the ion, including the algebraic sign. For example, at 25°C, S = 0.05916 V for monovalent cations, 0.02958 V for divalent cations, and S = −0.05916 V for monovalent anions. In practice, the experimental value of S may differ from the theoretical value.

If in addition to the ion A being measured (primary ion) the solution contains other ions, the measured potential of the selective electrode may be affected, in accordance with the modified Nernst equation

$$E = E' + S \log \left(a_A + \sum_{i=1}^{N} k^{pot}_{A,\,B_i} a_{B_i}^{z_A/z_{B_i}} \right) \tag{15.3}$$

where

$a_{B_1}, a_{B_2}, \ldots, a_{B_N}$ are the activities of the interfering ions with corresponding charges of $z_{B_1}, z_{B_2}, \ldots, z_{B_N}$

$k^{pot}_{A,\,B_1}, k^{pot}_{A,\,B_2}, \ldots, k^{pot}_{A,\,B_N}$ are coefficients that determine the contribution of each interfering ion to the value of E, known as *potentiometric selectivity coefficients* of the ion-selective electrode for the ions $B_1, B_2, \ldots, B_N$, respectively.

The value of the coefficient $k^{pot}_{A,\,B_i}$ is determined by many factors, mainly by the ratio of the relative mobilities of A and B_i in the membrane and by the equilibrium of the exchange reaction $B_{i(sol)} + A_{(mem)} \rightleftharpoons B_{i(mem)} + A_{(sol)}$. When an electrode is very selective towards the ion A in comparison with the ion B_i, then $k^{pot}_{A,\,B_i} << 1$. Generally,the interference of B_i increases with increasing $k^{pot}_{A,\,B_i}$ and a_{B_i}/a_A.

The most common calibration methods used in direct potentiometry are a) the method of the calibration curve, b) the method of standard addition, and c) the method of standard subtraction.

In the *method of the calibration curve*, the potential of the selective electrode in a series of standard solutions of the ion A is measured, and a calibration curve is constructed, E(ordinate) as a function of $\log a_A$ or ($\log C_A$). Subsequently, from the observed potential in an unknown solution, the activity (or concentration) of A is calculated by interpolation of the calibration curve (Example 15.2).

In the *method of standard addition*, the potential is measured before (E_1) and after (E_2) the addition of a known volume V_S of a standard solution of concentration C_s to a volume V_u of the unknown solution of concentration C_u. We have

$$E_1 = E' + S \log C_u \tag{15.4}$$

$$E_2 = E' + S \log \frac{C_u V_u + C_s V_s}{V_u + V_s} \tag{15.5}$$

Subtracting Equation (15.4) from Equation (15.5) we have

$$E_2 - E_1 = \Delta E = S \log \frac{C_u V_u + C_s V_s}{(V_u + V_s) C_u} \tag{15.6}$$

or

$$\frac{\Delta E}{S} = \log \frac{C_u V_u + C_s V_s}{(V_u + V_s) C_u} \tag{15.6a}$$

Solving Equation (15.6*a*) for C_u, we have

$$C_u = \frac{C_s V_s}{(V_u + V_s) 10^{\Delta E/S} - V_u} \tag{15.7}$$

If $V_s << V_u$, then $V_u + V_s \simeq V_u$ and $(C_s V_s / V_u) \simeq \Delta C$, and Equation (15.7) is simplified to

$$C_u = \frac{C_s V_s}{V_u} \cdot \frac{1}{10^{\Delta E/S} - 1} = \frac{\Delta C}{10^{\Delta E/S} - 1} \tag{15.8}$$

The slope S is positive for cations and negative for anions. Equation (15.8) is also used in the form

$$C_u = \frac{\Delta C}{\text{antilog}(\Delta E/S) - 1} \tag{15.8a}$$

The *method of standard subtraction* is similar to the method of standard addition, except that the standard solution added, of volume V_s and concentration C_s, acts as a binding agent and decreases the concentration of the unknown solution by ΔC, by forming a complex or a slightly soluble compound with the ion being measured. We have

$$C_u = \frac{n C_s V_s}{V_u - (V_u + V_s) 10^{\Delta E/S}} \tag{15.9}$$

where n is the number of ions bound by each molecule or ion of the binding agent. If $V_s << V_u$, Equation (15.9) is simplified to

$$C_u = \frac{nC_sV_s}{V_u} \cdot \frac{1}{1 - 10^{\Delta E/S}} = \frac{\Delta C}{1 - 10^{\Delta E/S}} \tag{15.10}$$

The methods of standard addition and standard subtraction are especially useful for solutions of high unknown ionic strength.

Example 15.1. If ΔE is the absolute error of potential measurement in mV, C is the concentration of the measured ion i and z_i is its charge (with algebraic sign), show that the relative error, at room temperature, is given roughly by the equation:

$$\%\text{error} = (\Delta C_i/C_i)100 \simeq 4z_i\Delta E. \tag{15.11}$$

Solution. If E is the actual value of the potential which corresponds io the concentration C_i, we have

$$E = E' + \frac{RT}{z_iF}\ln C_i. \tag{15.12}$$

Because of the error ΔE, the measured value of the potential is $E + \Delta E$ and corresponds to concentration $C_i + \Delta C_i$, and we have

$$E + \Delta E = E' + \frac{RT}{z_iF}\ln(C_i + \Delta C_i) \tag{15.13}$$

Subtracting Equation (15.12) from Equation (15.13) we have

$$\Delta E = \frac{RT}{z_iF}\ln\left(1 + \frac{\Delta C_i}{C_i}\right) \tag{15.14}$$

If $(\Delta C_i/C_i) << 1$, which is true for small values of ΔE, on the basis of the approximate relation $\ln(1 + x) \simeq x$ (the Taylor series for $|x| << 1$), we have

$$\Delta E \simeq \frac{RT}{z_iF}\frac{\Delta C_i}{C_i} \tag{15.15}$$

Therefore,

$$\%\text{error} = \frac{\Delta C_i}{C_i} \times 100 = \frac{Fz_i\Delta E}{RT} \times 100 = \frac{(96484.56\ \text{C/eq})(z_i\ \text{eq/mol})(\Delta E\ \text{mV})}{(8314.4\ \text{mV} \cdot \text{C} \cdot \text{K}^{-1}\ \text{mol}^{-1})(298.15\ \text{K})} \times 100$$
$$= 3.89z_i\Delta E \simeq 4z_i\Delta E. \tag{15.11}$$

Note. During the measurement of the potential with a common electrometer, the reading error is typically ± 1 mV; hence, the corresponding analytical error is 4% for monovalent ions (8% for divalent ions). Since the electrometers used in combination with ion-selective electrodes permit the measurement of the potential to ± 0.1 mV, the analytical error is reduced, theoretically, to a ± 0.4% for monovalent ions. However, in practice, the analytical error is larger, because there is an uncertainty of the order of ± 1 mV in all measurements with galvanic cells that contain a salt bridge, because of the liquid-junction potential which does not remain absolutely constant from measurement to measurement.

Example 15.2. To determine NH_4NO_3 in a fertilizer, the potential of a nitrate-selective electrode was measured in a series of standard KNO_3 solutions, containing in addition 0.10 M Na_2SO_4 to keep the ionic strength constant (the sulfate salt was chosen since sulfates do not interfere with the nitrate-selective electrode, $k^{pot}_{NO_3^-, SO_4^{2-}} \simeq 0$). The following data were obtained:

C_{KNO_3}, M:	1.00×10^{-4}	3.00×10^{-4}	1.00×10^{-3}	3.00×10^{-3}	1.00×10^{-2}
E, mV:	115.5	94.3	66.3	38.4	10.2

A 0.3312-g sample was dissolved in a 0.10 M Na_2SO_4 solution and diluted with the same solution to a final volume of 250.0 mL (solution A). The potential of the nitrate-selective electrode in solution A was 51.6 mV. Calculate the percentage of NH_4NO_3 in the fertilizer assuming that all other ingredients do not interfere with the nitrate-selective electrode.

Solution. On the basis of the measurements with the standard KNO_3 solutions, the following calibration curve is constructed:

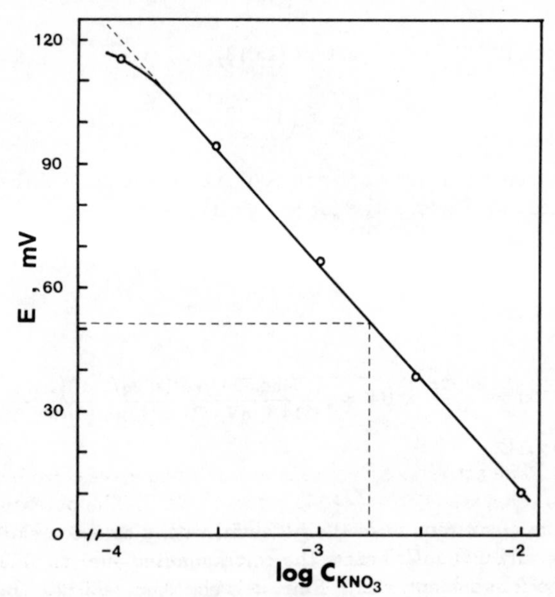

From the calibration curve we find that in solution A we have $\log C_{NH_4NO_3} = -2.75$. Hence, $C_{NH_4NO_3} = 1.78 \times 10^{-3}$ M (the fact that the cation of the nitrate salt is not the same in the standards and in the fertilizer is of no significance. KNO_3 was used, since NH_4NO_3 is not available as a standard substance). Hence, the fertilizer contains

$$\frac{(250.0 \text{ mL})(1.78 \times 10^{-3} \text{ mmol } NH_4NO_3/\text{mL})(80.5 \text{ mg/mmol})}{331.2 \text{ mg}} \times 100 = \mathbf{10.8\%}\ NH_4NO_3.$$

Note. The deviation from linearity of the last point used in the calibration curve is not symptomatic, but it is due to the fact that the nitrate concentration at this point is very close to the detection limit of the electrode.

Example 15.3. If E_{AB} is the potential of ion-selective electrode A in a solution containing ions A and B, and E_A is the potential of the electrode in a solution containing only ion A, prove the relation:

$$k_{A,\,B}^{pot} = \frac{a_A \cdot 10^{(E_{AB}-E_A)/S} - a'_A}{a_B^{z_A/z_B}} \tag{15.16}$$

where a'_A and a_B are the activities of ions A and B, respectively, in the solution containing both ions A and B, and a'_A is the activity of ion A in the solution containing only ion A.

Solution. We have

$$E_{AB} = E' + S \log(a'_A + k_{A,\,B}^{pot} a_B^{z_A/z_B}) \tag{15.17}$$

and

$$E_A = E' + S \log a_A \tag{15.18}$$

Subtracting Equation (15.18) from Equation (15.17), we have

$$E_{AB} - E_A = S[\log(a'_A + k_{A,\,B}^{pot} a_B^{z_A/z_B}) - \log a_A] = S \log\left(\frac{a'_A + k_{A,\,B}^{pot} a_B^{z_A/z_B}}{a_A}\right)$$

or

$$10^{(E_{AB}-E_A)/S} = \frac{a'_A + k_{A,\,B}^{pot} a_B^{z_A/z_B}}{a_A},$$

from which we find that

$$k_{A,\,B}^{pot} = \frac{a_A \cdot 10^{(E_{AB}-E_A)/S} - a'_A}{a_B^{z_A/z_B}} \tag{15.16}$$

Example 15.4. The activity of Ca^{2+} ions in a solution was determined by measurements with a calcium-ion selective electrode against a SCE. The potential was +0.0170 V in a 0.0100 $\underline{M}$ $CaCl_2$ solution and +0.0040 V in a solution 1.00×10^{-3} $\underline{M}$ in $CaCl_2$ and 1.00×10^{-2} $\underline{M}$ in $CuCl_2$. Calculate the calcium activity in an unknown solution A, if the potential of the calcium-electrode in solution A is +0.0120 V and the copper activity in solution A, determined with a copper-ion selective electrode, was found to be 5.00×10^{-3} $\underline{M}$.

Solution. The ionic strength of a 0.0100 M $CaCl_2$ solution is 0.0300. Using Equation (3.16), we find that the activity coefficient $f_{Ca^{2+}}$ is 0.50. Hence, we have

$$E_{Ca} = +0.0170 = E'_{Ca} + (0.05916/2)\log a_{Ca^{2+}} = E'_{Ca} + (0.05916/2)\log(0.50 \times 0.0100)$$

or

$$E'_{Ca} = +0.0851 \text{ V}.$$

The ionic strength of a solution 1.00×10^{-3} M in $CaCl_2$ and 1.00×10^{-2} M in $CuCl_2$ is 0.033. Using Equation (3.16) we find that the activity coefficients $f_{Ca^{2+}}$ and $f_{Cu^{2+}}$ are 0.49. Hence,

$$E_{CaCu} = E'_{Ca} + (0.05916/2)\log[a_{Ca^{2+}} + k^{pot}_{Ca^{2+},\ Cu^{2+}} a_{Cu^{2+}}] = +0.0040$$

$$= +0.0851 + (0.05916/2)\log[(0.49 \times 0.00100) + k^{pot}_{Ca^{2+},\ Cu^{2+}}(0.49 \times 0.0100)]$$

or

$$k^{pot}_{Ca^{2+},\ Cu^{2+}} = 0.27.$$

For the unknown solution we have

$$E_{CaCu} = +0.0120 = 0.0851 + (0.05916/2)\log(a_{Ca^{2+}} + 0.27 \times 0.00500)$$

or

$$a_{Ca^{2+}} = \mathbf{2.05 \times 10^{-3}} \text{ M}.$$

The problem can also be solved by using Equation (15.16) of Example (15.3).

Example 15.5. For the potentiometric determination of fluoride in drinking water, 50.0 mL of sample is mixed with an equal volume of TISAB (Total Ionic Strength Adjustment Buffer. The 1 M acetate −1 M NaCl buffer solution, pH 5.0–5.5, contains also *trans*-1,2-diaminocyclohexane-N, N, N', N'-tetraacetic acid (CDTA), 4 g/L. The CDTA displaces any F^- from fluorocomplexes of Fe(III) and Al(III)). The electrode system developed a potential of +0.1321 V when immersed in the mixture. After addition of 10.00 mL of 1.00×10^{-4} M NaF, containing the same concentration of TISAB, the potential decreased to +0.1191 V. Calculate the fluoride concentration of the sample, in ppm.

Solution. Using Equation (15.7), we find that in the unknown-TISAB solution we have

$$[F^-] = \frac{(1.00 \times 10^{-4})(10.00)}{(100.0 + 10.00)10^{(0.1191-0.1321)/(-0.05916)} - 100.0} = 1.213 \times 10^{-5} \text{ M}.$$

Hence, the concentration of F^- in the sample will be equal to $2 \times (1.213 \times 10^{-5}) = 2.426 \times 10^{-5}$ M or $(2.426 \times 10^{-5}$ mol $F^-/L)(18998$ mg $F^-/$mol $F^-) =$ **0.461** ppm F^-.

Example 15.6. What is the S^{2-} concentration in 50.00 mL of sample that gives a potential reading of −0.4521 V before the addition of 2.00 mL of 0.100 M $Pb(NO_3)_2$ and a reading of −0.4396 V after the addition (assume that $\alpha_2 = [S^{2-}]/C_s = 1.00$).

Solution. On the basis of Equation (15.19) we have

$$C_u = C_s = \frac{1 \times 0.100 \times 2.00}{50.00 - (50.00 + 2.00)10^{[-0.4396-(-0.4521)]/(-0.02958)}} = \mathbf{6.59 \times 10^{-3}}\ \underline{M}.$$

Example 15.7. Calculate the activity of interfering ions that can be tolerated for measurement of $a_{Ca^{2+}} = 0.00100\ \underline{M}$ with no more than a 3% error due to interference. The interfering ions and their potentiometric selectivity coefficients (in parentheses) are: Zn^{2+} (3.2), Fe^{2+} (0.8), Sr^{2+} (0.02), Ba^{2+} (0.01), Na^{+} (0.0016).

Solution. In order for the error not to exceed 3%, the second term inside the bracket of Equation (15.3) (contribution of foreign ions) should not exceed $0.03a_A$ (the error will always be positive). Hence, we have

- for Zn^{2+}: $0.03(1.00 \times 10^{-3}) = 3.2 \times a_{Zn^{2+}}^{2/2}$ and $a_{Zn^{2+}} = \mathbf{9.4 \times 10^{-6}}\ \underline{M}$
- for Fe^{2+}: $0.03(1.00 \times 10^{-3}) = 0.8 \times a_{Fe^{2+}}^{2/2}$ and $a_{Fe^{2+}} = \mathbf{3.8 \times 10^{-5}}\ \underline{M}$
- for Sr^{2+}: $0.03(1.00 \times 10^{-3}) = 0.02 \times a_{Sr^{2+}}^{2/2}$ and $a_{Sr^{2+}} = \mathbf{1.5 \times 10^{-3}}\ \underline{M}$
- for Ba^{2+}: $0.03(1.00 \times 10^{-3}) = 0.01 \times a_{Ba^{2+}}^{2/2}$ and $a_{Ba^{2+}} = \mathbf{3.0 \times 10^{-3}}\ \underline{M}$
- for Na^{+}: $0.03(1.00 \times 10^{-3}) = 0.0016 \times a_{Na^{+}}^{2/1}$ and $a_{Na^{+}} = \mathbf{0.14}\ \underline{M}$.

Example 15.8. If the potential of a calcium-selective electrode is +0.0443 V in a 0.0200 $\underline{M}$ $CaCl_2$ solution, what will the potential be in a 0.200 $\underline{M}$ $CaCl_2$ solution (use activities)?

Solution. The ionic strength of a 0.0200 M $CaCl_2$ solution is 0.0600. Using Equation (3.16), we find that the activity coefficient $f_{Ca^{2+}}$ is 0.397. Hence,

$$E_{Ca} = E'_{Ca} + (0.05916/2)\log a_{Ca^{2+}} = 0.0443 = E'_{Ca} + (0.05916/2)\log(0.397 \times 0.0200)$$

or

$$E'_{Ca} = +0.1064\ \text{V}.$$

The ionic strength of a 0.200 $\underline{M}$ $CaCl_2$ solution is 0.600. Using Equation (3.16) we find that $f_{Ca^{2+}} = 0.129$. Hence,

$$E_{Ca} = E'_{Ca} + (0.05916/2)\log a_{Ca^{2+}} = 0.1064 + (0.05916/2)\log(0.129 \times 0.200) = \mathbf{0.0594}\ \text{V}.$$

Example 15.9. The potential of a calcium-selective electrode in a 2.00×10^{-3} $\underline{M}$ $CaCl_2$ solution is +0.0147 V. If the potential of this electrode in an unknown Ca^{2+} solution is +0.0267 V, calculate the concentration of Ca^{2+} in the unknown solution. Both solutions have the same ionic strength.

Solution. We have

$$E_{Ca} = +0.0147 = E'_{Ca} + (0.05916/2)\log a_{Ca^{2+}} = E'_{Ca} + (0.05916/2)\log f_{Ca^{2+}}$$

$$+(0.05916/2)\log[Ca^{2+}] = E''_{Ca} + (0.05916/2)\log[Ca^{2+}] = E''_{Ca} + (0.05916/2)\log(0.00200)$$

or

$$E''_{Ca} = +0.0147 - (-0.0798) = +0.0945 \text{ V}.$$

If $[Ca^{2+}]_u$ is the concentration of the Ca^{2+} ions in the unknown solution, we have

$$+0.0267 = +0.0945 + (0.05916/2)\log[Ca^{2+}]_u$$

or

$$\log[Ca^{2+}]_u = -2.292, \text{ in which case } [Ca^{2+}]_u = \mathbf{5.10 \times 10^{-3}}\ \underline{M}.$$

Example 15.10. Calculate the relative error in the determination of nitrate in a nitrate-nitrite solution using a nitrate-selective electrode, if $a_{NO_3^-} = 0.00200\ \underline{M}$, $a_{NO_2^-} = 0.010\ \underline{M}$ and $k^{pot}_{NO_3^-, NO_2^-} = 0.06$.

Solution. *1st method.* Substituting the data in Equation (15.3), we have

$$E = E' - 0.05916\log[0.00200 + (0.06)(0.010)] = E' + 0.1529 \text{ V}.$$

Substituting the value (E′+0.529) V in Equation (15.2), we find the apparent (incorrect) activity of nitrate, $(a_{NO_3^-})_a$. That is, we have

$$E = E' + 0.1529 = E' - 0.05916\log(a_{NO_3^-})_a$$

or $(a_{NO_3^-})_a = 0.00260\ \underline{M}$. Hence, the relative error is

$$\left(\frac{0.00260 - 0.00200}{0.00200}\right) \times 100 = +\mathbf{30\%}.$$

2nd Method. Substituting the data in the relation

$$\%\text{error} = \frac{k^{pot}_{A,B} \cdot a_B^{z_B/z_A}}{a_A} \times 100,$$

we have

$$\%\text{error} = \frac{0.06 \cdot 0.010^{(-1)/(-1)}}{0.00200} \times 100 = \mathbf{30\%}.$$

Potentiometric determination of pH. The measurement of pH probably is the most common of all chemical measurements and is usually carried out potentiometrically with a pH meter. The measurement can be performed with a variety of electrodes (in the past, indicator electrodes based on redox system were used, such as the quinhydrone or the hydrogen electrode), but usually the calomel-glass electrode pair is used, which constitutes the cell

Ag | AgCl, HCl(0.100 M) | glass membrane | H^+(xM)‖KCl (satur.), **Hg_2Cl_2 | Hg**

The symbol | represents an interface and ‖ represents a liquid-junction (salt bridge).

The potential of the glass electrode as a function of pH is given by

$$E = k + S\log a_{H^+} = k - S \cdot pH \tag{15.19}$$

where k = a constant, which depends on the composition of the glass, the temperature, the internal reference electrode, and the a_{H^+} of the internal reference solution (k corresponds to the constant E′ of Equation (15.2)). Hence, the emf of the above cell is equal to

$$\begin{aligned} E_{cell} &= E_r - E_l = E_{SCE} - (k + S \log a_{H^+}) \\ &= (E_{SCE} - k) - S \log a_{H^+} = E' - S \log a_{H^+} = E' + S \cdot pH \end{aligned} \quad (15.20)$$

It should be noticed that Equation (15.20) differs from the general equation (15.2) in the sign of the logarithmic term, and this is due to the fact that in the literature (for historical reasons) the above cell is written with the glass electrode as the left electrode. On the other hand, the signs (+) and (−) on the voltage scale of the pH meter (or pIon meter) denote the relative polarity of the indicator electrode (measurement electrode) with respect to the reference electrode of the cell and will be of the opposite sign of Equation (15.20); that is, the measured cell potential will increase with increasing a_{H^+}, as predicted by Equation (15.19).

Since it is impossible to construct two glass electrodes that give exactly the same pH (or potential) reading when immersed in the same solution, and since the potential of a glass electrode does not remain constant for prolonged periods of time, the pH meter should be calibrated with a standard buffer prior to the pH measurement. Let $(E_{cell})_x$ and $(E_{cell})_s$ represent the emf of the cell with the unknown and the buffer solution, respectively (the experimental conditions are identical during the calibration of the pH meter and the pH measurement, and it is assumed that the junction potential remains constant in all measurements). If $(pH)_x$ and $(pH)_s$ are the corresponding pH values of the two solutions, the unknown pH is computed by the equation (operational definition of pH)

$$(pH)_x = (pH)_s + \frac{[(E_{cell})_x - (E_{cell})_s]F}{2.303RT} = (pH)_s + \frac{(E_{cell})_x - (E_{cell})_s}{S} \quad (15.21)$$

or

$$(pH)_x = (pH)_s + \frac{(E_{cell})_x - (E_{cell})_s}{0.05916} \text{ (for 25°C)} \quad (15.21a)$$

Similarly, for an ion-selective electrode that responds to the cation M^{n+} we have

$$(pM)_x = (pM)_s + \frac{(E_{cell})_x - (E_{cell})_s}{S} \quad (15.22)$$

where $(pM)_x$ and $(pM)_s$ represent the negative decimal logarithm of the activity of M^{n+} in the unknown and the standard solution, respectively. Similarly, for anions A^{n-} we have the equation

$$(pA)_x = (pA)_s + \frac{(E_{cell})_x - (E_{cell})_s}{S} \quad (15.23)$$

Example 15.11. The following cell was found to have a voltage of +0.2150 V when the solution in the left compartment was a buffer of pH 4.00:

$$\text{glass electrode} \mid H^+ \ (a_{H^+} = x\underline{M}) \| KCl \text{ (satur.)}, \ \mathbf{Hg_2Cl_2} \mid \mathbf{Hg}$$

a) Calculate the pH of the unknown solution for which the following voltage readings were obtained (the voltage of the cell as defined increases with increasing pH): 1) +0.2420 V, 2) +0.0810 V, 3) +0.5230 V. b) What voltage would be measured in a 0.100 M acetic acid solution (assume $a_{H^+} = [H^+]$)?

Solution. s) Substituting the data in Equation (15.21*a*), we have

1. pH = 4.00 + (0.2420 − 0.2150)/0.05916 = **4.46**

2. pH = 4.00 + (0.0810 − 0.2150)/0.05916 = **1.73**

3. pH = 4.00 + (0.5230 − 0.2150)/0.05916 = **9.21**.

b) We have $[H^+] \simeq \sqrt{1.8 \times 10^{-5} \times 0.100} = 1.34 \times 10^{-3}$ M and pH $= -\log(1.34 \times 10^{-3}) = 2.87$. Hence,

$$2.87 = 4.00 + \frac{(E_{cell})_x - 0.2150}{0.05916}$$

$$(E_{cell})_x = \mathbf{0.1482} \text{ V}.$$

Note. Equation (15.21*a*) applies only for the historical cell given before Equation (15.19). Since *measured* potentials are actually of the opposite sign (potential increased with increasing a_{H^+}), then in order to apply Equation (15.21*a*), the sign of the measured potential is reversed.

Example 15.12. What is the theoretical and the experimental pH of a solution 0.0100 M in NaOH and 0.100 M in NaCl, if the potentiometric selectivity coefficient $k^{pot}_{H^+, Na^+}$ for the glass electrode used is 5×10^{-11}?

Solution. The ionic strength of a solution 0.0100 M in NaOH and 0.100 M in NaCl is 0.110. Using Equation (3.16) we find that $f_{OH^-} = f_{Na^+} = 0.746$, then

$$a_{OH^-} = 0.746 \times 0.0100 = 0.00746 \text{ M}$$

and

$$a_{Na^+} = 0.746 \times 0.110 = 0.08206 \text{ M}.$$

Hence,

$$a_{H^+} = 1.00 \times 10^{-14}/7.46 \times 10^{-3} = 1.34 \times 10^{-12} \text{ M},$$

and the theoretical pH is equal to

$$(\text{pH})_{theor} = -\log(1.34 \times 10^{-12}) = \mathbf{11.87},$$

and this pH would have been found, if the Na^+ ions did not affect the potential of the glass electrode. In practice, because of the Na^+ ions, on the basis of Equations (15.3) and (15.20) we will have

$$E_{cell} = E' - S \log(a_{H^+} + k^{pot}_{H^+, Na^+} a_{Na^+}) = E' - S \log[1.34 \times 10^{-12} + (5 \times 10^{-11})(0.08206)].$$

Hence, the experimental pH, which will be found with the pH meter is equal to

$$(\text{pH})_{exp} = -\log[(1.34 \times 10^{-12} + (5 \times 10^{-11})(0.08206)] = \mathbf{11.26}.$$

Note. There is a negative error equal to 11.26 − 11.87 = −0.61 pH units, which is due to the contribution of the Na^+ ions to the emf of the cell. This kind of error in glass electrode pH measurements is known as *alkaline error*, and it is especially serious in strongly alkaline solutions.

Example 15.13. Derive an equation relating the potential of the following galvanic cell to pH (assume $a_{H^+} = [H^+]$):

$$(-)\ \mathbf{Pt},\ H_2\ (1\ \text{atm})\ |\ H^+\ (x\underline{M})||HCl\ (0.0100\ \underline{M}),\ \mathbf{AgCl}\ |\ \mathbf{Ag}\ (+) \qquad (15.24)$$

Solution. For the galvanic cell (15.24) we have

$$E_{cell} = E_{AgCl,\ Ag} - E_{H^+,\ H_2} = [0.2224 - 0.05916\log(0.0100)] - [0.0000-$$

$$\frac{0.05916}{2}\log\frac{1}{[H^+]^2}] = 0.3407 + 0.05916\ pH.$$

15.3 Potentiometric Titrations

In a potentiometric titration, the emf of a cell is measured after the addition of known volumes of a standard solution. The titration vessel with the titrand and the indicator electrode becomes one half-cell of the cell, while the other half-cell is a reference electrode. The indicator electrode responds either to the ions being titrated, whose activity is reduced during the titration because of their reaction with the titrant (a neutralization or redox or precipitation or complexometric reaction), or to the ions of the titrant. In the case of a redox titration, the potential of the indicator electrode, and consequently, the emf of the cell may depend on the ratio of the activities of the redox couple (redox electrode, e.g., Pt).

The titration curves (potential E vs. volume V) are constructed as previously described (Chapters 11 and 12) and in most cases they are sigmoid curves, whose inflection point is taken as the end point of the titration. The end point is determined more precisely from the first derivative curve as the point at which the first derivative has the maximal value, or even more precisely, from the second derivative curve as the point at which the second derivative equals zero (Figure 15.1). Only the points of the titration curve located in the vicinity of the end point are used for the drawing of the first and second derivative curves, and in these cases, the *average* volume between increments is plotted (see Example 15.14), which permits the extension of the abscissa scale, and consequently, results in a better precision in the determination of the end point.

The end point can also be determined from *Gran's plots* (antilog(E/S) or $\Delta V/\Delta E$ as a function of V). This provides a straight line plot since the antilog of E is linearly dependent on the concentration (or activity), neglecting dilution effects. To construct such a plot it is sufficient to obtain a few points near the start of the titration and

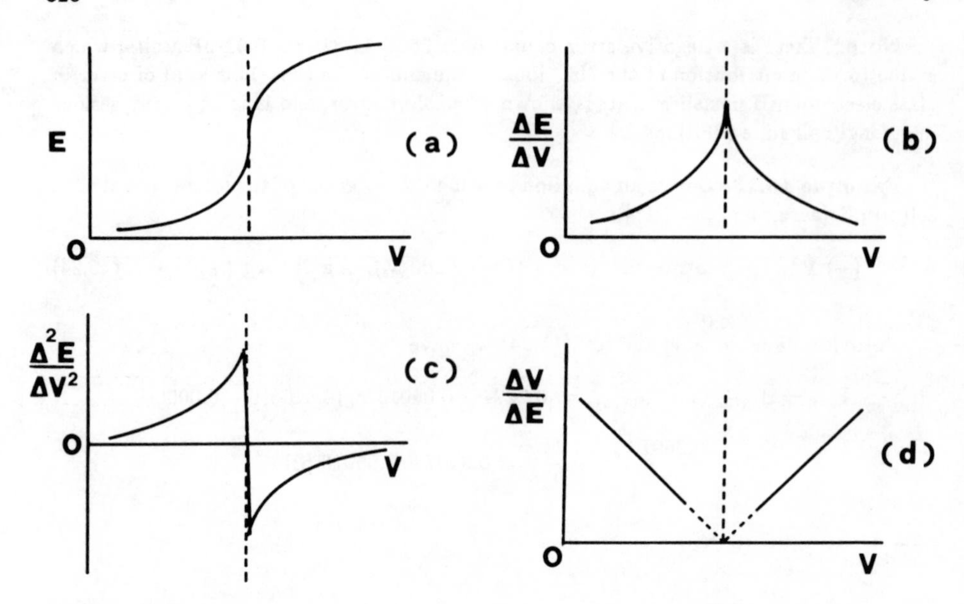

Figure 15.1. *(a). Potentiometric titration curve. (b) First derivative curve. (c) Second derivative curve.*

beyond the equivalence point, then the end point is easily and precisely determined as the intersection of the two straight lines defined by these points. For convenience, a plot of the reciprocal of a first derivative curve will represent a Gran's plot. That is, since in the former, $\Delta E/\Delta V$ goes to infinity at the equivalence point, the reciprocal will go to zero where the intersection of the two lines occurs. In this application of the Gran's plot, the average volume between the two increments is plotted, as in the first derivative plot.

In comparison to direct potentiometry, potentiometric titrations are characterized by better accuracy and precision, usually of the order of 0.1%, but they require more time and are more complex. They are more accurate and precise, because we are simply using the measured *changes* in potential to precisely indicate the end point to a few hundredths of a milliliter, as opposed to attempting to measure the absolute values of the potential. The potential changes are large and are little influenced by the liquid-junction potential or activity coefficients. In certain cases, titrations also permit the stepwise determination of two or more compounds in the same solution.

Example 15.14. The following data were obtained during the titration of 25.00 mL of a weak monoprotic acid with a standard 0.1000 M NaOH near the equivalence point:

V_{NaOH}, mL	pH	V_{NaOH}, mL	pH
28.00	5.97	28.50	7.25
28.10	6.04	28.60	9.97
28.20	6.14	28.70	10.45
28.30	6.30	28.80	10.67
28.40	6.52	28.90	10.81

Calculate a) the molarity of the acid solution, b) the pH at the equivalence point.

Solution. a) The titration data at the vicinity of the equivalence point and the values of the first and second derivative, $\Delta pH/\Delta V$ and $\Delta^2 pH/\Delta V^2$, which were calculated from these data, are given below:

V_{NaOH}, mL	pH	ΔpH	ΔV_{1st}	$\Delta pH/\Delta V_{1st}$	V'	$\Delta(\Delta pH/\Delta V_{1st})$	ΔV_{2nd}	$\Delta^2 pH/\Delta V^2$ $[\Delta(\Delta pH/\Delta V_{1st})/\Delta V_{2nd}]$	V''
28.00	5.97								
		0.07	0.10	0.7	28.05				
28.10	6.04					+0.3	0.10	+3	28.10
		0.10	0.10	1.0	28.15				
28.20	6.14					+0.6	0.10	+6	28.20
		0.16	0.10	1.6	28.25				
28.30	6.30					+0.6	0.10	+6	28.30
		0.22	0.10	2.2	28.35				
28.40	6.52					+5.1	0.10	+51	28.40
		0.73	0.10	7.3	28.45				
28.50	7.25					+19.9	0.10	+199	28.50
		2.72	0.10	27.2	28.55				
28.60	9.97					−22.4	0.10	−224	28.60
		0.48	0.10	4.8	28.65				
28.70	10.45								

From the above values it is apparent that the second derivative is equal to zero somewhere between 28.50 and 28.60 mL. Since 0.10 mL of NaOH results in a total second derivative change of $199-(-224) = 423$, it follows that the fraction $(199/423) \times 0.10$ mL is the volume of NaOH needed in addition to 28.50 mL for the second derivative to become zero. Therefore, the equivalence point volume is equal to

$$V_{eq} = 28.50 + 0.10[199/(199 + 224)] = 28.55 \text{ mL}.$$

Hence, the molarity of the acid solution is equal to

$$(28.55 \times 0.1000)/25.00 = \mathbf{0.1142}\ \underline{M}.$$

b) The pH at the equivalence point is equal to

$$pH_{EP} = 7.25 + 2.72[199/(199 + 224)] = \mathbf{8.53}.$$

Notes. 1. To facilitate the calculation of the first and the second derivative, equal increments of titrant are added near the equivalence point.

2. The volume V′ used for the first derivative curve is the average of the two successive volumes used to calculate $\Delta pH/\Delta V$. For example, the volume of (28.30 + 28.40)/2 = 28.35 mL corresponds to the value $\Delta pH/\Delta V = 2.2$. Similarly, the average of two successive volumes used in the first derivative curve, V″, is used for the second derivative curve. For example, the volume (28.35 + 28.45)/2 = 28.40 mL corresponds to the value $\Delta^2 pH/\Delta V^2 = +51$.

3. The point at which the first derivative is maximal (point of maximal slope of the titration curve, that is, inflection point) coincides with the theoretical equivalence point for a titration only if the titrant and titrand react at a stoichiometric ratio of 1:1 (e.g., $H^+ + OH^- \rightleftharpoons H_2O$, $Ag^+ + Cl^- \rightleftharpoons \mathbf{AgCl}$, $Ce^{4+} + Fe^{2+} \rightleftharpoons Fe^{3+} + Ce^{3+}$, $Ca^{2+} + H_2Y^{2-} \rightleftharpoons [CaY]^{2-} + 2H^+$), in which case the titration curve is symmetrical. If the titrant and titrand react at a different ratio, as for example in the titration of Fe^{2+} with MnO_4^-

$$MnO_4^- + 5Fe^{2+} + 8H^+ \rightleftharpoons Mn^{2+} + 5Fe^{3+} + 4H_2O,$$

the titration curve is asymmetric and the theoretical equivalence point does not coincide with the inflection point of the titration curve, so that we have an error, which, however, is so small that it can be neglected.

15.4 Problems

Note. In problems 1, 2, 3, and 18, the cell is written with the indicator electrode (glass or ion selective electrode) as the left electrode (as is done in the literature for historical reasons). This means that the voltages are of opposite sign of those actually obtained with a pH meter or a pIon meter (see Example 15.11). In the other problems, the general Equation (15.2) is applied.

Measurement of pH

15.1. The following cell was found to have a voltage of +0.3920 V when the solution in the left compartment was a buffer of pH 7.00:

$$\text{glass electrode} \mid H^+ \ (a_{H^+} = x\underline{M}) \| KCl \text{ (satur.)}, \ \mathbf{Hg_2Cl_2} \mid \mathbf{Hg}$$

a) Calculate the pH of the unknown solution for which a voltage reading of +0.5250 V was obtained. b) What voltage would be measured in a 0.100 M sodium acetate solution (assume $a_{H^+} = [H^+]$. The voltage of the cell as defined increases with increasing pH)?

15.2. The emf of the galvanic cell

$$(-)\ \mathbf{Pt},\ H_2\ (1\ \text{atm})\ |\ HCl\ (x\underline{M})\|KCl\ (\text{satur.}),\ \mathbf{Hg_2Cl_2}\ |\ \mathbf{Hg}\ (+)$$

is +0.3595 V. Calculate the pH of the HCl solution ($a_{H^+} = [H^+]$).

15.3. Derive an equation relating the emf of the following galvanic cell to pH (assume $a_{H^+} = [H^+]$).

$$(-)\ \mathbf{Pt},\ H_2\ (1\ \text{atm})\ |\ H^+\ (x\underline{M})\|KCl\ (\text{satur.}),\ \mathbf{Hg_2Cl_2}\ |\ \mathbf{Hg}\ (+)$$

Ion-selective electrodes

15.4. Solve the problem of Example (15.4), using Equation (15.16).

15.5. The calcium ion activity is measured with a calcium-selective electrode in solutions containing up to 0.50 M sodium ion. Calculate the minimum Ca^{2+} activity that can be measured under these conditions, so that the error does not exceed 3% ($k^{pot}_{Ca^{2+}, Na^+} = 0.0016$).

15.6. Calculate the activity of interfering ions that can be tolerated for measurement of $a_{Ca^{2+}} = 5.00 \times 10^{-3}$ M with no more than a 3% error due to interference. The interfering ions and their potentiometric selectivity coefficients (in parentheses) are: Cu^{2+} (0.27), K^+ (0.0001), Mg^{2+} (0.01).

15.7. What is the S^{2-} concentration in 50.00 mL of sample that gives a potential reading of −0.4621 V before the addition of 1.00 mL of 0.40 M $AgNO_3$ and a reading of −0.4496 V after the addition (assume that $\alpha_2 = [S^{2-}]/C_s = 1.00$)?

15.8. Calculate the relative error in the determination of calcium in a $CaCl_2$-NaCl solution, using a calcium-selective electrode, if $a_{Ca^{2+}} = 2.00 \times 10^{-3}$ M, $a_{Na^+} = 0.250$ M, and $k^{pot}_{Ca^{2+}, Na^+} = 0.0016$. What is the respective relative error in the determination of pCa?

15.9. The potential of a fluoride selective electrode against a SCE in a 1.00×10^{-5} M NaF solution is +0.1300 V. Solid NaF is added to 100 mL of this solution until the measured potential becomes +0.0120 V. Calculate the new concentration of F^- ions and the amount of NaF added (in mg) . Assume no change in the volume of the solution and $a_{F^-} = [F^-]$.

15.10. a) If an error of ±0.1 mV in the observed potential of the calcium-selective electrode is incurred, calculate the corresponding % error in the calcium activity. b) If the error in the calcium activity is not to exceed 1%, calculate the maximum permissible error in the potential of the electrode.

15.11. In a research laboratory the following experiment was carried out with a newly constructed acetate-selective electrode: Two solutions (A and B) were prepared, both 1.50×10^{-3} M in CH_3COOH and 0.0100 M in NaCl. The pH of solution A was adjusted at 4.60 (with the addition of a small volume of concentrated NaOH) and the pH of solution B at 5.30. The potential of the acetate electrode against a suitable reference electrode was +0.1830 V in solution A and +0.1739 V in solution B. Using these data calculate the potentiometric selectivity coefficient $k^{pot}_{CH_3COO^-, Cl^-}$ (assume that activities equal concentrations and that only Cl^- interferes).

15.12. Prove that for the monovalent A^+ and B^+ ions we have the following equation

$$\log k^{pot}_{A, B} = \frac{E^{0'}_B - E^{0'}_A}{S}$$

where $E^{0'}_A$ and $E^{0'}_B$ are the electrode potentials in solutions with $a_{A^+} = 1$ $\underline{M}$, and $a_{B^+} = 1$ $\underline{M}$, respectively.

15.13. A 2.00×10^{-3} $\underline{M}$ NH_4NO_3 solution is subjected to a bacterial action which results in the reduction of NO_3^- to NH_4^+. The reduction is monitored potentiometrically with ammonium- and nitrate-selective electrodes. The potential of the ammonium-selective electrode was +0.0401 V before the bacterial action and +0.0490 V afterwards, whereas the initial potential of the nitrate-selective electrode was +0.0886 V. Calculate the potential of the nitrate electrode after the bacterial action (assume activity equals concentration).

15.14. X mg of $Na_2SO_4 \cdot nH_2O$ is dissolved in water and the solution is diluted to 1000 mL (solution A). 2X mg of anhydrous Na_2SO_4 is dissolved and diluted to 1000 mL of solution B. The potential of a sodium-selective electrode against a suitable reference electrode is +0.1286 V in solution A and +0.1675 V in solution B. Calculate the number of water molecules n, in the salt $Na_2SO_4 \cdot nH_2O$ (assume activity equals concentration).

15.15. A 0.1172-g sample of black powder (a mixture of C, S and KNO_3) is extracted with water, filtered and the filtrate is diluted to 250.0 mL (solution A). A 20.00-mL portion of solution A is mixed with 10.00 mL of a 0.0100 $\underline{M}$ $NaNO_3$ solution (solution B). The potential of a nitrate-selective electrode is +0.1269 V and +0.1134 V in solutions A and B, respectively. Calculate the percentage of KNO_3 in the sample (assume activity equals concentration).

15.16. To determine $NaNO_3$ impurities in $NaNO_2$, a 0.2436-g sample of $NaNO_2$ (+ $NaNO_3$) is dissolved in water and diluted to 100.0 mL (solution A). An equal amount of sample is treated with an excess of hydroxylamine sulfate to destroy nitrite ($NO_2^- + 3H_2NOH \rightarrow 2N_2 + OH^- + 4H_2O$) and the solution is diluted to 100.0 mL (solution B). The potential of a nitrate-selective electrode was +0.2134 V and +0.2398 V, in solutions A and B, respectively. Calculate the percentage of $NaNO_3$ in the sample, if $k^{pot}_{NO_3^-, NO_2^-} = 0.021$. Assume that all other ions (sulfate, etc.) do not interfere and that activity equals concentration.

15.17. Prove Equation (15.9) for the known subtraction method with ion-selective electrodes.

15.18. The emf of the cell

$$\text{calcium-selective electrode} \mid Ca^{2+} \ (a_{Ca^{2+}} = x\underline{M}) \| SCE$$

is +0.4182 V in a solution with $a_{Ca^{2+}} = 2.00 \times 10^{-3}$ M. Calculate the pCa of a solution for which the emf of the cell is +0.4532 V (assume $a_{Ca^{2+}} = [Ca^{2+}]$).

15.19. The potential of a sodium-selective electrode against a SCE is +0.0320 V in a NaCl solution with $a_{Na^+} = 0.0100$ M, and +0.0720 V in a KCl solution with $a_{K^+} = 0.0100$ M. Calculate the potentiometric selectivity coefficient $k^{pot}_{Na^+, K^+}$.

15.20. The potential of a nitrate-selective electrode in a 2.50×10^{-3} M $NaNO_3$ solution is +0.1194 V vs. SCE. If solid $NaNO_3$ is added to 500 ml of this solution until the potential becomes +0.0826 V, what will the new nitrate concentration be and what is the quantity of $NaNO_3$ added? Assume negligible volume change upon the addition of the $NaNO_3$.

Chapter 16

Spectrophotometry

16.1 Introduction

Spectroscopic methods of analysis are based on the ability of various substances to interact with or emit radiation of characteristic frequencies. In spectrophotometry, the absorbance A or the transmittance T (Section 16.2) of the sample is measured, and a qualitative and quantitative analysis is carried out on the basis of these measurements. A plot of A or T versus wavelength λ or wavenumber (Equation 16.2) yields the *absorption spectrum*, which can be used to detect characteristic groups, to elucidate the structure of the absorbing substance X, and to identify X. The concentration of X can be calculated from A using Beer's law (Equation 16.3).

Characteristics of electromagnetic radiation. A radiation is characterized by the wavelength λ, the frequency ν, or the wavenumber $\bar{\nu}$, and the power P (or the intensity I). Each photon has a certain amount of energy, E, given by the relation (for a vacuum, where the index of refraction is 1)

$$E = h\nu = \frac{hc}{\lambda} = hc\bar{\nu} \tag{16.1}$$

where h is Planck's constant, equal to 6.6256×10^{-27} erg.s, c is the velocity of light, equal to 2.99792×10^{10} cm/s in vacuum, and $\bar{\nu}$ is the reciprocal of λ. The usual frequency unit is hertz, Hz, which corresponds to one cycle per second. Different units are used for λ in the various regions of the electromagnetic spectrum. We have the relation

$$\bar{\nu}(\text{cm}^{-1}) = 10^4/\lambda(\mu\text{m}) \tag{16.2}$$

Example 16.1. An optical filter permits the passage of only a red spectrum line of wavelength 6600 Å. Calculate a) the wavelength in nm and μm, b) the frequency, and c) the wavenumber.

Solution. a) We have

$$\lambda = 6600\text{Å} = 6600 \times 10^{-10}\ \text{m} = 660.0 \times 10^{-9}\ \text{m}$$

$$= \mathbf{660.0}\ nm = 0.6600 \times 10^{-6}\ m = \mathbf{0.6600}\ \mu m.$$

b) On the basis of Equation (16.1), we have

$$\nu = (2.99792 \times 10^{10} \text{ cm/s})/(6.600 \times 10^{-5} \text{ cm}) = \mathbf{4.542 \times 10^{14}} \text{ s}^{-1}$$

c) On the basis of Equation (16.2), we have

$$\bar{\nu} = 10^4 \ (\mu\text{m/cm})/0.6600 \ \mu\text{m} = \mathbf{1.515 \times 10^4} \text{ cm}^{-1}.$$

16.2 Direct (Absolute) Spectrophotometry

When monochromatic radiation passes through a solution of the absorbing substance X, the power of the beam is exponentially decreased along the optical path, because of its absorption by compound X. The decrease in radiant power depends on the concentration of substance X and the distance traversed by the beam in the solution. These relations are expressed by the Lambert-Beer law, which is usually referred to as Beer's law and is expressed in the form

$$A = \log \frac{P_0}{P} = -\log T = \log \left(\frac{100}{\%T}\right) = abc_{g/L} = \varepsilon bc_{mol/L} \qquad (16.3)$$

where

A = absorbance

P_0 = radiant power of incident radiation

P = power of transmitted radiation

T = transmittance, which is also expressed in per cent (%T)

a = proportionality constant, when the concentration c of the solution is expressed in g/L, called the *absorptivity*

b = path length through the solution

ε = proportionality constant, when c is expressed in mol/L, called the *molar absorptivity* [ε = a (MW), where MW = molecular weight of the absorbing substance].

The absorptivity a is used when the nature of the absorbing substance, and consequently, its molecular weight is unknown, whereas the molar absorptivity ε is preferred when we want to compare quantitatively the absorbance of various compounds of known molecular weight. In practice, the concentration is also expressed in other units, e.g., ppm, whereupon the proportionality constant in Beer's law (Equation (16.3)) has a different value from a or ε. When the concentration is expressed in weight-volume per cent, % w/v (Section 3.1), the proportionality constant is represented by the term $A_{1\,cm}^{1\%}$, which is numerically equal to the absorbance of 1% (w/v) solution of compound X in a 1-cm cell. The proportionality constant a or ε depends on the wavelength of the radiation, the solvent, the molecular structure of substance X, and to a small degree upon temperature.

Since the measurement of the absolute values P and P_0 is impractical, in practice the power of the beam transmitted through the sample is compared to the power of the beam which passes through a blank solution or solvent, whereupon

$$A = \varepsilon bc = \log \frac{P_0}{P} \simeq \log \frac{P_{blank}}{P_{sample}} \tag{16.4}$$

The calculation of concentration c is possible from Equation (16.4), on the basis of measured A and b and the value of ε taken from the literature (rarely) or determined from separate measurement with a standard. Such a procedure is applicable only if Beer's law holds over the concentration range of interest. In practice, a calibration curve is generally used.

Example 16.2. The absorbance of a 4.75×10^{-5} M $KMnO_4$ solution is 0.112 at 525 nm in a 1.000-cm cell. a) Calculate the molar absorptivity ε, and give its units. b) Calculate the proportionality constant k, when the concentration of $KMnO_4$ is expressed in ppm, and give its units. c) Calculate the per cent transmittance of the above solution at 525 nm in a 2.000-cm cell (the density of the solution is equal to 1.000 g/mL).

Solution. a) Substituting the data in Equation (16.3), we have

$$\varepsilon = \frac{A}{bc} = \frac{0.112}{(1.000\ \text{cm})(4.75 \times 10^{-5}\ \text{mol/L})} = 2.36 \times 10^3\ \text{L mol}^{-1}\ \text{cm}^{-1}.$$

b) We have ppm = mg/L. Hence, on the basis of Equation (16.3) we have for the proportionality constant k

$$k = \frac{A}{bc} = \frac{0.112}{(1.000\ \text{cm})(4.75 \times 10^{-5}\ \text{mol/L})(158.04 \times 10^{-3}\ \text{mg/mol})}$$
$$= \mathbf{1.49 \times 10^{-2}}\ \text{L mg}^{-1}\ \text{cm}^{-1} = \mathbf{1.49 \times 10^{-2}}\ \text{ppm}^{-1}\ \text{cm}^{-1}.$$

c) We have

$$A = 2.36 \times 10^3\ \text{L/mol cm} \times 2.000\ \text{cm} \times 4.75 \times 10^{-5}\ \text{mol/L} = 0.224.$$

Hence,

$$\%T = 100 \times 10^{-A} = 100 \times 10^{-0.224} = \mathbf{59.7}.$$

Example 16.3. A solution of substance X, that contains 0.1200 g X/L, has an absorbance of 0.126 at 257 nm in a 1.000-cm cell. a) Calculate the specific absorption coefficient $A_{1\,cm}^{1\%}$. b) Derive a formula for the interconversion of a, ε, and $A_{1\,cm}^{1\%}$.

Solution. a) We have the relations

$$0.126 = a \times 1.000\ \text{cm} \times 0.1200\ \text{g X/L} \tag{16.5}$$

$$A_{1\,cm}^{1\%} = a \times 1.000 \text{ cm} \times 1.000 \text{ g X}/0.1000 \text{ L} \tag{16.6}$$

Dividing Equations (16.5) and (16.6) and solving for $A_{1\,cm}^{1\%}$ we have

$$A_{1\,cm}^{1\%} = 0.126/0.0120 = \mathbf{10.5}.$$

b) We have

$$\varepsilon = a \cdot MW = (A_{1\,cm}^{1\%}/10) \cdot MW \tag{16.7}$$

Example 16.4. If A_1 is the absorbance of substance X of unknown concentration C_1, and A_2 is the absorbance of the solution that is obtained after the addition of a volume V_2 of a standard solution of X of concentration C_2 to a volume V_1 of the original solution, prove that

$$C_1 = \frac{A_1 V_2 C_2}{A_2(V_1 + V_2) - A_1 V_1} \tag{16.8}$$

Solution. We have the relations

$$A_1 = \varepsilon b C_1 \tag{16.9}$$

$$A_2 = \varepsilon b \left(\frac{V_1 C_1 + V_2 C_2}{V_1 + V_2} \right) \tag{16.10}$$

Dividing Equations (16.9) and (16.10) and solving for C_1 we find the sought-for relation (16.8).

Example 16.5. The transmittance of solutions B and D of substance X is 15.85% and 50.12%, respectively. At what volume ratio should solutions B and D be mixed so that the transmittance of the resulting solution would be 21.13% (all measurements are made in the same cell, and at the same wavelength)?

Solution. *1st method.* Assume that solutions B and D should be mixed at a volume ratio of

$$V_B/V_D = x, \tag{16.11}$$

whereupon a solution of absorbance A_M is obtained. We have the relations

$$A_B = -\log 0.1585 = 0.800 = \varepsilon b C_B \tag{16.12}$$

$$A_D = -\log 0.5012 = 0.300 = \varepsilon b C_D \tag{16.13}$$

$$A_M = -\log 0.2113 = 0.675 = \varepsilon b \left(\frac{V_B C_B + V_D C_D}{V_B + V_D} \right) \tag{16.14}$$

Dividing Equations (16.12) and (16.13), we have

$$C_B/C_D = 2.667, \tag{16.15}$$

whereas dividing Equations (16.14) and (16.13) we have

$$\frac{V_B C_B + V_D C_D}{(V_B + V_D) C_D} = 2.25 \tag{16.16}$$

Combining Equations (16.11), (16.15), and (16.16), we have

$$\frac{(C_B/C_D)V_B + V_D}{V_B + V_D} = 2.25 = \frac{2.667V_B + V_D}{V_B + V_D} = \frac{2.667(V_B/V_D) + 1}{(V_B + V_D) + 1} = \frac{2.667x + 1}{x + 1} \quad (16.17)$$

Solving Equation (16.17) for x, we find x = **3.00**.

2nd method. Equations (16.11), (16.12), (16.13), and (16.14) are also valid for this method. We have the relation

$$A_M = \frac{V_B A_B + V_D A_D}{V_B + V_D} \quad (16.18)$$

or, because of Equation (16.11),

$$A_M = \frac{xV_D A_B + V_D A_D}{xV_D + V_D} = \frac{xA_B + A_D}{x + 1}$$

or

$$xA_B + A_D = xA_M + A_M$$

or

$$x = \frac{A_M - A_D}{A_B - A_M} = \frac{0.675 - 0.300}{0.800 - 0.675} = \mathbf{3.00}.$$

Example 16.6. A 1.43×10^{-4} $\underline{M}$ solution of the pure substance B (MW 180.00) has an absorbance of 0.572, whereas solution D, which contains 0.1358 grams of a pharmaceutical preparation of substance B in 1 liter, has a transmittance of 0.362 at 284 nm, in a 1.00-cm cell. Calculate the %B in the sample.

Solution. Let c be the concentration of B in solution D. Substituting the data in Beer's Equation (16.3), we have

$$A = -\log 0.362 = 0.441 = \varepsilon \times 1.00 \times c \quad (16.19)$$

and

$$\varepsilon = 0.572/1.00 \times 1.43 \times 10^{-4} = 4.00 \times 10^3 \quad (16.20)$$

Combining Equations (16.19) and (16.20), we find $c = 0.441/4.00 \times 10^3 = 1.10 \times 10^{-4}$ $\underline{M}$. Hence, the preparation contains

$$\frac{(1.10 \times 10^{-4}\ \text{mol/L})(180.00\ \text{g/mol})}{0.1358\ \text{g}} \times 100 = \mathbf{14.6\%}\ \text{B}.$$

Example 16.7. A method for the determination of the molecular weight of organic bases, e.g., amines, is based on their conversion to their picrate salts by treatment with picric acid, 2,4,6-trinitrophenol (MW 229.11), according to the reaction

$$R - NH_2 + HOC_6H_2(NO_2)_3 \rightleftharpoons \mathbf{R\text{-}NH_3^+\ {}^-OC_6H_2(NO_2)_3} \quad (16.21)$$

For the determination of the molecular weight of an amine, 0.0302 g of its purified picrate salt was dissolved in 100.0 mL of water (solution B). A 2.00-mL aliquot of solution B was diluted to 100.0 mL (solution D). The absorbance of solution D was

0.270 in a 1.00-cm cell at 380 nm, where only picrate absorbs, with a molar absorptivity $\varepsilon_{380} = 1.35 \times 10^4$ L mol^{-1} cm^{-1}. Calculate the molecular weight of the amine.

Solution. If MW is the molecular weight of the amine picrate and c is its concentration in solution D, on the basis of Equation (16.3) we have

$$c = 0.270/1.00 \times 1.35 \times 10^4 = 2.00 \times 10^{-5} \text{ \underline{M}}.$$

The 0.0302 grams of amine picrate are equivalent to 5.00 liters of solution D. Hence, we have

$$\frac{0.0302 \text{ g}}{5.00 \text{ L} \times \text{MW (g/mol)}} = 2.00 \times 10^{-5} \text{ mol/L} \tag{16.22}$$

Solving Equation (16.22), we find MW = 302. Hence, from Equation (16.21) the molecular weight of the amine is equal to

$$302 - 229.11 = \mathbf{73}.$$

Example 16.8. The absorbance of solution B, 0.100 M in NaX and 1.00 M in NaOH, due only to X^- ion, was 0.276, whereas the absorbance of solution D, 0.200 M in HX, was 0.138 in the same cell and at the same wavelength. Calculate the dissociation constant of the acid HX.

Solution. Let c_1 and c_2 be the concentrations of ion X^- in solutions B and D, respectively. Then $c_1 = 0.100$ M (complete dissociation of the acid HX because of the strongly alkaline solution) and $c_2 = [X^-]$. According to Equation (16.3), we have the relations

$$0.276 = \varepsilon b c_1 = \varepsilon b 0.100 \tag{16.23}$$

$$0.138 = \varepsilon b c_2 = \varepsilon b [X^-] \tag{16.24}$$

Dividing Equations (16.23) and (16.24) and solving for $[X^-]$, we find that $[X^-] = 0.050$ M. Hence, we have $[HX] = 0.200 - 0.050 = 0.150$ M and

$$K_a = \frac{[H^+][X^-]}{[HX]} \simeq \frac{[X^-]^2}{[HX]} = \frac{(0.050)^2}{0.150} = \mathbf{0.0167}.$$

Example 16.9. The undissociated molecule of the acid-base indicator HX has a molar absorptivity of 500 L mol^{-1} cm^{-1} at 440 nm, where ion X^- does not absorb. Ion X^- has a molar absorptivity of 250 L mol^{-1} cm^{-1} at 650 nm, where the molecule HX does not absorb. The transmittance of an aqueous solution of the indicator was 0.398 at 440 nm and 0.316 at 650 nm, in a 1.00-cm cell. The dissociation constant of the indicator is 2.50×10^{-5}. Calculate the pH of the aqueous solution.

Solution. We have

$$A_{650} = -\log 0.316 = 0.500 = 250 \times 1.00 \times [X^-];$$

hence,

$$[X^-] = 2.00 \times 10^{-3} \text{ \underline{M}}$$

and

$$A_{440} = -\log 0.398 = 0.400 = 500 \times 1.00 \times [HX];$$

hence,

$$[HX] = 8.00 \times 10^{-4} \ \underline{M}.$$

Substituting the values of K_a, $[X^-]$ and $[HX]$ in the expression for the dissociation constant of the indicator, we have

$$2.50 \times 10^{-5} = \frac{[H^+] \times 2.00 \times 10^{-3}}{8.00 \times 10^{-4}} \tag{16.25}$$

Solving Equation (16.25), we find that $[H^+] = 1.00 \times 10^{-5}$ $\underline{M}$; hence, pH = **5.00**.

Example 16.10. Two colorless ions, M and L, form the colored complex ML (charges have been omitted for convenience), which absorbs at 525 nm and has a molar absorptivity $\varepsilon_{525} = 1.10 \times 10^3$ L mol^{-1} cm^{-1} and an instability constant equal to 1.00×10^{-4}. What would be the absorbance of solution X, prepared by mixing equal volumes of 0.00200 $\underline{M}$ solutions of M and L, in a 1.00-cm cell at 525 nm?

Solution. The initial concentrations of M and L in solution X (prior to the formation of the complex) are equal to 0.00100 $\underline{M}$. Assume that after equilibrium is established $[M] = [L] = y$, whereupon $[ML] = 0.00100 - y$. Hence, we have

$$1.00 \times 10^{-4} = \frac{[M][L]}{[ML]} = \frac{y^2}{0.00100 - y}$$

or

$$y = 270 \times 10^{-4} \ \underline{M} = [M] = [L]$$

and

$$[ML] = 0.00100 - 2.70 \times 10^{-4} = 7.30 \times 10^{-4} \ \underline{M}.$$

Hence, the absorbance of solution X is equal to

$$A = 1.10 \times 10^3 \times 1.00 \times 7.30 \times 10^{-4} = \mathbf{0.803}.$$

Example 16.11. For the spectrophotometric determination of the colorless ion M, the ion reacts with a controlled large excess of the colorless complexing reagent L (charges have been omitted for convenience) and is converted only partially to the colored complex ML, which has a formation constant K_f. Prove that the absorbance A is proportional to the total (analytical) concentration C_M.

Solution. Since L is in large excess over M, [L] remains practically constant. We have the relations

$$A = \varepsilon b[ML] \tag{16.26}$$

$$C_M = [ML] + [M] \tag{16.27}$$

$$K_f = [ML]/[M][L] \tag{16.28}$$

Combining Equations (16.27) and (16.28), we have

$$C_M = [ML] + \frac{[ML]}{K_f[L]} = [ML]\left(1 + \frac{1}{K_f[L]}\right) \tag{16.29}$$

Combining Equations (16.26) and (16.29), we have

$$A = \frac{\varepsilon b C_M}{1 + (1/K_f[L])} \tag{16.30}$$

From Equation (16.30) it can be seen that A is proportional to C_M, although the complexation of M is only partial, provided that the terms ε, b, K_f, and [L] are kept constant.

Example 16.12. The dissociation constant of the indicator HX is 2.50×10^{-6}. The following absorbance data are for 2.50×10^{-4} M solutions of the indicator in strongly acidic and strongly alkaline solutions, measured in 1.00-cm cells.

	Absorbance, A			Absorbance, A	
λ, nm	pH = 1.00	pH = 13.00	λ, nm	pH = 1.00	pH = 13.00
460	0.210	0.025	560	0.177	0.250
470	0.217	0.025	580	0.140	0.320
480	0.220	0.026	600	0.110	0.350
490	0.225	0.026	610	0.099	0.355
495	0.227	0.028	620	0.088	0.349
500	0.223	0.029	630	0.076	0.340
510	0.221	0.036	650	0.066	0.312
530	0.210	0.096	680	0.060	0.260
550	0.192	0.192			

a) What wavelength will be appropriate for the spectrophotometric determination of the indicator 1) in strongly acidic solutions, 2) in strongly alkaline solutions?

b) What would the absorbance be for an 8.00×10^{-5} M solution of the indicator in the alkaline form (X^-), at 570 nm in a 5.00-cm cell?

c) At what wavelength would the absorbance of an indicator solution be independent of pH?

Solution. a) On the basis of the data, the absorption spectra of X^- and HX are given in Figure 16.1.

Usually, the wavelength λ_{max} is selected, at the plateau of maximum absorbance, to achieve maximum sensitivity. Such wavelengths are **495** nm and **610** nm, for strongly acidic and strongly alkaline solutions, respectively.

b) From the spectrum we have $A_{X^-} = 0.292$, for b = 1.00 cm and $[X^-] \simeq C_{HX} = 2.50 \times 10^{-4}$ M. Hence, for the ion X^- we have

$$\varepsilon_{X^-,\,570} = 0.292/(1.00 \times 2.50 \times 10^{-4}) = 1168.$$

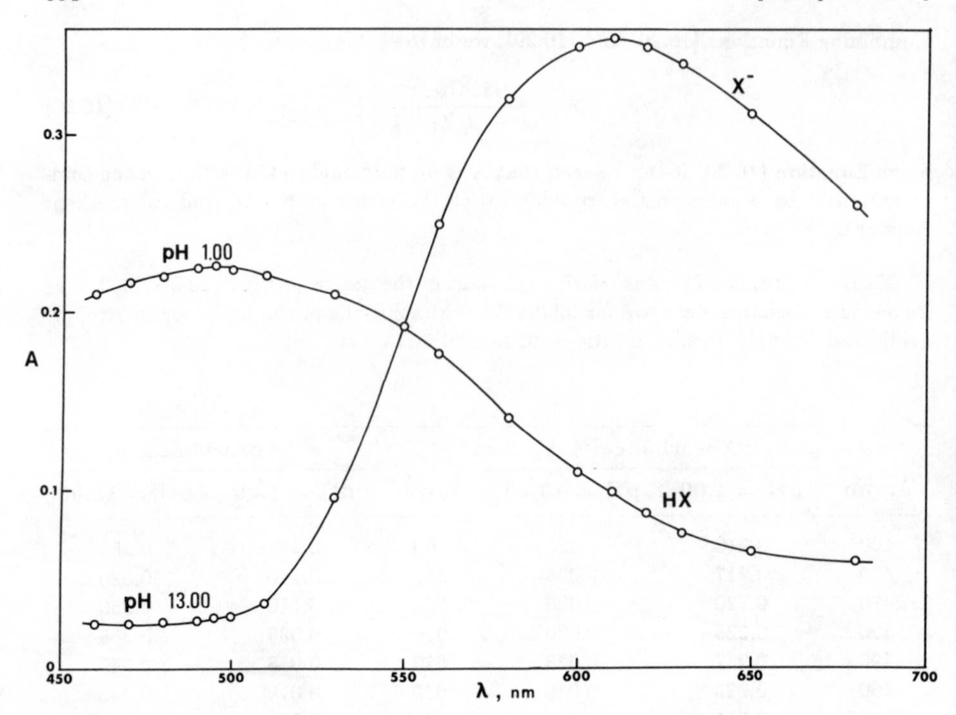

Figure 16.1. *Absorption spectra from the data in Example 16.12.*

Hence, for b = 5.00 cm and $[X^-] \simeq C_{HX} = 8.00 \times 10^{-5}$ M, we will have

$$A = 1168 \times 5.00 \times 8.00 \times 10^{-5} = \mathbf{0.467}.$$

c) The absorbance would be independent of pH at **550** nm, where A is the same for HX and X^-. This wavelength is called the *isosbestic point.*

Example 16.13. A 3.75×10^{-4} M solution of the indicator HX described in Example (16.12) has an absorbance of 0.453 at 600 nm in a 1.00-cm cell. a) What is the pH of the solution? b) What would the absorbance of the solution be at 550 nm, in the same cell?

Solution. a) From the data of Example (16.12), we find that

$$\varepsilon_{HX,\ 600} = \frac{0.110}{1.00 \times 2.50 \times 10^{-4}} = 440, \quad \varepsilon_{X^-,\ 600} = \frac{0.350}{1.00 \times 2.50 \times 10^{-4}} = 1400.$$

Hence, we have

$$1400 \times 1.00 \times [X^-] + 440 \times 1.00 \times (3.75 \times 10^{-4} - [X^-]) = 0.453 \qquad (16.31)$$

Solving Equation (16.31), we find that $[X^-] = 3.00 \times 10^{-4}$ M, whereupon $[HX] = 3.75 \times 10^{-4} - 3.00 \times 10^{-4} = 0.75 \times 10^{-4}$ M. Substituting the values of K_a, $[X^-]$ and $[HX]$

in the expression for the dissociation constant of the indicator, we have

$$2.50 \times 10^{-6} = \frac{[H^+] \times 3.00 \times 10^{-4}}{0.75 \times 10^{-4}} \tag{16.32}$$

Solving Equation (16.32), we find that $[H^+] = 6.25 \times 10^{-7}$ $\underline{M}$; hence, pH = **6.20.**

b) We have

$$\varepsilon_{HX,\ 550} = \varepsilon_{X^-,\ 550} = \frac{0.192}{1.00 \times 2.50 \times 10^{-4}} = 768.$$

Hence, the absorbance of the solution at 550 nm will be equal to

$$A = 768 \times 1.00 \times 3.75 \times 10^{-4} = \mathbf{0.288}.$$

Deviations from Beer's law. In many cases a positive or negative deviation is observed from the linear relation between A and c (Equation 16.3). The use of a polychromatic beam of radiation results in a deviation from Beer's law, unless the molar absorptivities of all radiations of the beam are equal. The same result is caused by *stray radiation*, that is, by the sum of radiations that reach the detector without being within the wavelength range which has been selected by the monochromator. If P_0, P, and P_s are the radiant powers of the incident, the transmitted, and the stray radiation, respectively, the real absorbance is $A_r = \log(P_0/P)$, whereas the experimentally measured absorbance is $A_{exp} = \log[(P_0 + P_s)/(P + P_s)]$. Since $A_{exp} < A_r$, a negative deviation from Beer's law is observed (Example (16.14). Apparent deviations from Beer's law are also observed when the absorbing species participate in chemical equilibria, such as equilibria involving polymerization, complex formation, and dissociation of weak acids or bases (Example (16.15)).

Example 16.14. If the stray radiation P_s is equal to a) 1.0%, or b) 2.0% of the radiant power P_0 that passes through the reference cell, and the real transmittance of samples D_1 and D_2 (in the absence of stray radiation) is 0.100 and 0.500, respectively, calculate the % error in the measured concentration of each sample that is caused by the stray radiation.

Solution. a) The real absorbance of sample D_1 is equal to $A_{1,r} = \log(100/10.0) = 1.00$, whereas its experimentally measured absorbance is equal to $A_{1,exp} = \log[(100 + 1.0)/(10.0 + 1.0)] = 0.963$. Hence, the % error in the concentration D_1 is equal to

$$\frac{0.963 - 1.00}{1.00} \times 100 = \mathbf{-3.7\%}.$$

Similarly, for sample D_2 we have $A_{2,r} = \log(100/50.0) = 0.301$ and $A_{2,exp} = \log[(100 + 1.0)/(50.0 + 1.0) = 0.297]$, and the % error in the concentration of D_2 is equal to

$$\frac{0.297 - 0.301}{0.301} \times 100 = \mathbf{-1.3\%}.$$

b) For solution D_1 we have $A_{1,exp} = \log[(100+2.0)/(10.0+2.0)] = 0.929$. Hence, the % error in the concentration of D_1 is equal to

$$\frac{0.929 - 1.00}{1.00} \times 100 = \mathbf{-7.1\%}.$$

For sample D_2, we have $A_{2,exp} = \log[(100 + 2.0)/(50.0 + 2.0)] = 0.293$. Hence, the % error in the concentration of D_2 is equal to

$$\frac{0.293 - 0.301}{0.301} \times 100 = \mathbf{-2.7\%}.$$

Note. From the values of the % error for samples D_1 and D_2 it can be seen that the % error that is due to the stray radiation decreases with increasing sample transmittance and increases with increasing stray radiation. Usually, a $P_s > 0.01P_0$ is unacceptable.

Example 16.15. The molar absorptivity of a weak acid HX ($K_a = 1.00 \times 10^{-5}$) is 1100 L mol^{-1} cm^{-1} at 305 nm, where X^- does not absorb. From the absorbance values for a) 1.000×10^{-3} $\underline{M}$, b) 5.00×10^{-4} $\underline{M}$, and c) 2.00×10^{-4} $\underline{M}$ HX solutions at 305 nm in a 1.000-cm cell, determine whether the system shows an apparent deviation from Beer's law. If an apparent deviation is observed, is it positive or negative?

Solution. Let C be the total (analytical) concentration of the HX solution. We have the relation

$$K_a = \frac{[H^+][X^-]}{[HX]} \simeq \frac{[H^+]^2}{C - [H^+]} = 1.00 \times 10^{-5} \tag{16.33}$$

For $C = 1.000 \times 10^{-3}$ $\underline{M}$, Equation (16.33) becomes

$$\frac{[H^+]^2}{1.000 \times 10^{-3} - [H^+]} = 1.00 \times 10^{-5} \tag{16.34}$$

Solving the quadratic Equation (16.34), we find $[H^+] = 9.5 \times 10^{-5}$ $\underline{M}$. Hence, we have $[HX] = 1.000 \times 10^{-3} - 9.5 \times 10^{-5} = 9.05 \times 10^{-4}$ $\underline{M}$. Hence, the absorbance of a 1.000×10^{-3} $\underline{M}$ HX solution is equal to $A = 1100 \times 1.000 \times 9.05 \times 10^{-4} = \mathbf{0.996}$. Similarly, for $C = 5.00 \times 10^{-4}$ $\underline{M}$, from Equation (16.33) we find that $[H^+] = 6.61 \times 10^{-5}$ $\underline{M}$. Hence, $[HX] = 5.00 \times 10^{-4} - 6.61 \times 10^{-5} = 4.34 \times 10^{-4}$ $\underline{M}$, and $A = 1100 \times 1.000 \times 4.34 \times 10^{-4} = \mathbf{0.477}$. Similarly, for $C = 2.00 \times 10^{-4}$ $\underline{M}$, we find that $[H^+] = 4.00 \times 10^{-5}$ $\underline{M}$, $[HX] = 1.60 \times 10^{-4}$ $\underline{M}$, and $A = 1100 \times 1.000 \times 1.60 \times 10^{-4} = \mathbf{0.176}$. From the three calculated absorbance values (or from a plot, A = f(C)) it can be seen that there is no linear relation between A and C, and that the system shows an apparent positive deviation from Beer's law.

Note. The deviation from Beer's law is apparent, because when the concentration C increases, the degree of dissociation of the acid decreases, and a larger percentage of the acid remains in the undissociated form. There is a linear relation between absorbance and the *real* concentration of the absorbing species at the equilibrium state, that is, Beer's law is valid in the form $A = \varepsilon b[HX]$.

Colorimetry. In colorimetry the color intensity of a sample solution is compared visually with the color intensities of a series of standard solutions of the substance being determined, in identical tubes. The following relation is the basis of colorimetry

$$b_s c_s = b_u c_u \qquad (16.35)$$

where c_s and c_u are the concentrations of a standard solution and the unknown solution, respectively, and b_s and b_u are the corresponding path lengths of the tubes (cells). If the path lengths are held constant, then the concentrations are directly determined by matching the color intensities.

Spectrophotometric analysis of mixtures of absorbing substances. In the case of a mixture of absorbing substances, Beer's law is expressed by the relation

$$A_t = A_1 + A_2 + \ldots + A_n = \varepsilon_1 bc_1 + \varepsilon_2 bc_2 + \ldots + \varepsilon_n bc_n \qquad (16.36)$$

where A_t is the total absorbance of the mixture, and A_1, A_2, ... A_n are the absorbances of the individual absorbing components 1, 2, ... n. Equation (16.36) is the basis for the quantitative analysis of mixtures of absorbing substances. The principle of the additivity of absorbances, which is expressed by this equation, is an indispensable condition for the spectrophotometric analysis of a mixture of n components with partially overlapping spectra. In such an analysis, the absorbance of the mixture is measured at n wavelengths, and the concentrations of n components are calculated by solving the system of n equations with n unknowns.

Example 16.16. A 0.9934-g sample of steel is dissolved in acid and the solution diluted to 250.0 mL (solution B). A 50.00-mL aliquot of solution B is treated with potassium persulfate in the presence of Ag^+ ions acting as a catalyst, and potassium periodate, whereupon Mn and Cr are oxidized to MnO_4^- and $Cr_2O_7^{2-}$ and diluted to 100.0 mL (solution D). The absorbance of solution D at 440 nm and 545 nm was 0.204 and 0.170, respectively, in a 1.00-cm cell. Calculate the per cent Mn and Cr in steel, also taking into account the following data from the literature:

λ, nm	$\varepsilon_{MnO_4^-}$	$\varepsilon_{Cr_2O_7^{2-}}$
440	95	369
545	2350	11.0

Solution. Let $[MnO_4^-] = y$ and $[Cr_2O_7^{2-}] = z$ in solution D, whereupon we have

$$0.204 = 95y + 369z \qquad (16.37)$$

$$0.170 = 2350y + 11.0z \qquad (16.38)$$

Solving the system of Equations (16.37) and (16.38), we find that $y = 6.98 \times 10^{-5}$ and $z = 5.35 \times 10^{-4}$. Hence, the steel contains

$$\frac{6.98 \times 10^{-5}\ \text{mmol/mL} \times 100.0\ \text{mL}\ (250.0/50.00) \times 54.94\ \text{mg Mn/mmol}}{993.4\ \text{mg}} \times 100$$

$$= \mathbf{0.193\%}\ \text{Mn},$$

and

$$\frac{5.35 \times 10^{-4}\ \text{mmol/mL} \times 100.0\ \text{mL}\ (250.0/50.00) \times (2 \times 52.00\ \text{mg Cr/mmol})}{993.4\ \text{mg}} \times 100$$

$$= \mathbf{2.80\%}\ \text{Cr.}$$

Photometric error. The photometric error of an analysis is the error in the determination of concentration that results from the reading error ΔT of the transmittance scale in the spectrophotometer, which is usually constant and approximated by the width of the instrument needle or the width of the recorded line. If Δc is the respective absolute error in the concentration c of the substance being determined, the relation between the relative error $\Delta c/c$ and ΔT and T is given by the equation

$$\frac{\Delta c}{c} = \left(\frac{0.4343}{T \log T}\right) \Delta T \tag{16.39}$$

(Equation (16.39) is obtained by differentiating the equation $c = -\log T/\varepsilon b$ with respect to T and combining the result with this equation.) The term ΔT is usually in the range 0.002–0.01 and is called the *absolute photometric error.* A plot of the relative error $\Delta c/c$ as a function of transmittance T gives a curve with a minimum at $T = 0.368$ ($A = 0.434$) (Example 16.17).

Example 16.17. Calculate the per cent relative error in the concentration (or absorbance), $(\Delta c/c)\ 100$, as a function of transmittance T, for an absolute photometric error (ΔT) of 0.01 and draw the corresponding curve.

Solution. Using Equation (16.39) the following data are calculated and the corresponding curve is drawn (Figure 16.2). For example, for $T = 0.05$ we have

$$\frac{\Delta c}{c} \times 100 = \frac{0.4343}{(0.05) \log(0.05)} = -6.68.$$

T	$-(\Delta c/c)100$	T	$-(\Delta c/c)100$
0.01	21.72	0.50	2.89
0.05	6.68	0.55	3.04
0.10	4.34	0.60	3.26
0.15	3.51	0.65	3.57
0.20	3.11	0.70	4.01
0.25	2.89	0.75	4.63
0.30	2.77	0.80	5.60
0.35	2.72	0.85	7.24
0.40	2.73	0.90	10.55
0.45	2.78	0.95	20.52

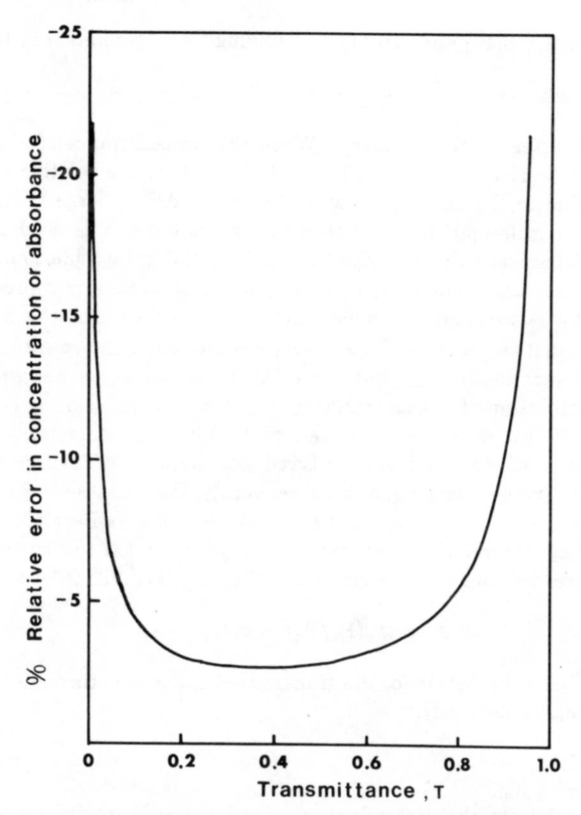

Figure 16.2. *Per cent relative error in the calculated concentration (or in A, Equation (16.3)) as a function of transmittance, for an absolute photometric error (ΔT) of 0.01.*

Example 16.18. a) Calculate the per cent relative error in the computed concentration of a substance, resulting from the photometric error, if the transmittance of the solution is 0.500 and the absolute error in the transmittance reading is 0.002. b) If the preceding transmittance measurement was made in a 1.00-cm cell, calculate what the path length should have been to ensure the smallest error in the calculation of the concentration.

Solution. a) Substituting the data in Equation (16.39), we have

$$\frac{\Delta c}{c} \times 100 = \left(\frac{0.4343}{0.500 \log 0.500}\right) 0.2 = \mathbf{-0.58\%}.$$

b) The smallest error is ensured for A = 0.434 (Figure 16.2). We have the relations

$$A = -\log 0.500 = 0.301 = \varepsilon \times 1.00 \times c \tag{16.40}$$

$$0.434 = \varepsilon bc \tag{16.41}$$

Dividing Equations (16.40) and (16.41) and solving for b, we find that b = **1.44** cm.

Differential Spectrophotometry. When the transmittance of a solution is very low ($T < 0.1$, $A > 1$) or very high ($T > 0.8$, $A < 0.1$), the relative analytical error $\Delta c/c$, which is due to the absolute photometric error ΔT, is large (Example (16.17)). In such cases, the technique of differential spectrophotometry is used to decrease the relative error. There are three methods of differential spectrophotometry: the high absorbance method, the trace analysis method, and the maximum precision method. The *high absorbance method* is applied when the absorbance of the solution is high ($A > 1$). In this method, the 0% T ($A = \infty$) reading on the spectrophotometer is set as in classical spectrophotometry, but the 100% T ($A = 0$) is set not with a blank but with a standard solution of concentration c_2 ($c_2 < c_x$, where c_x is the concentration of the unknown solution). In this way the scale of the spectrophotometer is expanded, and consequently, the relative analytical error is reduced, because ΔT still remains the same. The extent of the scale expansion, and consequently, the increase in precision, depend upon the transmittance of the standard solution. As long as Beer's law is obeyed for solutions c_2 and c_x, the relative absorbance A′ is proportional to the difference $c_x - c_2$ between the concentrations of the sample and the standard. That is, we have

$$A' = \log(P_2/P_x) = \varepsilon b(c_x - c_2) \tag{16.42}$$

where P_2 and P_x are the powers of the transmitted radiations through the standard c_2 and the sample c_x, respectively.

Example 16.19. In the determination of substance K by differential spectrophotometry (high absorbance method) the 0% T was set with a closed shutter (without light), whereas the 100% T was set with a 1.224×10^{-4} $\underline{M}$ solution of pure substance K (solution B). During the successive filling of the cell with a 1.388×10^{-4} $\underline{M}$ standard solution of substance K (solution D) and an unknown solution of K (solution E), the absorbance was 0.205 and 0.335, respectively. Calculate the concentration c_x of substance K in solution E.

Solution. *1st method.* Substituting the data in Equation (16.42), we have

$$0.205 = \varepsilon b(1.388 \times 10^{-4} - 1.224 \times 10^{-4}) \tag{16.43}$$

$$0.335 = \varepsilon b(c_x - 1.224 \times 10^{-4}) \tag{16.44}$$

Dividing Equations (16.43) and (16.44) and solving for c_x, we find that c_x = $\mathbf{1.492 \times 10^{-4}}$ $\underline{M}$.

2nd method. Let A be the absorbance that solution B would have had, if the 100% T was set with a blank (classic spectrophotometry), whereupon the absorbances of solutions D and E would have been A + 0.205 and A + 0.335, respectively. We have the relation

$$\frac{A}{A + 0.205} = \frac{1.224 \times 10^{-4}}{1.388 \times 10^{-4}}$$

or A = 1.530. Similarly, we have

$$\frac{A + 0.0205}{A + 0.335} = \frac{1.530 + 0.205}{1.530 + 0.335} = \frac{1.338 \times 10^{-4}}{c_x}$$

or

$$c_x = \mathbf{1.492 \times 10^{-4}} \ \underline{M}.$$

Spectrophotometric determination of the formula of a complex ion and its instability constant. The spectrophotometric determination of the ratio of ligand to metal ion, s/r, in the complex M_rL_s, which is formed according to the reaction (charges have been omitted for convenience)

$$rM + sL \rightleftharpoons M_rL_s \qquad (16.45)$$

can be performed by several methods.

In the *mole-ratio method*, a series B of solutions is prepared, in which the analytical concentration of one of the reactants (usually of the metal M) is kept constant, while that of the other is varied, so that the [L]/[M] varies, for example, in the range 0.1 to 10. The absorbances of these solutions are measured, usually at the wavelength of maximum absorbance of the complex M_rL_s and a plot is prepared, A(ordinate) as a function of ([L]/[M]). Absorbance A increases linearly until near the stoichiometric point, at which M is almost quantitatively in the form M_rL_s and [L]/[M] = s/r, whereas beyond this point $[M_rL_s]$, and hence its absorbance, remains practically constant and the curve becomes parallel to the x-axis. In the region of the stoichiometric point a curvature is observed, which increases with increasing instability constant K_{inst} of the complex. Despite the curvature, the value of the ratio s/r and the stoichiometric formula of the complex can be found by extrapolating the straight line portions of the curve. When the curvature is not too small or too large, it can be used for the calculation of K_{inst} (Example (16.20)).

In the *method of continuous variations* (Job's method) the absorbance is measured in a series of solutions which have the same *total* molarity but different concentrations [M] and [L]. In practice, two *equimolar* solutions of the metal ion M and the ligand L are prepared, and from these solutions, a series C of solutions is prepared by mixing V_L mL of solution L and $(V_T - V_L)$ mL of solution M, where V_T is the total volume, fixed and the same for all solutions, and V_L is the variable volume $(0 \le V_L \le V_T)$. The absorbances of all solutions of series C are measured, usually at the wavelength of maximum absorbance of M_rL_s and a plot of absorbance versus the mole fraction of the ligand, $A = f(V_L/V_T)$, or of the metal ion, $A = f[(V_T - V_L)/V_T]$, is prepared. The position of the maximum absorbance, A_{max}, relative to the axis of the molar fractions (abscissa) gives the stoichiometry of the complex (Example (16.22)). In the region of A_{max} a curvature is observed, and the stoichiometric point is located by extrapolating the straight-line portions of the curve. The curvature also permits the determination of K_{inst}.

Example 16.20. The ion M^{2+} reacts with the ion L^-, forming a colored complex. To determine the formula of the complex, a series of solutions was prepared in which the analytical concentration of M^{2+} was 1.00×10^{-4} M, while the concentration of L^- varied. The following data were obtained by measuring the absorbance at 510 nm, where only the complex absorbs, in a 1.00-cm cell:

C_L, M	A_{510}	C_L, M	A_{510}
0.250×10^{-4}	0.148	1.50×10^{-4}	0.589
0.500×10^{-4}	0.294	1.75×10^{-4}	0.592
0.750×10^{-4}	0.434	2.00×10^{-4}	0.594
1.00×10^{-4}	0.543	2.50×10^{-4}	0.597
1.25×10^{-4}	0.580	3.00×10^{-4}	0.599

a) What is the formula of the complex? b) What is the molar absorptivity of the complex at 510 nm? c) What is the instability constant of the complex?

Solution. a) The data are plotted in Figure 16.3, from which it can be seen that M^{2+} and L^- react at a 1:1 ratio. Hence, the formula of the complex is $[ML]^+$ (it is assumed that complexes of the formula $[M_nL_n]^{n+}$, where n is an integer larger than 1, are not formed in such dilute solutions).

b) The absorbance A_{max} corresponds to practically quantitative complexation of M^{2+}. Hence, we have

$$[ML^+] \simeq 1.00 \times 10^{-4}\ \underline{M} \text{ and } \varepsilon_{ML^+,\,510} = 0.599/(1.00 \times 1.00 \times 10^{-4}) = \mathbf{5.99 \times 10^3}.$$

c) If C is the analytical concentration of M^{2+}, we have

$$[ML^+] = (A/A_{max})C \tag{16.46}$$

$$[M^{2+}] = [L^-] = C - [ML^+] = C - (A/A_{max})C = C[1 - (1/A_{max})] \tag{16.47}$$

$$K_{inst} = \frac{[M^{2+}][L^-]}{[ML+]} = \frac{[1-(A/A_{max})]^2 C}{A/A_{max}} \tag{16.48}$$

Substituting the data corresponding to the stoichiometric point in Equation (16.48), we have

$$K_{inst} = \frac{[1-(0.543/0.599)]^2 \times 1.00 \times 10^{-4}}{0.543/0.599} = \mathbf{9._6 \times 10^{-7}}$$

Note. The value of K_{inst} can also be found from other experimental points of the plot. For example, for $C_M = 1.00 \times 10^{-4}$ M, $C_L = 0.500 \times 10^{-4}$ M and $A = 0.294$, we have $[ML^+] = 0.294/5.99 \times 10^3 = 4.91 \times 10^{-5}$ M, $[M^{2+}] = 1.00 \times 10^{-4} - 4.91 \times 10^{-5} = 5.1 \times 10^{-5}$ M, $[L^-] = 0.500 \times 10^{-4} - 4.91 \times 10^{-5} = 0.09 \times 10^{-5}$ M. Hence

$$K_{inst} = \frac{(5.1 \times 10^{-5})(0.09 \times 10^{-5})}{4.91 \times 10^{-5}} = \mathbf{9._3 \times 10^{-7}}.$$

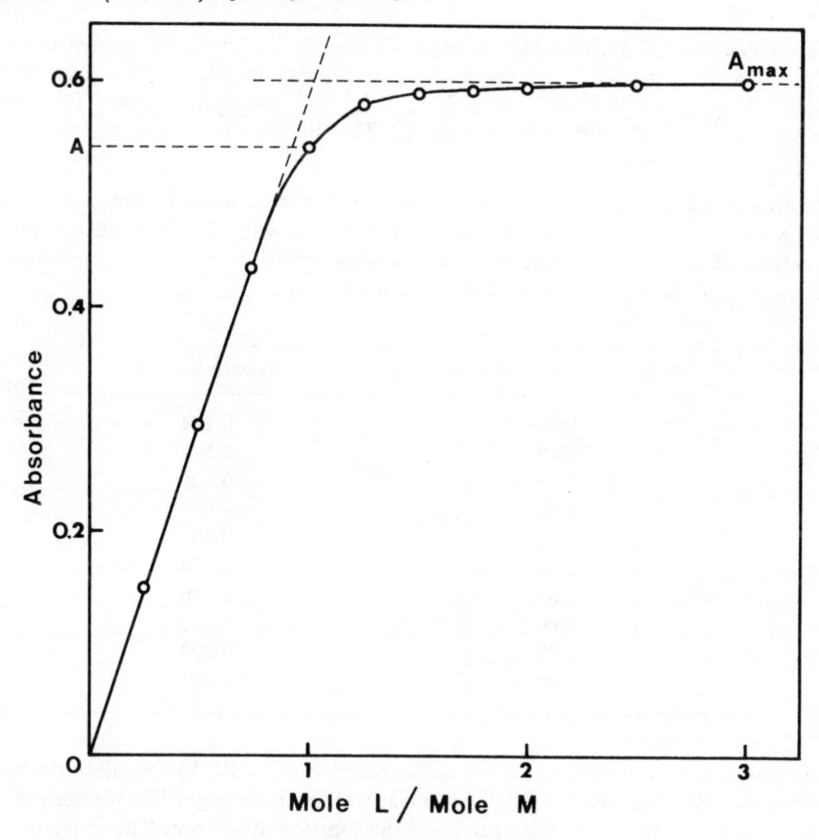

Figure 16.3. *Mole-ratio plot.*

Example 16.21. The ion M^{n+} and ligand L form the colored complex $[ML_3]^{n+}$. A 25.0-mL aliquot of 2.00×10^{-5} $\underline{M}$ M^{n+} solution is mixed with 25.0 mL of 1.00×10^{-4} $\underline{M}$ L solution, and the resulting solution B has an absorbance of 0.565. When a 25.0-mL aliquot of 1.00×10^{-5} $\underline{M}$ M^{n+} solution is mixed with 25.0 mL of 1.00 $\underline{M}$ L solution, the ion M^{n+} is complexed practically quantitatively, and the resulting solution D has an absorbance of 0.315. Calculate the formation constant K_f of the complex $[ML_3]^{n+}$. All measurements are made at a wavelength at which only the complex absorbs, in a 1.00-cm cell.

Solution. In solution D we have practically complete conversion of M^{n+} to $[ML_3]^{n+}$; therefore, $[ML_3{}^{n+}] = 5.00 \times 10^{-6}$ $\underline{M}$ and $\varepsilon_{ML_3^{n+}} = 0.315/(1.00 \times 5.00 \times 10^{-6}) = 6.30 \times 10^4$. Hence, in solution B we have $[ML_3{}^{n+}] = 0.565/(1.00 \times 6.30 \times 10^4) = 8.97 \times 10^{-6}$ $\underline{M}$, $[M^{n+}] = 1.00 \times 10^{-5} - 8.97 \times 10^{-6} = 1.03 \times 10^{-6}$ $\underline{M}$ and $[L] = 5.00 \times 10^{-5} - 3(8.97 \times 10^{-6}) = 2.31 \times 10^{-5}$ $\underline{M}$. Substituting the preceding values of solution B in the expression for the formation constant of the complex, we have

$$K_f = \frac{[ML_3{}^{n+}]}{[M^{n+}][L]^3} = \frac{8.97 \times 10^{-6}}{(1.03 \times 10^{-6})(2.31 \times 10^{-5})^3} = \mathbf{7.1 \times 10^{14}}.$$

Example 16.22. During the determination of the formula of the complex ion $[Fe(1,10\text{-phenanthroline})_n]^{2+}$, which the ion Fe^{2+} forms with 1,10-phenanthroline, by the method of continuous variations, the following experimental data were obtained, when measuring the absorbance at 510-nm in a 1.00-cm cell:

Molar fraction of Fe(II)	Absorbance
0.06	0.174
0.12	0.347
0.18	0.520
0.22	0.637
0.28	0.699
0.40	0.583
0.50	0.486
0.60	0.388
0.70	0.293
0.80	0.194

a) Plot absorbance as a function of the molar fraction of Fe(II). b) Calculate the value of n (coordination number of Fe(II)), by extrapolating the straight-line portions of the curve. c) Calculate the molar absorptivity of the complex at 510 nm. The constant sum of the Fe(II) and 1,10-phenanthroline concentrations was 2.64×10^{-4} M.

Solution. a) See Figure 16.4.

b) From Figure 16.4, it can be seen that $n = (1.0{-}0.25)/0.25 = \mathbf{3}$.

c) The maximum absorbance of the complex, obtained by extrapolating the linear straight-line portions, is 0.724, and corresponds to a complex concentration of $2.64 \times 10^{-4}/4 = 6.60 \times 10^{-5}$ M. Hence, the molar absorptivity of the complex at 510 nm is equal to

$$\varepsilon_{510} = 0.724/(1.00 \times 660 \times 10^{-5}) = \mathbf{1.10 \times 10^4}\ \text{L mol}^{-1}\ \text{cm}^{-1}.$$

Spectrophotometric determination of the dissociation constant of an acid-base indicator. For the acid-base indicator HX, which is a weak organic acid, we have (see Equation (5.37))

$$pH = pK_{HX} - \log \frac{[HX]}{[X^-]} \qquad (16.49)$$

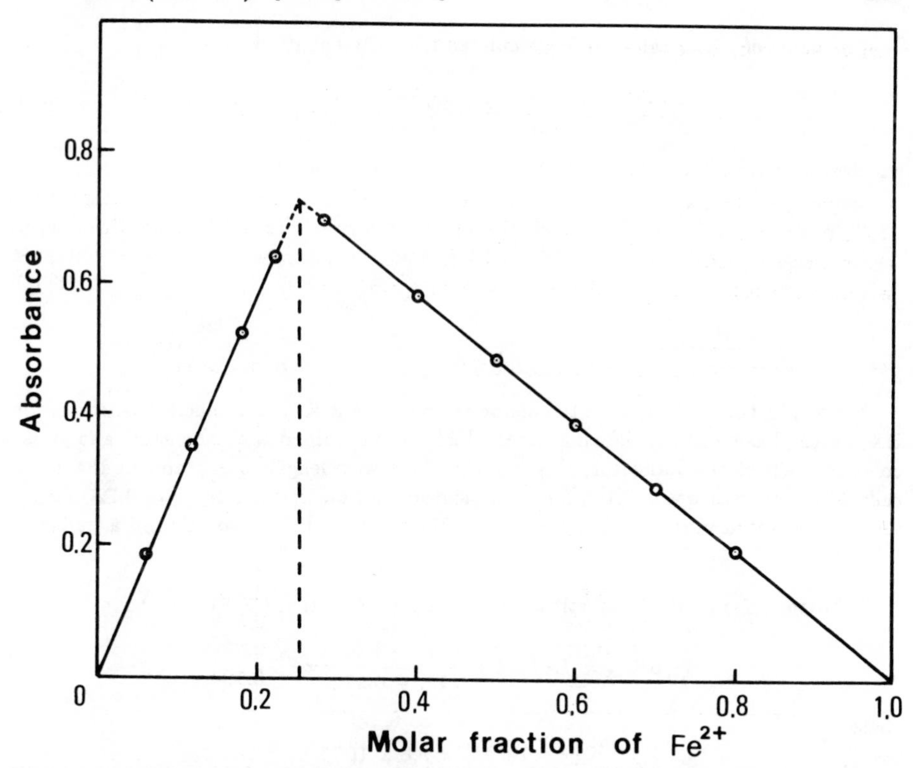

Figure 16.4. *Plot taken by the method of continuous variations (Job's method) for the complex* $[Fe(1,10\text{-phenanthroline})_3]^{2+}$.

or

$$\log([X^-]/[HX]) = pH - pK_{HX} \tag{16.49a}$$

The undissociated molecule HX has a different color from the anion X^-. A plot of Equation (16.49*a*), $\log([X^-]/[HX])$ versus pH (abscissa), gives a straight line with a slope of +1, which intercepts the ordinate at $-pK_{HX}$, and the pH-axis at a point where $\log([X^-]/[HX]) = 0$, that is, at a point where $pH = pK_{HX}$. The value of the ratio $[X^-]/[HX]$ is found spectrophotometrically, as follows: A series of solutions is prepared which contain the same amount of the indicator and have different pH values (lower or higher than pK_{HX}). Their absorbance is measured at a certain wavelength, say $\lambda_{X^-, max}$, and then absorbance A_{X^-} is plotted as a function of pH. From this graph we have

$$\frac{[X^-]}{[HX]} = \frac{A - A_{min}}{A_{max} - A} \tag{16.50}$$

where A_{max} is the absorbance of X^- (in a solution of very high pH), A_{min} is the absorbance of HX (in a solution of very low pH), and A is the absorbance of a mixture of X^- and HX (in a solution of intermediate pH).

The value of pK_{HX} can also be calculated from the equation

$$pK_{HX} = pH - \log \frac{A - A_{min}}{A_{max} - A} \tag{16.51}$$

which results from a combination of Equations (16.49) and (16.50).

If the absorbance is measured at $\lambda_{HX,max}$, we have the relation

$$pK_{HX} = pH - \log \frac{A_{max} - A}{A - A_{min}} \tag{16.52}$$

Example 16.23. Calculate the dissociation constant K_{HX} of the acid-base indicator HX, on the basis of the following data, which were obtained with the same analytical concentration of the indicator, C_{HX}, at the same wavelength (λ_{X^-}) and in the same cell: The absorbance was 0.165 for an indicator solution in 0.100 $\underline{M}$ HCl, 0.720 for an indicator solution in 0.100 $\underline{M}$ NaOH, and 0.372 for an indicator solution in a buffer of pH 7.88.

Solution. *1st method.* Substituting the data in Equation (16.51), we have

$$pK_{HX} = 7.88 - \log \frac{0.372 - 0.165}{0.720 - 0.372} = 8.11$$

Hence,

$$K_{HX} = 10^{-8.11} = \mathbf{7.8 \times 10^{-9}}.$$

2nd method. In the HCl solution the indicator is practically quantitatively in the form HX, whereas in the NaOH solution is in the form X^-. Therefore, $\varepsilon_{HX} = 0.165/bC_{HX}$, and $\varepsilon_{X^-} = 0.720/bC_{HX}$. Hence, for pH 7.88 we have the equations

$$0.165[HX]/C_{HX} + 0.720[X^-]/C_{HX} = 0.372 \tag{16.53}$$

$$[HX]/C_{HX} + [X^-]/C_{HX} = 1 \tag{16.54}$$

Solving the system of Equations (16.53) and (16.54), we find $[HX]/C_{HX} = 0.627$ and $[X^-]/C_{HX} = 0.373$, whereupon we have

$$K_{HX} = \frac{[H^+][X^-]}{[HX]} = \frac{10^{-7.88} \times 0.373\ C_{HX}}{0.627 C_{HX}} = \mathbf{7.8 \times 10^{-9}}.$$

Infrared spectrophotometry. In infrared spectra the wavenumber (Section 16.1) for a diatomic molecule is calculated by the equation

$$\overline{\nu} = \frac{1}{2\pi c}\sqrt{\frac{k}{\mu}} \tag{16.55}$$

where

c = light velocity in cm/s

π = a constant = 3.14159...

k = force constant (a constant related to the strength of the chemical bond) in dyn/cm (the stronger the bond, the larger the force constant)

μ = reduced mass

The reduced mass is given by the equation

$$\mu = \frac{m_1 m_2}{m_1 + m_2} \tag{16.56}$$

where m_1 and m_2 are the masses of the two atoms, in grams.

Example 16.24. Calculate the wavenumber $\overline{\nu}$ and the wavelength of the stretching vibration of the group C–H (the force constant of the bond C–H is equal to 5.0×10^5 dyn/cm).

Solution. Substituting the data and the values of the various constants in Equation (16.55), we have

$$\begin{aligned}\overline{\nu}_{C-H} &= \frac{1}{2 \times 3.1416 \times 2.998 \times 10^{10}} \sqrt{\frac{5.0 \times 10^5[(1.008 + 12.011)/6.023 \times 10^{23}]}{(1.008 \times 12.011)/(6.023 \times 10^{23})^2}} \\ &= \mathbf{3.0 \times 10^3}\ \mathrm{cm^{-1}}.\end{aligned}$$

Hence, according to Equation (16.2) we have

$$\lambda = 10^4/3.0 \times 10^3 = \mathbf{3.3}\ \mu\mathrm{m}.$$

16.3 Spectrophotometric Titrations

In spectrophotometric titrations, increments of titrant (T) are added to a solution of the substance (S) being titrated and the absorbance of the solution is measured after each increment. The titration curve, in which the absorbance A is plotted on the ordinate as a function of titrant volume V_T, usually consists of two straight-line portions, the extrapolations of which intersect at a point that corresponds to the end point (Example (16.25). It is presupposed that two of the three substances that take part in the titration reaction

$$S + T \rightleftharpoons P \tag{16.57}$$

do not have the same molar absorptivity at the wavelength of the measurements. The selection of the appropriate wavelength is based on the absorption spectra of solutions that have a composition similar to that of the titrated solution before and after the equivalence point. For best accuracy, the corrected (real) absorbance value A_c should be used for the construction of the titration curve ($A < A_c$, because the solution is diluted during the titration).

Example 16.25. A 1.00×10^{-3} M solution of substance S is titrated spectrophotometrically with a 0.100 M standard solution of substance T (Equation (16.57)). Sketch the expected spectrophotometric titration curves, if the following substances absorb at the wavelength at which the absorbance of the solution is measured: a) S, b) T, c) P, d) S and T, e) S and P ($\varepsilon_S < \varepsilon_P$), f) S and P ($\varepsilon_S > \varepsilon_P$), g) T and P ($\varepsilon_T < \varepsilon_P$), h) T and P ($\varepsilon_T > \varepsilon_P$).

Solution.

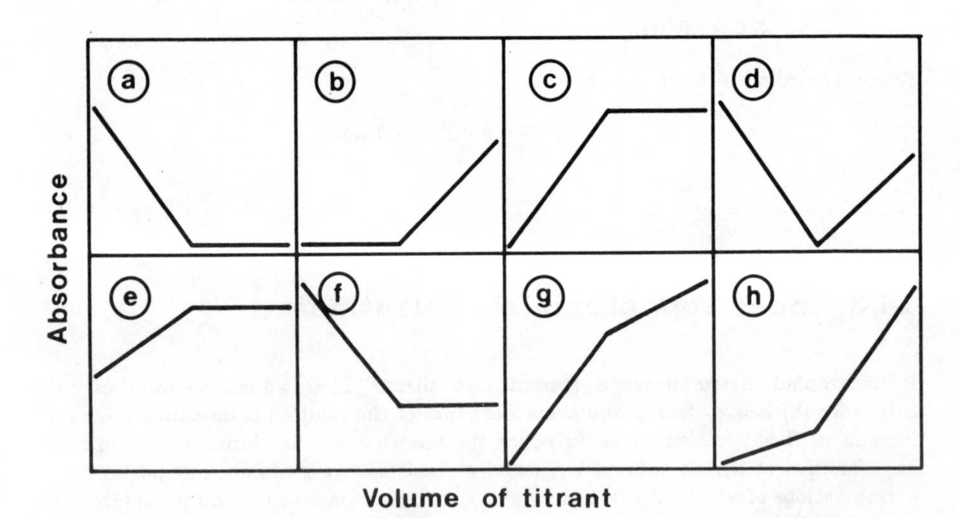

Example 16.26. During the spectrophotometric titration of 50.00 mL of solution X which contains Fe^{3+} and Cu^{2+} ions, in concentrations of the same order of magnitude, at pH 1.8 and a wavelength of 748 nm, with a standard 0.0800 M EDTA solution, the following titration curve was obtained (A_c = corrected absorbance value). Calculate

the concentrations of Fe^{3+} and Cu^{2+} ions in solution X.

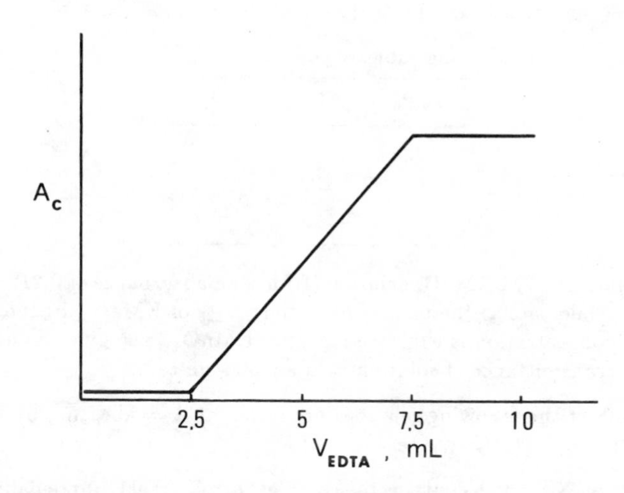

Solution. The stability constants of the Fe-EDTA and Cu-EDTA complexes are $10^{25.1}$ and $10^{18.80}$, respectively. Therefore, Fe^{3+} is titrated first (the initial horizontal part of the titration curve) and then Cu^{2+} (ascending part of the titration curve due to the absorbance of the Cu-EDTA complex at 748 nm). Hence, the concentrations of Fe^{3+} and Cu^{2+} ions in solution X are

$$[Fe^{3+}] = (2.50 \text{ mL} \times 0.0800 \text{ mmol/mL})/50.00 \text{ mL} = \mathbf{0.00400} \; \underline{\text{M}}$$

$$[Cu^{2+}] = [(7.50 - 2.50) \text{ mL} \times 0.0800 \text{ mmol/mL}]/50.00 \text{ mL} = \mathbf{0.00800} \; \underline{\text{M}}.$$

16.4 Problems

Wavelength-frequency

16.1. Complete the following table:

λ, Å	λ, nm	λ, μm	λ, cm	ν, Hz	$\overline{\nu}$, cm^{-1}
5890					
	440				
		6.54			
			3.00×10^{-8}		
				7.50×10^{9}	
					5.00×10^{3}

Absorbance

16.2. Prove that A = 2.00 − log % T.

16.3. Complete the following table:

P/P_0	% T	A
1.00		
	10.0	
		2.00

16.4. A 9.00×10^{-5} M $KMnO_4$ solution (B) has an absorbance of 0.212 at 525 nm in a 1.00-cm cell. Calculate a) the molar absorptivity ε_{525} of $KMnO_4$, b) the absorptivity k_{525} when the concentration is expressed in ppm $KMnO_4$ (also give the units of ε and k), and c) the transmittance of solution B in a 0.500-cm cell.

16.5. Which of the following relations are false? a) $A = abc_{mol/L}$, b) $T = 10^{-A}$, c) $A = \varepsilon bc_{g/L}$, d) $A = \log(P/P_0)$, e) $a = \varepsilon \times MW$.

16.6. Solution X_1, which contains the complex ferroin, $[Fe(1,10\text{-phenanthroline})_3]^{2+}$, has a transmittance of 45.0%. A 10.00-mL aliquot of this is mixed (to form solution X_2) with 5.00 mL of a 5.00 ppm Fe^{2+} solution and 5.00 mL of a 1,10-phenanthroline solution, which oversuffices for the complexation of the Fe^{2+}. Calculate the absorbance of solution X_2 and the analytical concentration of the complex in this solution. All measurements were made in a 1.00-cm cell, at 510 nm, where the molar absorptivity of the complex is 1.10×10^4 L mol^{-1} cm^{-1}.

16.7. The molar absorptivity of the complex $[FeL_3]^{2+}$ at 538 nm, ε_{538}, is 2.24×10^4 L mol^{-1} cm^{-1}, whereas ε_{538} for the ligand L is 12. A 20.00-mL aliquot of 5.00×10^{-5} M Fe(II) is mixed with 20.00 mL of 5.00×10^{-3} M L and diluted to 100.0 mL (solution B), whereupon Fe(II) is complexed quantitatively. Calculate the absorbance of solution B at 538 nm, in a 1.00-cm cell.

16.8. Calculate the absorbance of a solution 0.010 M in Na_2HPO_4, 0.050 M in NaH_2PO_4, and 2.50×10^{-4} M in the indicator HX of Example 16.12, at a) 460 nm, b) 680 nm, in a 1.00-cm cell. Ignore the effect of ionic strength upon the values of the dissociation constant.

16.9. For the spectrophotometric determination of the colorless ion M, the ion reacts with a controlled large excess of the colorless complexing reagent L (charges have been omitted for convenience) and is converted only partially to the colored complex ML_2, which has a formation constant K_f. Derive an expression which relates the absorbance A to the total concentration C_M.

Transmittance

16.10. When setting 100% T with a water blank, by mistake the transmittance scale reading was set at 90 instead of 100. After this setting, the transmittance of an unknown solution was 30.8%. Calculate the real transmittance of the unknown solution.

Absorptivity

16.11. If the molecular weight of substance X is 250 and the specific absorption coefficient $A_{1\,cm}^{1\%}$ is 340, calculate a) the absorptivity a and b) the molar absorptivity ε of substance X.

Dilution

16.12. The transmittance of solution B is 0.100. What is the required dilution so that a) the absorbance is halved, b) the absorbance is decreased by a factor of four, c) the transmittance is doubled, d) the transmittance is increased by a factor of four?

Analysis

16.13. For the determination of iron in magnesia (almost pure MgO), a 2.451-g sample is dissolved in HCl and the solution is diluted to 100 mL (solution B). A 25.00-mL aliquot of solution B is treated with an excess of the appropriate reagents for the spectrophotometric determination of iron (formation of ferroin) and diluted to 100.0 mL (solution D). The % transmittance of solution D at 510 nm in a 5.00-mL cell is 79.2% and the molar absorptivity of ferroin at 510 nm is 1.10×10^4 L mol^{-1} cm^{-1}. Calculate the iron content of the magnesia, in ppm.

16.14. A solution B, 3.31×10^{-4} M in pure substance K (MW 273) has a transmittance of 0.178 at 285 nm in a 1.00-cm cell. A 0.3122-g sample of a pharmaceutical preparation of compound K is dissolved in 500 mL of water, giving solution D, which has a transmittance of 0.222 under the same experimental conditions. Calculate % K in the sample.

16.15. During the spectrophotometric determination of iron by the 1,10-phenanthroline method, it was found that when the colored solution B of the complex (ferroin) was diluted with twice its volume of water, the transmittance of the solution was doubled. Calculate the concentration of iron in solution B, in ppm, given that the molar absorptivity of ferroin is 1.10×10^4 L mol^{-1} cm^{-1} and a 1.00-cm cell was used.

16.16. y grams of limestone ($CaCO_3$ + trace impurities) is dissolved in HCl, the solution is treated with hydroxylamine hydrochloride (to reduce Fe(III)), 1,10-phenanthroline is added and the mixture is diluted to 100 mL (solution X). If the iron content in the limestone is in the 0.1%–0.2% range, calculate the range within which y should lie, so that the absorbance of solution X falls in the range 0.200–0.800. All measurements are carried out at 510 nm, at which the molar absorptivity of the complex being formed (ferroin) is 1.10×10^4 L mol^{-1} cm^{-1}, in a 1.00-cm cell.

16.17. A 0.0782 sample of butylamine is dissolved in water and diluted to 250.0 mL (solution B). A 10.00-mL aliquot of solution B is reacted with picric acid and diluted to 100.0 mL (solution D). A 10.00-mL aliquot of solution D is diluted to 100.0 mL, and the absorbance of the final solution at 380 nm in a 1.00-cm cell is 0.515. Calculate the percent purity of butylamine, given that the molar absorptivity ε_{380} of the absorbing compound (butylamine picrate) is 1.35×10^4 L mol^{-1} cm^{-1}.

16.18. The transmittance of solution X which contains M^{3+} and M^{2+} ions was 0.182 in a 1.00 cm cell at 431 nm, where only M^{3+} ions absorb with $\varepsilon_{431} = 450$ L mol^{-1} cm^{-1}. The potential of a platinum indicator electrode immersed in solution X, which depends on the ratio $[M^{3+}]/[M^{2+}]$, was +0.0558 V versus the saturated calomel electrode. Calculate the concentrations of M^{3+} and M^{2+} ions in solution X, given that $E^0_{M^{3+}, M^{2+}} = +0.402$ V.

16.19. A 1.00-mL aliquot of a standard $KMnO_4$ solution that contains 4.00 μg Mn/mL is added to 10.00 mL of $KMnO_4$ solution B which has an absorbance of 0.326 and the absorbance becomes 0.312. Calculate the concentration of solution B, in μg Mn/mL.

16.20. A 1.00×10^{-4} $\underline{M}$ solution of substance B has absorbances of 0.840 and 0.360 at 320 and 278 nm, respectively, whereas a 1.00×10^{-4} $\underline{M}$ solution of substance C has absorbances of 0.480 and 0.432 at these wavelengths, respectively. An unknown solution has $A_{320} = 0.386$ and $A_{278} = 0.347$. All measurements were carried out under the same experimental conditions. a) Find which compound (B or C) exists in the unknown solution, and b) calculate its concentration.

16.21. The transmittance of solutions A and B of a colored substance X is 0.565 and 0.295, respectively. a) How many milliliters of each solution should be mixed so as to obtain 250.0 mL of solution D which has a transmittance of 0.500? b) If the molecular weight of X is 323.1 and its molar absorptivity at the wavelength of the measurements is 6.85×10^3 L mol^{-1} cm^{-1}, how many milligrams of X does solution D contain? All measurements were made in a 1.00-cm cell.

16.22. A 0.3421-g sample of manganese steel is dissolved in acid, and after converting Mn to MnO_4^-, the solution is diluted to 1000 mL (solution B). The transmittance of solution B at a wavelength where permanganate absorbs was 21.5%, whereas that of a 2.00×10^{-4} $\underline{N}$ $KMnO_4$ solution (standardized against $Na_2C_2O_4$ in a strongly acidic solution) was 46.0%. Calculate the percent manganese in the sample.

16.23. When reading the transmittance of an unknown solution of the Fe(II)-1,10-phenanthroline complex at 510 nm in a 1.00-cm cell, there was a reading error of +0.02 transmittance unit, and the iron concentration was found to be 1.62 ppm, by using a correct calibration curve. From this result, the iron content of an ore was calculated to be 5.25%. Calculate the real content of iron in the ore, given that the molar absorptivity of the complex at 510 nm is 1.10×10^4 L mol^{-1} cm^{-1}.

16.24. For the spectrophotometric determination of the colored substance K (MW 250) in a mixture which contains about 1% K and other colored substances, each of three samples, weighing a) 0.3535 g, b) 1.6821 g, and c) 5.0050 g, was dissolved in water and diluted to a final volume of 500 mL. The absorbance was measured at 530 nm, where only K absorbs, with a molar absorptivity $\varepsilon_{530} = 3.10 \times 10^3$ L mol^{-1} cm^{-1}, in a 1.00-cm cell. Which of the three analyses is expected to yield the most accurate result?

16.25. If the molar absorptivity of substance X at 5.3 μm (infrared) is 2.50×10^3 L mol^{-1} cm^{-1} and the minimum absorbance that can be measured in a 0.100-mm cell is 0.01, calculate the minimum concentration of substance X that can be measured.

16.26. In the spectrophotometric determination of iron by forming the red complex $[Fe(C_{12}H_8N_2)_3]^{2+}$ (ferroin), the system obeys Beer's law in the concentration range 0.1–5 ppm Fe. If the absorbance was 0.591 and 0.357 for a 3.00 ppm standard Fe^{2+} solution and an unknown solution, respectively, calculate the concentration of iron in the unknown solution, in ppm.

16.27. In the spectrophotometric study of solutions of substances B and C, the following results were obtained in a 1.00-cm cell:

	Solution		
	(1)	(2)	(3)
Concentration of B	1.20×10^{-4} $\underline{M}$	0	?
Concentration of C	0	1.50×10^{-4} $\underline{M}$	?
A at 470 nm (λ_{max} of B)	0.912	0.132	0.607
A at 530 nm (isosb. point)	0.216	0.270	0.252
A at 630 nm (λ_{max} of C)	0.054	0.975	?

Calculate a) the concentrations of B and C in solution (3), and b) the absorbance of solution (3) at 630 nm.

16.28. The molar absorptivity of substance B (MW 160) at 310 nm is 4.3×10^3 L mol^{-1} cm^{-1}. How many grams of a pharmaceutical preparation containing 1.10% B should be taken for analysis, so that maximum accuracy may be achieved in the spectrophotometric determination of B? The sample is finally diluted to 250.0 mL and a 1.00-cm cell is used.

Analysis of mixtures

16.29. The determination of iron and aluminum in a calcite ore (almost pure $CaCO_3$) was carried out as follows: A 6.8520-g sample was dissolved in HCl and diluted to 100.0 mL (solution A). A 25.00-mL aliquot of solution A was treated with HNO_3 and then an excess of ammonia was added. The precipitate formed was ignited to a constant weight of 0.0568 g. A 2.00-mL aliquot of solution A was treated with an excess of hydroxylamine hydrochloride ($Fe^{3+} \rightarrow Fe^{2+}$), an excess of 1,10-phenanthroline was added and the solution was diluted to 500.0 mL (solution B). The transmittance of solution B at 510 nm was 44.5% in a 1.00-cm cell. Calculate the per cent content of Fe and Al in the ore, given that the molar absorptivity of the complex $[Fe(1,10\text{-phenanthroline})_3]^{2+}$ at 510 nm is 1.10×10^4 L mol^{-1} cm^{-1}.

Calibration graph

16.30. The molecule of the weak electrolyte MX exhibits an absorption maximum at 510 nm, whereas the M^+ and X^- ions absorb only in the ultraviolet region of the electromagnetic spectrum. Which of the following calibration curves, $A_{510} = f(C_{MX})$, represents the actual calibration curve?

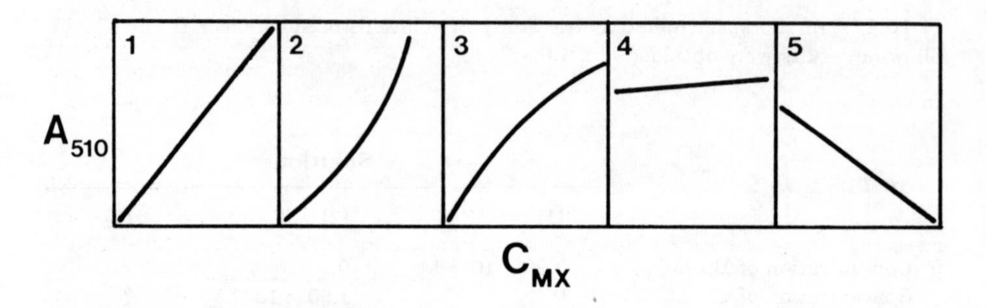

Deviations from Beer's law

16.31. The indicator HX, with a dissociation constant $K_a = 1.00 \times 10^{-5}$, has a molar absorptivity at 480 nm of 6.00×10^3 L mol^{-1} cm^{-1}, whereas the molar absorptivity ε_{480} of the ion X^- is 120 L mol^{-1} cm^{-1}. a) From the absorbance values for 1) 5.00×10^{-5} M, 2) 1.00×10^{-4} M, 3) 2.00×10^{-4} M, and 4) 3.00×10^{-4} M HX solutions at 480 nm in a 1.00-cm cell, find whether the system shows an apparent deviation from Beer's law, and if it does, determine whether it is positive or negative. b) If the pH of the preceding indicator solutions is maintained constant at 1) 4.00, 2) 5.00, with a buffer solution, would there be any deviation from Beer's law?

Stray radiation

16.32. If the stray radiation P_S is equal to 1.0% of the radiant power P_0 that passes through the reference cell, and the real transmittance of samples D_1 and D_2 (in the absence of stray radiation) is 0.040 and 0.400, respectively, calculate the per cent error in the measured concentration of each sample that is caused by the stray radiation.

Photometric error

16.33. Calculate the per cent relative error in the concentration (or absorbance), $(\Delta c/c)100$, as a function of transmittance T, for an absolute photometric error (ΔT) of 0.005, and draw the corresponding curve.

16.34. If there is a systematic reading error of +0.005 transmittance units in the transmittance reading, calculate the per cent analytical error during the spectrophotometric analysis of samples having a real absorbance of a) 0.090, b) 0.450, c) 1.050. What conclusion can be drawn from the results of these calculations?

Differential spectrophotometry

16.35. In the determination of substance S by differential spectrophotometry (high absorbance method) the 0% T was set with a closed shutter (without light), whereas the 100% T was set with a 1.000×10^{-4} $\underline{M}$ solution of pure substance S. During the successive filling of the cell with a 1.150×10^{-4} $\underline{M}$ standard solution of substance S and an unknown solution of S (solution D) the absorbance was 0.165 and 0.550, respectively. Calculate the concentration c_X of substance S in solution D.

16.36. A 6.00×10^{-5} $\underline{M}$ standard solution of pure substance B and an unknown solution D of substance B have absorbances of 0.660 and 1.210, respectively, when measured by the method of classical spectrophotometry at 510 nm in a 1.00-cm cell. If the standard solution is used as blank (A = 0, high absorbance method), calculate a) the absorbance of the unknown solution D, and b) the differences in %T of the two solutions by the two methods.

16.37. A 1.000×10^{-4} $\underline{M}$ $KMnO_4$ solution, which has an absorbance of 0.236, is used to set the 100% T of the spectrophotometer scale. Subsequently, under the same experimental conditions, the absorbance of an unknown $KMnO_4$ solution was found equal to 0.223. Calculate the concentration of the unknown solution, in ppm Mn.

Mole-ratio method

16.38. The ion M^{2+} reacts with ion L^-, forming a colored complex. To determine the formula of the complex, a series of solutions was prepared in which the analytical concentration of M^{2+} was 2.50×10^{-4} $\underline{M}$, while the concentration of L^- varied. The following data were obtained by measuring the absorbance at 495 nm, where only the complex absorbs, in a 1.00-cm cell:

C_L, $\underline{M}$	A_{495}	C_L, $\underline{M}$	A_{495}
6.25×10^{-5}	0.124	3.75×10^{-4}	0.496
1.25×10^{-4}	0.248	5.00×10^{-4}	0.498
1.875×10^{-4}	0.375	6.25×10^{-4}	0.499
2.50×10^{-4}	0.469	1.25×10^{-3}	0.499_5
3.25×10^{-4}	0.493		

a) What is the formula of the complex? b) What is the molar absorptivity of the complex at 495 nm? c) What is the instability constant of the complex?

Stability constant

16.39. The ion M^{n+} and the ligand L form the colored complex $[ML_3]^{n+}$. A 10.00-mL aliquot of 2.00×10^{-5} $\underline{M}$ M^{n+} solution is mixed with 10.00 mL of 1.00×10^{-4} $\underline{M}$ L solution, and the resulting solution has an absorbance of 0.565. When a 10.00-mL aliquot of 1.00×10^{-5} $\underline{M}$ M^{n+} solution is mixed with 10.00 mL of 1.00 $\underline{M}$ L solution, the ion M^{n+} is complexed practically quantitatively, and the resulting solution has an absorbance of 0.315. All measurements are made at a wavelength at which only the complex absorbs, in a 1.00-cm cell. Calculate the stability (formation) constant of the complex.

Instability constant

16.40. The ion M^{n+} and the ligand L, form the complex $[ML_4]^{n+}$. A 10.00-mL aliquot of 2.00×10^{-5} $\underline{M}$ M^{n+} solution is mixed with 10.00 mL of 1.00×10^{-4} $\underline{M}$ L solution, and the resulting solution has an absorbance of 0.465. When a 10.00-mL aliquot of 2.00×10^{-5} $\underline{M}$ M^{n+} solution is mixed with 10.00 mL of 1.00 $\underline{M}$ L solution, the ion M^{n+} is complexed practically quantitatively, and the resulting solution has an absorbance of 0.515. All measurements are made at a wavelength at which only the complex absorbs, in a 1.00-cm cell. Calculate the instability constant of the complex.

16.41. The ion M^{n+} and the ligand L form the colored complex $[ML]^{n+}$. Solution B, in which $C_M = 4.00 \times 10^{-4}$ $\underline{M}$ and $C_L = 3.12 \times 10^{-4}$ $\underline{M}$, has an absorbance of 0.375, exactly equal to the absorbance of solution D, which has $C_M = 6.00 \times 10^{-4}$ $\underline{M}$ and $C_L = 3.04 \times 10^{-4}$ $\underline{M}$. All measurements are made at 550 nm, at which only the complex absorbs, in a 1.00-cm cell. Calculate the molar absorptivity ε_{550} and the instability constant of the complex.

16.42. The ion M^{n+} and the ligand L form the complex $[ML]^{n+}$. At the wavelength of maximum absorption by the complex, where M^{n+} and L do not absorb, the absorbance of solution B, which has $C_M = 3.60 \times 10^{-4}$ $\underline{M}$ and $C_L = 2.48 \times 10^{-4}$ $\underline{M}$, is exactly equal to that of solution D, which has $C_M = 2.80 \times 10^{-4}$ $\underline{M}$ and $C_L = 2.62 \times 10^{-4}$ $\underline{M}$. Calculate the instability constant of the complex.

Dissociation constant

16.43. The molecule of the weak acid HX is colorless, whereas the ion X^- absorbs at 560 nm, where it has a molar absorptivity of 3.36×10^4 L mol^{-1} cm^{-1}. A 2.00×10^{-5} $\underline{M}$ HX solution has a pH of 5.30 and an absorbance of 0.444 at 560 nm, in a 2.00-cm cell. Calculate the dissociation constant of the acid HX.

16.44. The spectrophotometric determination of the dissociation constant of the acid-base indicator HX was carried out as follows: A known quantity of the indicator (always the same) was added to each of a series of buffer solutions, and the absorbance of each solution was measured at 530 nm, where only the acid form (HX) of the indicator absorbs. The following experimental data were obtained:

pH	A_{530}	pH	A_{530}
2.00	0.599	5.60	0.120
3.00	0.594	6.30	0.029
4.00	0.545	7.00	0.006
4.30	0.500	8.00	0.001
4.70	0.400		

Calculate the dissociation constant of the indicator.

16.45. Calculate the dissociation constant of the acid-base indicator HX, on the basis of the following spectrophotometric data. All measurements were made at the same wavelength in a 1.00-cm cell. The total (analytical) concentration of the indicator was 4.50×10^{-4} $\underline{M}$.

Solution	Absorbance
4.50×10^{-4} $\underline{M}$ HX in 0.1 $\underline{M}$ HCl	0.090
4.50×10^{-4} $\underline{M}$ HX in a buffer of pH 4.70	0.285
4.50×10^{-4} $\underline{M}$ HX in 0.1 $\underline{M}$ NaOH	0.675

Titration curve

16.46. For the molar absorptivities of the molecule of the weak acid HX and the ion X^- at 225, 330 and 420 nm, we have the following relations:

$$\varepsilon_{HX,225} >> \varepsilon_{X^-,225}, \varepsilon_{HX,330} = \varepsilon_{X^-,330} \text{ and } \varepsilon_{HX,420} << \varepsilon_{X^-,420}.$$

Draw the spectrophotometric titration curve for the titration of an HX solution with an NaOH solution, when the absorbance is measured at a) 330, b) 225, and c) 420 nm. In which of the three cases is the determination of the equivalence point from the titration curve impossible? It is assumed that the volume of the solution remains constant throughout the titration.

Chapter 17

Separation Methods

17.1 Introduction to Separation Methods

Chemical analysis encompasses various kinds of measurements, such as weighing and measurement of absorbance or conductivity. In a small number of analytical methods the measurement required for the determination of a component can be performed in the presence of other substances, because these methods are *selective*, or very rarely, completely *specific*. However, in most cases such a measurement is hindered by other substances present in the sample, and therefore a *separation step* precedes the measurement.

In the separation step, either the interfering substances are removed or the analyte is isolated. For a successful separation, the substances to be separated must differ in a chemical or physical property, which will form the basis of the separation method. A separation does not necessarily result in quantitative removal. The extent of separation required in a particular case depends on many factors, such as the selected method of measurement, the nature and the quantity of the sample, the ratio of the components in the sample, the required accuracy, and the available time. The most important methods of separation are precipitation, extraction, ion exchange, distillation, electrodeposition and the various chromatographic methods (Chapter 18).

Recovery coefficient and separation coefficient. In any analytical chemical separation we are mainly interested in the completeness of the recovery of the analyte A from the sample, and in the extent of separation of A from the other components of the sample. The completeness of recovery for A is expressed by the *recovery coefficient*, R_A, which is defined by

$$R_A = Q_A/(Q_A)_0 \tag{17.1}$$

where Q_A is the quantity of component A that is recovered and $(Q_A)_0$ is the initial quantity of A in the sample. The degree of separation of two components A and B is given by the *separation coefficient*, $S_{B/A}$, which is defined by

$$S_{B/A} = \frac{Q_B/Q_A}{(Q_B)_0/(Q_A)_0} = \frac{R_B}{R_A} \tag{17.2}$$

where Q_A and Q_B are the quantities of A and B that are recovered and $(Q_A)_0$ and $(Q_B)_0$ their corresponding initial quantities in the sample. If $R_A \simeq 1$, which is the usual case in quantitative analysis, Equation (17.2) simplifies to

$$S_{B/A} \simeq \frac{Q_B}{(Q_B)_0} = R_B \tag{17.3}$$

Errors resulting from the separation process. During the separation of component A that is being determined from the undesirable component B, incomplete recovery of A results in a negative error, whereas the error caused by incomplete removal of B will be either positive or negative, depending on whether B increases or decreases the value of the measured parameter X (X may be the weight of a precipitate, the rate of a reaction, the intensity of the diffusion current, the potential of an electrode, etc.). For a better understanding of these errors, let us assume that the analysis is based on the measurement of parameter X, which depends on the quantities Q_A and Q_B present in the solution after the separation. We have

$$X_A = \kappa_A Q_A \tag{17.4}$$

$$X_B = \kappa_B Q_B \tag{17.5}$$

where κ_A and κ_B are proportionality constants. The term κ_A is always positive, whereas the term κ_B becomes negative whenever the presence of B causes a decrease in the value of the parameter X.

If both components A and B are present in the solution, the measured value of X is given by

$$X = X_A + X_B = \kappa_A Q_A + \kappa_B Q_B \tag{17.6}$$

If component B is absent, a separation will not be necessary and the measured value of X will be given by

$$X_0 = \kappa_A (Q_A)_0 \tag{17.7}$$

The error resulting from the separation is equal to $X - X_0$. Therefore, the relative error caused by the separation is given by

$$E_r = \frac{X - X_0}{X_0} \tag{17.8}$$

Combining Equations (17.6), (17.7) and (17.8), we have

$$E_r = \frac{\kappa_A Q_A + \kappa_B Q_B - \kappa_A (Q_A)_0}{\kappa_A (Q_A)_0} \tag{17.9}$$

Setting $Q_A = R_A (Q_A)_0$ and $Q_B = R_B (Q_B)_0$ in Equation (17.9), we have

$$E_r = (R_A - 1) + \frac{\kappa_B}{\kappa_A} \frac{(Q_B)_0}{(Q_A)_0} R_B \tag{17.10}$$

The first term of Equation (17.10) gives the error caused by the incomplete recovery of A, whereas the second term gives the error caused by the incomplete removal of B during the separation process. To minimize the relative error, E_r, the values R_A and R_B, which depend on the separation method, should tend toward 1 and 0, respectively, and the ratios κ_B/κ_A and $(Q_B)_0/(Q_A)_0$, which depend on the nature of the measurement and the composition of the sample, respectively, should have smaller (absolute) values as much as possible. Occasionally, the two terms in the right side of equation (17.10) cancel out and no error or a negligible error is observed. Therefore, the method of separation should be checked by analyzing samples over a large composition range.

Example 17.1. The gravimetric determination of magnesium is based on its precipitation as $MgNH_4PO_4$, which is ignited to $Mg_2P_2O_7$, and weighed as such. Zn interferes because it gives the same reactions. If Mg is to be determined in a Mg-Zn alloy, what is the maximum permissible recovery coefficient for Zn, if the initial weight ratio Zn/Mg is 1:2 and the error from the presence of Zn is not to exceed 0.2%? Assume that the recovery of Mg is virtually complete.

Solution. The weights of $Mg_2P_2O_7$ and $Zn_2P_2O_7$ are related to the corresponding initial quantities of the metals in the sample by the expressions

$$W_{Mg_2P_2O_7} = \frac{Mg_2P_2O_7}{2Mg}(Q_{Mg})_0 = 4.579(Q_{Mg})_0$$

$$W_{Zn_2P_2O_7} = \frac{Zn_2P_2O_7}{2Zn}(Q_{Zn})_0 = 2.330(Q_{Zn})_0$$

(the terms of the fractions refer to atomic and molecular weights). Substituting the above values in Equation (17.10), we have

$$E_r = (1.000 - 1) + \frac{2.330}{4.579}\frac{1}{2}R_{Zn} \leq 0.0020$$

or

$$R_{Zn} \leq \mathbf{0.0079}.$$

Example 17.2. The determination of element A (AW 63.55) in an alloy containing a number of other elements and the interfering element B (AW 207.19) is performed spectrophotometrically, by adding an excess of reagent X to a solution of this alloy, which results in the formation of compounds AX_2 and BX_2 with molar absorptivities, ε, of 12550 and 22700, respectively. The weight ratio A/B is 8:1. a) What is the relative error if the determination is carried out in the original solution? b) What is the relative error if the determination is carried out after the separation of A and B with recovery coefficients of 0.98 for A and 0.12 for B? c) What is the maximum permissible recovery coefficient for B, R_B, in another separation scheme, with quantitative recovery of A and a maximum permissible error of 0.50%?

Solution. a) The molar ratio of A and B is

$$\text{mol A/mol B} = (8/63.55)/(1/207.19) = 26.08.$$

Substituting in Equation (17.9), where $Q_A = (Q_A)_0$ and $Q_B = (Q_B)_0$, because no separation is carried out, we have

$$E_r = \frac{\kappa_B}{\kappa_A}\frac{(Q_B)_0}{(Q_A)_0} = \frac{22700}{12550} \times \frac{1}{26.08} = +0.0694 \text{ or } +\mathbf{6.94\%}.$$

b) Substituting in Equation (17.10), we have

$$E_r = (0.98 - 1) + \frac{22700}{12550} \times \frac{1}{26.08} \times 0.12 = -0.0117 \text{ or } -\mathbf{1.17\%}.$$

c) Since the recovery coefficient of A, R_A, is practically equal to 1, substituting in Equation (17.10), we have

$$\frac{0.50}{100} \geq (1.000 - 1) + \frac{22700}{12550} \cdot \frac{1}{26.08} \cdot R_B,$$

or

$$R_B \leq \mathbf{0.072}.$$

17.2 Separation by Extraction

Extraction is one of the most common separation methods and is based on the distribution equilibria of substances between two liquids that are immiscible. The more usual case in separation by extraction is the extraction of an aqueous solution with an organic solvent, whereupon the inorganic ions and the polar organic compounds are found mainly in the aqueous phase and the non-polar organic compounds distribute in the organic phase. Inorganic ions may be reacted with an appropriate reagent to yield a non-polar compound that distributes in the organic solvent. The densities of the two phases should differ appreciably so that they are separated easily after mixing. The extraction is facilitated by using a solvent that has a small viscosity and is therefore easily mixed with the aqueous layer.

Extraction can be very selective for the removal of a substance. A multitude of parameters, such as pH, the choice of the extracting agent, the concentration of reagents, the extraction rate, and so forth, make possible the selection of the proper conditions for a rapid and quantitative separation of substances which otherwise are separated with difficulty.

Distribution equilibria and distribution law. When a substance A is distributed between two solvents 1 and 2, at equilibrium we have

$$K_D = a_2/a_1 \tag{17.11}$$

where a_1 and a_2 are the activities of the substance in phases 1 and 2, and K_D is the *distribution coefficient*, which is an expression of the *distribution law.* For a given system of solvents, K_D depends practically solely on temperature.

Equation (17.11) is valid only if the solute is present in both phases in the same form. Otherwise, if dissociation or dimerization or polymerization or complexation of the solute takes place, the *distribution ratio*, D, is used instead, which is given by

$$D = C_2/C_1 \tag{17.12}$$

where C_1 and C_2 are the analytical concentrations of substance A in phases 1 and 2 (rather than equilibrium concentrations of given species). By convention, when one of the two solvents is water, Equation (17.12) is written with the aqueous concentration in the denominator and the organic solvent concentration in the numerator. The distribution ratio depends on the experimental conditions, and many separation methods are based on controlling its value by varying the conditions.

For a given volume of extractant, successive extractions with smaller individual volumes of extractant are more efficient than a single extraction with all the volume of the extractant. If W_0 g of substance A is initially present in V_1 mL of solvent 1 and A is extracted successively with equal fractions of V_2 mL of solvent 2, the quantity W_n of substance A that remains in layer 1 after n such extractions is given by

$$W_n = \left(\frac{V_1}{DV_2 + V_1}\right)^n W_0. \tag{17.13}$$

Equation (17.13) is also used in the form

$$f_n = \frac{W_n}{W_0} = \left(\frac{V_1}{DV_2 + V_1}\right)^n \tag{17.14}$$

where f_n is the fraction of A that remains in solvent A after n extractions. One can derive from equations (17.13) and (17.14) that for a given total volume V of extractant, the extraction efficiency increases with increasing number of extractions, n, each with the same volume $V_2 = V/n$ (Table 4.2). Mechanical complications limit the value of the ratio V_1/V_2 to between 0.01 and 100.

Example 17.3. The distribution ratio of substance A between toluene and water is 4.70. If 100 mL of an aqueous solution of A is extracted with a) one 120-mL, and b) four 30-mL portions of toluene, calculate the percentage of substance A which is extracted in each case.

Solution. a) According to Equation (17.13) we have

$$W_1 = \left(\frac{100}{(4.70 \times 120) + 100}\right) W_0 = 0.1506 W_0.$$

Therefore, the percentage of A that is extracted is equal to

$$\%E = \left(\frac{W_0 - W_1}{W_0}\right) 100 = \left(\frac{1 - 0.1506}{1}\right) 100 = \mathbf{84.9\%}.$$

b) After four extractions, according to Equation (17.13), we have

$$W_4 = \left(\frac{100}{(4.70 \times 30) + 100}\right)^4 W_0 = 0.0296 W_0.$$

Then

$$\%E = \left(\frac{1 - 0.0296}{1}\right)100 = \mathbf{97.0\%}.$$

Example 17.4. If the distribution ratio for substance A is 9.0, what is the minimum number of 5.00-mL portions of ether that must be used in order to extract 99.9% of substance A from 5.00 mL of an aqueous solution that contains 0.0400 g of substance A? What weight of substance A is removed with each extraction?

Solution. Substituting the data in Equation (17.14), we have

$$\frac{W_n}{W_0} = \frac{100 - 99.9}{100} = \left(\frac{5.00}{(9.0 \times 5.00) + 5.00}\right)^n$$

or

$$\frac{1}{1000} = \left(\frac{1}{10}\right)^n \quad \text{or } n = \mathbf{3}.$$

For $n = 1$,

$$W_1 = \left(\frac{5.00}{(9.0 \times 5.00) + 5.00}\right) W_0 = 0.100 W_0.$$

Hence, 90% of substance A, that is, **0.0360** g, is removed with the first extraction. In a similar way, it is found that **0.0036** and **0.00036** g of substance A is removed with the second and third extraction, respectively.

Example 17.5. A 5.00-mL portion of solution A containing Zn^{2+} and HCl is titrated with 15.96 mL of 0.01003 $\underline{M}$ EDTA using Eriochrome Black T as indicator, at pH 10.0. A 100.0-mL portion of A is extracted with four 25.00-mL portions of a 20% (v/v) tributyl phosphate solution in benzene. A 50.00-mL portion of the aqueous phase (solution B) is titrated with 19.20 mL EDTA. Calculate the distribution ratio, D, of Zn between the two solvents.

Solution. The zinc concentration in solution A, prior to the extraction, is equal to

$$[Zn]_0 = \frac{(0.01003 \text{ mmol/mL})(15.96 \text{ mL})}{5.00 \text{ mL}} = 0.03202 \ \underline{M},$$

whereas its concentration in solution B, after 4 extractions, is equal to

$$[Zn]_n = \frac{(0.01003 \text{ mmol/mL})(19.20 \text{ mL})}{50.00 \text{ mL}} = 0.003852 \ \underline{M}.$$

Substituting these values in Equation (17.14), we have

$$\frac{W_n}{W_0} = \frac{[Zn]_n V_1}{[Zn]_0 V_1} = \frac{[Zn]_n}{[Zn]_0} = \left(\frac{100}{25.00D + 100}\right)^4 = \frac{0.003852}{0.03202} = 0.1203$$

or

$$\sqrt[4]{0.1203} = 0.5889 = \frac{100}{25.00D + 100} \quad \text{or } D = \mathbf{2.79}.$$

Apparent deviations from the distribution law. The expression for the distribution coefficient K_D (Equation (17.11)) describes the distribution equilibrium only if the solute is present in one and the same form in both solvents. However, if the solute is present in more than one form in one or both solvents, that is, if we have a system of multiple equilibria, then we have *apparent deviations* from the distribution law. These deviations are eliminated if the distribution ratio D (Equation (17.12)) is used instead of the distribution coefficient K_D.

Since during extractions we are more interested in the distribution of the sum of all species than in the distribution of one species, it is useful to derive expressions that give the value of D in cases of multiple equilibria as a function of known concentrations and equilibrium constants, as in the following examples.

Example 17.6. The weak acid HA with a dissociation constant K_a (in water) is distributed between an organic solvent and water. If the only extractable species is the undissociated species HA with a distribution coefficient K_D and this is the only existing form of the acid in the organic phase, derive an expression showing the dependence of the distribution ratio D on $[H^+]$ of the aqueous phase and draw conclusions from this expression.

Solution. The equilibria are represented schematically as follows:

$$\begin{array}{ccc} & HA & \\ K_D & \Big\Downarrow\!\!\Big\Uparrow & \text{org. phase} \\ \hline & K_a & \text{aq. phase} \\ & HA \rightleftharpoons H^+ + A^- & \end{array}$$

We have the expressions

$$D = \frac{(C_{HA})_o}{(C_{HA})_w} = \frac{[HA]_o}{[HA]_w + [A^-]_w} \tag{17.15}$$

$$K_D = [HA]_o/[HA]_w \tag{17.16}$$

$$K_a = \frac{[H^+]_w[A^-]_w}{[HA]_w} \tag{17.17}$$

The subscripts o and w denote concentrations of species in the organic and aqueous phase, respectively. Combining Equations (17.15), (17.16) and (17.17), we have the sought-for expression

$$D = \frac{[HA]_o}{[HA]_w + \left(\dfrac{K_a[HA]_w}{[H^+]_w}\right)} = \frac{[HA]_o}{[HA]_w\left(1 + \dfrac{K_a}{[H^+]_w}\right)} = \frac{K_D[H^+]_w}{[H^+]_w + K_a} \tag{17.18}$$

Equation (17.18) predicts that if $[H^+]_w >> K_a$ (strongly acidic aqueous phase), $D \simeq K_D$ and the acid is extracted in the organic phase. On the other hand, if $[H^+] << K_a$ (strongly alkaline aqueous phase), we have $D \simeq K_D[H^+]_w/K_a$, and because of the small value of D the acid is then extracted in the aqueous phase. In this way, by regulating the pH, the course of extraction is shifted towards the desired direction. By regulating the pH of the aqueous layer, the separation of a mixture of acids or bases becomes feasible (Problem 15.10).

Example 17.7. Molecular iodine has a small solubility in water (0.3 g I_2/L at 20°C) but a large one in aqueous solutions of KI because of the formation of triiodide anion, $[I_3]^-$, which has a formation constant of $K_f = 710$. If the only extractable species between an organic solvent and water is the I_2 species, with a distribution coefficient K_D, and this is the only existing form of iodine in the organic phase, derive an expression showing the dependence of the distribution ratio D on the concentration of iodide ion in the aqueous phase, $[I^-]_w$, and draw conclusions from this expression.

Solution. The equilibria are represented schematically as follows:

$$\begin{array}{c} I_2 \\ K_D \updownarrow \quad \dfrac{\text{org. phase}}{\text{aq. phase}} \\ [I_3]^- \underset{K_f}{\rightleftharpoons} I_2 + I^- \end{array}$$

We have the expressions

$$D = \frac{(C_{I_2})_o}{(C_{I_2})_w} = \frac{[I_2]_o}{[I_2]_w + [I_3^-]_w} \tag{17.19}$$

$$K_D = [I_2]_o/[I_2]_w \tag{17.20}$$

$$K_f = \frac{[I_3^-]_w}{[I_2]_w[I^-]_w} \tag{17.21}$$

Combining Equations (17.19), (17.20) and (17.21), we have the sought-for expression

$$D = \frac{[I_2]_o}{[I_2]_w + K_f[I_2]_w[I^-]_w} = \frac{[I_2]_o}{[I_2]_w} \cdot \frac{1}{1 + K_f[I^-]_w} = \frac{K_D}{1 + K_f[I^-]_w} \tag{17.22}$$

Equation (17.22) predicts that at very low iodide concentrations we have $K_f[I^-]_w << 1$, then $D \simeq K_D$ and the iodine is extracted in the organic layer with maximum efficiency. On the other hand, at high iodide concentrations we have $K_f[I^-]_w >> 1$, and therefore, the value of D and the quantitativeness of the extraction are reduced (i.e., the aqueous equilibrium is shifted toward $[I_3]^-$, which does not extract).

Example 17.8. 8-Hydroxyquinoline, $C_9H_6(OH)N$, often referred by the trivial name "oxine," forms in acidic solutions the cation $C_9H_6(OH)NH^+$ and in alkaline solutions the anion $C_9H_6(O^-)N$. If $CHCl_3$ extracts only the neutral molecule of oxine with a distribution coefficient $K_D = 720$, derive an expression showing the dependence of the distribution ratio, D, of 8-hydroxyquinoline on $[H^+]$ of the aqueous phase and calculate the pH value at which D is maximized. The consecutive dissociation constants of the cationic acid $C_9H_6(OH)NH^+$ are: $K_1 = 1 \times 10^{-5}$, $K_2 = 2 \times 10^{-10}$.

Solution. For convenience, the molecule of 8-hydroxyquinoline is symbolized as BH. The equilibria are represented schematically as follows:

$$\begin{array}{c} \mathrm{BH} \\ \overline{\mathrm{K_1}}\,\rightleftharpoons\mathrm{K_D}\,\frac{\mathrm{CHCl_3}}{\mathrm{H_2O}} \\ \mathrm{BH_2^+} \rightleftharpoons \mathrm{BH} + \mathrm{H^+} \\ \mathrm{K_2}\, \rightleftharpoons \\ \mathrm{B^-} + \mathrm{H^+} \end{array}$$

We have the expressions

$$D = \frac{(C_{BH})_o}{(C_{BH})_w} = \frac{[BH]_o}{[BH_2^+]_w + [BH]_w + [B^-]_w} \tag{17.23}$$

$$K_D = [BH]_o/[BH]_w = 720 \tag{17.24}$$

$$K_1 = \frac{[BH]_w[H^+]_w}{[BH_2^+]_w} = 1 \times 10^{-5} \tag{17.25}$$

$$K_2 = \frac{[B^-]_w[H^+]_w}{[BH]_w} = 2 \times 10^{-10} \tag{17.26}$$

Combining Equations (17.23), (17.24), (17.25) and (17.26), we have the sought-for expression

$$\begin{aligned} D &= \frac{[BH]_o}{\dfrac{[BH^+]_w[H^+]_w}{K_1} + [BH]_w + \dfrac{K_2[BH]_w}{[H^+]_w}} \\ &= \frac{[BH]_o}{[BH]_w} \cdot \frac{1}{\dfrac{[H^+]_w}{K_1} + 1 + \dfrac{K_2}{[H^+]_w}} = \frac{K_D}{\dfrac{[H^+]_w}{K_1} + 1 + \dfrac{K_2}{[H^+]_w}} \end{aligned} \tag{17.27}$$

We calculate the first and the second derivatives of the denominator, i.e.,

$$\phi([H^+]_w) = \frac{[H^+]_w}{K_1} + 1 + \frac{K_2}{[H^+]_w}$$

whereupon we have for the first derivative

$$\phi'([H^+]_w) = \frac{1}{K_1} - \frac{K_2}{[H^+]_w^2}$$

and for the second derivative

$$\phi''([H^+]_w) = \frac{2K_2}{[H^+]_w^3}.$$

Since always $\phi''([H^+]_w) > 0$, then when $\phi'([H^+]_w) = 0$, $\phi([H^+]_w)$ is minimum under these conditions. Consequently, the distribution ratio is maximum when

$$\frac{1}{K_1} - \frac{K_2}{[H^+]^2} = 0$$

or

$$[H^+]_w = \sqrt{K_1K_2} = \sqrt{(1 \times 10^{-5})(2 \times 10^{-10})} = 4.5 \times 10^{-8}\ \underline{M}$$

or

$$pH = \mathbf{7.35}.$$

Example 17.9. A 20.00-mL portion of a 1.000×10^{-4} $\underline{M}$ Zn^{2+} solution is mixed with 20.00 mL of a 0.0800 $\underline{M}$ solution of the complexing reagent L^-, which forms with zinc the complex $[ZnL_2]$ with a formation constant $K_f = 455$. The mixture is extracted with 20.00 mL of $CHCl_3$ and after the separation of phases the analytical concentration of Zn in the aqueous phase is found to be equal to 1.85×10^{-5} $\underline{M}$, using atomic absorption spectroscopy. Assuming that the only extractable species is the complex $[ZnL_2]$, calculate its distribution coefficient, K_D, between $CHCl_3$ and water.

Solution. The equilibria are represented schematically as follows:

$$\begin{array}{ll} [ZnL_2] & \\ \quad\Big\Updownarrow K_D & \overline{CHCl_3} \\ & H_2O \\ [ZnL_2] \overset{K_f}{\rightleftharpoons} Zn^{2+} + 2L^- & \end{array}$$

After the extraction, we have $(C_{Zn})_w = 1.85 \times 10^{-5}$ $\underline{M}$, whereas the concentration of the complex in the organic phase is equal to

$$[ZnL_2]_o = [(20.00 \times 1.000 \times 10^{-4}) - (40.00 \times 1.85 \times 10^{-5})]/20.00 = 6.30 \times 10^{-5}\ \underline{M}.$$

We have the expressions

$$D = \frac{(C_{Zn})_o}{(C_{Zn})_w} = \frac{[ZnL_2]_o}{[Zn^{2+}]_w + [ZnL_2]_w} \tag{17.28}$$

$$K_D = [ZnL_2]_o/[ZnL_2]_w \tag{17.29}$$

$$K_f = \frac{[ZnL_2]_w}{[Zn_2]_w[L^-]_w^2} = 455 \tag{17.30}$$

Combining Equations (17.28), (17.29) and (17.30), we have

$$D = \frac{[ZnL_2]_o}{\dfrac{[ZnL_2]_w}{K_f[L^-]_w^2} + [ZnL_2]_w} = \frac{[ZnL_2]_o}{[ZnL_2]_w} \cdot \frac{K_f[L^-]_w^2}{1 + K_f[L^-]_w^2} = \frac{K_DK_f[L^-]_w^2}{1 + K_f[L^-]_w^2} \tag{17.31}$$

Because of the large excess of complexing reagent with respect to zinc, $[L^-]$ is considered constant and equal to $(20.00 \times 0.0800)/40.00 = 0.0400$ $\underline{M}$. Substituting the data in Equation (17.31), we have

$$K_D = D\frac{1 + K_f[L^-]_w^2}{K_f[L^-]_w^2} = \frac{6.30 \times 10^{-5}}{1.85 \times 10^{-5}} \cdot \frac{1 + 455(0.0400)^2}{455(0.0400)^2} = \mathbf{8.08}.$$

Extraction of metal ions with organic solvent containing a chelating agent. The extraction of metal ions with an organic solvent (immiscible with water) containing a chelating agent is the most common technique for their separation in microquantities. Dissolving the chelating agent in the organic phase overcomes various problems, such as the limited solubility and the chemical instability (because of hydrolysis or oxidation) of many chelating agents in water. If an organic solution that contains an excess of the chelating agent HL is shaken with an aqueous solution of the metal ion M^{n+}, the extractable complex $[ML_n]$ is formed, with the equilibria represented schematically as follows:

$$\begin{array}{ccc} HL & & [ML_n] \quad \text{org. phase} \\ \Updownarrow K_{D_{HL}} & & \Updownarrow K_{D_{ML_n}} \quad \text{aq. phase} \\ HL & & \\ K_a \Updownarrow & & \\ M^{n+} + nL^- & \overset{K_f}{\rightleftharpoons} & [ML_n] \\ + & & \\ H^+ & & \end{array}$$

We have the expressions

$$K_{D_{HL}} = [HL]_o/[HL]_w \tag{17.32}$$

$$K_{D_{ML_n}} = [ML_n]_o/[ML_n]_w \tag{17.33}$$

$$K_a = \frac{[H^+]_w[L^-]_w}{[HL]_w} \tag{17.34}$$

$$K_f = \frac{[ML_n]_w}{[M^{n+}]_w[L^-]_w^n} \tag{17.35}$$

$$D = (C_M)_o/(C_M)_w \tag{17.36}$$

If we ignore other intermediate complexes ($[ML_{n-1}]^+$, $[ML_{n-2}]^{2+}$, etc.) in both the organic and the inorganic phases, because of the large excess of chelating agent with respect to the metal ion, we have

$$(C_M)_o = [ML_n]_o \tag{17.37}$$

and

$$(C_M)_w = [M^{n+}]_w + [ML_n]_w \tag{17.38}$$

If the distribution of the complex favors the organic phase, which is often the case, the term $[ML_n]_w$ can be omitted in Equation (17.38), as too small in comparison with the term $[M^{n+}]_w$, and Equation (17.36) becomes

$$D = [ML_n]_o/[M^{n+}]_w \tag{17.39}$$

Combining Equations (17.33), (17.35) and (17.39), we have

$$D = K_{D_{ML_n}} K_f [L^-]_w^n \tag{17.40}$$

Combining Equations (17.32), (17.34) and (17.40), we have

$$D = \frac{K_{D_{ML_n}} K_f K_a^n [HL]_o^n}{K_{D_{HL}}^n [H^+]_w^n} \tag{17.41}$$

Equation (17.41) gives the distribution ratio of the metal as a function of the constant $K_{D_{ML_n}}$, $K_{D_{HL}}$, K_f and K_a, which depend on the particular chemical compounds involved in the system, and of the variables $[HL]_o$ and $[H^+]$, which can be regulated experimentally.

If we define the *conditional extraction constant* as equal to

$$K'_{ex} = \frac{K_{D_{ML_n}} K_f K_a^n}{K_{D_{HL}}^n}, \tag{17.42}$$

Equation (17.41) becomes

$$D = K'_{ex} \frac{[HL]_o^n}{[H^+]_w^n} \tag{17.43}$$

Taking logarithms in Equation (17.43), we have

$$\log D = \log K'_{ex} + n \log[HL]_o - n \log[H^+]_w \tag{17.44}$$

or

$$\log D = \log K'_{ex} + n \log[HL]_o + n\text{pH} \tag{17.44a}$$

Equation (17.44*a*) predicts that a plot of $\log D$ vs pH (when $[HL]_o$ is maintained constant) should be a straight line with a slope of n (method of finding the formula of a complex) and an intercept on the $\log D$ axis equal to $\log K'_{ex} + n \log[HL]_o$. The slope of the curve increases with increasing charge on the metal ion. A change in the concentration of the chelating agent results in a shifting of the curve along the pH axis.

The constants K_f and $K_{D_{ML_n}}$ have different values for the various metal ions and this is the basis of the separation of metal ions by extracting their aqueous solutions with organic solvents that contain chelating agents.

Example 17.10. A chelating agent, HL, dissolved in an organic solvent extracts the metal ion M^{2+} from an aqueous solution according to the reaction

$$M^{2+}\text{ (w)} + 2HL\text{ (o)} \rightleftharpoons [ML_2]\text{ (o)} + 2H^+\text{ (w)} \tag{17.45}$$

The equilibrium constant of this reaction, K, is 0.0025. a) Identify K in terms of other constants of the system. b) Calculate the pH values at which 1, 20, 50, 80, 99, and 99.9% of the metal is extracted into the organic phase if 10.0 mL of the aqueous solution is shaken with 10.0 mL of a 0.010 M solution of HL (assume that $[M^{2+}]_w << 0.010$ M). c) Plot the results, percent extracted vs. pH. d) If for the metal cation N^{2+} the equilibrium constant for the extraction reaction is 2.5×10^{-7}, calculate the pH at which metal ions M^{2+} and N^{2+} could be separated quantitatively by solvent extraction of M^{2+} (extraction of $M^{2+} \geq 99\%$ with coextraction of $N^{2+} \leq 1\%$).

Solution. a) The equilibrium constant of Reaction (17.45) is given by

$$K = \frac{[ML_2]_o[H^+]_o^2}{[M^{2+}]_w[HL]_o^2} \tag{17.46}$$

Multiplying the numerator and the denominator by the product $[ML_2]_w[L^-]_w^2[HL]_w^2$, we have

$$\begin{aligned} K &= \frac{[ML_2]_o[H^+]_w^2}{[M^{2+}]_w[HL]_o^2} \cdot \frac{[ML_2]_w[L^-]_w^2[HL]_w^2}{[ML_2]_w[L^-]_w^2[HL]_w^2} \\ &= \frac{\left(\frac{[ML_2]_o}{[ML_2]_w}\right)\left(\frac{[ML_2]_w}{[M^{2+}]_w[L^-]_w^2}\right)\left(\frac{[H^+]_w[L^-]_w}{[HL]_w}\right)^2}{([HL]_o/[HL]_w)^2} \\ &= \frac{K_{D_{ML_2}} K_f K_a^2}{K_{D_{HL}}^2} = K'_{ex} \end{aligned} \tag{17.47}$$

It can be seen from Equation (17.47) that the equilibrium constant K of Reaction (17.45) coincides with the conditional extraction constant K'_{ex} (see Equation (17.42)).

b) From Equation (17.44*a*) we have

$$pH = \frac{\log D - \log K'_{ex} - n \log[HL]_o}{n} \tag{17.44b}$$

Since 1, 20, 50, 80, 99 and 99.9% of the metal is extracted and the volumes of the two phases are equal, the corresponding distribution ratios are 1/99, 20/80, 50/50, 80/20, 99/1 and 99.9/0.1. Substituting these values of D and the other data in Equation (17.44*b*), we find the sought-for values of pH, which are **2.30, 3.00, 3.30, 3.60, 4.30** and **4.80**, respectively.

c)

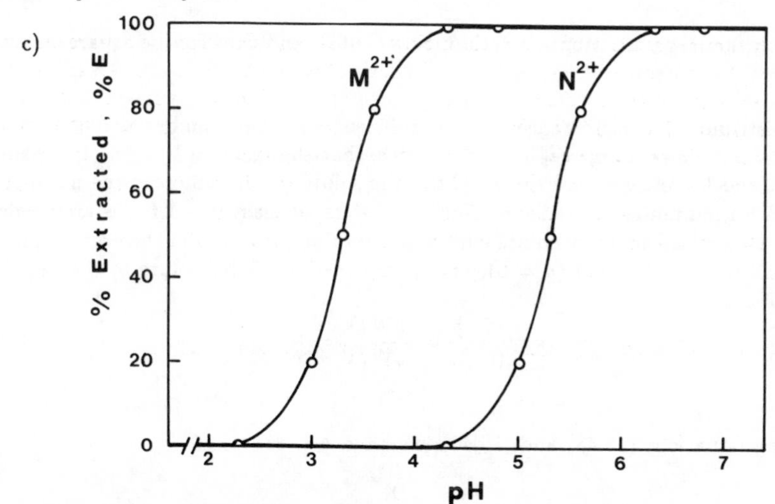

d) In a similar way as in (b) above we find that the pH values for the ion N^{2+} are two units higher than the corresponding values for the ion M^{2+}, that is, 4.30, 5.00, 5.30, 5.60, 6.30 and 6.80, respectively (the curve is shifted by 2 pH units). At pH **4.30**, 99% of M^{2+} is extracted and 1% of N^{2+} is coextracted.

Note. In general, a metal ion that forms a complex with a very large formation constant can be extracted at a low pH and thus be separated from another metal ion that forms a complex with a small formation constant.

Countercurrent extraction

A method of multiple liquid-liquid extractions is *countercurrent extraction*, which permits the separation of substances with similar distribution coefficients (ratios). The Craig apparatus is used for countercurrent extractions. It consists of a series of connected glass tubes, 0, 1, 2, 3... t, so arranged that the lighter liquid is transferred from each tube to the next tube in the series. A large number of extractions takes place simultaneously in all the tubes of the apparatus.

The separation process is carried out as follows: At the start of a run all tubes are partly filled with the heavier solvent (e.g., aqueous phase, stationary phase), whereas tube 0 contains an aqueous solution of the sample. The lighter solvent (organic phase, mobile phase) is added to tube 0 and the two solvents are mixed and allowed to separate. The upper phase of tube 0 is transferred to tube 1, fresh organic solvent is added to tube 0, and equilibrium is reached again in the two tubes (mixing-separation). Then the upper layers of tubes 1 and 0 are simultaneously transferred to tubes 2 and 1, respectively, fresh solvent is added to tube 0, the whole process is repeated and so on. Since only the organic phases are transferred, the solutes of the sample "move" along the series of tubes with different speeds, which depend on their distribution ratios (the solute with the larger distribution ratio moves faster), and thus can be separated from each other.

The distribution of a substance in countercurrent extraction can be calculated mathematically, as follows: Suppose, for convenience, that the volumes V of the two phases are equal, and let W represent the weight of the solute in the sample, t the number of tubes and n the number of transfers. For n transfers, at least $n + 1$ tubes are needed. Let p represent the fraction of solute with a distribution ratio of D in the organic phase, after equilibration in tube 0 ($n = 0$), and q the fraction of this solute in the aqueous phase. Then we have

$$\mathrm{D} = \frac{(\mathrm{C_A})_o}{(\mathrm{C_A})_w} = \frac{p\mathrm{W/V}}{q\mathrm{W/V}} = \frac{p}{q} \tag{17.48}$$

$$p + q = 1 \tag{17.49}$$

Combining Equations (15.48) and (15.49), we have

$$p = \frac{\mathrm{D}}{\mathrm{D}+1} \tag{17.50}$$

$$q = \frac{1}{\mathrm{D}+1} \tag{17.51}$$

If $\mathrm{V_o}$ and $\mathrm{V_w}$ are the volumes of the organic and the aqueous phases, respectively, then

$$p = \frac{\mathrm{D}}{\mathrm{D} + (\mathrm{V_o}/\mathrm{V_w})} \tag{17.52}$$

$$q = \frac{1}{(\mathrm{V_w}/\mathrm{V_o})\mathrm{D} + 1}. \tag{17.53}$$

After the transfer of the organic phase from tube 0 to tube 1, the addition of fresh solvent in tube 0 and the new equilibration in tubes 0 and 1 ($n = 1$), then in tube 0 the organic phase contains a fraction p of the initial aqueous fraction q, that is, the fraction pq, and the aqueous phase a fraction q of the initial fraction q, that is, q^2. In tube 1 the organic phase contains the fraction p^2 and the aqueous phase the fraction pq. After the repetition and termination of the whole process ($n = 2$), in tube 0 the organic phase contains a fraction $p \cdot q^2 = pq^2$ and the aqueous phase the fraction q^3, in tube 1 the organic phase contains a fraction $p(pq + pq) = 2p^2q$ and the aqueous phase the fraction $2pq^2$, whereas in tube 2 the organic phase contains the fraction p^3 and the aqueous phase the fraction p^2q. In a similar way the fractions of solute in the two phases of the various tubes are calculated for $n = 3$ and $n = 4$, and are given in Figure 17.1. We observe that the numerical coefficients of the fractions in the various tubes are identical to those of the binomial expansion $(p + q)^n$. After n transfers, the quantities present in successive organic and aqueous phases are equal to $p\mathrm{f}_{n,r}$ and $q\mathrm{f}_{n,r}$, respectively, where $\mathrm{f}_{n,r}$ is the fraction of solute in tube r after n transfers.

The fraction of solute in tube r after n transfers is given by

$$\mathrm{f}_{n,r} = \frac{n!}{r!(n-r)!}p^r q^{n-r} \tag{17.54}$$

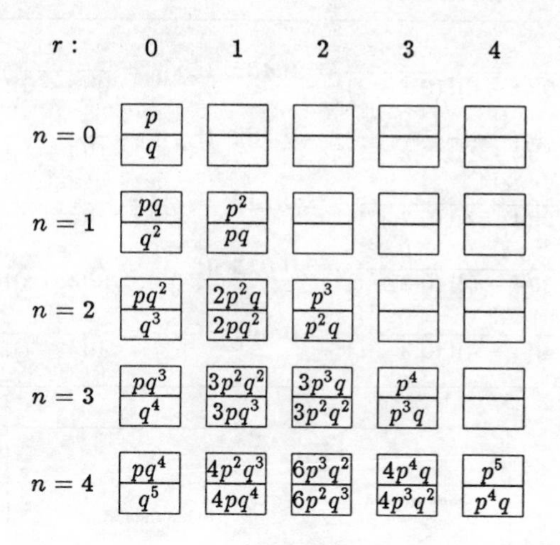

Figure 17.1. *Fractions of solute in successive tubes in countercurrent extraction with equal volumes of solvents.*

(Reminder: By definition $0! = 1$). Combining Equations (17.50), (17.51) and (17.54), we have

$$f_{n,r} = \frac{n!}{r!(n-r)!} = \frac{D^r}{(D+1)^n} \tag{17.55}$$

The possibility of separating two substances by countercurrent extraction is examined in the following example.

Example 17.11. Calculate the fraction of substances A and B remaining in tubes $r = 0, 1, 2, 3, 4$ of a Craig extraction apparatus after 4 transfers ($n = 4$), if the distribution ratios D_A and D_B are 0.2 and 3, respectively. Plot the results, f vs. r. Assume that the volumes of the two phases in each tube are equal.

Solution. Substituting the data in Equation (17.55), we have the following results, which are also represented in a diagram.

Tube number	Fraction of substance A	Fraction of substance B
0	$(f_{4,0})_A = \frac{4!}{0!(4-0)!}\frac{0.2^0}{(0.2+1)^4} = \mathbf{0.482}$	$(f_{4,0})_B = \frac{4!}{0!(4-0)!}\frac{3^0}{(3+1)^4} = \mathbf{0.004}$
1	$(f_{4,1})_A = \frac{4!}{1!(4-1)!}\frac{0.2^1}{(0.2+1)^4} = \mathbf{0.386}$	$(f_{4,1})_B = \frac{4!}{1!(4-1)!}\frac{3^1}{(3+1)^4} = \mathbf{0.047}$
2	$(f_{4,2})_A = \frac{4!}{2!(4-2)!}\frac{0.2^2}{(0.2+1)^4} = \mathbf{0.116}$	$(f_{4,2})_B = \frac{4!}{2!(4-2)!}\frac{3^2}{(3+1)^4} = \mathbf{0.211}$
3	$(f_{4,3})_A = \frac{4!}{3!(4-3)!}\frac{0.2^3}{(0.2+1)^4} = \mathbf{0.015}$	$(f_{4,3})_B = \frac{4!}{3!(4-3)!}\frac{3^3}{(3+1)^4} = \mathbf{0.422}$
4	$(f_{4,4})_A = \frac{4!}{4!(4-4)!}\frac{0.2^4}{(0.2+1)^4} = \mathbf{0.001}$	$(f_{4,4})_B = \frac{4!}{4!(4-4)!}\frac{3^4}{(3+1)^4} = \mathbf{0.316}$

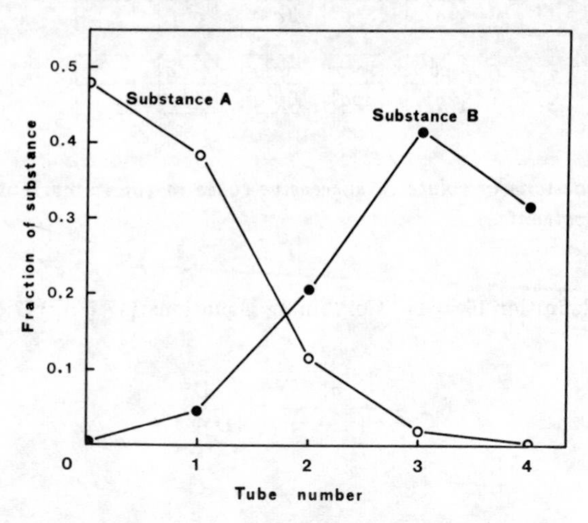

It can be seen that substance B moves faster than substance A. The tubes 0 and 4 contain almost pure substance A and substance B, respectively. A gradient of the fractions is observed in the intermediate tubes. If the number of tubes and transfers is increased, the separation will be more complete. In practice, this is accomplished with the Craig apparatus, which contains 600–1000 tubes.

Equations (17.54) and (17.55) are inconvenient for large values for n and r. For $n > 25$, the binomial distribution curve of the fraction, f, in relation to the number of tubes approximates the normal distribution curve (Gauss' distribution, Section 1.2). In such cases we can use the equation

$$f_{n,r} = \frac{1}{\sqrt{2\pi npq}} e^{-\frac{(r_{\max} - r)^2}{2npq}} \tag{17.56}$$

where r_{max} is the serial number of the tube with the maximum concentration of a substance and is given by

$$r_{max} = np = \frac{nD}{D+1} \tag{17.57}$$

For $r = r_{max}$, Equation (17.56) yields

$$f_{n,r} = 1/\sqrt{2\pi npq} \tag{17.58}$$

The standard deviation of the binomial distribution is equal to

$$\sigma = \sqrt{npq} = \sqrt{nD}/(D+1) \tag{17.59}$$

as can also be seen by comparison of Equations (1.2) and (17.56). Combining Equations (17.50), (17.51) and (17.56), we have

$$f_{n,r} = \frac{D+1}{\sqrt{2\pi nD}} e^{-(r_{max}-r)^2(D+1)^2/2nD} \tag{17.60}$$

The advantage of the above equations is that by using them and statistical tables we can calculate the total fraction of a substance among selected tubes of the Craig apparatus.

Example 17.12. In the countercurrent extractive separation of substances A, B and C, with distribution ratios of 0.75, 0.95 and 1.70, respectively, among which tubes is 95% of each substance found after 500 transfers? Assume that the volumes of the two phases in each tube are equal.

Solution. Using Equation (17.57), we find for each substance the tube that contains the maximum fraction:

$$(r_{max})_A = \frac{500 \times 0.75}{0.75 + 1} \simeq 214$$

$$(r_{max})_B = \frac{500 \times 0.95}{0.95 + 1} \simeq 244$$

$$(r_{max})_C = \frac{500 \times 1.70}{1.70 + 1} \simeq 315$$

The standard deviations for each distribution are calculated by Equation (17.59). They are

$$\sigma_A = \frac{\sqrt{500 \times 0.75}}{0.75 + 1} = 11.1$$

$$\sigma_B = \frac{\sqrt{500 \times 0.95}}{0.95 + 1} \simeq 11.2$$

$$\sigma_C = \frac{\sqrt{500 \times 1.70}}{1.70 + 1} \simeq 10.8$$

95% of the area under a normal distribution curve lies between the values $\mu - 1.96\sigma$ and $\mu + 1.96\sigma$ (Figure 1.3). Consequently, 95% of each substance is found among the tubes:

For substance A : $(214 - 1.96 \times 11.1)$ to $(214 + 1.96 \times 11.1)$, that is **192–236**
For substance B : $(224 - 1.96 \times 11.2)$ to $(244 + 1.96 \times 11.2)$, that is **224–266**
For substance C : $(315 - 1.96 \times 10.8)$ to $(315 + 1.96 \times 10.8)$, that is **294–336**

17.3 Separation by Ion Exchange

The separation by ion exchange is based on the reversible exchange of ions between an outer liquid phase and an ionic solid phase, the *ion exchanger*. From a variety of ion exchange materials we are mainly interested in synthetic organic ion exchanges known as *ion exchange resins*.

The ion exchange resins are high molecular weight polymeric organic compounds (usually copolymers of styrene and divinylbenzene), insoluble in water and the common organic solvents, that contain ionizable functional groups (cationic or anionic), e.g., $-SO_3H$, $-COOH$, $-N(CH_3)_3OH$, etc. The functional groups are ionized to ions permanently attached to the polymer chain, plus counter ions in the liquid phase that can be exchanged reversibly with like-charge ions.

Ion exchange resins are classified as cation exchange resins and anion exchange resins, depending on the type of the ionizable groups. On the basis of the strength of the ionizable groups, that is, their degree of ionization, the ion exchange resins are classified as strongly acid resins, weak acid resins, strongly base resins and weak base resins. Typical examples of ion exchange are given below:

$$RSO_3H + NaCl \rightleftharpoons RSO_3Na + HCl \tag{17.61}$$

$$2RSO_3Na + CaCl_2 \rightleftharpoons (RSO_3)_2Ca + 2NaCl \tag{17.62}$$

$$RN(CH_3)_3OH + HCl \rightleftharpoons RN(CH_3)_3Cl + H_2O \tag{17.63}$$

$$RN(CH_3)_3Cl + NaNO_3 \rightleftharpoons RN(CH_3)_3NO_3 + NaCl \tag{17.64}$$

(R represents the unchanged part of the resin polymer molecule.)

Characteristic properties of resins

The *exchange capacity* of a resin is the number of milliequivalents of H^+ or OH^- per unit weight of dry resin (meq/g) (or per unit volume of the liquid-swollen resin, meq/mL), which can be adsorbed by ion exchange.

The *selectivity* of a resin refers to its increased preference for a certain ion over another ion. The *selectivity coefficient* is equal to the equilibrium constant of the ion-exchange reaction. For example, for the following general ion-exchange reactions

$$\text{(cation exchange) } M^+ + X^+R^- \rightleftharpoons X^+ + M^+R^- \tag{17.65}$$

$$\text{(anion exchange) } M^- + R^+X^- \rightleftharpoons X^- + R^+M^- \tag{17.66}$$

the selectivity coefficient is given by

$$K_X^M = \frac{[X][MR]}{[M][XR]} = \frac{[X][M]_r}{[M][X]_r} \tag{17.67}$$

where [X] and [M] represent the concentrations of the like-charged ions in the solution (meq/mL) and [XR] or $[X]_r$ and [MR] or $[M]_r$ their concentrations in the resin (meq/g) (charges are omitted for convenience). The selectivity coefficient is constant only if activities are used.

The distribution of an ion between the resin and the solution is described by the *distribution coefficient*, K_D. For the general ion-exchange reaction

$$M + XR \rightleftharpoons X + MR, \tag{17.68}$$

if the concentration of M is relatively small, so that the amount of M is not larger than 5% of the exchange capacity of the resin, then at the equilibrium state we have

$$K_D = \frac{[MR]}{[M]} = \frac{\text{amount of ion/g of dry resin}}{\text{amount of ion/mL of solution}} \tag{17.69}$$

If an element is not present in the same form in the resin and the solution, we use the distribution ratio, D.

Ion exchange resins are used either as a batch process or in a column.

Example 17.13. A column contains 0.500 g of a strong acid ion-exchange resin in the H^+ form. A 1.00 $\underline{M}$ NaCl solution is passed through the column until the eluent shows a neutral reaction. The total eluent is titrated with 17.50 mL of 0.0500 $\underline{M}$ NaOH. A 100-mL portion of a 1.00×10^{-3} $\underline{M}$ NaCl solution is equilibrated with 0.150 g of the same resin, and the sodium concentration in the supernatant solution is found to be 2.20×10^{-4} $\underline{M}$ by flame spectroscopy. Calculate a) the exchange capacity of the resin, b) the selectivity coefficient K_H^{Na} of the resin and c) the distribution coefficient of sodium (the concentration is expressed in meq/g for the resin and in meq/mL for the solution).

Solution. a) When the NaCl solution is passed through the column, the following reaction takes place

$$Na^+ + H^+R^- \rightleftharpoons H^+ + Na^+R^-.$$

Since the eluent finally shows a neutral reaction, it is concluded that the sum of the available H^+ in the resin has been substituted by Na^+ ions. Therefore, we have

$$\text{exchange capacity} = \text{meq eluted } H^+/\text{weight of resin in g}$$

$$= (17.50 \text{ mL} \times 0.0500 \text{ meq/mL})/0.500 \text{ g} = \mathbf{1.75} \text{ meq/g}.$$

b) The selectivity coefficient is equal to

$$K_H^{Na} = \frac{[H^+][Na^+]_r}{[H^+]_r[Na^+]} \tag{17.70}$$

We have $[Na^+] = 2.20 \times 10^{-4}$ meq/mL, $[Na^+]_r$ = (total meq Na^+ – meq Na^+ in the solution)/g of resin = $[(100 \text{ mL} \times 1.00 \times 10^{-3} \text{ meq/L}) - (100 \text{ mL} \times 2.20 \times 10^{-4} \text{ meq/mL})]/0.150 \text{ g} = 0.52$ meq/g, $[H^+] = [Na^+]_{initial} - [Na^+]_{final} = (1.00 \times 10^{-3} - 2.20 \times 10^{-4})$ meq/mL $= 7.80 \times 10^{-4}$ meq/mL and $[H^+]_r$ = (total meq H^+ – meq Na^+ in the solution)/g of resin = [(0.150 ×1.75)meq - (100 mL ×7.80 × 10^{-4} meq/mL)]/0.150 g = 1.23 meq/g. Substituting the above values in Equation (17.70), we have

$$K_H^{Na} = \frac{(7.80 \times 10^{-4})(0.52)}{(1.23)(2.20 \times 10^{-4})} = \mathbf{1.50}.$$

c) The distribution coefficient is equal to

$$K_D = 0.52/(2.20 \times 10^{-4}) = \mathbf{2.36 \times 10^3}.$$

Example 17.14. V mL of a solution of ion M^+ with an initial concentration $[M^+]_0$ is shaken with W g of a cation exchange resin in the X^+R^- form, having an exchange capacity of C meq/g, and a selectivity coefficient K_X^M. Derive a general expression that gives the final concentration of M^+, $[M^+]$, in terms of the parameters V, $[M^+]_0$, W, C, and K_X^M.

Solution. The selectivity coefficient is equal to

$$K_X^M = \frac{[X^+][M^+]_r}{[M^+][X^+]_r} \tag{17.71}$$

and symbolizes the equilibrium constant of the ion exchange reaction

$$M^+ + X^+R^- \rightleftharpoons X^+ + M^+R^-. \tag{17.65}$$

We have the relations

$$[M^+]_r = (\text{total meq } M^+ - \text{meq } M^+ \text{ in the solution})/\text{weight of resin} = (V[M^+]_0 - V[M^+])/W \tag{17.72}$$

$$[X^+] = [M^+]_0 - [M^+] \tag{17.73}$$

$$[X^+]_r = (\text{total meq } X^+ - \text{meq } X^+ \text{ in the solution})/\text{weight of resin} = (CW - [X^+]V)/W = [CW - ([M^+]_0 - [M^+])V]/W \tag{17.74}$$

Substituting the values of $[M^+]_r$, $[X^+]$, and $[X^+]_r$ from Equations (17.72), (17.73) and (17.74), respectively, in (17.71) we have

$$K_X^M = \frac{([M^+]_0 - [M^+])[(V[M^+]_0 - V[M^+])/W]}{[M^+]([CW - ([M^+]_0 - [M^+])V]/W)}$$

or

$$K_X^M = \frac{([M^+]_0 - [M^+])^2}{[M^+]\left(\frac{CW}{V} - [M^+]_0 + [M^+]\right)} \tag{17.75}$$

From Equation (17.75) we have the second order equation for $[M^+]$

$$(K_X^M - 1)[M^+]^2 + \left[\frac{K_X^M CW}{V} + [M^+]_0\left(2 - K_X^M\right)\right][M^+] - [M^+]_0^2 = 0 \tag{17.76}$$

If a practically quantitative exchange of ion M^+ by ion X^+ has occurred, then $[M^+] << [M^+]_0$, $[M^+]_0 - [M^+] \simeq [M^+]_0$, $[X^+] \simeq [M^+]_0$ and Equation (17.75) is simplified to

$$K_X^M = \frac{[M^+]_0^2}{[M^+]\left(\frac{CW}{V} - [M^+]_0\right)} \tag{17.77}$$

from which we obtain the relation

$$[M^+] = \frac{[M^+]_0^2}{K_X^M\left(\frac{CW}{V} - [M^+]_0\right)} \tag{17.77a}$$

Note. Whenever Equation (17.77*a*) is used, we should verify that the relation $[M^+] << [M^+]_0$ holds true, since it was used in deriving this equation. Usually, it is sufficient to have $[M^+] \leq 0.05[M^+]_0$.

Example 17.15. A 2.00-g portion of the resin described in Example 17.13 is shaken with 100 mL of 1.00×10^{-3} $\underline{M}$ Na^+ solution. a) What percentage of sodium ions will be adsorbed by the resin? b) If 1.00 g of resin is equilibrated with the solution, and the stripped solution is decanted and equilibrated with the remaining 1.00 g of fresh resin, what percentage of sodium ions will be adsorbed by the resin?

Solution. a) On the basis of Equation (17.77*a*), we have

$$[Na^+] = \frac{(1.00 \times 10^{-3})^2}{1.50\left(\frac{1.75 \times 2.00}{100} - 1.00 \times 10^{-3}\right)} = 1.96 \times 10^{-5}.$$

(Now, $1.96 \times 10^{-5} < 0.05 \times 100 \times 10^{-3} = 5 \times 10^{-5}$, therefore, we correctly selected the simplified Equation (17.77*a*), instead of (17.76), because the relation $[M^+] << [M^+]_0$ is valid.) Consequently,

$$\left(\frac{1.00 \times 10^{-3} - 1.96 \times 10^{-5}}{1.00 \times 10^{-3}}\right) 100 = \mathbf{98.0\%}$$

of sodium ions will be adsorbed by the resin.

b) For 1.00 g of resin, on the basis of Equation (17.77*a*), we have

$$[Na^+] = \frac{(1.00 \times 10^{-3})^2}{1.50\left(\frac{1.75 \times 1.00}{100} - 1.00 \times 10^{-3}\right)} = 4.04 \times 10^{-5}.$$

(Now, $4.04 \times 10^{-5} < 5 \times 10^{-5}$; therefore, we correctly selected Equation (17.77*a*).) After the removal of the initial resin from the solution and the addition of the remaining 1.00 g of resin, we have

$$[Na^+] = \frac{(4.04 \times 10^{-5})^2}{1.50\left(\frac{1.75 \times 1.00}{100} - 4.04 \times 10^{-5}\right)} = 6.2 \times 10^{-8}.$$

Consequently,

$$\left(\frac{1.00 \times 10^{-3} - 6.2 \times 10^{-8}}{1.00 \times 10^{-3}}\right) 100 = \mathbf{99.99\%}$$

of sodium ions will be adsorbed by the resin.

From the determined values, 98.0 and 99.99%, it can be seen that the ion exchange is more complete when the available resin is added sequentially and each new addition is done after the removal of the previous resin, but this procedure requires more time. This difficulty is overcome by using the resin in a column.

Note. Using the precise Equation (17.76) we find that 98.1 and 99.99% of the sodium ions are adsorbed by the resin in the cases a and b, respectively, i.e., practically the same results we found using the simplified Equation (17.77*a*).

17.4 Problems

All organic solvents in the extraction problems are considered perfectly immiscible with water.

Errors resulting from the separation process

17.1. A gravimetric procedure for the determination of zirconium in ores is based on its precipitation with mandelic acid, $C_6H_5CH(OH)COOH$, ignition of the precipitate to ZrO_2 and weighing of the oxide. Hafnium always accompanies zirconium in its ores, giving exactly the same reactions. If a particular sample of zirconium ore contains 15% Zr and 0.8% Hf, calculate the error in the determination of Zr, if the determination takes place a) without any prior separation, b) after a separation by which Zr is recovered by 99% and Hf by 12%.

17.2. Iron is to be determined gravimetrically in a series of ores containing 20–30% Fe and 5–15% Al. The analyst has to devise a separation scheme, otherwise Al as Al_2O_3 will be coweighed with Fe_2O_3, causing a positive error. A relative error smaller than 0.1% is required. Calculate the maximum percentage of Al that can remain after the separation. The recovery of iron is virtually complete.

17.3. A method for the determination of chloride in a chloride-bromide mixture is based on the separation of Cl^- from Br^- and subsequent precipitation as AgCl and weighing. If the analytical error is −1.45% for a 0.150 $\underline{M}$ NaCl-0.100 $\underline{M}$ NaBr solution and +4.05% for a 0.050 $\underline{M}$ NaCl-0.200 M NaBr, and these errors are caused solely by incomplete separation, calculate a) the separation coefficient of bromide with respect to chloride and b) the expected percent analytical error for an equimolar chloride-bromide solution.

17.4. The gasometric determination of urea is based on the measurement of the volume of nitrogen released upon the reaction of urea with an excess of hypobromite in alkaline solution, according to the reaction $H_2NCONH_2 + 3\,OBr^- + 2\,OH^- \rightarrow N_2 + 3Br^- + CO_3{}^{2-} + 3H_2O$. Ammonia interferes in this analysis because it reacts in a similar manner, $2NH_3 + 3\,OBr^- \rightarrow N_2 + 3Br^- + 3H_2O$. For the determination of urea in its mixture with $(NH_4)_2SO_4$ by this method, find the maximum tolerable weight ratio $(NH_4)_2SO_4/H_2NCONH_2$ for which the analytical error is not to exceed 3% a) without any prior separation, b) after a separation scheme, by which 95% of the ammonium salt is removed.

17.5. In a solution containing Ni and Co in a weight ratio Ni/Co of 1:20, Ni is to be determined gravimetrically after its separation from Co. Assuming that Ni will be recovered quantitatively and that Co has the same effect as an equal weight of Ni, what should be the value of the separation factor $S_{Co/Ni}$ so that the error due to the presence of Co will not exceed 0.15%?

17.6. The determination of element A (AW55.85) in an alloy containing a number of other elements and the interfering element B (AW95.94) is performed spectrophotometrically, by adding an excess of reagent X to a solution of this alloy. This results in the formation of compound AX_3 and BX_3 with molar absorptivities of 2850 and 5650, respectively. The weight ratio A/B is 20:1. a) What is the relative error if the determination is carried out in the alloy solution, without prior separation of A? b) What is the relative error if the determination is carried out after the separation of A and B, with recovery coefficients of 0.99 for A and 0.18 for B? c) What is the maximum permissible recovery coefficient for B, R_B, in another separation scheme, with quantitative recovery of A and a maximum permissible error of 0.30%?

Distribution equilibria and distribution law

17.7. What is the minimum number of extractions required for the removal of at least 99.5% of the substance A from 100 mL of an aqueous solution containing 0.7500 g of A, if each extraction is carried out with 50.0 mL of CCl_4 and the distribution coefficient is 8? What is the amount of substance A removed by each extraction?

17.8. The distribution coefficient of substance X in the system CCl_4-H_2O is 5.00. An aqueous solution contains 20.0 mg X/mL. If 50.0 mL of this solution is equilibrated with 50.0 mL of CCl_4, how many grams of X will remain in the aqueous phase?

17.9. A 25.0-mL aliquot of an aqueous 0.400 $\underline{M}$ solution of the weak organic acid HX is transferred to each of three 100-mL volumetric flasks, A, B, and C. All flasks are filled to the mark, flask A with 1.00 $\underline{M}$ HCl, flask B with 1.00 $\underline{M}$ NaOH, and flask C with water. Aliquots of 25.0 mL each were extracted with 25.0 mL of an organic solvent. It was found that the extract from flask A contained no Cl^- or HCl but it was 0.0580 $\underline{M}$ in HX (by extraction with standard NaOH and back-titration for the unreacted NaOH with standard HCl). The extract from flask B contained no HX or X, whereas the extract from flask C was 0.0370 $\underline{M}$ in HX. Assume HX does not dissociate in the organic solvent, and calculate the dissociation constant K_a of HX in water.

17.10. The dissociation constants K_{HA} and K_{HB}, of acids HA and HB, are 1.0×10^{-3} and 1.0×10^{-9}, respectively, whereas their distribution coefficients $K_{D_{HA}}$ and $K_{D_{HB}}$ are 10 and 1000. Calculate the distribution ratios D_{HA} and D_{HB} of the two acids at pH values 4.0, 5.0, 6.0, 7.0, and 8.0. Assuming that the ratio D_{HA}/D_{HB} needs to be 10^4 for the quantitative extraction of HB ($\geq$ 99.9%) without appreciable coextraction of HA ($\leq$ 1%), what is the lowest pH (approximately) at which such a separation is feasible?

17.11. The anion of the weak acid HL ($K_a = 1.00 \times 10^{-5}$) forms the complexes $[CuL_2]$ and $[ZnL_2]$ with formation constants 3200 and 25, respectively. Given that both complexes are equally extractable with $CHCl_3$ ($K_D = 400$), calculate the percent of the extracted metal, if 20 mL of a 0.50 M HL solution containing 1.00×10^{-3} M Cu or Zn are extracted with 20 mL of $CHCl_3$ and the pH of the aqueous phase is adjusted to 2.0, 3.0, 4.0, and 5.0.

17.12. An aqueous solution of NaA is extracted with an organic solvent. If K_a is the dissociation constant of weak acid HA, V_w is the volume of the solution, C is the analytical concentration of NaA, V_o is the volume of the organic solvent, K_D is the distribution coefficient for HA between organic solvent and water, and NaA is not soluble in the organic solvent, show that $[H^+]$ for the aqueous phase can be calculated by solving the equation

$$\lambda[H^+]^3 + (K_a + \lambda C)[H^+]^2 - \lambda K_w[H^+] - K_w K_a = 0$$

where K_w is the ion product of water and $\lambda = 1 + (V_o/V_w)K_D$.

17.13. The dissociation constant for HA is $K_a = 5.0 \times 10^{-5}$, and its distribution coefficient between carbon tetrachloride and water is $K_{D_{HA}} = 1000$. A 100-mL portion of 0.0100 M NaA is equilibrated with 50 mL of CCl_4. Calculate the pH increase of the aqueous solution after this equilibration.

17.14. An aqueous 0.200 M solution of the organic compound A is extracted with an equal volume of an organic solvent. After the determination of the concentration of A in the aqueous phase, the distribution ratio $D = (C_A)_o/(C_A)_w$ is found equal to 3.9. If the same extraction is repeated with an aqueous 0.0100 M solution of A, the distribution ratio is found equal to 2.3. If the deviation from the distribution law is attributed to a dimerization reaction, $2A \rightleftharpoons A_2$, that takes phase solely in the organic phase, calculate the equilibrium constant of the dimerization reaction and the distribution coefficient $K_D = [A]_o/[A]_w$.

17.15. When 100 mL of an aqueous solution of substance A is extracted with four successive 25-mL portions of $CHCl_3$, 5% of A remains in the aqueous phase. If, in order to speed up the above separation, the solution is extracted with two successive V-mL portions of $CHCl_3$, what is the value of V?

17.16. Niobium is to be determined gravimetrically as Nb_2O_5 in 50 mL of an aqueous Nb(V)-Ta(V) solution. The weight ratio Nb/Ta is 5:1. Tantalum interferes because it is coweighed as Ta_2O_5. a) What is the relative error if the determination is carried out in the original solution? b) What is the relative error if the determination is carried out after 1 or 2 or 3 extractions of the solution, each time with 25 mL of an organic solution of a complexing agent? The distribution ratios of Nb and Ta are 0.040 and 2.1, respectively.

17.17. A 100-mL portion of a 0.1000 M aqueous solution of the weak acid HA is extracted with 50.00 mL of CCl_4. After the extraction, a 25.0-mL aliquot of the aqueous phase was titrated with 10.00 mL of a 0.1000 M NaOH. Calculate the distribution ratio of HA.

17.18. The distribution ratio of organic acid HA between an organic solvent and water is 40. At pH 7.0, 25% of the acid is extracted into the organic solvent. Calculate the dissociation constant of HA.

17.19. A 100-mL portion of an aqueous solution of substance A is extracted with two 45-mL portions of CCl_4. If 1.00% of A remained in the aqueous phase, calculate the distribution ratio of A.

17.20. The distribution ratio of substance A between CCl_4 and H_2O is 20. A 100-mL portion of a 2.00 M aqueous solution of A is extracted with CCl_4 until the concentration of A in the aqueous phase is reduced to 0.00200 M or less. What total volume of CCl_4 is required if the extraction is done with a) 20-mL, and b) 10-mL portions of CCl_4?

17.21. A 100-mL portion of a 0.100 M aqueous solution of substance A is extracted with $CHCl_3$ until the concentration of A in the aqueous phase is reduced to 0.00100 M or less. Calculate the minimum value of the distribution coefficient if the extraction is done with a) two 20-mL and b) four 10-mL portions of $CHCl_3$.

17.22. What is the minimum value of K_D that would allow the extraction of 99.9% of a solute from 25 mL of water with four 25-mL portions of $CHCl_3$?

17.23. If two extractions with 50-mL portions of ether extract 90% of substance A from an aqueous solution, what percentage of A will five similar extractions remove?

17.24. The metal ion X^{2+} is extracted by a chelating reagent as in Example 17.10, but using $[HL]_o = 0.020$ M. The following results are obtained:

pH	2.0	2.5	3.0	3.5	4.0
D	2.4×10^{-4}	7.6×10^{-3}	0.24	7.6	240

From a plot of log D vs. pH, calculate the conventional extraction constant and the number of ligands in the molecule of the extracted complex.

17.25. Derive an expression similar to Equation (17.18) for the distribution of weak base B between an organic solvent and water.

Countercurrent extraction

17.26. The volume for the two phases in all the tubes of a Craig extraction apparatus is the same, and the distribution ratio of A is 3.0. Calculate the total percentage of A remaining in the first two tubes after a) 20 transfers, b) 50 transfers.

17.27. The distribution ratio of a substance between $CHCl_3$ and H_2O is 4.0. A 0.0200-g sample is placed in tube 0 of a Craig extraction apparatus. Each tube holds 2.00 mL of each phase. Calculate the quantity of the substance present in each tube after four transfers.

17.28. Calculate the fractions f of substances A and B remaining in tubes r = 0, 1, 2, 3, 4, 5 of a Craig extraction apparatus after 5 transfers, if the distribution ratios D_A and D_B are 1.0 and 2.0, respectively. Plot the results, f vs. r. Assume that the volumes of the two phases in each tube are equal.

Characteristic properties of resins

17.29. A 100-mL portion of a 2.00×10^{-3} $\underline{M}$ Na^+ solution is shaken with 2.00 g of a strong acid resin in the H^+ form. The exchange capacity of the resin is 1.75 meq/g and its selectivity coefficient, K_H^{Na}, is 1.50. a) What percentage of sodium ions will be adsorbed by the resin? b) If 1.00 g of resin is equilibrated with the solution, and the stripped solution is decanted and equilibrated with the remaining 1.00 g of fresh resin, what percentage of sodium ions will be adsorbed by the resin?

17.30. An aqueous solution containing 0.1401 g of a KCl-KNO_3 mixture is passed through a column of a strong acid cation-exchange resin in the H^+ form. The column is eluted until the eluent shows no acidic reaction. The total eluent is neutralized with 16.30 mL of 0.0997 M NaOH. Calculate the percent of KCl and KNO_3 in the mixture.

17.31. How many grams of a strongly acid cation-exchange resin in the H^+ form, having an exchange capacity of 3.50 meq/g should be added to 100 mL of 2.00×10^{-3} $\underline{M}$ CsCl, in order to retain 99% of Cs, if the selectivity coefficient K_H^{Cs} is 2.70?

17.32. A 100-mL portion of a 0.00100 $\underline{M}$ Cs^+ solution is shaken with 2.0 g of Dowex 50-X8 resin in the H^+ form. The exchange capacity of the resin is 4.00 meq/g and its selectivity coefficient K_H^{Cs} is 2.56. a) Calculate the distribution coefficient of Cs and the percentage of Cs remaining in the solution. b) If the same amount of resin is used successively in two 1.0 g fractions, calculate the percentage of Cs remaining in the solution.

17.33. The total content of natural water is determined by passing the sample through a strongly acid cationic resin in the H^+ form. All cations are exchanged with an equivalent amount of hydrogen ions. The eluent collected from a 100.0-mL sample of natural water and washings were titrated with 10.82 mL of 0.1002 $\underline{M}$ NaOH. Calculate the number of milliequivalents of cation present in one liter of sample.

17.34. A 50.00-mL sample of y <u>M</u> $MgCl_2$ was passed through a strongly acid cationic resin in the H^+ form. The eluent and washings were titrated with 30.70 mL of 0.0998 <u>M</u> NaOH. Calculate the value of y.

17.35. An aqueous solution A, containing HCl and $CaCl_2$, is analyzed as follows: 50.00 mL of A is titrated with 11.85 mL of 0.1004 <u>M</u> NaOH. Another 20.00-mL aliquot of A is passed through a strong acid cation-exchange resin in the H^+ form and the eluent is titrated with 15.64 mL of the same NaOH solution (the same indicator is used in both titrations). Calculate the molar concentrations of HCl and $CaCl_2$ in solution A.

Chapter 18

Chromatographic Methods of Analysis

18.1 Introduction

Chromatographic analysis, usually known as chromatography, * comprises a series of methods for the separation and determination of the components of mixtures of inorganic and organic substances. The separation is achieved by distribution of the components between two phases, a *stationary* phase and a *mobile* phase, which are present in the *chromatographic column.* The mobile phase moves down the column, in effect washing (eluting) analyte components down the column at different rates. The separation is based on differences in certain properties of the components of a mixture, such as the boiling point, the polarity, the electric charges (for ionic compounds), the size of the molecule, and so forth. These differences result in varying physicochemical affinities of sample components towards the two phases of the chromatographic column. For example, in moving along the column, the sample components move at different rates, depending on their boiling points and their affinity for the packing material of the column, and therefore are separated from each other and eluted from the column at different times. If at the column outlet there is a system for detecting and measuring the quantity of each component, then a quantitative determination of the separated components is achieved.

The various chromatographic methods differ from each other in the nature of the mobile phase (liquid or gaseous) or the stationary phase (solid or a liquid on a solid support), the mechanism of the separation (adsorption, ion-exchange, distribution), and the form of the stationary phase. Although the general principles of the various chromatographic methods are the same, the laws governing the movement and the separation of the components of a mixture cannot be generalized. Because of the special importance of *gas-liquid chromatography* (GLC) (gaseous mobile phase, liquid stationary phase supported on a solid support, gas distribution mechanism), this method is the main subject of the present chapter. It is often referred to simply as *gas chromatography* (GC), and is the most widely used of all chromatographic methods.

---

*From the Greek words $\chi\rho\omega\mu\alpha$ = color, and $\gamma\rho\alpha\phi\omega$ = to write.

18.2 Theories of Chromatography. Nomenclature. Basic Relations

Plate theory. According to the plate theory, the movement of substance A through a chromatographic column can be considered as a movement through successive equilibration chambers, called *theoretical plates.* A theoretical plate is the volume of the column required for an equilibration between the stationary and the mobile phases. The equilibrium is described by the *distribution ratio* or *distribution coefficient*, K, defined by the relation

$$K = C_S/C_M \tag{18.1}$$

where C_S and C_M are the concentrations of substance A in the stationary and the mobile (carrier gas in gas chromatography) phases, respectively (in general, the subscripts S and M refer to the stationary and mobile phases, respectively). The equilibrium is also described by the *capacity factor* or *mass distribution ratio*, k′, which is equal to the ratio of the moles of a substance in the stationary phase to its moles in the mobile phase. That is, we have

$$k' = \frac{C_S V_S}{C_M V_M} = K\frac{V_S}{V_M} \tag{18.2}$$

where V_S is the volume of the stationary liquid phase and V_M is the dead (void) volume of the column.

The movement of compound A is governed by the same laws as for countercurrent extraction (Section 17.2), but in this case the equilibration chambers do not have well defined limits among each other (a theoretical plate is an imaginary concept that is not related to real column parameters, but greatly facilitates their study). Each theoretical plate has a concrete length, which is called the *height equivalent to a theoretical plate* (HETP), and is equal to

$$h = L/n \tag{18.3}$$

where h is the HETP, L is the length of the column and n is its theoretical plate number. The fact is that there is a finite value of h for a given column and substance is due to the inertia of the equilibria involved and the inability, therefore, to reestablish instantaneously the distribution equilibrium at each point of the column. It is evident, as it can also be seen from Examples 17.11 and 17.12, that a chromatographic column should have the largest possible number n of theoretical plates, in order to achieve a better separation of components differing only slightly in their distribution coefficients.

Basic chromatographic terms. When there is a detector at the outlet of the chromatographic column giving a signal proportional to the quantity of the substance eluted per unit time, then a chromatographic peak of the substance is obtained, which in the ideal case has the shape of a normal distribution curve (Gaussian curve, Section 1.2). In Figure 18.1 a *chromatogram* is shown, which is a graphic presentation of the detector response or the concentration in the gaseous phase at the end of the column as a function of time or volume. Substance O is inert and is not distributed between the

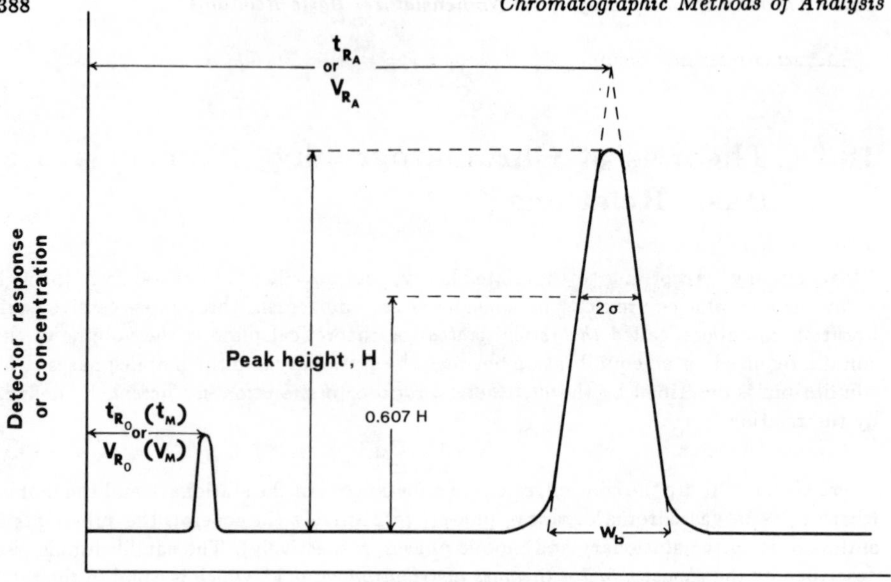

Figure 18.1. *Chromatogram and plot of concentrations of substances O and A in the order of elution from the chromatographic column.*

stationary and the gaseous phases, i.e., it remains in the gaseous phase (K = 0), whereas substance A is distributed normally and is eluted from the column after substance O.

Some fundamental chromatographic terms shown in Figure 18.1 are defined below, and the pertinent chromatographic relations are given.

Dead time, t_M: This is the time elapsed between the introduction (injection) of an inert substance which is not distributed between the two phases, such as air, and the appearance of the maximum of the peak. (In Figure 18.1, $t_M = t_{R_O}$).

Retention time, t_R: This is the time elapsed between the injection of the sample at the inlet of the column and the appearance of the maximum of the peak. We have the relation

$$t_R = t_M(1 + k') \tag{18.4}$$

(for an inert gas, we have $k' = 0$; hence, $t_R = t_M$).

Retention volume, V_R: This is the volume of a carrier gas required for the elution of a substance, and is given by the equation

$$V_R = t_R F_C \tag{18.5}$$

where F_C is the flow rate of the carrier gas, measured at column temperature and outlet pressure (usually, atmospheric pressure).

Adjusted retention time, t'_R: This is given by the relation

$$t'_R = t_R - t_M. \tag{18.6}$$

Adjusted retention volume, V'_R: This is given by the relation

$$V'_R = V_R - V_M, \tag{18.7}$$

where V_M is the dead volume of the column, given by the relation

$$V_M = t_M F_C. \tag{18.8}$$

Corrected retention volume, V^0_R: This is given by the relation

$$V^0_R = jV_R, \tag{18.9}$$

where j is the *pressure drop correction factor*, given by the relation

$$j = 1.5\left(\frac{p^2 - 1}{p^3 - 1}\right), \tag{18.10}$$

where $p = p_i/p_o$ (p_i and p_o are the pressures of the carrier gas at the inlet and the outlet of the column, respectively.)

For the identification of substances in qualitative analysis, we mainly use the adjusted retention time and the adjusted retention volume.

Retardation factor, R_F: This is the ratio of velocities of movement of the sample and the carrier gas, and is given by the relation

$$R_F = \frac{L/t_R}{L/t_M} = \frac{t_M}{t_R} = \frac{V^0_M}{V^0_R} \tag{18.11}$$

Peak height, H: This is the distance between the peak maximum and the baseline corresponding to zero concentration or quantity of the eluted substance A.

Peak width, w_b: This is the distance between the points of intersection of the tangents, drawn at the inflection points of the peak and the baseline. As long as the peak has the shape of a Gaussian curve, then we have the relation

$$w_b = 4\sigma, \tag{18.12}$$

where σ is the standard deviation of the distribution represented by the Gaussian curve, which is equal to the half-width of the peak at a height equal to $H/\sqrt{e} = 0.607H$.

Relative retention or *selectivity factor* or *separation factor*, α or $r_{B/A}$: This is the ratio of the *adjusted* retention times or volumes of the substance A being examined and a reference substance B, that is

$$\alpha = r_{B/A} = \frac{t'_{R_B}}{t'_{R_A}} = \frac{t_{R_B} - t_M}{t_{R_A} - t_M} \tag{18.13}$$

or

$$\alpha = r_{B/A} = \frac{V'_{R_B}}{V'_{R_A}} = \frac{V_{R_B} - V_M}{V_{R_A} - V_M} \tag{18.14}$$

The value of α is an indicative measure of the possibility of separating A and B, and it is independent of the length of the column, the flow rate of the carrier gas and the quantity of the liquid stationary phase.

Theoretical plate number, n. For substance A, the number of theoretical plates can be calculated by the equations

$$n = 16(t_{R_A}/w_b)^2 \tag{18.15}$$

$$n = 5.54(t_{R_A}/w_{1/2})^2, \tag{18.16}$$

where $w_{1/2}$ is the width of the peak at half the peak height (the terms t_{R_A} and w are measured in the same units, e.g., cm or mm of recorder chart). The higher the number n, the narrower is the peak and the more efficient the column for the separation of a mixture of substances with similar retention times (volumes). If the adjusted retention time t'_{R_A} is used instead of t_{R_A} in Equations (18.15) and (18.16), the *effective plate number* N is calculated. The numbers n and N are interrelated by the relation

$$N = n\left(\frac{k'}{1+k'}\right)^2 \tag{18.17}$$

The real separating power of a column is more accurately described by N. At any rate, the best measure of the column efficiency is the plate height (HETP) and not the theoretical plate number, because the height is practically independent of the length of the column.

Resolution. The separation efficiency of two substances A and B by gas chromatography depends on the degree of overlap of the respective peaks. A measure of the degree of separation is the *resolution*, R, defined by the equations

$$R = \frac{\Delta t_R}{4\sigma} = \frac{t_{R_B} - t_{R_A}}{0.5(w_A + w_B)} = \frac{2(t_{R_B} - t_{R_A})}{w_A + w_B} \tag{18.18}$$

$$R = \left(\frac{\sqrt{N}}{4}\right)\left(\frac{\alpha - 1}{\alpha}\right)\left(\frac{k'_B}{1 + k'_B}\right) \tag{18.19}$$

[It is assumed that the two peaks are eluted in similar times so that their standard deviations are almost equal. Actually, $N = (N_A + N_B)/2$.] If the width of the peaks is measured at half the peak height, which often is more easily done, we have the relation

$$R = \frac{2(t_{R_B} - t_{R_A})}{1.699(w_{A_{1/2}} + w_{B_{1/2}})} = \frac{1.177(t_{R_B} - t_{R_A})}{w_{A_{1/2}} + w_{B_{1/2}}} = \tag{18.20}$$

Figure 18.2. shows the peaks of substances A and B for three different values of R. The separation is practically complete for $R > 1.5$ (Example 18.3).

We have the following relations

$$h = \sigma^2/L \tag{18.21}$$

$$v = L/t_M \tag{18.22}$$

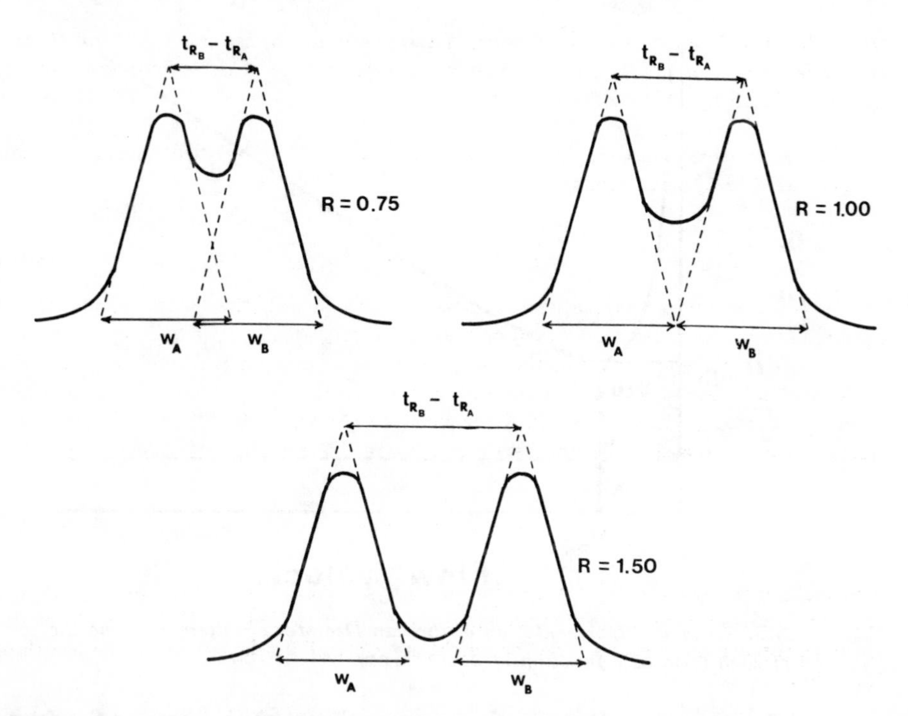

Figure 18.2. *Two peaks of equal height for three different values of resolution R. The dashed lines are the triangular approximations of the individual peaks.*

where v is the mean linear velocity of the carrier gas.

$$k' = \frac{t_R - t_M}{t_M} = \frac{V_R - V_M}{V_M} \tag{18.23}$$

$$t_R = t_M(1 + k') = t_M\left(1 + K\frac{V_S}{V_M}\right) \tag{18.23a}$$

$$V_R = V_M(1 + k') = V_M\left(1 + K\frac{V_S}{V_M}\right) = V_M + KV_S \tag{18.23b}$$

$$V'_R = V_R - V_M = KV_S \tag{18.24}$$

$$V^0_R = V^0_M + KV_S \tag{18.25}$$

$$t_{R_B} = \left(\frac{16R^2h}{v}\right)\left(\frac{\alpha}{\alpha - 1}\right)^2\left(\frac{1 + k'_B}{k'_B}\right)^3 \tag{18.26}$$

Rate theory. In gas chromatography the velocity, v, of the carrier gas greatly affects the values of the parameters h and n, and therefore the column efficiency as well, as can be seen from the van Deemter equation

$$h = A + \frac{B}{v} + Cv. \tag{18.27}$$

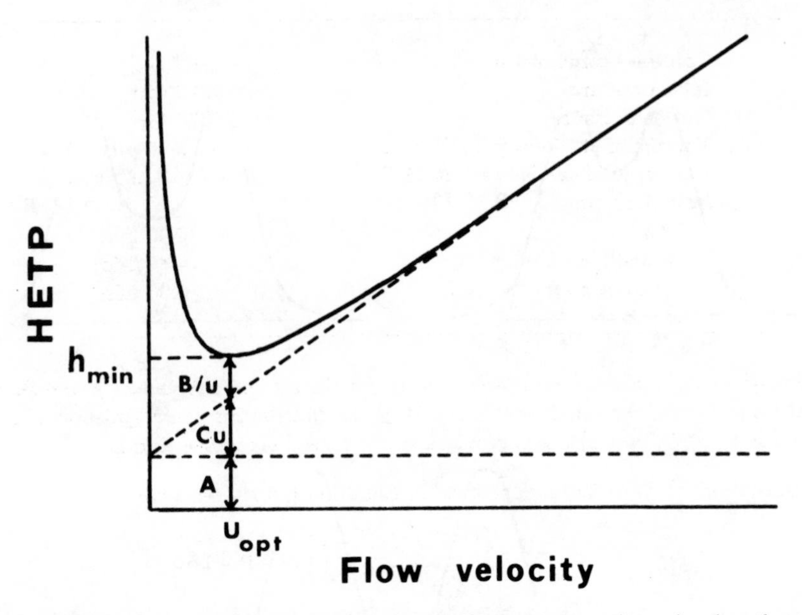

Figure 18.3. *Graphical representation of the van Deemter equation, showing the contribution of each term as a function of flow velocity.*

A, B, and C are constants characteristic of a given column, which depend on the nature of the static and gaseous phases, and on the filling material and the packing of the chromatographic column. (Actually, the first term of the right side of Equation (18.27) is $Av^{1/3}$, but in the literature the dependence of the first term on v is ignored as negligible, and the constant A is considered as the first term.)

The graphic presentation of Equation (18.27) is shown in Figure 18.3, from which it can be seen that there is an optimum velocity value, v_{opt}, given by the relation

$$v_{opt} = \sqrt{B/C}. \tag{18.28}$$

The corresponding minimum value of HETP, h_{min}, is given by the relation

$$h_{min} = A + 2\sqrt{BC}, \tag{18.29}$$

for which the parameter n, and therefore the column efficiency as well are maximized. The values of A, B, and C in the van Deemter equation are determined either graphically or by calculation (Example 18.5).

In the other chromatographic methods there are respective equations corresponding to Equation (18.27), but different in format. For example, in liquid chromatography we have a continuous increase in h with increasing velocity v, because diffusion in a liquid is small compared to a gas at low velocities.

Example 18.1. The following data were obtained with a GLC column:

Column temperature	60°C
Inlet pressure	1270 torr
Outlet pressure	770 torr
Volume of stationary liquid phase	3.00 mL
Flow rate of carrier gas, at 25°C	18.0 mL/min
Retention times:	
Air	0.30 min
Substance A	4.40 min
Substance B	5.00 min

Calculate a) the corrected retention volumes for the air and substances A and B, b) the retardation factors for substances A and B, c) the distribution coefficients for substances A and B at 60°C, and d) the separation factor for substances A and B.

Solution. a) Substituting the data in Equation (18.10) we have

$$j = 1.5\left[\frac{(1270/770)^2 - 1}{(1270/770)^3 - 1}\right] = 0.740.$$

Hence, combining Equations (18.5), (18.9), and (18.10), and adjusting the flow rate of the carrier gas to the temperature of the column, we have

$$V^0_{air} = V^0_M = jt_RF_c(T_{col}/T_{air}) = 0.740 \times 0.30 \times 18.0(333/298) = \mathbf{4.5}\ \text{mL}$$

$$V^0_{R_A} = 0.740 \times 4.40 \times 18.0(333/298) = \mathbf{65.5}\ \text{mL}$$

$$V^0_{R_B} = 0.740 \times 5.00 \times 18.0(333/298) = \mathbf{74.4}\ \text{mL}.$$

b) Substituting the data in Equation (18.11), we have

$$R_{F_A} = 4.5/65.5 = \mathbf{0.069}$$
$$R_{F_B} = 4.5/74.4 = \mathbf{0.060}.$$

c) Substituting the data in Equation (18.25), we have

$$K_A = \frac{V^0_{R_A} - V^0_M}{V_s} = \frac{65.5 - 4.5}{3.00} = \mathbf{20.3}$$

$$K_B = \frac{V^0_{R_B} - V^0_M}{V_s} = \frac{74.4 - 4.5}{3.00} = \mathbf{23.3}.$$

d) Substituting the data in Equation (18.13), we have

$$\alpha = \frac{5.00 - 0.30}{4.40 - 0.30} = \mathbf{1.15}.$$

Example 18.2. The retention times for substances A and B were 14.80 and 16.14 min, respectively, in a 40.0-cm column, whereas the widths of the respective peaks were

1.28 and 1.40 min. Calculate a) the average theoretical plate number of the column, b) the resolution (R_1) of the column, c) the height equivalent to a theoretical plate, d) the length that the column should have in order to achieve a resolution (R_2) of 1.50, and e) the time required for the elution of substance B from the longer column.

Solution. a) Substituting the data in Equation (18.15), we have

$$n_A = 16(14.80/1.28)^2 = 2139$$

and

$$n_B = 16(16.14/1.40)^2 = 2127.$$

Hence,

$$\overline{n} = (2139 + 2127)/2 = 2133 \simeq \mathbf{2.13 \times 10^3}.$$

b) Substituting the data in Equation (18.18), we have

$$R_1 = \frac{2(16.14 - 14.80)}{1.28 + 1.40} = \mathbf{1.00}.$$

c) Substituting the data in Equation (18.3), we have

$$h = 40.0 \text{ cm}/(2.13 \times 10^3) = \mathbf{0.0188} \text{ cm}.$$

d) Combining Equations (18.17) and (18.19), we have

$$\frac{R_1}{R_2} = \frac{\sqrt{n_1}}{\sqrt{n_2}} = \frac{1.00}{1.50} = \frac{\sqrt{2133}}{\sqrt{n_2}}$$

or

$$n_2 = (1.50/1.00)^2 2133 = \mathbf{4.80 \times 10^3}.$$

Hence, the length that the column should have is equal to

$$L = nh = 4.80 \times 10^3 \times 0.0188 = \mathbf{90.2} \text{ cm}.$$

e) Substituting the values of R_1 and R_2 in Equation (18.26) and dividing the resulting equations, we have

$$\frac{(t_{R_B})_1}{(t_{R_B})_2} = \frac{R_1{}^2}{R_2{}^2} = \frac{16.14}{(t_{R_B})_2} = \frac{1.00^2}{(1.50)^2}$$

or

$$(t_{R_B})_2 = \mathbf{36.3} \text{ min}.$$

Note. From the above results [$(t_{R_B})_1 = 16.14$ min, $(t_{R_B})_2 = 36.3$ min], it can be seen that in order to achieve a quantitative separation of substances A and B ($R \geq 1.5$, Example 18.3), the time required for the separation should be more than doubled.

Example 18.3. a) Calculate the percent overlap of two chromatographic peaks, when the resolution R is 1) 0.75, 2) 1.00, and 3) 1.50 (assume that the peaks follow

Gaussian normal distribution). b) Calculate the resolution, when the overlap of the two peaks is 20.0% (assume that both peaks have practically the same width).

Solution. a) 1) For the calculation of the resolution, the peak width is defined as equal to 4σ (σ = standard deviation). Substituting the data in Equation (18.18), we have

$$0.75 = \frac{2(t_{R_2} - t_{R_1})}{4\sigma + 4\sigma}$$

or

$$t_{R_2} - t_{R_1} = 3.00\sigma.$$

The distance between t_{R_1} and the point of intersection of the peaks is equal to 1.50σ. From a statistical table of the normal distribution curve (see Figures 1.2 and 1.3), it is found that the area between t_{R_1} and 1.50σ is equal to 0.4332A (A = total area of the one peak). Given that the area between t_{R_1} and the limit of the curve is equal to 0.5000A, it follows that the area of the overlap is equal to $2(0.5000 - 0.4332)\text{A} = 0.1336\text{A}$. Hence, the per cent overlap of the two peaks is equal to **13.4**.

2) Working as above, we have

$$1.00 = \frac{2(t_{R_2} - t_{R_1})}{4\sigma + 4\sigma}$$

or $t_{R_2} - t_{R_1} = 4.00\sigma$. The distance between t_{R_1} and the point of intersection is 2.00σ, the area between t_{R_1} and 2.00σ is 0.4773A, whereas the area of the overlap is equal to $2(0.5000 - 0.4773)\text{A} = 0.0454\text{A}$. Hence, the overlap amounts to **4.54%**.

3) We have $1.50 = 2(t_{R_2} - t_{R_1})/(4\sigma + 4\sigma)$ or $t_{R_2} - t_{R_1} = 6.00\sigma$. The distance between t_{R_1} and the point of intersection is 3.00σ, the area between t_{R_1} and 3.00σ is 0.4987A, whereas the area of the overlap is equal to $2(0.5000 - 0.4987)\text{A} = 0.0026\text{A}$. Hence, the overlap amounts to **0.26%**. From the preceding results, it can be seen that for R = 1.50 the separation of the two substances corresponding to the two peaks is practically complete.

b) The area between t_{R_1} and the point of intersection of the peaks is equal to $(0.5000 - 0.1000)\text{A} = 0.4000\text{A}$, and this corresponds to a distance between t_{R_1} and the intersection point of the peaks equal to 1.28σ (from the statistical table of the normal distribution curve). Hence, $t_{R_2} - t_{R_1} = 2 \times 1.28\sigma = 2.56\sigma$, and $\text{R} = 2(2.56\sigma)/(4\sigma + 4\sigma)$ = **0.64**.

Example 18.4. Derive Equation a) (18.28), b) (18.29).

Solution. a) In Equation (18.27) we have h_{min} when

$$\frac{dh}{dv} = 0 = -\frac{B}{v_{opt}^2} + C$$

or

$$v_{opt} = \sqrt{B/C} \tag{18.28}$$

b) In Equations (18.27), for $v = v_{opt}$, we have $h = h_{min}$. For $v_{opt} = \sqrt{B/C}$, Equation (18.27) becomes

$$h_{min} = A + B\sqrt{\frac{C}{B}} + C\sqrt{\frac{B}{C}} = A + 2\sqrt{BC} \tag{18.29}$$

Example 18.5. The theoretical plate numbers for a gas chromatographic system are found equal to 55.3, 96.9, and 89.8, for linear carrier gas velocities of 0.55, 1.65, and 3.10 cm/s, respectively. Calculate the maximum possible theoretical plate number for this system, and the carrier gas velocity required to achieve it.

Solution. Substituting the data in the van Deemter Equation (18.27), and taking into account Equation (18.3), we have

$$A + \frac{B}{0.55} + 0.55C = \frac{L}{55.3} \tag{18.30}$$

$$A + \frac{B}{1.65} + 1.65C = \frac{L}{96.9} \tag{18.31}$$

$$A + \frac{B}{3.10} + 3.10C = \frac{L}{89.8} \tag{18.32}$$

Solving the system of Equations (18.30), (18.31), and (18.32), we find that $A = 0.001587L$, $B = 0.008404L$, and $C = 0.002206L$. We have $n = n_{max}$, for $h = h_{min}$. Hence, substituting the values of A, B, and C in Equation (18.29), we have

$$h_{min} = \frac{L}{n_{max}} = 0.001587L + 2\sqrt{(0.008404L)(0.002206L)} = 0.0101984L$$

or $n_{max} = \mathbf{98}$. Substituting the values of B and C in Equation (18.28), we have

$$v_{opt} = \sqrt{0.008404L/0.002206L} = \mathbf{1.952}.$$

Example 18.6. The effective plate number N of a column is 690, whereas its relative retention (α) for substances A and B is 1.22. Calculate the resolution R achieved with this column for the two substances, a) when $k'_A = 1.08$, $k'_B = 1.18$, and b) when $k'_A = 4.97$, $k'_B = 5.43$.

Solution. a) Substituting the data in Equation (18.19), we have

$$R = \left(\frac{\sqrt{690}}{4}\right)\left(\frac{1.22 - 1}{1.22}\right)\left(\frac{1.18}{1 + 1.18}\right) = \mathbf{0.64}.$$

b) Similarly, we have

$$R = \left(\frac{\sqrt{690}}{4}\right)\left(\frac{1.22 - 1}{1.22}\right)\left(\frac{5.43}{1 + 5.43}\right) = \mathbf{1.00}.$$

Note. Although the ratio of the capacity factors k' is the same it both cases ($k'_B/k'_A = 1.18/1.08 = 1.093$, in the first case; $k'_B/k'_A = 5.43/4.97 = 1.093$ in the second case), the separation is more effective in the second case, in which the capacity factors are larger (peak overlap = 4.5%., Example (18.3a2), than in the first case (peak overlap = 20%, Example 18.3b).

18.3 Qualitative and Quantitative Analysis by GLC

Qualitative analysis. The identification of the components of a mixture by gas chromatography can be made on the basis of the adjusted retention time t'_R or the adjusted retention volume V'_R. However, since the parameters t'_R and V'_R depend on the experimental conditions (temperature, flow velocity of carrier gas, quantity of liquid phase, etc.), the identification of a substance is usually made by applying the *Kováts retention index* or simply *Kováts index*, I, which is independent from these factors. The Kováts index expresses the relative retention (Section 18.2) of a substance with respect to linear alkanes, which are taken as reference materials, and it is calculated by the relation

$$I = 100\frac{\log t'_{R,X} - \log t'_{R,Z}}{\log t'_{R,Z+1} - \log t'_{R,Z}} + 100z, \qquad (18.33)$$

where $t'_{R,X}$, $t'_{R,Z}$, $t'_{R,Z+1}$ are the adjusted retention times for substance X and the alkanes with z and z + 1 carbon atoms, respectively. The choice of z should be such that $t'_{R,Z} < t'_{R,X} < t'_{R,Z+1}$ (Example 18.7). For example, I = 650 for benzene. Hence, benzene behaves like a linear alkane with 6.50 carbon atoms and is eluted between n-hexane and n-heptane. If the experimental conditions are changed, the values of t'_R or V'_R are also changed, but the value of I for a substance remains constant.

Another fact which assists in identification is that for a homologous series of compounds and a certain column and temperature, the plot of $\log t'_R$ or $\log V'_R$ versus the number of carbon atoms is a straight line.

Example 18.7. The following adjusted retention times (t'_R), expressed in cm, were measured on a chromatogram: n-hexane, 8.4; cyclohexane, 14.6; n-heptane, 15.7; toluene, 18.3; and n-octane, 30.8. a) Calculate the Kováts index for cyclohexane and for toluene. b) Does the value of I/100 have any physical significance?

Solution. a) Substituting the data in Equation (18.33), we have for cyclohexane

$$I = 100\left(\frac{\log 14.6 - \log 8.4}{\log 15.7 - \log 8.4}\right) + (100 \times 6) = \mathbf{688},$$

and for toluene

$$I = 100\left(\frac{\log 18.3 - \log 15.7}{\log 30.8 - \log 15.7}\right) + (100 \times 7) = \mathbf{723}.$$

b) It only signifies that the corresponding compound behaves in this system as if it were a linear alkane with (I/100) carbon atoms.

Example 18.8. In a gas-chromatographic column the retention time t_R is 19.6 s for ethane, 46.7 s for n-butane, and 1.9 s for air. Calculate the expected retention times in the same column for propane and n-hexane.

Solution. *1st method.* Let z be the number of carbon atoms in the hydrocarbon molecule, whereupon we have the relation

$$\log t'_R = az + b \tag{18.34}$$

Substituting the data in Equation (18.34), we have the relations

$$\log(19.6 - 1.9) = 1.2480 = 2a + b \tag{18.35}$$

$$\log(46.7 - 1.9) = 1.6513 = 4a + b \tag{18.36}$$

Solving the system of Equations (18.35) and (18.36), we find that $a = 0.2016_5$ and $b = 0.8447$. Hence, if t_x is the sought-for retention time for propane, we have the relation

$$\log(t_x - 1.9) = 3a + b = (3 \times 0.2016_5) + 0.8447 = 1.4495$$

or

$$t_x - 1.9 = 28.2, \text{ and } t_x = \mathbf{30.1} \text{ s.}$$

Similarly, if t_y is the sought-for time for n-hexane, we have the relation

$$\log(t_y - 1.9) = 6a + b = (6 \times 0.2016_5) + 0.8447 = 2.0546,$$

or

$$t_y - 1.9 = 113.4, \text{ and } t_y = \mathbf{115.3} \text{ s.}$$

2nd method (graphical). A graphical solution is shown in Figure 18.4, where $\log t'_R = f(z)$. From this graph we have for propane $\log t'_R = 1.45$, $t'_R = 28.2 = t_x - 1.9$, and $t_x =$ **30.1** s. Similarly, for n-hexane we have $\log t'_R = 2.056$, $t'_R = 113.8 = t_y - 1.9$, and $t_y =$ **115.7** s.

Quantitative analysis. Under certain conditions (region of linear detector response, constant experimental conditions, and so forth), the area of a chromatographic peak is proportional to the quantity of a component, and it is used for the construction of the calibration curve. For small t'_R values, whereupon the peaks are sharp and narrow, the peak height can be used instead of the area, but in this way the precision is reduced.

If the detector responded equally well to all components of a mixture, the *relative* areas of the peaks would be equal to the corresponding *relative* quantities of the components. However, often this is not the case and the sensitivity of the detector is different for different compounds (the sensitivity of a thermal conductivity detector depends on the nature of the substance, the response of a flame ionization detector varies, depending on the number of carbon atoms in the molecule of the compound, etc.) In these cases, an empirical *response factor* (F) is determined experimentally for each component, which is calculated by the relation

$$F = \frac{C_x/C_s}{A_x/A_s} \tag{18.37}$$

where C_x and C_s are the concentrations of component X and the reference substance (standard), and A_x and A_s are the areas of the corresponding peaks.

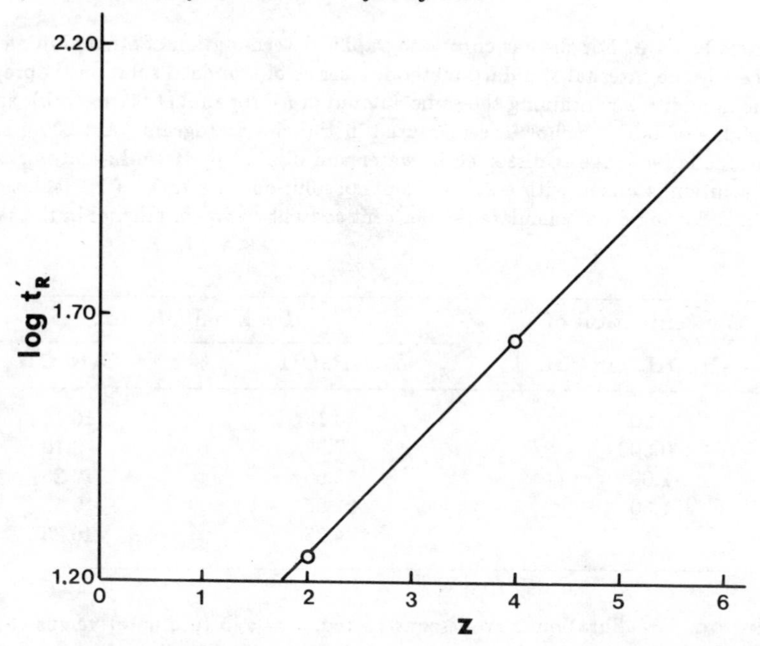

Figure 18.4. *Plot of* $\log t'_R$ *versus number z of carbon atoms.*

When fluctuations in the experimental conditions are expected, the method of internal standard is used, whereupon the *relative* areas of the peaks for two compounds remain constant and independent of the flow velocity of the carrier gas. In this method we have the relation

$$C_{x,u} = \left(\frac{A_{x,u}}{A_{x,s}}\right)\left(\frac{A_{is,s}}{A_{is,u}} C_{x,s}\right) \tag{18.38}$$

where

$C_{x,u}$ = concentration of substance X in the sample

$C_{x,s}$ = concentration of substance X in the standard solution

$A_{x,u}$ = peak areas of substance X in the sample

$A_{x,s}$ = peak areas of substance X in the standard solution

$A_{is,u}$ = peak area of internal standard in the sample

$A_{is,s}$ = peak area of internal standard in the standard solution

The measurement of the area of a peak is usually made by one of the following methods: 1) by triangulating the peak and multiplying the width $w_{1/2}$ by the peak height h (the peak must be Gaussian), 2) with a planimeter, 3) with an integrator, 4) by cutting out the peak and weighing the paper. The area can be calculated automatically with a computer.

Example 18.9. For the gas chromatographic determination of ethanol in alcoholic beverages by the internal standard method, a series of standard solutions is prepared, each one in addition containing the same amount of n-propanol (1.00 mg/mL), and the peak heights of both alcohols are measured in the chromatogram. A 1.230-g sample of an alcoholic beverage is dissolved in water and diluted to 100 mL, and part of the diluted solution is mixed with a 2‰ n-propanol solution, at a ratio of 1:1 (solution A). From the following data, calculate the per cent content (w/w) of ethanol in the sample.

Concentration of	Peak height, cm	
C_2H_5OH, mg/mL	C_2H_5OH	C_3H_7OH
0.200	3.15	10.14
0.500	7.55	9.70
1.00	16.00	10.25
1.50	21.95	9.50
A	9.25	10.85

Solution. A calibration curve is constructed, y = a/b (ordinate) versus C_2H_5OH concentration, where a and b are the peak heights of ethanol and n-propanol, respectively. From the calibration graph we find that solution A contains 0.548 mg C_2H_5OH/mL. Hence, the alcoholic beverage contains

$$\frac{0.548 \text{ mg/mL} \times 200 \text{ mL}}{1230 \text{ mg}} \times 100 = \mathbf{8.91\%} \text{ w/w } C_2H_5OH.$$

Notes. 1. The C_2H_5OH concentration can also be found from the regression equation of the calibration curve, which in the present case is: $y = 1.540x + 0.007$.

2. If we assume that the ratio of peak areas equals the ratio of the corresponding heights and for the calculations we use Equation (18.38), then on the basis of the ethanol standard solution containing 0.200 mg C_2H_5OH/mL we find that solution A contains

$$\left(\frac{9.25}{3.15}\right)\left(\frac{10.14}{10.85}\right) 0.200 = 0.549 \text{ mg } C_2H_5OH/\text{mL},$$

a value that practically coincides with the one found by the calibration curve method. Similarly, if in Equation (18.38) we use the data for solution A and the standards which contain 0.500, 1.00 and 1.50 mg C_2H_5OH/mL, we find that solution A contains 0.548, 0.546, and 0.553 mg C_2H_5OH/mL, respectively. From these results it can be seen that there is satisfactory agreement between the values calculated by the two methods.

Example 18.10. For the determination of methanol in methanol-water samples by gas chromatography, the chromatograms of four standard mixtures containing 20.0, 40.0, 60.0, and 80.0% v/v methanol are taken. The methanol and water peaks are cut out with scissors and the respective weights w_m and w_w of the recorder paper are measured. From the following data, calculate the per cent content of methanol in solution A.

Concentration of CH_3OH, % v/v	Weight of paper, g		$\frac{w_m}{w_m + w_w}$
	w_m	w_w	
20.0	0.0436	0.1955	0.182
40.0	0.0692	0.1485	0.318
60.0	0.1098	0.1190	0.480
80.0	0.1107	0.0700	0.613
A	0.1124	0.1124	0.500

Solution. A calibration curve is constructed, $y = w_m/(w_m + w_w)$ (ordinate) versus per cent methanol (x = % v/v CH_3OH), and from this curve we find that solution A contains 64.0% v/v CH_3OH.

Note. For better accuracy, the equation of the calibration curve is found by the method of least squares (Section 1.9), which gives $y = 0.007275x + 0.0345$. From this equation, for $y = 0.500$, we find $x =$ **64.0%** v/v CH_3OH.

Example 18.11. In the determination of free ethanol in a commercial ethyl acetate sample, a 2.0-μL sample gave an ethanol peak having an area of 0.85 cm^2. Twenty-five microliters of ethanol were mixed with 5.00 mL of sample (solution A). A 2.0-μL aliquot of solution A gave an ethanol peak having an area of 1.45 cm^2. Calculate the per cent content (w/w) of ethanol in the sample (density of ethanol = 0.789 g/mL, of ethyl acetate = 0.901 g/mL).

Solution. Assume that 2.0 μL of sample contains y μg of ethanol, whereupon 5.00 mL of the sample contains (5.00/0.0020)y = 2500y μg C_2H_5OH. Therefore, 2.0 μL of solution A contains

$$[2500y\ \mu g + (0.0250\ mL\ C_2H_5OH \times 0.789\ g/mL \times 10^6\ \mu g/g]\ (0.0020/5.025)$$
$$= (0.995y + 7.851)\ \mu g\ C_2H_5OH.$$

Hence, we have the relation

$$\frac{0.995y + 7.851}{y} = \frac{1.45\ cm^2}{0.85\ cm^2},$$

from which we find that y = 11.04 μg. Hence, the sample contains

$$\frac{11.04\ \mu g\ C_2H_5OH}{0.0020\ mL\ ester \times 0.901\ g/mL \times 10^6 \mu g/g} \times 100 = \mathbf{0.61\%}\ w/w\ C_2H_5OH.$$

Example 18.12. The GLC analysis of a mixture of n-pentane, n-hexane, n-heptane, and n-octane, using a thermal conductivity detector, is based on the oxidation of the eluted hydrocarbons to CO_2 and H_2O and the subsequent passing of CO_2 through the detector *after* the removal of H_2O. The following data were obtained:

Compound	Area, arbitrary units
n-pentane	10.0
n-hexane	24.0
n-heptane	42.0
n-octane	64.0

Calculate a) the mole per cent composition of the sample, b) the weight per cent composition of the sample.

Solution. The detector measures the carbon content. Since each carbon atom produces one CO_2 molecule, regardless of its origin, we have the relation

$$\text{relative number of moles} = \text{relative area/number of C atoms} \tag{18.39}$$

On the basis of the data and Equation (18.39), we construct the following table:

Substance	Area, arbitrary units	Area/number of C atoms = relative number of moles	% mol	% mol x MW	% w/w
n − pentane	10.0	2.00	**10.0**	722	**7.2**
n − hexane	24.0	4.00	**20.0**	1723	**17.2**
n − heptane	42.0	6.00	**30.0**	3006	**30.0**
n − octane	64.0	8.00	**40.0**	4569	**45.6**
		20.00	100.0	10020	100.0

Example 18.13. The determination of Cr(III) in blood serum (in cases of poisoning with chromium compounds) by gas chromatography is performed as follows: One milliliter of serum is treated with acetylacetone, and the complex formed, [Cr(acetylacetone)$_3$], is extracted with an organic solvent and diluted to 5.00 mL (solution A). A 5.00-μL aliquot of solution A gives a peak for the complex having an area of 6.30 cm^2. One milliliter of a 10.0 ppm Cr(III) solution is treated similarly and diluted to 5.00 mL (solution B); 1.00 μL of solution B gives a peak having an area of 5.45 cm^2. Also, 1.00 mL of water is treated similarly and gives 5.00 mL of solution C, 10.0 μL of which gives a peak having an area of 1.10 cm^2 (blank). Calculate the chromium content in the serum, in ppm.

Solution. Suppose that the serum contains x μg Cr/mL and the blank b μg Cr/mL. If S represents the peak area and V the volume of the solution used to obtain the chromatogram, we have the relation

$$\mu\text{g Cr/mL} = k \cdot \frac{S(\text{cm}^2)}{V(\mu\text{L})} \tag{18.40}$$

Substituting the data for solutions A, B, and C in Equation (18.40), we have

$$x + b = k(6.30/5.00) \tag{18.41}$$

$$10.0 + b = k(5.45/1.00) \tag{18.42}$$

$$b = k(1.10/10.0) \tag{18.43}$$

Solving the system of Equations (18.41), (18.42), and (18.43), we find that x = 2.15 μg Cr/mL $\equiv$ **2.15** ppm Cr.

Example 18.14. Trace amounts of cyanide are determined by gas chromatography after converting to CNCl with an aqueous solution of chloramine T (the sodium salt of N-chloro-p-toluenesulfonamide). The CNCl is extracted with hexane (C_6H_{14}) and injected on the column. The hexane peak serves as an internal standard. The calibration curve, i.e., the ratio of peak areas (CNCl/hexane) versus cyanide concentration (abscissa), has an average slope S equal to 2.48 mL/μg. In the analysis of a 1.00-mL blood sample (from the coroner's office) the areas of the CNCl and n-hexane peaks were 32.95 and 0.2433 units, respectively. Calculate the cyanide concentration in the blood, in μg/mL.

Solution. We have the relation

$$S = \frac{A_{CNCl}/A_{C_6H_{14}}}{[CN^-]} = \frac{32.95/0.2433}{[CN^-]} = 2.48 \text{ mL}/\mu\text{g}, \tag{18.44}$$

where A_{CNCl} and $A_{C_6H_{14}}$ are the areas of the peaks of CNCl and hexane, expressed in the same (arbitrary) units. From Equation (18.44) we find that $[CN^-]$ = **54.6** μg/mL.

Example 18.15. In an experimental GC system, the eluent from the column is directed to a chamber for pyrolytic cracking and hydrogenation, so that, if there are sulfur atoms in the eluted compounds, they are finally converted to an equal number of H_2S molecules. From the chamber, the gases are directed to a beaker containing a solution that traps the H_2S. An automatic buret delivers an iodine solution which reacts with H_2S according to the reaction $H_2S + [I_3]^- \rightarrow 2H^+ + 3I^- + \mathbf{S}$. The reaction is monitored potentiometrically and an appropriate amount of iodine solution is delivered whenever needed to oxidize quantitatively the H_2S. The following figure shows the recording taken for a mixture of ethyl mercaptan (C_2H_5SH), diethylthioether ($C_2H_5SC_2H_5$), and diethyldisulfide ($C_2H_5SSC_2H_5$). Calculate the per cent (w/w) content of each component in the mixture.

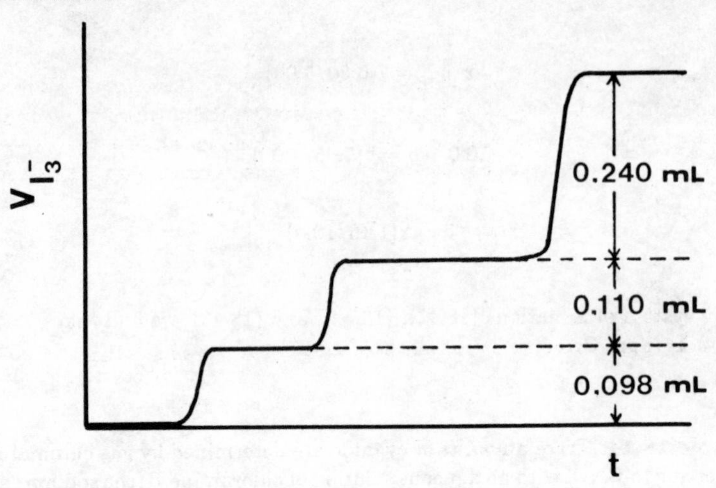

Solution. Let $\underline{N}$ be the normality of the iodine solution. The injected mixture contains

(0.098 mL)($\underline{N}$ meq/mL)(0.5 mmol H_2S/meq)(1 mmol C_2H_5SH/mmol H_2S)
(62.13 mg/mmol C_2H_5SH) = 3.044 $\underline{N}$ mg C_2H_5SH,

(0.110 mL)($\underline{N}$ meq/mL)(0.5 mmol H_2S/meq)[1 mmol $(C_2H_5)_2S$/mmol H_2S]
[90.19 mg/mmol $(C_2H_5)_2S$] = 4.960 $\underline{N}$ mg $(C_2H_5)_2S$

and

(0.240 mL)($\underline{N}$ meq/mL)(0.5 mmol H_2S/meq)[0.5 mmol$(C_2H_5)_2S_2$/mmol H_2S]

[122.25 mg/mmol $(C_2H_5)_2S_2$] = 7.315 $\underline{N}$ mg $(C_2H_5)_2S_2$.

Therefore, the mixture weighs 3.044 $\underline{N}$ + 4.960 $\underline{N}$ + 7.315 $\underline{N}$ = 15.319 $\underline{N}$ mg. Hence, the mixture contains

(3.044 $\underline{N}$/15.319 $\underline{N}$)100 = **19.9**% w/w C_2H_5SH,
(4.960 $\underline{N}$/15.319 $\underline{N}$)100 = **32.4**% w/w $(C_2H_5)_2S$,

and

(7.315 $\underline{N}$/15.319 $\underline{N}$)100 = **47.8**% w/w $(C_2H_5)_2S_2$.

Example 18.16. A standard mixture that contains equal weights of 2-methylbutane, n-pentane, and n-hexane, is injected into a gas chromatograph and gives three peaks having areas of 4.70 cm^2, 4.20 cm^2, and 3.60 cm^2, respectively. Sample A containing the above hydrocarbons gives corresponding peaks having areas of 3.20 cm^2, 1.20 cm^2, and 4.50 cm^2. Calculate the per cent content of each hydrocarbon in sample A.

Solution. We have the relation

$$w_n = k_n S_n, \tag{18.45}$$

where w_n is the weight (mg) of a component in the injected mixture, S_n is the area of the corresponding peak, and k_n is a proportionality constant. Let w_1, w_2, and w_3 be the weights of 2-methyl-butane, n-pentane, and n-hexane in the standard mixture, and S_1, S_2, and S_3 be the areas of the corresponding peaks, whereupon we have

$$w_1 = k_1 S_1 \tag{18.46}$$

$$w_2 = k_2 S_2 \tag{18.47}$$

$$w_3 = k_3 S_3 \tag{18.48}$$

Since $w_1 = w_2 = w_3$, from Equations (18.46) and (18.48), we have that

$$k_1 = (S_3/S_1)k_3 = (3.60/4.70)k_3 = 0.766k_3 \tag{18.49}$$

and from Equations (18.47) and (18.48) that

$$k_2 = (S_3/S_2)k_3 = (3.60/4.20)k_3 = 0.857k_3. \tag{18.50}$$

Let w'_1, w'_2, and w'_3 be the weights of 2-methyl-butane, n-pentane, and n-hexane in sample A, and S'_1, S'_2, and S'_3 be he areas of the corresponding peaks, whereupon we have

$$w'_1 = k_1 S'_1 = 0.766k_3 \cdot 3.20 = 2.451k_3$$

$$w'_2 = k_2 S'_2 = 0.857k_3 \cdot 1.20 = 1.028k_3$$

$$w'_3 = k_3 S'_3 = 4.50k_3.$$

Therefore, sample A weighs $2.451k_3 + 1.028k_3 + 4.50k_3 = 7.979k_3$ mg. Hence, sample A contains

$(2.451k_3/7.979k_3)100 =$ **30.7%** w/w 2-methyl-butane
$(1.028k_3/7.979k_3)100 =$ **12.9%** w/w n-pentane

and

$(4.50k_3/7.979k_3)100 =$ **56.4%** w/w n-hexane.

Example 18.17. A mixture containing 1.25 mmol of n-pentanol and 1.20 mmol of n-hexanol is dissolved in the appropriate solvent and the solution is analyzed by GC. The area of the n-pentanol peak is 2.08 cm^2, whereas the area of the n-hexanol peak is 2.32 cm^2. A sample containing 1.10 mmol of n-pentanol and y mmol of n-hexanol is treated similarly and gives peaks having areas of 1.83 cm^2 and 1.12 cm^2, respectively. Calculate a) the response factor for n-hexanol, relative to n-pentanol (reference compound, internal standard), and b) the value of y.

Solution. a) Substituting the data in Equation (18.37) (the volumes cancel), we find that the response factor for n-hexanol is equal to

$$F = \frac{1.20/1.25}{2.32/2.08} = \mathbf{0.861}.$$

b) Again, using Equation (18.37), we have the relation

$$F = \frac{\text{mmol n-hexanol/mmol n-pentanol}}{\text{area of n-hexanol/area of n-pentanol}} = 0.861 = \frac{y/1.10}{1.12/1.83},$$

from which we find that y = **0.580**.

18.4 Problems

Basic chromatographic relations

18.1. Prove that for a chromatographic column we have the relation $n = L^2/\sigma^2$.

18.2. The distribution coefficients of substances A and B in a chromatographic column are 180 and 225, respectively. Which of the two substances is eluted first from the column?

18.3. Calculate the retention volumes of substances A and B, if $V_S = 1.3$ mL, $V_M = 2.1$ mL, $K_A = 10.0$, $K_B = 40.0$. If the column contains 24 theoretical plates, is the separation of substances A and B satisfactory?

18.4. After injecting 3μL of n-hexane in a gas chromatograph with a 2.48-m column, the retention time was 1.29 min for air and 6.05 min for n-hexane, whereas the peak width for n-hexane was 0.50 min. Calculate a) the theoretical plate number n, b) the effective plate number N, and c) the HETP (h).

18.5. What will the change in peak width w be, if the column length is tripled?

Resolution

18.6. a) For the column C having $t_M << t_R$, and $w_A \simeq w_R$, prove the relation

$$R = \left(\frac{\sqrt{n}}{4}\right)\left(\frac{\alpha - 1}{\alpha}\right)$$

b) If the resolution R for a mixture of substances A and B in a 20-cm column C_1 is 0.90, what will the resolution be for the same mixture in a 40-cm column C_2? (With respect to other characteristics, column C_2 is exactly like columns C and C_1.)

18.7. In the chromatographic separation of a mixture of substances A (eluted first) and B, how many times should the retention time t_{R_B} be increased for the resolution of the column to be doubled?

18.8. A satisfactory chromatographic separation ($R \geq 1.5$) is achieved with various combinations of column parameters, e.g., for $N = 3136$, $\alpha = 1.20$, and $k' = 1.80$. What will the effect on the resolution of the above column be, if a) N is decreased to 1394 (the other parameters remain constant), b) if k' is decreased to 0.75?

18.9. Calculate the length that a column must have for the quantitative separation of substances A and B ($R = 1.50$), when $V_M = 8.0$ mL, $V_{R_B} = 96$ mL, $\alpha = 1.20$, and $h = 0.110$ cm.

18.10. Two chromatographic columns, C_1 with the liquid phase on a solid support, and C_2, a capillary one with the liquid phase on the walls of the capillary tube, have the following characteristics:

	Column C_1	Column C_2
Dead time, t_M, min	2.0	0.40
Adjusted retention time, t'_R, min :		
t'_{R_A}	21.0	0.12
t'_{R_B}	22.0	0.13
Theoretical plate number, n	4900	19600

With which of the two columns is a better separation achieved (higher R)? Interpret the results on the basis of the effective plate number N.

van Deemter equation

18.11. In what units are the constants A, B, and C of the van Deemter Equation (18.27) expressed, if the height is expressed in cm and the linear velocity of the carrier gas in cm/s?

18.12. The constants of the van Deemter Equation (18.27) for a particular chromatographic column at 180°C are: $A = 0.07$ cm, $B = 0.50$ cm^2/s, $C = 0.04$ s. Calculate the values of h at several values of the linear velocity of the carrier gas (v), and from the graphical presentation of the results [$h = f(v)$, Figure 18.3], calculate the optimum velocity (v_{opt}) and the corresponding HETP (h_{min}). How do these values compare with those obtained by using Equations (18.28) and (18.29)?

18.13. The following results were obtained with a chromatographic column:

v, cm/s	1.0	2.0	3.0	4.0	5.0	6.0	8.0	10.0	15.0
R, cm	0.63	0.44	0.42	0.44	0.47	0.51	0.61	0.72	1.00

a) Construct a plot of h vs. v, and from this determine the best linear velocity of the carrier gas (v_{opt}), and the corresponding minimum h_{min}. b) From the plot determine the values of the constants A, B, and C, at v_{opt}, and compare them with the values of the constants calculated by the van Deemter Equation (18.27), on the basis of data corresponding to three points located close to v_{opt} (*Hint*: system of three equations with three unknowns).

18.14. The constants A, B, and C of the van Deemter Equation (18.27) have the following values for two columns of equal length:

	A	B	C
Column 1	0.10 cm	0.40 cm^2/s	0.06s
Column 2	0.130	0.32	0.13

a) Which of the two columns gives the larger theoretical plate number, if the flow velocity of the carrier gas is 1.0 cm/s? b) Which is the best velocity (v_{opt}) for each of the two columns?

Qualitative analysis

18.15. The retention times, in seconds, in a GLC column are the following: air, 5.6; n-butane, 48.7; substance A, 85.4; n-hexane, 103.0; substance B, 152.0. Which one of substances A and B cannot be a linear alkane?

18.16. In a gas chromatogram, the adjusted retention times t'_R, in minutes, were the following: ethane, 0.20; n-pentane, 1.48; n-octane, 11.2; ethylene, 0.11; butene-1, 0.66; benzene, 3.31. Calculate the Kováts indices for the above compounds.

18.17. Below, the adjusted retention volumes, V'_R in mL, are given for a series of homologous compounds:

Number of C atoms	1	2	3	4
V'_R, mL	6.0	12.0	24.0	48.0

Calculate the expected adjusted retention volume, with the same column, for a compound of the same homologous series having five carbon atoms.

18.18. The unknown compounds X_1 and X_2 were co-injected into a gas chromatograph with their homologous compounds A and B, having 5 and 8 carbon atoms in their molecules, respectively. The retention times in minutes were the following: air, 0.5; compound A, 1.98; compound B, 11.7; compound X_1, 0.7; compound X_2, 3.4. How many carbon atoms are there in the molecules of compounds X_1 and X_2?

Quantitative analysis

18.19. Virgin olive oil A is subjected to methanolysis, in order to convert the nonvolatile glycerin esters of the organic acids to volatile methyl esters. The mixture of methyl esters is injected into a gas chromatograph, giving peaks having areas of 2.20, 5.86, and 1.21 cm^2, for the palmitic, oleic, and stearic methyl esters, respectively. Olive oil B of inferior quality is subjected to the same treatment and gives respective peaks of 1.82, 6.90, and 1.05 cm^2. Olive oil C, obtained by adulteration of olive oil A with olive oil B, is subjected to the same treatment and gives respective peaks of 1.97, 5.90, and 1.10 cm^2. Calculate the per cent (w/w) content of olive oil B in olive oil C (in all cases, the same amount of sample is injected. Assume that the weight factor (weight/area unit) is the same for all esters.)

18.20. In a (gas) chromatogram obtained in the determination of the purity of substance A using a 10.0-mg sample, besides the peak of substance A there was a second peak of 0.50 cm^2, corresponding to an impurity B. When 3.0 mg of substance B was injected, the peak area was 4.17 cm^2. Calculate the *chromatographic* purity of the sample.

18.21. A 3.00-mg sample containing n-pentane, n-hexanol, and toluene gave (gas) chromatographic peaks having areas of 2.40, 2.15, and 3.12 cm^2, respectively. a) If the mole response of the detector (area/mole) is the same for all three components of the sample, calculate the per cent mole composition and the per cent weight composition of the sample. b) If the weight factor (weight/cm^2) is 0.87, 1.15, and 1.23, respectively, calculate the per cent weight composition of the sample.

18.22. The determination of cyclohexane in cyclohexane-benzene mixtures is performed by obtaining the (gas) chromatogram of a standard and of an unknown mixture. The standard mixture was prepared by mixing 1.525 g of cyclohexane with 1.783 g of benzene, and a 2-μL sample of this mixture gave peaks having areas of 15.41 and 18.32 cm^2, respectively. The respective peaks in the unknown mixture had areas of 13.98 and 12.12 cm^2. Calculate the per cent cyclohexane in the unknown mixture.

Appendix A

Equilibrium Constants

The equilibrium constants given in the following four tables are for a temperature of 25°C. They appear in two forms, the exponential, K, and the logarithmic, pK.

Table A.1. *Dissociation constants for acids.*

Acid	Equilibrium equation	K_a	pK_a
Acetic	$CH_3COOH \rightleftharpoons H^+ + CH_3COO^-$	1.8×10^{-5}	4.74
Aluminum hydroxide	$Al(OH)_3 \rightleftharpoons H^+ + AlO_2^- + H_2O$	4×10^{-13}	12.4
Aluminum ion	$[Al(H_2O)_6]^{3+} \rightleftharpoons H^+ + [Al(H_2O)_5(OH)]^{2+}$	1.1×10^{-5}	4.96
Ammonium ion	$NH_4^+ \rightleftharpoons H^+ + NH_3$	5.6×10^{-10}	9.25
Antimony(III) hydroxide	$Sb(OH)_3 \rightleftharpoons H^+ + SbO_2^- + H_2O$	1×10^{-11}	11.0
Arsenic	$H_3AsO_4 \rightleftharpoons H^+ + H_2AsO_4^-$	6.0×10^{-3} (K_{a_1})	2.22
	$H_2AsO_4^- \rightleftharpoons H^+ + HAsO_4^{2-}$	1×10^{-7} (K_{a_2})	7.0
	$HAsO_4^{2-} \rightleftharpoons H^+ + AsO_4^{3-}$	3×10^{-12} (K_{a_3})	11.5
Benzoic	$C_6H_5COOH \rightleftharpoons H^+ + C_6H_5COO^-$	6.6×10^{-5}	4.18
Boric	$H_3BO_3 \rightleftharpoons H^+ + H_2BO_3^-$	6.0×10^{-10}	9.22
Carbonic	$H_2CO_3 \rightleftharpoons H^+ + HCO_3^-$	4.2×10^{-7} (K_{a_1})	6.38
	$HCO_3^- \rightleftharpoons H^+ + CO_3^{2-}$	4.8×10^{-11} (K_{a_2})	10.32
Chloroacetic	$ClCH_2COOH \rightleftharpoons H^+ + ClCH_2COO^-$	1.4×10^{-3}	2.85
Chromic	$H_2CrO_4 \rightleftharpoons H^+ + HCrO_4^-$	$\simeq 10^{-1}$ (K_{a_1})	1.0
	$HCrO_4^- \rightleftharpoons H^+ + CrO_4^{2-}$	3.2×10^{-7} (K_{a_2})	6.49
Copper(II) hydroxide	$Cu(OH)_2 \rightleftharpoons H^+ + HCuO_2^-$	1.5×10^{-16} (K_{a_1})	15.82
	$HCuO_2^- \rightleftharpoons H^+ + CuO_2^{2-}$	8×10^{-14} (K_{a_2})	13.1
Dichloroacetic	$Cl_2CHCOOH \rightleftharpoons H^+ + Cl_2CHCOO^-$	5.5×10^{-2}	1.26
Formic	$HCOOH \rightleftharpoons H^+ + HCOO^-$	2.1×10^{-4}	3.68
Hydrocyanic	$HCN \rightleftharpoons H^+ + CN^-$	4×10^{-10}	9.4
Hydrofluoric	$HF \rightleftharpoons H^+ + F^-$	6.9×10^{-4}	3.16
Hydrogen peroxide	$H_2O_2 \rightleftharpoons H^+ + HO_2^-$	2.4×10^{-12}	11.62

Table A.1. *Dissociation constants for acids (continued).*

Acid	Equilibrium equation	K_a	pK_a
Hydrogen sulfide	$H_2S \rightleftharpoons H^+ + HS^-$	$1.0 \times 10^{-7} (K_{a_1})$	7.00
	$HS \rightleftharpoons H^+ + S^{2-}$	$1.0 \times 10^{-14} (K_{a_2})$	14.00
Hypochlorous	$HClO \rightleftharpoons H^+ + ClO^-$	3.2×10^{-8}	7.49
Iron(III) ion	$[Fe(H_2O)_6]^{3+} \rightleftharpoons H^+ + [Fe(H_2O)_5(OH)]^{2+}$	8.9×10^{-4}	3.05
Lead(II) hydroxide	$\mathbf{Pb(OH)_2} \rightleftharpoons H^+ + HPbO_2^-$	2×10^{-16}	15.7
Nitrous	$HNO_2 \rightleftharpoons H^+ + NO_2^-$	4.5×10^{-4}	3.35
Oxalic	$H_2C_2O_4 \rightleftharpoons H^+ + HC_2O_4^-$	$3.8 \times 10^{-2} (K_{a_1})$	1.42
	$HC_2O_4^- \rightleftharpoons H^+ + C_2O_4^{2-}$	$5.0 \times 10^{-5} (K_{a_2})$	4.30
Periodic	$HIO_4 \rightleftharpoons H^+ + IO_4^-$	2.3×10^{-2}	1.64
Phenol	$C_6H_5OH \rightleftharpoons H^+ + C_6H_5O^-$	1×10^{-10}	10.0
Phosphoric	$H_3PO_4 \rightleftharpoons H^+ + H_2PO_4^-$	$7.5 \times 10^{-3} (K_{a_1})$	2.12
	$H_2PO_4^- \rightleftharpoons H^+ + HPO_4^{2-}$	$6.2 \times 10^{-8} (K_{a_2})$	7.21
	$HPO_4^{2-} \rightleftharpoons H^+ + PO_4^{3-}$	$1 \times 10^{-12} (K_{a_3})$	12.0
Phosphorous	$H_3PO_3 \rightleftharpoons H^+ + H_2PO_3^-$	$1.6 \times 10^{-2} (K_{a_1})$	1.80
	$H_2PO_3^- \rightleftharpoons H^+ + HPO_3^{2-}$	$7 \times 10^{-7} (K_{a_2})$	6.2
Silicic (meta)	$H_2SiO_3 \rightleftharpoons H^+ + HSiO_3^-$	$3.2 \times 10^{-10} (K_{a_1})$	9.49
	$HSiO_3^- \rightleftharpoons H^+ + SiO_3^{2-}$	$6.3 \times 10^{-12} (K_{a_2})$	11.80
Sulfamic	$HNH_2SO_3 \rightleftharpoons H^+ + NH_2SO_3^-$	1.1×10^{-1}	0.96
Sulfuric	$H_2SO_4 \rightleftharpoons H^+ + HSO_4^-$	$1.0 \times 10^{2} (K_{a_1})$	−2.00
	$HSO_4^- \rightleftharpoons H^+ + SO_4^{2-}$	$1.2 \times 10^{-2} (K_{a_2})$	1.92
Sulfurous	$H_2SO_3 \rightleftharpoons H^+ + HSO_3^-$	$1.3 \times 10^{-2} (K_{a_1})$	1.89
	$HSO_3^- \rightleftharpoons H^+ + SO_3^{2-}$	$5.6 \times 10^{-8} (K_{a_2})$	7.25
Tartaric	$H_2C_4H_4O_6 \rightleftharpoons H^+ + HC_4H_4O_6^-$	$1.1 \times 10^{-3} (K_{a_1})$	2.96
	$HC_4H_4O_6^- \rightleftharpoons H^+ + C_4H_4O_6^{2-}$	$6.9 \times 10^{-5} (K_{a_2})$	4.16
Thiocyanic	$HSCN \rightleftharpoons H^+ + SCN^-$	1.4×10^{-1}	0.85
Thiosulfuric	$H_2S_2O_3 \rightleftharpoons H^+ + HS_2O_3^-$	$2.0 \times 10^{-2} (K_{a_1})$	1.70
	$HS_2O_3^- \rightleftharpoons H^+ + S_2O_3^{2-}$	$3.2 \times 10^{-3} (K_{a_2})$	2.49
Tin(II) hydroxide	$\mathbf{Sn(OH)_2} \rightleftharpoons H^+ + HSnO_2^-$	4×10^{-15}	14.4
Trichloroacetic	$Cl_3CCOOH \rightleftharpoons H^+ + Cl_3CCOO^-$	1.3×10^{-1}	0.89
Zinc hydroxide	$\mathbf{Zn(OH)_2} \rightleftharpoons H^+ + HZnO_2^-$	$1 \times 10^{-17} (K_{a_1})$	17.0
	$HZnO_2^- \rightleftharpoons H^+ + ZnO_2^{2-}$	$2 \times 10^{-13} (K_{a_2})$	12.7
Zinc ion	$[Zn(H_2O)_4]^{2+} \rightleftharpoons H^+ + [Zn(H_2O)_3(OH)]^+$	2.5×10^{-10}	9.60

Table A.2. *Dissociation constants for bases.*

Base	Equilibrium equation	K_b	pK_b
Ammonia	$NH_3 + H_2O \rightleftharpoons NH_4^+ + OH^-$	1.8×10^{-5}	4.74
Aniline	$C_6H_5NH_2 + H_2O \rightleftharpoons C_6H_5NH_3^+ + OH^-$	3.8×10^{-10}	9.42
Dimethylamine	$(CH_3)_2NH + H_2O \rightleftharpoons (CH_3)_2NH_2^+ + OH^-$	5.1×10^{-4}	3.29
Ethylamine	$C_2H_5NH_2 + H_2O \rightleftharpoons C_2H_5NH_3^+ + OH^-$	5.6×10^{-4}	3.25
Hydrazine	$N_2H_4 + H_2O \rightleftharpoons N_2H_5^+ + OH^-$	9.8×10^{-7}	6.01
Methylamine	$CH_3NH_2 + H_2O \rightleftharpoons CH_3NH_3^+ + OH^-$	4.4×10^{-4}	3.36
Pyridine	$C_5H_5N + H_2O \rightleftharpoons C_5H_5NH^+ + OH^-$	1.4×10^{-9}	8.85
Trimethylamine	$(CH_3)_3N + H_2O \rightleftharpoons (CH_3)_3NH^+ + OH^-$	5.3×10^{-5}	4.28

Table A.3. *Solubility product constants.*

Substance	Equilibrium equation	K_{sp}	pK_{sp}
Acetates			
Silver acetate	$\mathbf{CH_3COOAg} \rightleftharpoons Ag^+ + CH_3COO^-$	4×10^{-3}	2.4
Arsenates			
Silver arsenate	$\mathbf{Ag_3AsO_4} \rightleftharpoons 3Ag^+ + AsO_4^{3-}$	1×10^{-22}	22.0
Bromates			
Silver bromate	$\mathbf{AgBrO_3} \rightleftharpoons Ag^+ + BrO_3^-$	6×10^{-5}	4.2
Bromides			
Copper(I) bromide	$\mathbf{CuBr} \rightleftharpoons Cu^+ + Br^-$	6×10^{-9}	8.2
Lead bromide	$\mathbf{PbBr_2} \rightleftharpoons Pb^{2+} + 2Br^-$	4.6×10^{-6}	5.34
Mercury(I) bromide	$\mathbf{Hg_2Br_2} \rightleftharpoons Hg_2^{2+} + 2Br^-$	1.3×10^{-22}	21.89
Silver bromide	$\mathbf{AgBr} \rightleftharpoons Ag^+ + Br^-$	5×10^{-13}	12.3
Carbonates			
Barium carbonate	$\mathbf{BaCO_3} \rightleftharpoons Ba^{2+} + CO_3^{2-}$	1.6×10^{-9}	8.80
Cadmium carbonate	$\mathbf{CdCO_3} \rightleftharpoons Cd^{2+} + CO_3^{2-}$	5.2×10^{-12}	11.28
Calcium carbonate	$\mathbf{CaCO_3} \rightleftharpoons Ca^{2+} + CO_3^{2-}$	6.9×10^{-9}	8.16
Cobalt(II) carbonate	$\mathbf{CoCO_3} \rightleftharpoons Co^{2+} + CO_3^{2-}$	8×10^{-13}	12.1
Copper(II) carbonate	$\mathbf{CuCO_3} \rightleftharpoons Cu^{2+} + CO_3^{2-}$	2.5×10^{-10}	9.60
Iron(II) carbonate	$\mathbf{FeCO_3} \rightleftharpoons Fe^{2+} + CO_3^{2-}$	2.1×10^{-11}	10.68
Lead carbonate	$\mathbf{PbCO_3} \rightleftharpoons Pb^{2+} + CO_3^{2-}$	1.5×10^{-13}	12.82

Table A.3. *Solubility product constants (continued).*

Substance	Equilibrium equation	K_{sp}	pK_{sp}
Carbonates (continued)			
Magnesium carbonate	$MgCO_3 \rightleftharpoons Mg^{2+} + CO_3^{2-}$	4×10^{-5}	4.4
Manganese(II) carbonate	$MnCO_3 \rightleftharpoons Mn^{2+} + CO_3^{2-}$	9×10^{-11}	10.1
Mercury(I) carbonate	$Hg_2CO_3 \rightleftharpoons Hg_2^{2+} + CO_3^{2-}$	9×10^{-17}	16.1
Nickel carbonate	$NiCO_3 \rightleftharpoons Ni^{2+} + CO_3^{2-}$	7×10^{-9}	8.2
Silver carbonate	$Ag_2CO_3 \rightleftharpoons 2Ag^{+} + CO_3^{2-}$	8.2×10^{-12}	11.09
Strontium carbonate	$SrCO_3 \rightleftharpoons Sr^{2+} + CO_3^{2-}$	7×10^{-10}	9.2
Zinc carbonate	$ZnCO_3 \rightleftharpoons Zn^{2+} + CO_3^{2-}$	2×10^{-11}	10.7
Chlorides			
Copper(I) chloride	$CuCl \rightleftharpoons Cu^{+} + Cl^{-}$	3.2×10^{-7}	6.49
Lead chloride	$PbCl_2 \rightleftharpoons Pb^{2+} + 2Cl^{-}$	1.6×10^{-5}	4.80
Mercury(I) chloride	$Hg_2Cl_2 \rightleftharpoons Hg_2^{2+} + 2Cl^{-}$	1.1×10^{-18}	17.96
Silver chloride	$AgCl \rightleftharpoons Ag^{+} + Cl^{-}$	1.8×10^{-10}	9.74
Thallium(I) chloride	$TlCl \rightleftharpoons Tl^{+} + Cl^{-}$	3.5×10^{-4}	3.46
Chromates			
Barium chromate	$BaCrO_4 \rightleftharpoons Ba^{2+} + CrO_4^{2-}$	1.2×10^{-10}	9.92
Calcium chromate	$CaCrO_4 \rightleftharpoons Ca^{2+} + CrO_4^{2-}$	7.1×10^{-4}	3.15
Copper(II) chromate	$CuCrO_4 \rightleftharpoons Cu^{2+} + CrO_4^{2-}$	3.6×10^{-6}	5.44
Lead chromate	$PbCrO_4 \rightleftharpoons Pb^{2+} + CrO_4^{2-}$	2×10^{-14}	13.7
Mercury(I) chromate	$Hg_2CrO_4 \rightleftharpoons Hg_2^{2+} + CrO_4^{2-}$	2×10^{-9}	8.7
Silver chromate	$Ag_2CrO_4 \rightleftharpoons 2Ag^{+} + CrO_4^{2-}$	1.9×10^{-12}	11.72
Strontium chromate	$SrCrO_4 \rightleftharpoons Sr^{2+} + CrO_4^{2-}$	3.6×10^{-5}	4.44
Cyanides			
Mercury(I) cyanide	$Hg_2(CN)_2 \rightleftharpoons Hg_2^{2+} + 2CN^{-}$	5×10^{-40}	39.3
Silver cyanide	$AgCN \rightleftharpoons Ag^{+} + CN^{-}$	1.6×10^{-14}	13.80
Ferrocyanides			
Copper(II) ferrocyanide	$Cu_2[Fe(CN)_6] \rightleftharpoons 2Cu^{2+} + [Fe(CN)_6]^{4-}$	1.3×10^{-16}	15.89
Silver ferrocyanide	$Ag_4[Fe(CN)_6] \rightleftharpoons 4Ag^{+}[Fe(CN)_6]^{4-}$	1.6×10^{-41}	40.80
Zinc potassium ferrocyanide	$K_2Zn_3[Fe(CN)_6]_2 \rightleftharpoons 2K^{+} + 3Zn^{2+} + 2[Fe(CN)_6]^{4-}$	1×10^{-95}	95.0
Fluorides			
Barium fluoride	$BaF_2 \rightleftharpoons Ba^{2+} + 2F^{-}$	2.4×10^{-5}	4.62
Calcium fluoride	$CaF_2 \rightleftharpoons Ca^{2+} + 2F^{-}$	1.7×10^{-10}	9.77
Lead fluoride	$PbF_2 \rightleftharpoons Pb^{2+} + 2F^{-}$	2.7×10^{-8}	7.57
Magnesium fluoride	$MgF_2 \rightleftharpoons Mg^{2+} + 2F^{-}$	6.5×10^{-9}	8.19
Strontium fluoride	$SrF_2 \rightleftharpoons Sr^{2+} + 2F^{-}$	7.9×10^{-10}	9.10

Table A.3. *Solubility product constants (continued).*

Substance	Equilibrium equation	K_{sp}	pK_{sp}
Hydroxides			
Aluminum hydroxide	$Al(OH)_3 \rightleftharpoons Al^{3+} + 3\,OH^-$	5×10^{-33}	32.3
Cadmium hydroxide	$Cd(OH)_2 \rightleftharpoons Cd^{2+} + 2\,OH^-$	2.0×10^{-14}	13.70
Chromium(III) hydroxide	$Cr(OH)_3 \rightleftharpoons Cr^{3+} + 3\,OH^-$	7×10^{-31}	30.2
Cobalt(III) hydroxide	$Co(OH)_3 \rightleftharpoons Co^{3+} + 3\,OH^-$	2.5×10^{-43}	42.60
Cobalt(II) hydroxide	$Co(OH)_2 \rightleftharpoons Co^{2+} + 2\,OH^-$	2.5×10^{-16}	15.60
Copper(II) hydroxide	$Cu(OH)_2 \rightleftharpoons Cu^{2+} + 2\,OH^-$	1.6×10^{-19}	18.80
Iron(III) hydroxide	$Fe(OH)_3 \rightleftharpoons Fe^{3+} + 3\,OH^-$	6×10^{-38}	37.2
Iron(II) hydroxide	$Fe(OH)_2 \rightleftharpoons Fe^{2+} + 2\,OH^-$	1.8×10^{-15}	14.74
Lead hydroxide	$Pb(OH)_2 \rightleftharpoons Pb^{2+} + 2\,OH^-$	4.2×10^{-15}	14.38
Magnesium hydroxide	$Mg(OH)_2 \rightleftharpoons Mg^{2+} + 2\,OH^-$	8.9×10^{-12}	11.05
Manganese(II) hydroxide	$Mn(OH)_2 \rightleftharpoons Mn^{2+} + 2\,OH^-$	2×10^{-13}	12.7
Mercury(II) hydroxide	$HgO + H_2O \rightleftharpoons Hg^{2+} + 2\,OH^-$	3×10^{-26}	25.5
Nickel hydroxide	$Ni(OH)_2 \rightleftharpoons Ni^{2+} + 2\,OH^-$	1.6×10^{-16}	15.80
Silver hydroxide	$1/2Ag_2O + 1/2H_2O \rightleftharpoons Ag^+ + OH^-$	2×10^{-8}	7.7
Tin(IV) hydroxide	$Sn(OH)_4 \rightleftharpoons Sn^{4+} + 4\,OH^-$	1×10^{-56}	56.0
Tin(II) hydroxide	$Sn(OH)_2 \rightleftharpoons Sn^{2+} + 2\,OH^-$	3×10^{-27}	26.5
Zinc hydroxide	$Zn(OH)_2 \rightleftharpoons Zn^{2+} + 2\,OH^-$	5×10^{-17}	16.3
Iodates			
Barium iodate	$Ba(IO_3)_2 \rightleftharpoons Ba^{2+} + 2IO_3^-$	1.3×10^{-9}	8.89
Calcium iodate	$Ca(IO_3)_2 \rightleftharpoons Ca^{2+} + 2IO_3^-$	1.7×10^{-6}	5.77
Lead iodate	$Pb(IO_3)_2 \rightleftharpoons Pb^{2+} + 2IO_3^-$	2.6×10^{-13}	12.59
Mercury(I) iodate	$Hg_2(IO_3)_2 \rightleftharpoons Hg_2^{2+} + 2IO_3^-$	1.9×10^{-14}	13.72
Mercury(II) iodate	$Hg(IO_3)_2 \rightleftharpoons Hg^{2+} + 2IO_3^-$	3×10^{-13}	12.5
Silver iodate	$AgIO_3 \rightleftharpoons Ag^+ + IO_3^-$	3×10^{-8}	7.5
Iodides			
Copper(I) iodide	$CuI \rightleftharpoons Cu^+ + I^-$	1×10^{-12}	12.0
Lead iodide	$PbI_2 \rightleftharpoons Pb^{2+} + 2I^-$	8.3×10^{-9}	8.08
Mercury(I) iodide	$Hg_2I_2 \rightleftharpoons Hg_2^{2+} + 2I^-$	4×10^{-29}	28.4
Mercury(II) iodide	$HgI_2 \rightleftharpoons Hg^{2+} + 2I^-$	4×10^{-18}	17.4
Silver iodide	$AgI \rightleftharpoons Ag^+ + I^-$	8.5×10^{-17}	16.07
Thallium(I) iodide	$TlI \rightleftharpoons Tl^+ + I^-$	2.5×10^{-8}	7.60
Nitrites			
Silver nitrite	$AgNO_2 \rightleftharpoons Ag^+ + NO_2^-$	1.2×10^{-4}	3.92
Oxalates			
Barium oxalate	$BaC_2O_4 \rightleftharpoons Ba^{2+} + C_2O_4^{2-}$	1.5×10^{-8}	7.82
Cadmium oxalate	$CdC_2O_4 \rightleftharpoons Cd^{2+} + C_2O_4^{2-}$	1.5×10^{-8}	7.82

Table A.3. *Solubility product constants (continued).*

Substance	Equilibrium equation	K_{sp}	pK_{sp}
Oxalates (continued)			
Calcium oxalate	$CaC_2O_4 \rightleftharpoons Ca^{2+} + C_2O_4{}^{2-}$	1.3×10^{-9}	8.89
Iron(II) oxalate	$FeC_2O_4 \rightleftharpoons Fe^{2+} + C_2O_4{}^{2-}$	2×10^{-7}	6.7
Magnesium oxalate	$MgC_2O_4 \rightleftharpoons Mg^{2+} + C_2O_4{}^{2-}$	8.6×10^{-5}	4.07
Manganese(III) oxalate	$Mn_2(C_2O_4)_3 \rightleftharpoons 2Mn^{3+} + 3C_2O_4{}^{2-}$	7×10^{-20}	19.2
Manganese(II) oxalate	$MnC_2O_4 \rightleftharpoons Mn^{2+} + C_2O_4{}^{2-}$	1.1×10^{-15}	14.96
Silver oxalate	$Ag_2C_2O_4 \rightleftharpoons 2Ag^{+} + C_2O_4{}^{2-}$	1×10^{-11}	11.0
Strontium oxalate	$SrC_2O_4 \rightleftharpoons Sr^{2+} + C_2O_4{}^{2-}$	5.6×10^{-8}	7.25
Zinc oxalate	$ZnC_2O_4 \rightleftharpoons Zn^{2+} + C_2O_4{}^{2-}$	1.5×10^{-9}	8.82
Phosphates			
Barium phosphate	$Ba_3(PO_4)_2 \rightleftharpoons 3Ba^{2+} + 2PO_4{}^{3-}$	6×10^{-39}	38.2
Calcium phosphate	$Ca_3(PO_4)_2 \rightleftharpoons 3Ca^{2+} + 2PO_4{}^{3-}$	1×10^{-25}	25.0
Iron(III) phosphate	$FePO_4 \rightleftharpoons Fe^{3+} + PO_4{}^{3-}$	1.3×10^{-22}	21.89
Magnesium ammonium phosphate	$MgNH_4PO_4 \rightleftharpoons Mg^{2+} + NH_4{}^{+} + PO_4{}^{3-}$	2×10^{-13}	12.7
Magnesium phosphate	$Mg_3(PO_4)_2 \rightleftharpoons 3Mg^{2+} + 2PO_4{}^{3-}$	2.6×10^{-13}	12.59
Silver phosphate	$Ag_3PO_4 \rightleftharpoons 3Ag^{+} + PO_4{}^{3-}$	1.8×10^{-18}	17.74
Strontium phosphate	$Sr_3(PO_4)_2 \rightleftharpoons 3Sr^{2+} + 2PO_4{}^{3-}$	1×10^{-31}	31.0
Zirconium phosphate	$Zr_3(PO_4)_4 \rightleftharpoons 3Zr^{4+} + 4PO_4{}^{3-}$	10^{-132}	132
Sulfates			
Barium sulfate	$BaSO_4 \rightleftharpoons Ba^{2+} + SO_4{}^{2-}$	1.5×10^{-9}	8.82
Calcium sulfate	$CaSO_4 \rightleftharpoons Ca^{2+} + SO_4{}^{2-}$	2.4×10^{-5}	4.62
Lead sulfate	$PbSO_4 \rightleftharpoons Pb^{2+} + SO_4{}^{2-}$	1.3×10^{-8}	7.89
Mercury(I) sulfate	$Hg_2SO_4 \rightleftharpoons Hg_2{}^{2+} + SO_4{}^{2-}$	1×10^{-6}	6.0
Silver sulfate	$Ag_2SO_4 \rightleftharpoons 2Ag^{+} + SO_4{}^{2-}$	1.6×10^{-5}	4.80
Strontium sulfate	$SrSO_4 \rightleftharpoons Sr^{2+} + SO_4{}^{2-}$	2.8×10^{-7}	6.55
Sulfides			
Bismuth sulfide	$Bi_2S_3 \rightleftharpoons 2Bi^{3+} + 3S^{2-}$	10^{-100}	100
Cadmium sulfide	$CdS \rightleftharpoons Cd^{2+} + S^{2-}$	6×10^{-27}	26.2
Cobalt(II) sulfide	$CoS \rightleftharpoons Co^{2+} + S^{2-}$	5×10^{-22} (α)	21.3
		6×10^{-29} (β)	28.2
Copper(I) sulfide	$Cu_2S \rightleftharpoons 2Cu^{+} + S^{2-}$	1.2×10^{-49}	48.92
Copper(II) sulfide	$CuS \rightleftharpoons Cu^{2+} + S^{2-}$	4×10^{-36}	35.4
Iron(III) sulfide	$Fe_2S_3 \rightleftharpoons 2Fe^{3+} + 3S^{2-}$	1×10^{-88}	88.0
Iron(II) sulfide	$FeS \rightleftharpoons Fe^{2+} + S^{2-}$	5×10^{-18}	17.3
Lead sulfide	$PbS \rightleftharpoons Pb^{2+} + S^{2-}$	8×10^{-28}	27.1
Manganese(II) sulfide	$MnS \rightleftharpoons Mn^{2+} + S^{2-}$	8×10^{-14}	13.1
Mercury(I) sulfide	$Hg_2S \rightleftharpoons Hg_2{}^{2+} + S^{2-}$	1×10^{-45}	45.0

Table A.3. *Solubility product constants (continued).*

Substance	Equilibrium equation	K_{sp}	pK_{sp}
Sulfides (continued)			
Mercury(II) sulfide	$HgS \rightleftharpoons Hg^{2+} + S^{2-}$	1×10^{-50}	50.0
Nickel sulfide	$NiS \rightleftharpoons Ni^{2+} + S^{2-}$	1×10^{-22} (α)	22.0
		3×10^{-28} (β)	27.5
		7×10^{-30} (γ)	29.2
Silver sulfide	$Ag_2S \rightleftharpoons 2Ag^{+} + S^{2-}$	1×10^{-50}	50.0
Tin(II) sulfide	$SnS \rightleftharpoons Sn^{2+} + S^{2-}$	1×10^{-26}	26.0
Zinc sulfide	$ZnS \rightleftharpoons Zn^{2+} + S^{2-}$	1.6×10^{-23}	22.80
Sulfites			
Barium sulfite	$BaSO_3 \rightleftharpoons Ba^{2+} + SO_3{}^{2-}$	1.0×10^{-8}	8.00
Calcium sulfite	$CaSO_3 \rightleftharpoons Ca^{2+} + SO_3{}^{2-}$	1.0×10^{-4}	4.00
Mercury(I) sulfite	$Hg_2SO_3 \rightleftharpoons Hg_2{}^{2+} + SO_3{}^{2-}$	9×10^{-28}	27.0
Silver sulfite	$Ag_2SO_3 \rightleftharpoons 2Ag^{+} + SO_3{}^{2-}$	1.9×10^{-11}	10.72
Strontium sulfite	$SrSO_3 \rightleftharpoons Sr^{2+} + SO_3{}^{2-}$	3.9×10^{-8}	7.41
Thiocyanates			
Copper(I) thiocyanate	$CuSCN \rightleftharpoons Cu^{+} + SCN^{-}$	4×10^{-14}	13.4
Mercury(I) thiocyanate	$Hg_2(SCN)_2 \rightleftharpoons Hg_2{}^{2+} + 2SCN^{-}$	3×10^{-20}	19.5
Silver thiocyanate	$AgSCN \rightleftharpoons Ag^{+} + SCN^{-}$	1×10^{-12}	12.0

Table A.4. *Instability (dissociation) constants for complex ions.*

Ligand	Equilibrium equation	K_{inst}	pK_{inst}
Ammonia	$[Cd(NH_3)_4]^{2+} \rightleftharpoons Cd^{2+} + 4NH_3$	1.9×10^{-7}	6.72
	$[Co(NH_3)_6]^{3+} \rightleftharpoons Co^{3+} + 6NH_3$	2.2×10^{-34}	33.66
	$[Co(NH_3)_4]^{2+} \rightleftharpoons Co^{2+} + 4NH_3$	9×10^{-6}	5.0
	$[Co(NH_3)_6]^{2+} \rightleftharpoons Co^{2+} + 6NH_3$	1.3×10^{-5}	4.89
	$[Cu(NH_3)_4]^{2+} \rightleftharpoons Cu^{2+} + 4NH_3$	1×10^{-12}	12.0
	$[Ni(NH_3)_4]^{2+} \rightleftharpoons Ni^{2+} + 4NH_3$	1×10^{-8}	8.0
	$[Ni(NH_3)_6]^{2+} \rightleftharpoons Ni^{2+} + 6NH_3$	6×10^{-9}	8.2
	$[Zn(NH_3)_4]^{2+} \rightleftharpoons Zn^{2+} + 4NH_3$	2×10^{-9}	8.7 (at 30°C)
	$[Ag(NH_3)]^{+} \rightleftharpoons Ag^{+} + NH_3$	5.0×10^{-4}	3.30
	$[Ag(NH_3)_2]^{+} \rightleftharpoons Ag^{+} + 2NH_3$	5.9×10^{-8}	7.23

Table A.4. *Instability (dissociation) constants for complex ions (continued).*

Ligand	Equilibrium equation	K_{inst}	pK_{inst}
Bromide ion	$[HgBr_4]^{2-} \rightleftharpoons Hg^{2+} + 4Br^-$	1×10^{-21}	21.0
Chloride ion	$[CdCl_4]^{2-} \rightleftharpoons Cd^{2+} + 4Cl^-$	9.1×10^{-4}	3.04
	$[FeCl]^{2+} \rightleftharpoons Fe^{3+} + Cl^-$	3×10^{-2}	1.5
	$[FeCl_2]^+ \rightleftharpoons Fe^{3+} + 2Cl^-$	0.222	0.654
	$[HgCl_4]^{2-} \rightleftharpoons Hg^{2+} + 4Cl^-$	1×10^{-15}	15.0
	$[AgCl_2]^- \rightleftharpoons Ag^+ + 2Cl^-$	7×10^{-6}	5.2
Cyanide ion	$[Cd(CN)_4]^{2-} \rightleftharpoons Cd^{2+} + 4CN^-$	8×10^{-18}	17.1
	$[Cu(CN)_3]^{2-} \rightleftharpoons Cu^+ + 3CN^-$	5×10^{-28}	27.3
	$[Cu(CN)_4]^{3-} \rightleftharpoons Cu^+ + 4CN^-$	5×10^{-30}	29.3
	$[Fe(CN)_6]^{3-} \rightleftharpoons Fe^{3+} + 6CN^-$	1×10^{-42}	42.0
	$[Fe(CN)_6]^{4-} \rightleftharpoons Fe^{2+} + 6CN^-$	1×10^{-35}	35.0
	$[Hg(CN)_4]^{2-} \rightleftharpoons Hg^{2+} + 4CN^-$	3×10^{-42}	41.5
	$[Ni(CN)_4]^{2-} \rightleftharpoons Ni^{2+} + 4CN^-$	1×10^{-22}	22.0
	$[Ag(CN)_2]^- \rightleftharpoons Ag^+ + 2CN^-$	1×10^{-20}	20.0
	$[Zn(CN)_4]^{2-} \rightleftharpoons Zn^{2+} + 4CN^-$	1×10^{-19}	19.0
Ethylenediamine-	$[CaY]^{2-} \rightleftharpoons Ca^{2+} + Y^{4-}$	2×10^{-11}	10.7
tetraacetate ion	$[FeY]^- \rightleftharpoons Fe^{3+} + Y^{4-}$	1×10^{-25}	25.0
(= Y^{4-})(at 20°C)	$[MgY]^{2-} \rightleftharpoons Mg^{2+} + Y^{4-}$	2×10^{-9}	8.7
	$[ZnY]^{2-} \rightleftharpoons Zn^{2+} + Y^{4-}$	3.1×10^{-17}	16.51
Fluoride ion	$[AlF_6]^{3-} \rightleftharpoons Al^{3+} + 6F^-$	2×10^{-21}	20.7
	$[FeF_6]^{3-} \rightleftharpoons Fe^{3+} + 6F^-$	1×10^{-16}	16.0
Hydroxide ion	$[Al(OH)_4]^- \rightleftharpoons Al^{3+} + 4\,OH^-$	1.2×10^{-34}	33.92
	$[Zn(OH)_4]^{2-} \rightleftharpoons Zn^{2+} + 4\,OH^-$	2.5×10^{-15}	14.60
Iodide ion	$[CdI_4]^{2-} \rightleftharpoons Cd^{2+} + 4I^-$	7×10^{-7}	6.2
	$[HgI_4]^{2-} \rightleftharpoons Hg^{2+} + 4I^-$	5.3×10^{-31}	30.28
Oxalate ion	$[Al(C_2O_4)_3]^{3-} \rightleftharpoons Al^{3+} + 3C_2O_4{}^{2-}$	5×10^{-17}	16.3
	$[Fe(C_2O_4)_3]^{3-} \rightleftharpoons Fe^{3+} + 3C_2O_4{}^{2-}$	6×10^{-21}	20.2
Sulfide ion	$[HgS_2]^{2-} \rightleftharpoons Hg^{2+} + 2S^{2-}$	2.0×10^{-55}	54.70
Sulfite ion	$[Ag(SO_3)_2]^{3-} \rightleftharpoons Ag^+ + 2SO_3{}^{2-}$	3×10^{-9}	8.5
Thiocyanate ion	$[Fe(SCN)]^{2+} \rightleftharpoons Fe^{3+} + SCN^-$	9.4×10^{-4}	3.03
	$[Fe(SCN)_6]^{3-} \rightleftharpoons Fe^{3+} + 6SCN^-$	1×10^{-4}	4.0
	$[Hg(SCN)_4]^{2-} \rightleftharpoons Hg^{2+} + 4SCN^-$	1×10^{-22}	22.0
Thiosulfate ion	$[Ag(S_2O_3)_2]^{3-} \rightleftharpoons Ag^+ + 2S_2O_3{}^{2-}$	1×10^{-13}	13.0

Appendix B

Standard and formal potentials, E^0 and $E^{0'}$, at 25° C

Half reaction	E^0, volts	$E^{0'}$, volts	Conditions for the formal potentials
1. Acidic solutions			
$F_2 + 2H^+ + 2e \rightleftharpoons 2HF$	+3.06		
$F_2 + 2e \rightleftharpoons 2F^-$	+2.85		
$S_2O_8^{2-} + 2e \rightleftharpoons 2SO_4^{2-}$	+2.01		
$Co^{3+} + e \rightleftharpoons Co^{2+}$	+1.82		
$H_2O_2 + 2H^+ + 2e \rightleftharpoons 2H_2O$	+1.77		
$MnO_4^- + 4H^+ + 3e \rightleftharpoons \mathbf{MnO_2} + 2H_2O$	+1.695		
$Ce^{4+} + e \rightleftharpoons Ce^{3+}$		+1.70	1 $\underline{M}$ $HClO_4$
		+1.61	1 $\underline{M}$ HNO_3
		+1.44	1 $\underline{M}$ H_2SO_4
		+1.28	1 $\underline{M}$ HCl
$2HClO + 2H^+ + 2e \rightleftharpoons Cl_2 + 2H_2O$	+1.63		
$\mathbf{NaBiO_3} + 6H^+ + 2e \rightleftharpoons Na^+ + Bi^{3+} + 3H_2O$	+1.6		
$H_5IO_6 + H^+ + 2e \rightleftharpoons IO_3^- + 3H_2O$	+1.6		
$2BrO_3^- + 12H^+ + 10e \rightleftharpoons Br_2 + 6H_2O$	+1.52		
$MnO_4^- + 8H^+ + 5e \rightleftharpoons Mn^{2+} + 4H_2O$	+1.51		
$Mn^{3+} + e \rightleftharpoons Mn^{2+}$	+1.51		
$\mathbf{PbO_2} + 4H^+ + 2e \rightleftharpoons Pb^{2+} + 2H_2O$	+1.455		
$Cl_2 + 2e \rightleftharpoons 2Cl^-$	+1.359		
$Cr_2O_7^{2-} + 14H^+ + 6e \rightleftharpoons 2Cr^{3+} + 7H_2O$	+1.33		
		+1.09	1 $\underline{M}$ HCl
$\mathbf{MnO_2} + 4H^+ + 2e \rightleftharpoons Mn^{2+} + 2H_2O$	+1.23		
		+1.24	1 $\underline{M}$ $HClO_4$
$O_2 + 4H^+ + 4e \rightleftharpoons 2H_2O$	+1.229		
$2IO_3^- + 12H^+ + 10e \rightleftharpoons \mathbf{I_2} + 6H_2O$	+1.195		
$Br_{2(aq)} + 2e \rightleftharpoons 2Br^-$	+1.087[a]		
$Br_2(l) + 2e \rightleftharpoons 2Br^-$	+1.065[a]		

a. E^0 for $Br_2(l)$ is used for saturated solutions of Br_2, while E^0 for $Br_2(aq)$ is used for unsaturated solutions.

b. E^0 for $\mathbf{I_2}$ is used for saturated solutions of I_2, while E^0 for $I_2(aq)$ is used for unsaturated solutions.

Half reaction	E^0, volts	$E^{0'}$, volts	Conditions for the formal potentials
1. Acidic solutions (continued)			
$2ICl_2^- + 2e \rightleftharpoons I_2 + 4Cl^-$	+1.06		
$HNO_2 + H^+ + e \rightleftharpoons NO + H_2O$	+1.00		
$NO_3^- + 4H^+ + 3e \rightleftharpoons NO + 2H_2O$	+0.96		
$NO_3^- + 3H^+ + 2e \rightleftharpoons HNO_2 + H_2O$	+0.94		
$NO_3^- + 10H^+ + 8e \rightleftharpoons NH_4^+ + 3H_2O$	+0.87		
$2Hg^{2+} + 2e \rightleftharpoons Hg_2^{2+}$	+0.920		
$Cu^{2+} + I^- + e \rightleftharpoons \mathbf{CuI}$	+0.86		
$Ag^+ + e \rightleftharpoons \mathbf{Ag}$	+0.7994		
$Hg_2^{2+} + 2e \rightleftharpoons \mathbf{2Hg}$	+0.789		
$Fe^{3+} + e \rightleftharpoons Fe^{2+}$	+0.771		
		+0.732	1 $\underline{M}$ $HClO_4$
		+0.700	1 $\underline{M}$ HCl
		+0.674	1 $\underline{M}$ H_2SO_4
		+0.46	2 $\underline{M}$ H_3PO_4
$O_2 + 2H^+ + 2e \rightleftharpoons H_2O_2$	+0.682		
$I_2(aq) + 2e \rightleftharpoons 2I^-$	+0.6197[b]		
$MnO_4^- + e \rightleftharpoons MnO_4^{2-}$	+0.564		
$H_3AsO_4 + 2H^+ + 2e \rightleftharpoons H_3AsO_3 + H_2O$	+0.559		
		+0.577	1 $\underline{M}$ HCl, 1 $\underline{M}$ $HClO_4$
$[I_3]^- + 2e \rightleftharpoons 3I^-$	+0.536		
$\mathbf{I_2} + 2e \rightleftharpoons 2I^-$	+0.5355[b]		
$Cu^+ + e \rightleftharpoons \mathbf{Cu}$	+0.521		
$H_2SO_3 + 4H^+ + 4e \rightleftharpoons \mathbf{S} + 3H_2O$	+0.45		
$\mathbf{Ag_2CrO_4} + 2e \rightleftharpoons \mathbf{2Ag} + CrO_4^{2-}$	+0.446		
$2H_2SO_3 + 2H^+ + 4e \rightleftharpoons S_2O_3^{2-} + 3H_2O$	+0.40		
$[Fe(CN)_6]^{3-} + e \rightleftharpoons [Fe(CN)_6]^{4-}$	+0.356		
		+0.71	1 $\underline{M}$ HCl
$Cu^{2+} + 2e \rightleftharpoons \mathbf{Cu}$	+0.337		
$\mathbf{Hg_2Cl_2} + 2e \rightleftharpoons \mathbf{2Hg} + 2Cl^-$	+0.2680		
		+0.3337	0.1 $\underline{M}$ KCl
		+0.2801	1 $\underline{M}$ KCl
		+0.2412	KCl satur.
$\mathbf{AgCl} + e \rightleftharpoons \mathbf{Ag} + Cl^-$	+0.2224		
$SO_4^{2-} + 4H^+ + 2e \rightleftharpoons H_2SO_3 + H_2O$	+0.17		
$Cu^{2+} + e \rightleftharpoons Cu^+$	+0.153		
$Sn^{4+} + 2e \rightleftharpoons Sn^{2+}$	+0.15		
$[SnCl_6]^{2-} + 2e \rightleftharpoons [SnCl_4]^{2-} + 2Cl^-$		+0.14	1 $\underline{M}$ HCl
$\mathbf{S} + 2H^+ + 2e \rightleftharpoons H_2S$	+0.141		
$\mathbf{AgBr} + e \rightleftharpoons \mathbf{Ag} + Br^-$	+0.095		
$S_4O_6^{2-} + 2e \rightleftharpoons 2S_2O_3^{2-}$	+0.08		
$[Ag(S_2O_3)_2]^{3-} + e \rightleftharpoons \mathbf{Ag} + 2S_2O_3^{2-}$	+0.01		
$2H^+ + 2e \rightleftharpoons H_2$	0.000		

Half reaction	E^0, volts	$E^{0\prime}$, volts	Conditions for the formal potentials
1. Acidic solutions (continued)			
$Pb^{2+} + 2e \rightleftharpoons \mathbf{Pb}$	−0.126		
$Sn^{2+} + 2e \rightleftharpoons \mathbf{Sn}$	−0.136		
$\mathbf{AgI} + e \rightleftharpoons \mathbf{Ag} + I^-$	−0.151		
$\mathbf{CuI} + e \rightleftharpoons \mathbf{Cu} + I^-$	−0.185		
$Ni^{2+} + 2e \rightleftharpoons \mathbf{Ni}$	−0.250		
$V^{3+} + e \rightleftharpoons V^{2+}$	−0.255		
$Co^{2+} + 2e \rightleftharpoons \mathbf{Co}$	−0.277		
$[Ag(CN)_2)]^- + e \rightleftharpoons \mathbf{Ag} + 2CN^-$	−0.31		
$\mathbf{PbSO_4} + 2e \rightleftharpoons \mathbf{Pb} + SO_4{}^{2-}$	−0.356		
$Cd^{2+} + 2e \rightleftharpoons \mathbf{Cd}$	−0.403		
$Cr^{3+} + e \rightleftharpoons Cr^{2+}$	−0.41		
$Fe^{2+} + 2e \rightleftharpoons \mathbf{Fe}$	−0.440		
$Cr^{3+} + 3e \rightleftharpoons \mathbf{Cr}$	−0.74		
$Zn^{2+} + 2e \rightleftharpoons \mathbf{Zn}$	−0.763		
$Mn^{2+} + 2e \rightleftharpoons \mathbf{Mn}$	−1.18		
$Al^{3+} + 3e \rightleftharpoons \mathbf{Al}$	−1.66		
$Mg^{2+} + 2e \rightleftharpoons \mathbf{Mg}$	−2.37		
$Na^+ + e \rightleftharpoons \mathbf{Na}$	−2.71		
$Ca^{2+} + 2e \rightleftharpoons \mathbf{Ca}$	−2.87		
$Sr^{2+} + 2e \rightleftharpoons \mathbf{Sr}$	−2.89		
$Ba^{2+} + 2e \rightleftharpoons \mathbf{Ba}$	−2.90		
$K^+ + e \rightleftharpoons \mathbf{K}$	−2.92		
$Li^+ + e \rightleftharpoons \mathbf{Li}$	−3.04		
2. Alkaline solutions			
$ClO^- + H_2O + 2e \rightleftharpoons Cl^- + 2\,OH^-$	+0.89		
$O_2{}^{2-} + 2H_2O + 2e \rightleftharpoons 4\,OH^-$	+0.88		
$BrO^- + H_2O + 2e \rightleftharpoons Br^- + 2\,OH^-$	+0.76		
$MnO_4{}^- + 2H_2O + 3e \rightleftharpoons \mathbf{MnO_2} + 4\,OH^-$	+0.59		
$O_2 + 2H_2O + 4e \rightleftharpoons 4\,OH^-$	+0.401		
$[Ag(NH_3)_2]^+ + e \rightleftharpoons \mathbf{Ag} + 2NH_3$	+0.373		
$ClO_3{}^- + H_2O + 2e \rightleftharpoons ClO_2{}^- + 2\,OH^-$	+0.33		
$\mathbf{Co(OH)_3} + e \rightleftharpoons \mathbf{Co(OH)_2} + OH^-$	+0.17		
$[Co(NH_3)_6]^{3+} + e \rightleftharpoons [Co(NH_3)_6]^{2+}$	+0.1		
$NO_3{}^- + H_2O + 2e \rightleftharpoons NO_2{}^- + 2\,OH^-$	+0.01		
$\mathbf{MnO_2} + 2H_2O + 2e \rightleftharpoons \mathbf{Mn(OH)_2} + 2\,OH^-$	−0.05		
$[Cu(NH_3)_4]^{2+} + 2e \rightleftharpoons \mathbf{Cu} + 4NH_3$	−0.11		
$NO_3{}^- + 6H_2O + 8e \rightleftharpoons NH_3 + 9\,OH^-$	−0.13		
$NO_2{}^- + 5H_2O + 6e \rightleftharpoons NH_3 + 7\,OH^-$	−0.18		
$[Ag(CN)_2]^- + e \rightleftharpoons \mathbf{Ag} + 2CN^-$	−0.31		
$NO_2{}^- + H_2O + e \rightleftharpoons 2\,OH^- + NO$	−0.46		

Half reaction	E^0, volts	$E^{0'}$, volts	Conditions for the formal potentials
1. Alkaline solutions (continued)			
$S + 2e \rightleftharpoons S^{2-}$	−0.48		
$[Pb(OH)_3]^- + 2e \rightleftharpoons Pb + 3\,OH^-$	−0.54		
$Fe(OH)_3 + e \rightleftharpoons Fe(OH)_2 + OH^-$	−0.56		
$[Cd(NH_3)_4]^{2+} + 2e \rightleftharpoons Cd + 4NH_3$	−0.597		
$AsO_4^{3-} + H_2O + 2e \rightleftharpoons AsO_3^{3-} + 2\,OH^-$	−0.67		
$HgS + 2e \rightleftharpoons Hg + S^{2-}$	−0.72		
$2H_2O + 2e \rightleftharpoons H_2 + 2\,OH^-$	−0.828		
$[Sn(OH)_6]^{2-} + 2e \rightleftharpoons [Sn(OH)_3]^- + 3\,OH^-$	−0.90		
$[Zn(NH_3)_4]^{2+} + 2e \rightleftharpoons Zn + 4NH_3$	−0.103		
$[Cd(CN)_4]^{2-} + 2e \rightleftharpoons Cd + 4CN^-$	−1.03		
$[Cr(OH)_4]^- + 3e \rightleftharpoons Cr + 4\,OH^-$	−1.2		
$[Zn(OH)_4]^{2-} + 2e \rightleftharpoons Zn + 4\,OH^-$	−1.22		
$ZnS + 2e \rightleftharpoons Zn + S^{2-}$	−1.44		
$Mn(OH)_2 + 2e \rightleftharpoons Mn + 2\,OH^-$	−1.55		
$[Al(OH)_4]^- + 3e \rightleftharpoons Al + 4\,OH^-$	−2.35		

Appendix C

Common logarithms

N	0	1	2	3	4	5	6	7	8	9	1	2	3	4	5	6	7	8	9
10	0000	0043	0086	0128	0170	0212	0253	0294	0334	0374	4	8	12	17	21	25	29	33	37
11	0414	0453	0492	0531	0569	0607	0645	0682	0719	0755	4	8	11	15	19	23	26	30	34
12	0792	0828	0864	0899	0934	0969	1004	1038	1072	1106	3	7	10	14	17	21	24	28	31
13	1139	1173	1206	1239	1271	1303	1335	1367	1399	1430	3	6	10	13	16	19	23	26	29
14	1461	1492	1523	1553	1584	1614	1644	1673	1703	1732	3	6	9	12	15	18	21	24	27
15	1761	1790	1818	1847	1875	1903	1931	1959	1987	2014	3	6	8	11	14	17	20	22	25
16	2041	2068	2095	2122	2148	2175	2201	2227	2253	2279	3	5	8	11	13	16	18	21	24
17	2304	2330	2355	2380	2405	2430	2455	2480	2504	2529	2	5	7	10	12	15	17	20	22
18	2553	2577	2601	2625	2648	2672	2695	2718	2742	2765	2	5	7	9	12	14	16	19	21
19	2788	2810	2833	2856	2878	2900	2923	2945	2967	2989	2	4	7	9	11	13	16	18	20
20	3010	3032	3054	3075	3096	3118	3139	3160	3181	3201	2	4	6	8	11	13	15	17	19
21	3222	3243	3263	3284	3304	3324	3345	3365	3385	3404	2	4	6	8	10	12	14	16	18
22	3424	3444	3464	3483	3502	3522	3541	3560	3579	3598	2	4	6	8	10	12	14	15	17
23	3617	3636	3655	3674	3692	3711	3729	3747	3766	3784	2	4	6	7	9	11	13	15	17
24	3802	3820	3838	3856	3874	3892	3909	3927	3945	3962	2	4	5	7	9	11	12	14	16
25	3979	3997	4014	4031	4048	4065	4082	4099	4116	4133	2	3	5	7	9	10	12	14	15
26	4150	4166	4183	4200	4216	4232	4249	4265	4281	4298	2	3	5	7	8	10	11	13	15
27	4314	4330	4346	4362	4378	4393	4409	4425	4440	4456	2	3	5	6	8	9	11	13	14
28	4472	4487	4502	4518	4533	4548	4564	4579	4594	4609	2	3	5	6	8	9	11	12	14
29	4624	4639	4654	4669	4683	4698	4713	4728	4742	4757	1	3	4	6	7	9	10	12	13
30	4771	4786	4800	4814	4829	4843	4857	4871	4886	4900	1	3	4	6	7	9	10	11	13
31	4914	4928	4942	4955	4969	4983	4997	5011	5024	5038	1	3	4	6	7	8	10	11	12
32	5051	5065	5079	5092	5105	5119	5132	5145	5159	5172	1	3	4	5	7	8	9	11	12
33	5185	5198	5211	5224	5237	5250	5263	5276	5289	5302	1	3	4	5	6	8	9	10	12
34	5315	5328	5340	5353	5366	5378	5391	5403	5416	5428	1	3	4	5	6	8	9	10	11
35	5441	5453	5465	5478	5490	5502	5514	5527	5539	5551	1	2	4	5	6	7	9	10	11
36	5563	5575	5587	5599	5611	5623	5635	5647	5658	5670	1	2	4	5	6	7	8	10	11
37	5682	5694	5705	5717	5729	5740	5752	5763	5775	5786	1	2	3	5	6	7	8	9	10
38	5798	5809	5821	5832	5843	5855	5866	5877	5888	5899	1	2	3	5	6	7	8	9	10
39	5911	5922	5933	5944	5955	5966	5977	5988	5999	6010	1	2	3	4	5	7	8	9	10
40	6021	6031	6042	6053	6064	6075	6085	6096	6107	6117	1	2	3	4	5	6	8	9	10
41	6128	6138	6149	6160	6170	6180	6191	6201	6212	6222	1	2	3	4	5	6	7	8	9
42	6232	6243	6253	6263	6274	6284	6294	6304	6314	6325	1	2	3	4	5	6	7	8	9
43	6335	6345	6355	6365	6375	6385	6395	6405	6415	6425	1	2	3	4	5	6	7	8	9
44	6435	6444	6454	6464	6474	6484	6493	6503	6513	6522	1	2	3	4	5	6	7	8	9
45	6532	6542	6551	6561	6571	6580	6590	6599	6609	6618	1	2	3	4	5	6	7	8	9
46	6628	6637	6646	6656	6665	6675	6684	6693	6702	6712	1	2	3	4	5	6	7	7	8
47	6721	6730	6739	6749	6758	6767	6776	6785	6794	6803	1	2	3	4	5	5	6	7	8
48	6812	6821	6830	6839	6848	6857	6866	6875	6884	6893	1	2	3	4	4	5	6	7	8
49	6902	6911	6920	6928	6937	6946	6955	6964	6972	6981	1	2	3	4	4	5	6	7	8
50	6990	6998	7007	7016	7024	7033	7042	7050	7059	7067	1	2	3	3	4	5	6	7	8
51	7076	7084	7093	7101	7110	7118	7126	7135	7143	7152	1	2	3	3	4	5	6	7	8
52	7160	7168	7177	7185	7193	7202	7210	7218	7226	7235	1	2	2	3	4	5	6	7	7
53	7243	7251	7259	7267	7275	7284	7292	7300	7308	7316	1	2	2	3	4	5	6	6	7
54	7324	7332	7340	7348	7356	7364	7372	7380	7388	7396	1	2	2	3	4	5	6	6	7

N	0	1	2	3	4	5	6	7	8	9	1	2	3	4	5	6	7	8	9
55	7404	7412	7419	7427	7435	7443	7451	7459	7466	7474	1	2	2	3	4	5	5	6	7
56	7482	7490	7497	7505	7513	7520	7528	7536	7543	7551	1	2	2	3	4	5	5	6	7
57	7559	7566	7574	7582	7589	7597	7604	7612	7619	7627	1	2	2	3	4	5	5	6	7
58	7634	7642	7649	7657	7664	7672	7679	7686	7694	7701	1	1	2	3	4	4	5	6	7
59	7709	7716	7723	7731	7738	7745	7752	7760	7767	7774	1	1	2	3	4	4	5	6	7
60	7782	7789	7796	7803	7810	7818	7825	7832	7839	7846	1	1	2	3	4	4	5	6	6
61	7853	7860	7868	7875	7882	7889	7896	7903	7910	7917	1	1	2	3	4	4	5	6	6
62	7924	7931	7938	7945	7952	7959	7966	7973	7980	7987	1	1	2	3	3	4	5	6	6
63	7993	8000	8007	8014	8021	8028	8035	8041	8048	8055	1	1	2	3	3	4	5	5	6
64	8062	8069	8075	8082	8089	8096	8102	8109	8116	8122	1	1	2	3	3	4	5	5	6
65	8129	8136	8142	8149	8156	8162	8169	8176	8182	8189	1	1	2	3	3	4	5	5	6
66	8195	8202	8209	8215	8222	8228	8235	8241	8248	8254	1	1	2	3	3	4	5	5	6
67	8261	8267	8274	8280	8287	8293	8299	8306	8312	8319	1	1	2	3	3	4	5	5	6
68	8325	8331	8338	8344	8351	8357	8363	8370	8376	8382	1	1	2	3	3	4	4	5	6
69	8388	8395	8401	8407	8414	8420	8426	8432	8439	8445	1	1	2	2	3	4	4	5	6
70	8451	8457	8463	8470	8476	8482	8488	8494	8500	8506	1	1	2	2	3	4	4	5	6
71	8513	8519	8525	8531	8537	8543	8549	8555	8561	8567	1	1	2	2	3	4	4	5	5
72	8573	8579	8585	8591	8597	8603	8609	8615	8621	8627	1	1	2	2	3	4	4	5	5
73	8633	8639	8645	8651	8657	8663	8669	8675	8681	8686	1	1	2	2	3	4	4	5	5
74	8692	8698	8704	8710	8716	8722	8727	8733	8739	8745	1	1	2	2	3	4	4	5	5
75	8751	8756	8762	8768	8774	8779	8785	8791	8797	8802	1	1	2	2	3	3	4	5	5
76	8808	8814	8820	8825	8831	8837	8842	8848	8854	8859	1	1	2	2	3	3	4	5	5
77	8865	8871	8876	8882	8887	8893	8899	8904	8910	8915	1	1	2	2	3	3	4	4	5
78	8921	8927	8932	8938	8943	8949	8954	8960	8965	8971	1	1	2	2	3	3	4	4	5
79	8976	8982	8987	8993	8998	9004	9009	9015	9020	9025	1	1	2	2	3	3	4	4	5
80	9031	9036	9042	9047	9053	9058	9063	9069	9074	9079	1	1	2	2	3	3	4	4	5
81	9085	9090	9096	9101	9106	9112	9117	9122	9128	9133	1	1	2	2	3	3	4	4	5
82	9138	9143	9149	9154	9159	9165	9170	9175	9180	9186	1	1	2	2	3	3	4	4	5
83	9191	9196	9201	9206	9212	9217	9222	9227	9232	9238	1	1	2	2	3	3	4	4	5
84	9243	9248	9253	9258	9263	9269	9274	9279	9284	9289	1	1	2	2	3	3	4	4	5
85	9294	9299	9304	9309	9315	9320	9325	9330	9335	9340	1	1	2	2	3	3	4	4	5
86	9345	9350	9355	9360	9365	9370	9375	9380	9385	9390	1	1	2	2	3	3	4	4	5
87	9395	9400	9405	9410	9415	9420	9425	9430	9435	9440	0	1	1	2	2	3	3	4	4
88	9445	9450	9455	9460	9465	9469	9474	9479	9484	9489	0	1	1	2	2	3	3	4	4
89	9494	9499	9504	9509	9513	9518	9523	9528	9533	9538	0	1	1	2	2	3	3	4	4
90	9542	9547	9552	9557	9562	9566	9571	9576	9581	9586	0	1	1	2	2	3	3	4	4
91	9590	9595	9600	9605	9609	9614	9619	9624	9628	9633	0	1	1	2	2	3	3	4	4
92	9638	9643	9647	9652	9657	9661	9666	9671	9675	9680	0	1	1	2	2	3	3	4	4
93	9685	9689	9694	9699	9703	9808	9713	9717	9722	9727	0	1	1	2	2	3	3	4	4
94	9731	9736	9741	9745	9750	9754	9759	9763	9768	9773	0	1	1	2	2	3	3	4	4
95	9777	9782	9786	9791	9795	9800	9805	9809	9814	9818	0	1	1	2	2	3	3	4	4
96	9823	9827	9832	9836	9841	9845	9850	9854	9859	9863	0	1	1	2	2	3	3	4	4
97	9868	9872	9877	9881	9886	9890	9894	9899	9903	9908	0	1	1	2	2	3	3	4	4
98	9912	9917	9921	9926	9930	9934	9939	9943	9948	9952	0	1	1	2	2	3	3	4	4
99	9956	9961	9965	9969	9974	9978	9983	9987	9991	9996	0	1	1	2	2	3	3	3	4

Appendix D

Answers to Odd-Numbered Problems

Chapter 1

1. const. det. error. −0.0022 g. **3.** 0.1000 g. **5.** −3.0 ppt. **7.** grav.: 26.47 ± 0.05, 26.47 ± 0.07, 26.47 ± 0.11; vol.: 26.50 ± 0.07, 26.50 ± 0.09, 26.50 ± 0.15. **9.** (a) N = 2, 65.5%; N = 4, 81.6%; N = 9, 95.5%; (b) 9.2%; (c) N = 2, ± 4.16 mg/dL; N = 4, ± 2.94 mg/dL; N = 9, ± 1.96 mg/dL; (d) ± 6, 2; ± 3, 7: ± 1, 60. **11.** yes, at 95% level. **13.** no. at 90 and 99% levels. **15.** no. **17.** 6.02 and 7.92 rejected. **19.** 0.400 mL. **21.** (a) $A = 0.2110\,[Fe]_{ppm} + 0.0320$; (b) 1.99.

Chapter 2

1. all agree.

Chapter 3

1. (a) MW/3; (b) 2 MW; (c) MW/6; (d) MW/2; (e) MW/2; (f) MW/2; (g) MW/2; (h) MW; (i) MW/4; (j) 2 MW. **3.** A: 1) 6.99 g, 2) 9.37 g, 3) 31.3 g, 4) 22.2 g; B: 1) 7.6 g, 2) 9.5 g, 3) 7.4 g, 4) 7.8 g. **5.** (a) 368; (b) 24.0; (c) 250. **7.** 2.21 g. **9.** 1.64 L + 8.36 L 0.500 $\underline{N}$ H_2SO_4. **11.** (a) 15.8; (b) 10.0 mL; (c) 5.00 mL conc. HNO_3 diluted to 500 mL. **13.** 0.614. **15.** 37.62. **17.** (a) 1.960; (b) 0.3350; (c) 1.060; (d) 0.5300. **19.** $\underline{M}_{Ca^{2+}} = 0.0400$, $\underline{M}_{Na^+} = 0.120$, $\underline{M}_{Cl^-} = 0.200$; $f_{Ca^{2+}} = 0.213$, $f_{Na^+} = f_{Cl^-} = 0.680$. **21.** 40.0. **23.** 0.125 $\underline{F}$, 0.250 $\underline{N}$, 24. **25.** (a) 79.6; (b) 74.5. **27.** 1.4 mg.

Chapter 4

1. (a) $v = k[A][B]^2$, third order; (b) 5.12×10^{-2} $\underline{M} \cdot min^{-1}$; (c) 4.0×10^{-4} $\underline{M} \cdot min^{-1}$. **3.** (a) none; (b) equil. shifted to the right; (c) equil. shifted to the right; (d) none. **5.** $v = k[A][B]^2[C]$, fourth order, $k = 0.30\ M^{-3} \cdot s^{-1}$. **7.** 3.7×10^{-4}. **9.** 4.

Chapter 5

1. $1.6_2 \times 10^{-3}$ $\underline{M}$. **5.** 1.01×10^{-14}. **7.** 1.6×10^{-6}. **9.** 6.3×10^{-5}. **11.** $2.2_2 \times 10^{-5}$. **13.** (a) 0.4%; (b) 99.45%; (c) 0.55%. **15.** 1.56 $\underline{M}$. **17.** (a) $\log K_a = \log K_a^\circ + 2A\sqrt{\mu}$; (b) $\log K_a = \log K_a^\circ + 4A\sqrt{\mu}$; (c) $\log K_a = \log K_a^\circ$. Largest change for $H_2PO_4^-$, minim. for $CH_3NH_3^+$. **19.** 154. **21.** (a) 1.7×10^{-9} $\underline{M}$; (b) 3.9×10^{-12} $\underline{M}$. **23.** 1.79. **25.** 30.3. **27.** 6.50. **29.** 2.08. **31.** 7.51. **33.** 6.38. **35.** 0.0065 $\underline{M}$. **37.** (a) 7.21; (b) 7.02; (c) 6.98. **39.** (a) 10.96; (b) 9.13; (c) 9.50; (d) 7.00. **41.** 1.61. **43.** (a) 2.50; (b) 18.7. **53.** 2.0×10^{-4} $\underline{M}$. **55.** pH = 9.00, $[H_2A] = 0.00100$ $\underline{M}$, $[HA^-] = 0.0085$ $\underline{M}$, $[A^{2-}] = 0.00050$ $\underline{M}$, $K_1 = 8.5 \times 10^{-9}$, $K_2 = 5.9 \times 10^{-11}$. **57.** 9.26. **59.** (a) incr. 0.024 units; (b) decr. 0.024 units; (c) 0.576. **61.** (a) $[H^+]$ decr. from 1.8×10^{-5} to 1.47×10^{-5} $\underline{M}$, pH incr. 0.087 units; (b) $[H^+]$ incr. to 2.20×10^{-5} $\underline{M}$. pH decr. 0.087 units. **63.** 26.7_4 g NH_4Cl + 185 mL conc. NH_3. **65.** 339 mL CH_3COOH, 61 mL CH_3COONa. **67.** 1.34. **69.** 2.95. **71.** 33.3. **73.** 6.91. **75.** 8.78. **77.** 4.44. **79.** 32.1. **81.** 56.8. **83.** 0.470. **85.** 375 mL B + 625 mL HCl. **87.** 40.0 g $NaHSO_4$, 23.6 g Na_2SO_4. **89.** 4.56. **91.** $C_{NaX} = 0.358$ $\underline{M}$, $C_{HX} = 0.716$ $\underline{M}$. **93.** (a) 1.100; (b) 10.63; (c) 17.0. **95.** $\alpha = 2.19 \times 10^{-5}$, $K_a = 2.09 \times 10^{-4}$. **97.** 8.92, $\alpha_{NH_4^+} = 0.32$, $\alpha_{CN^-} = 0.75$. **99.** (a) 3.34; (b) 3.32.

Chapter 6

1. 1.9×10^{-12}. **3.** 2.8×10^{-7}. **5.** (a) $K_{sp(Ag_2S)} = 4S^3$; (b) $K_{sp(PbCl_2)} = 4S^3$; (c) $K_{sp(PbSO_4)} = S^2$; (c) $K_{sp(PbClF)} = S^3$; (e) $K_{sp[Ca_3(PO_4)_2]} = 108S^5$. **7.** 3.6×10^{-4} mol/L. **11.** 9.8. **13.** 2.3_8. **15.** (a) yes; (b) Cr. **17.** 2.1×10^{-8}%. **19.** 0.0652. **21.** 5.7%. **23.** $[Ag^+] = 5.5 \times 10^{-6}$ $\underline{M}$, $[C_2O_4^{2-}] = 0.330$ $\underline{M}$, $[Na^+] = 0.896$ $\underline{M}$, $[NO_3^-] = 0.235$ M. **25.** $[Mg^{2+}] = 0.020$ $\underline{M}$, $[Cl^-] = 0.0150$ $\underline{M}$, $[Ag^+] = 1.2 \times 10^{-8}$ $\underline{M}$, $[NO_3^-] = 0.025$ $\underline{M}$. **27.** 1.72×10^{-4} mol/L. **29.** (a) 6.7×10^{-5} mol/L; (b) 4.2×10^{-5} mol/L; (c) 3.9×10^{-5} mol/L. Solubility incr. with incr. pH. **31.** 0.0065. **33.** 8.9×10^{-8} mol/L. **35.** 0.065 mg/100 mL. **37.** > 0.65 $\underline{M}$. **39.** 1.8×10^{-11} $\underline{M} < [OH^-] < 3.0 \times 10^{-5}$ $\underline{M}$. **41.** 4.5×10^{-9} $\underline{M}$. **43.** (a) Ag_2CrO_4; (b) AgCl. **45.** 1.7×10^{-7}. **47.** yes. **49.** $[Ba^{2+}] = 4.5 \times 10^{-9}$ $\underline{M}$, $[Ca^{2+}] = 2.66 \times 10^{-2}$ $\underline{M}$, $[CrO_4^{2-}] = 2.66 \times 10^{-2}$ $\underline{M}$. **51.** 2.3×10^{-10} mg Cu^{2+}/L, 9.4×10^2 mg Zn^{2+}/L. **53.** 1.9×10^{-14} g. **55.** 0.89 $\underline{M}$.

Chapter 7

1. 0.053. **3.** $4._4 \times 10^{-25}$ $\underline{M}$. **9.** 1.1×10^{-6}, 1.6×10^{-7}, 8.3×10^{-8}, 2.3×10^{-7}, 2.0×10^{-6}, 5.0×10^{-5} $\underline{M}$; 7.66×10^{-5} $\underline{M}$. **11.** 12.2. **13.** 3.7×10^{-8} $\underline{M}$. **15.** 0.24 mol/L.

Chapter 8

5. (a) −0.487 V; (b) −1.56 V. **7.** −0.151 V. **9.** −0.680 V. **11.** −0.219 V. **13.** (a) +0.7330 V; (b) +0.7298 V. **15.** (a) −0.015 V; (b) +0.6942 V; (c) −0.046 V; (d) −0.14 V. **17.** (a) − 1.004 V; (b) +1.04 V; (c) −0.377 V. **19.** (a) −0.406 V; (b) +0.540 V; (c) −0.134 V; (d) +0.177 V; (e) +2.13 V; (f) +0.233 V; (g) +0.522 V. **21.** 0.98. **23.** 5.18×10^{-8} $\underline{M}$. . 25 +0.062 V. **27.** 2:1. **29.** 0.212. **31.** −0.59 V. **33.** (a) 0.167; (b) 2.0×10^{21}; (c) 3×10^{62}; (d) 4.2×10^{13}; (e) 4.4×10^{40}; (f) 4.9×10^{20}; (g) 9×10^{40}. **35.** 1.2×10^{83}. **37.** 5×10^{56}. **39.** 1.00×10^{-14}. **41.** 1.6×10^{-19}. **43.** 8.6×10^{-17}. **45.** 7.5×10^{-3}. **47.** 1.6×10^{-15}. **49.** 4.8×10^{-9}. **51.** $[Ce^{3+}] = [Fe^{3+}] \simeq 0.0500$ $\underline{M}$; $[Ce^{4+}] = [Fe^{2+}] = 1.68 \times 10^{-8}$ $\underline{M}$. **53.** 1.00%. **55.** 0.825. **57.** $[Fe^{2+}] \simeq 0.050$ $\underline{M}$; $[Ag^+] = 2.5 \times 10^{-22}$ $\underline{M}$. **59.** (a) 9.0; (b) 1.94×10^{-2}.

Chapter 9

1. f = $As_2S_3/6AgCl$. **3.** (a) 0.2180 g; (b) 0.2805 g; (c) 0.4684 g. **5.** (a) 98.75%; (b) 14.06. **7.** 4. **9.** 23.5. **11.** 2.46. **13.** 5. **15.** 55.00% CaO, 45.00% MgO. **17.** 20.40. **19.** 53.36% CaC_2O_4, 46.64% CaO. **21.** 29.40% K, 17.29% Na. **23.** 37.83% KCl, 62.17% $KClO_3$. **25.** 56.51. **27.** 34.31% phenol, 65.69% glucose. **29.** 0.137 g. **31.** (a) 0.187 – 0.304 g; (b) 37.7y. **33.** 21.31. **35.** (a) 23.86; (b) 23.56. **37.** 12.36. **39.** +2.0%. **41.** −0.59. **43.** 0.34.

Chapter 10

1. 1.4610 g. **3.** 0.09837. **5.** 0.1006. **7.** 36.21. **9.** 17.06. **11.** 75.6. **13.** 6.25. **15.** (a) 0.19–0.24; (b) 0.37–0.48. **17.** 8.79. **19.** 35.13. **21.** −0.2. **23.** 27.42.

Chapter 11

1. 0.530 g. **3.** 3002. **5.** 0.2296. **7.** (a) 0.0791; (b) 0.1363. **13.** (a) −5.35%; (b) +0.2%. **15.** 9.24. **17.** 3.92 g NaOH, 0.1060 g Na_2CO_3. **19.** 0.0863. **21.** V_p = 10.61 mL, V_m = 53.07 mL. **23.** 3.18% Na_2CO_3, 96.82% $NaHCO_3$. **25.** 60.0% Na_2CO_3, 23.22% $NaHCO_3$. **27.** (a) $V_p = 0$, V_m = 14.2 mL; (b) $V_p = V_m$ = 23.1 mL; (c) V_p = 21.5 mL, V_m = 43.0 mL; (d) V_p = 14.8 mL, V_m = 39.6 mL; (e) V_p = 24.8 mL, V_m = 39.6 mL; (f) V_p = 14.8 mL, V_m = 29.6 mL; (g) V_p = 11.9 mL, V_m = 32.1 mL; (h) V_p = 35.6 mL, V_m = 45.6 mL. **29.** (a) $[H_3PO_4]$ = 0.0765 $\underline{M}$, $[NaH_2PO_4]$ = 0.0448 $\underline{M}$; (b) [HCl] = 0.1494 $\underline{M}$; (c) $[NaH_2PO_4]$ = 0.1566 $\underline{M}$; (d) $[H_3PO_4]$ = 0.0474 $\underline{M}$, [HCl] = 0.0203 $\underline{M}$; (e) $[H_3PO_4]$ = 0.0898 $\underline{M}$. **31.** 11.7. **33.** 0.3204 g. **35.** 51.1% CH_3OH, 48.9% C_2H_5OH. **37.** (a) 93.6; (b) 56.9 mL. **39.** 88.7. **41.** 5.00. **43.** (a) 1; (b) 81.3. **45.** 150.1. **47.** 97.0.

Chapter 12

3. $[Mn^{2+}] \simeq 0.0400$ $\underline{M}$, $[MnO_4^-] = 5.8 \times 10^{-35}$ $\underline{M}$, $[Sn^{2+}] = 1.45 \times 10^{-34}$ $\underline{M}$, $[Sn^{4+}] \simeq 0.100$ $\underline{M}$, 1.12 V. **5.** 53.36. **7.** 3. **9.** 11.30. **11.** 17.86% Fe, 8.36% FeO, 16.25% Fe_2O_3. **13.** (a) 86.8; (b) 90.00 mL. **15.** 0.1011. **17.** 37.37. **19.** (a) 0.0986; (b) 0.0980. **21.** 49.02. **23.** (a) 7.1 mL; (b) 25.0 mL; (c) 28.6 mL. **25.** 632.5. **27.** 38.66. **29.** 0.1011. **31.** 44.51. **33.** 95.67. **35.** 2.04. **37.** 0.01000. **39.** 0.0500. **41.** 92.8. **43.** 2.045. **47.** 9.2. **49.** 0.1102.

Chapter 13

1. 0.43–0.53 g. **5.** 25.76% KCl, 74.24% $KClO_4$. **7.** 0.2559. **9.** 20.76. **11.** 24.30. **13.** (a) 24.00; (b) 32.00. **15.** 0.2821. **17.** 29.2. **19.** 0.369. **21.** $2.1 \times 10^{-12} < K_{sp(AgA)} < 1.3 \times 10^{-9}$. **23.** (a) −0.03%; (b) $+0.009_4$. **25.** 1.27, 9.18.

Chapter 14

1. 42.96% NaCN, 57.04% KCN. **3.** 1.001. **5.** (a) 3.6×10^{-5} M; (b) 8.2×10^{-9} M. **7.** 2.0×10^{8}. **9.** 5.0×10^{11}. **11.** 99.9%. **15.** $[Ca^{2+}] = 0.00552$ M, $[Zn^{2+}] = 0.00427$ M. **17.** 18.46% MgO, 76.93% $NaHCO_3$, 4.61% inert mat. **19.** 29.91% Ca, 1.64% Mg. **21.** 80.7% Fe, 18.90% Zn. **23.** (a) 0.01105; (b) 45.40. **25.** (a) −1.80%; (b) −0.02%; (c) +1.57%.

Chapter 15

1. (a) 9.25; (b) +0.5026 V. **3.** $E_{cell} = +0.2412 + 0.05916$ pH. **5.** 0.0133 M. **7.** 6.51×10^{-3} M. **9.** 9.88×10^{-4} M, 4.11. **11.** 0.0654. **13.** +0.1023 V. **15.** 70.2. **19.** 4.7.

Chapter 16

5. all but (b). **7.** 0.236. **11.** (a) 34.0; (b) 8.50×10^{3}. **13.** 168. **15.** 2.29. **17.** 89.2. **19.** 7.58. **21.** (a) 203.0 mL A, 47.0 mL B; (b) 3.55. **23.** 5.55%. **25.** 4.0×10^{-4} M. **27.** (a) $[B] = 7.2 \times 10^{-5}$ M, $[C] = 6.8 \times 10^{-5}$ M; (b) 0.474. **29.** 0.65% Fe, 1.26% Al. **31.** (a) positive; (b) no. **35.** 1.500×10^{-4} M. **37.** 10.69. **39.** 7.1×10^{14}. **41.** $\varepsilon = 1.25 \times 10^{3}$, 4.0×10^{-6}. **43.** 2.5×10^{-6}. **45.** 1.0×10^{-5}.

Chapter 17

1. (a) +4.66%; (b) −0.44%. **3.** (a) 0.0129; (b) −0.90%. **5.** 7.5×10^{-5}. **7.** 4: 0.6000, 0.1200, 0.0240, 0.0048 g. **9.** 0.0489. **11.** % Cu = 24.2, 96.7, 99.7, 99.7; % Zn = 0.25, 19.7, 95.1, 99.6. **13.** 1.35. **15.** 77.9 mL. **17.** 3.00. **19.** 20.0. **21.** (a) 45; (b) 21.6. **23.** 99.7%. **27.** 0.008192, 0.008192, 0.003072, 0.000512, 0.000032 g. **29.** (a) 96.0%; (b) 99.96%. **31.** 2.13 g. **33.** 10.84. **35.** 0.02379, 0.02736 M.

Chapter 18

3. 15.1, 54.1 mL. **5.** increased 1.73 times. **7.** 4. **9.** 202 cm. **11.** A in cm, B in cm^2/s, C in s. **13.** (a) $V_{opt} = 3.0$ cm/s, $h_{min} = 0.42$ cm. **15.** A. **17.** 95.5 mL. **19.** 31.0% B. **21.** (a) Mole % = 31.3 pen, 28.0 hex, 40.7 tol. % (w/w): 25.5 pen, 32.3 hex, 42.3 tol; (b) 24.9% pen, 29.4% hex, 45.7% tol.

Index